VOLUME FIVE HUNDRED AND SEVENTY SEVEN

METHODS IN
ENZYMOLOGY

Computational Approaches for
Studying Enzyme Mechanism
Part A

METHODS IN ENZYMOLOGY

Editors-in-Chief

ANNA MARIE PYLE
*Departments of Molecular, Cellular and Developmental Biology and Department of Chemistry
Investigator, Howard Hughes Medical Institute
Yale University*

DAVID W. CHRISTIANSON
*Roy and Diana Vagelos Laboratories
Department of Chemistry
University of Pennsylvania
Philadelphia, PA*

Founding Editors

SIDNEY P. COLOWICK and NATHAN O. KAPLAN

VOLUME FIVE HUNDRED AND SEVENTY SEVEN

METHODS IN
ENZYMOLOGY

Computational Approaches for Studying Enzyme Mechanism
Part A

Edited by

GREGORY A. VOTH

Department of Chemistry
The University of Chicago
Chicago, Illinois, United States

AMSTERDAM • BOSTON • HEIDELBERG • LONDON
NEW YORK • OXFORD • PARIS • SAN DIEGO
SAN FRANCISCO • SINGAPORE • SYDNEY • TOKYO

Academic Press is an imprint of Elsevier

Academic Press is an imprint of Elsevier
50 Hampshire Street, 5th Floor, Cambridge, MA 02139, United States
525 B Street, Suite 1800, San Diego, CA 92101-4495, United States
The Boulevard, Langford Lane, Kidlington, Oxford OX5 1GB, United Kingdom
125 London Wall, London, EC2Y 5AS, United Kingdom

First edition 2016

Copyright © 2016 Elsevier Inc. All rights reserved.

No part of this publication may be reproduced or transmitted in any form or by any means, electronic or mechanical, including photocopying, recording, or any information storage and retrieval system, without permission in writing from the publisher. Details on how to seek permission, further information about the Publisher's permissions policies and our arrangements with organizations such as the Copyright Clearance Center and the Copyright Licensing Agency, can be found at our website: www.elsevier.com/permissions.

This book and the individual contributions contained in it are protected under copyright by the Publisher (other than as may be noted herein).

Notices
Knowledge and best practice in this field are constantly changing. As new research and experience broaden our understanding, changes in research methods, professional practices, or medical treatment may become necessary.

Practitioners and researchers must always rely on their own experience and knowledge in evaluating and using any information, methods, compounds, or experiments described herein. In using such information or methods they should be mindful of their own safety and the safety of others, including parties for whom they have a professional responsibility.

To the fullest extent of the law, neither the Publisher nor the authors, contributors, or editors, assume any liability for any injury and/or damage to persons or property as a matter of products liability, negligence or otherwise, or from any use or operation of any methods, products, instructions, or ideas contained in the material herein.

ISBN: 978-0-12-805347-8
ISSN: 0076-6879

For information on all Academic Press publications
visit our website at https://www.elsevier.com/

Publisher: Zoe Kruze
Acquisition Editor: Zoe Kruze
Editorial Project Manager: Helene Kabes
Production Project Manager: Magesh Kumar Mahalingam
Cover Designer: Greg Harris

Typeset by SPi Global, India

CONTENTS

Contributors   xi
Preface   xv

1. The Role of Molecular Dynamics Potential of Mean Force Calculations in the Investigation of Enzyme Catalysis   **1**
Y. Yang, L. Pan, F.C. Lightstone, and K.M. Merz Jr.

1. Introduction   2
2. Method   3
3. Applications   4
4. Summary   24
Acknowledgments   25
References   26

2. Empirical Force Fields for Mechanistic Studies of Chemical Reactions in Proteins   **31**
A.K. Das and M. Meuwly

1. Introduction   31
2. Computational Approaches   35
3. Applications   41
4. Outlook   49
Acknowledgments   51
References   51

3. Generalized Ensemble Sampling of Enzyme Reaction Free Energy Pathways   **57**
D. Wu, M.I. Fajer, L. Cao, X. Cheng, and W. Yang

1. Introduction   58
2. Collective Variable and Reaction Order Parameter   59
3. Traditional Importance Sampling vs GE Sampling   62
4. Dimensionality Limit   64
5. One of the First Metadynamics-Based Enzyme Reaction Studies   65
6. GE-Based String Optimization: The OTPRW Method   67
7. OTPRW Study of a Substrate-Assisted Glycosylation Reaction   69
8. Final Remarks   70
Acknowledgments   71
References   71

4. Methods for Efficiently and Accurately Computing Quantum Mechanical Free Energies for Enzyme Catalysis 75
F.L. Kearns, P.S. Hudson, S. Boresch, and H.L. Woodcock

1. Introduction 76
2. Alchemical FES 86
3. Reaction Profiles 92
Appendix 97
Acknowledgment 98
References 98

5. Born–Oppenheimer Ab Initio QM/MM Molecular Dynamics Simulations of Enzyme Reactions 105
Y. Zhou, S. Wang, Y. Li, and Y. Zhang

1. Introduction 106
2. Methods 107
3. Examples 112
4. Enzyme Simulation Protocol 114
5. Conclusion 116
Acknowledgments 116
References 116

6. QM/MM Calculations on Proteins 119
U. Ryde

1. Introduction 120
2. Methods 122
3. Applications 135
4. Suggested Approach for QM/MM Investigations 147
Acknowledgments 150
References 150

7. Enzymatic Cleavage of Glycosidic Bonds: Strategies on How to Set Up and Control a QM/MM Metadynamics Simulation 159
L. Raich, A. Nin-Hill, A. Ardèvol, and C. Rovira

1. Introduction 160
2. Carbohydrate Structures and Glycoside Hydrolases 160
3. QM/MM Molecular Dynamics Simulation of the Catalytic Mechanism 167
4. Conclusions 179
Acknowledgments 180
References 180

8. Toward Determining ATPase Mechanism in ABC Transporters: Development of the Reaction Path–Force Matching QM/MM Method   185
Y. Zhou, P. Ojeda-May, M. Nagaraju, and J. Pu

1. Introduction   186
2. Biological Questions   187
3. QM/MM Simulations of the ABC-Transporter HlyB   193
4. Computational Challenges   196
5. A Multiscale QM/MM Method: RP–FM   200
6. Concluding Remarks   207
Acknowledgments   208
References   208

9. QM/MM Analysis of Transition States and Transition State Analogues in Metalloenzymes   213
D. Roston and Q. Cui

1. Introduction   214
2. Background on Computational Methods   215
3. Case Study: The Transition State of AP   223
4. Summary/Conclusions   241
Acknowledgments   242
References   243

10. Practical Aspects of Multiscale Classical and Quantum Simulations of Enzyme Reactions   251
M. Dixit, S. Das, A.R. Mhashal, R. Eitan, and D.T. Major

1. Introduction   252
2. Enzyme System Modeling   253
3. The Potential Energy Surface   260
4. Reaction Coordinates and Classical Free Energy Simulations   267
5. Concluding Words   274
Acknowledgments   277
References   277

11. Examinations of the Chemical Step in Enzyme Catalysis   287
P. Singh, Z. Islam, and A. Kohen

1. Introduction   288
2. KIE_{obs} vs KIE_{int} 290
3. The Northrop Method   292

4. Case Study 1: TSase	296
5. Case Study 2: DHFR	305
6. Summary	312
Acknowledgments	313
References	313

12. Use of QM/DMD as a Multiscale Approach to Modeling Metalloenzymes — 319

N.M. Gallup and A.N. Alexandrova

1. Introduction	320
2. Overview of QM/DMD	321
3. Setting Up QM/DMD	324
4. Running QM/DMD and Details of the Procedure	331
5. Conclusions	338
References	338

13. Adaptive Partitioning QM/MM Dynamics Simulations for Substrate Uptake, Product Release, and Solvent Exchange — 341

A. Duster, C. Garza, and H. Lin

1. Introduction	342
2. Methodology	344
3. Implementation	349
4. Applications of AP Schemes	351
5. Summary	353
Acknowledgments	354
References	354

14. Enzymatic Kinetic Isotope Effects from Path-Integral Free Energy Perturbation Theory — 359

J. Gao

1. Introduction	360
2. Methods	362
3. Illustrative Examples	371
4. Concluding Remarks	382
Acknowledgment	382
References	383

15. Simulating Nuclear and Electronic Quantum Effects in Enzymes 389
L. Wang, C.M. Isborn, and T.E. Markland

1. Introduction 390
2. Electronic Quantum Effects in Biological Systems 392
3. Incorporating NQEs in AIMD Simulations 401
4. Outlook 409
Acknowledgments 410
References 410

16. Using Molecular Simulation to Study Biocatalysis in Ionic Liquids 419
K.G. Sprenger and J. Pfaendtner

1. Introduction 420
2. Methods for Simulating Biomolecules in ILs 425
3. Perspective: Challenges and Future Directions 436
References 437

17. The MOD-QM/MM Method: Applications to Studies of Photosystem II and DNA G-Quadruplexes 443
M. Askerka, J. Ho, E.R. Batista, J.A. Gascón, and V.S. Batista

1. Introduction 444
2. Methods 449
3. EXAFS Simulations 461
4. MOD-QM/MM Models of DNA Quadruplexes 464
5. Conclusions 468
Acknowledgments 469
References 469

Author Index *483*
Subject Index *513*

CONTRIBUTORS

A.N. Alexandrova
University of California, Los Angeles; California NanoSystems Institute, Los Angeles, CA, United States

A. Ardèvol
Max-Planck Institut für Biophysik, Frankfurt am Main, Germany

M. Askerka
Yale University, New Haven, CT, United States

E.R. Batista
Los Alamos National Laboratory, Los Alamos, NM, United States

V.S. Batista*
Yale University, New Haven, CT, United States

S. Boresch
Faculty of Chemistry, University of Vienna, Vienna, Austria

L. Cao
Florida State University, Tallahassee, FL, United States

X. Cheng
UT-ORNL Center for Molecular Biophysics, Oak Ridge National Laboratory, Oak Ridge; The University of Tennessee, Knoxville, TN, United States

Q. Cui
Theoretical Chemistry Institute, University of Wisconsin–Madison, Madison, WI, United States

A.K. Das
University of Basel, Basel, Switzerland

S. Das
Lise Meitner-Minerva Center of Computational Quantum Chemistry, Bar-Ilan University, Ramat Gan, Israel

M. Dixit
Lise Meitner-Minerva Center of Computational Quantum Chemistry, Bar-Ilan University, Ramat Gan, Israel

A. Duster
University of Colorado Denver, Denver, CO, United States

---

*Institute of High Performance Computing, 1 Fusionopolis Way, #16-16 Connexis North, Singapore 138632.

R. Eitan
Lise Meitner-Minerva Center of Computational Quantum Chemistry, Bar-Ilan University, Ramat Gan, Israel

M.I. Fajer
UT-ORNL Center for Molecular Biophysics, Oak Ridge National Laboratory, Oak Ridge, TN; Florida State University, Tallahassee, FL, United States

N.M. Gallup
University of California, Los Angeles, Los Angeles, CA, United States

J. Gao
Theoretical Chemistry Institute, Jilin University, Changchun, Jilin Province, PR China; University of Minnesota, Minneapolis, MN, United States

C. Garza
University of Colorado Denver, Denver, CO, United States

J.A. Gascón
University of Connecticut, Storrs, CT, United States

J. Ho
Yale University, New Haven, CT, United States

P.S. Hudson
University of South Florida, Tampa, FL, United States

C.M. Isborn
School of Natural Sciences, University of California, Merced, CA, United States

Z. Islam
University of Iowa, Iowa City, IA, United States

F.L. Kearns
University of South Florida, Tampa, FL, United States

A. Kohen
University of Iowa, Iowa City, IA, United States

Y. Li
International Center of Quantum and Molecular Structures, and Shanghai Key Laboratory of High Temperature Superconductors, Shanghai University, Shanghai, China

F.C. Lightstone
Physical and Life Sciences Directorate, Lawrence Livermore National Laboratory, Livermore, CA, United States

H. Lin
University of Colorado Denver, Denver, CO, United States

D.T. Major
Lise Meitner-Minerva Center of Computational Quantum Chemistry, Bar-Ilan University, Ramat Gan, Israel

T.E. Markland
Stanford University, Stanford, CA, United States

K.M. Merz Jr.
Michigan State University, East Lansing, MI, United States

M. Meuwly
University of Basel, Basel, Switzerland

A.R. Mhashal
Lise Meitner-Minerva Center of Computational Quantum Chemistry, Bar-Ilan University, Ramat Gan, Israel

M. Nagaraju
Indiana University-Purdue University Indianapolis, Indianapolis, IN, United States

A. Nin-Hill
Institut de Química Teòrica i Computacional (IQTCUB), Universitat de Barcelona, Barcelona, Spain

P. Ojeda-May
Indiana University-Purdue University Indianapolis, Indianapolis, IN, United States

L. Pan
Michigan State University, East Lansing, MI, United States

J. Pfaendtner
University of Washington, Seattle, WA, United States

J. Pu
Indiana University-Purdue University Indianapolis, Indianapolis, IN, United States

L. Raich
Institut de Química Teòrica i Computacional (IQTCUB), Universitat de Barcelona, Barcelona, Spain

D. Roston
Theoretical Chemistry Institute, University of Wisconsin–Madison, Madison, WI, United States

C. Rovira
Institut de Química Teòrica i Computacional (IQTCUB), Universitat de Barcelona; Institució Catalana de Recerca i Estudis Avançats (ICREA), Barcelona, Spain

U. Ryde
Chemical Centre, Lund University, Lund, Sweden

P. Singh
University of Iowa, Iowa City, IA, United States

K.G. Sprenger
University of Washington, Seattle, WA, United States

L. Wang
Rutgers University, Piscataway, NJ, United States

S. Wang
New York University, New York, NY, United States

H.L. Woodcock
University of South Florida, Tampa, FL, United States

D. Wu
Institute of Molecular Biophysics, Florida State University, Tallahassee, FL, United States

W. Yang
Institute of Molecular Biophysics, Florida State University; Florida State University, Tallahassee, FL, United States

Y. Yang
Physical and Life Sciences Directorate, Lawrence Livermore National Laboratory, Livermore, CA, United States

Y. Zhang
New York University, New York, NY, United States; NYU-ECNU Center for Computational Chemistry at NYU Shanghai, Shanghai, China

Y. Zhou
Laboratory of Mesoscopic Chemistry, Collaborative Innovation Center of Chemistry for Life Sciences, Institute of Theoretical and Computational Chemistry, School of Chemistry and Chemical Engineering, Nanjing University, Nanjing, China; Indiana University-Purdue University Indianapolis, Indianapolis, IN, United States

PREFACE

The computational study of enzyme structure and function has reached exciting and unprecedented levels. This state of affairs is due to a combination of powerful new computational methods, a critical evolution of ideas and insights, the ever-increasing power of computers, and important new experimental results for validation. In these two volumes of *Methods in Enzymology*, many of the leading computational researchers present their latest work, representing a range of state-of-the-art topics in computational enzymology. Generally speaking, the two volumes are divided into two general areas, the first being mostly devoted to the calculation of the free energy barriers and reaction pathways for enzymes—often using powerful quantum mechanics/molecular mechanics (QM/MM) methods—while the second volume contains a broader range of topics, including the role of enzyme dynamics and allostery, electrostatics, ligand binding, and several specific case studies.

The topic of computational enzymology—and the field of computational biophysics and biochemistry in general—has grown enormously since its inception in the early 1970s, with some of these original work being recognized by the 2013 Nobel Prize in Chemistry. It is tempting to conclude that the field is now "mature" and all that remains is for researchers in it to carry out increasingly detailed and accurate set of applications of the powerful computational methods presented herein. However, nothing could be farther from the truth. Real enzymes exist and function in highly complex, multiscale biological environments. Often several enzymes function cooperatively, and they can be influenced by, or respond to, their local environment, whether be it a lipid membrane or the crowded cellular interior. In the venerable theories of activated dynamics for condensed-phase chemical kinetics, such as transition state theory, it is tacitly assumed that the chemical reaction dictates the slowest dynamical time scale of the system, thus corresponding to the highest free energy barrier. This basic assumption also allows one to apply these simpler theories to calculate quantities such as the free energy profile, ie, free energy barrier, for a reaction along a chemical pathway in an enzyme (the so-called potential of mean force). Yet, in complex biological systems there are a wide range of timescales associated with *numerous* processes, some of which may be intrinsically coupled to the reactive process of the enzyme. In that light a key question then arises for the future.

Can we better understand enzyme kinetics in the larger biological context of the living cell through computation? This challenge awaits us.

There is also the important fact that real biological systems are *not in a state of equilibrium*. Indeed, they can be rather far from it. Much of the standard condensed-phase kinetic theory developed in the last century—and applied in present-day computational enzymology—relies on the key notion of the famous fluctuation-dissipation theorem, ie, the behavior of systems perturbed out of equilibrium can be understood from studies of ones that are actually *in equilibrium*. This so-called linear response assumption leads us to powerful mathematical formulas for observables such as kinetic rate constants, as well as the algorithms one uses to compute them, which are based on *equilibrium* molecular dynamics simulation. However, much work remains to be done to develop theories and computational algorithms for enzymes functioning in a nonequilibrium biological context, albeit some important work in that direction, motivated by experiments, has already been initiated.

The great degree of progress to date on the topic of computational enzymology is reflected in these two volumes of *Methods in Enzymology*. Moreover, this field of research continues to evolve at an increasingly rapid pace. The scope of the enzyme systems presently under study, and the elaboration of their complex behaviors, is remarkable. As an example, I can point to some of the research completed by talented young theorists as they passed through my own research group [see McCullagh, M., Saunders, M. G., & Voth, G. A. (2014). Unraveling the mystery of ATP hydrolysis in actin filaments. *Journal of the American Chemical Society, 136*, 13053–13058.]. In this work, QM/MM, molecular dynamics, advanced free energy sampling, and insights from coarse-grained modeling were all combined to explain the origins of the $>10^4$ acceleration of ATP hydrolysis in actin filaments (F-actin) over the free monomeric form (G-actin). This ATP hydrolysis by F-actin, which has been a mystery for years, is critical to the functioning of the actin-based eukaryotic cellular cytoskeleton and now computation has solved it.

Nevertheless, in light of the great remaining challenges described earlier, it is abundantly clear that much remains to be done to further advance computational enzymology, in some cases even at a qualitative level of basic understanding. It will certainly be both important and fascinating to survey future volume(s) of *Methods in Enzymology* devoted to this topic—perhaps 10 or even 20 years from now—and to celebrate what I am sure will the outcomes from an exciting and continual evolution of this important field of research.

G.A. VOTH
The University of Chicago

CHAPTER ONE

The Role of Molecular Dynamics Potential of Mean Force Calculations in the Investigation of Enzyme Catalysis

Y. Yang*, L. Pan[†], F.C. Lightstone*, K.M. Merz Jr.[†,1]

*Physical and Life Sciences Directorate, Lawrence Livermore National Laboratory, Livermore, CA, United States
[†]Michigan State University, East Lansing, MI, United States
[1]Corresponding author: e-mail address: kmerz1@gmail.com

Contents

1. Introduction   2
2. Method   3
   2.1 Umbrella Sampling   3
   2.2 Steered MD   4
3. Applications   4
   3.1 Protein Farnesyltransferase   4
   3.2 Aromatic Prenyltransferase NphB   11
   3.3 Aspergillus Fumigatus Prenyltransferase   16
   3.4 Cytochrome P450s   20
4. Summary   24
Acknowledgments   25
References   26

Abstract

The potential of mean force simulations, widely applied in Monte Carlo or molecular dynamics simulations, are useful tools to examine the free energy variation as a function of one or more specific reaction coordinate(s) for a given system. Implementation of the potential of mean force in the simulations of biological processes, such as enzyme catalysis, can help overcome the difficulties of sampling specific regions on the energy landscape and provide useful insights to understand the catalytic mechanism. The potential of mean force simulations usually require many, possibly parallelizable, short simulations instead of a few extremely long simulations and, therefore, are fairly manageable for most research facilities. In this chapter, we provide detailed protocols for applying the potential of mean force simulations to investigate enzymatic mechanisms for several different enzyme systems.

1. INTRODUCTION

Molecular dynamics (MD) has been applied to study protein structure and biochemical processes since the 1970s (Mccammon, Gelin, & Karplus, 1977). Using molecular mechanics (MM) force fields or hybrid quantum mechanics/molecular mechanics (QM/MM) potentials, the MD algorithm is frequently applied to simulate motions of biological systems, including peptides, proteins, and nucleic acids with the resultant trajectories aiding in the experimental interpretation of biomolecular structure, function, and dynamics. With ever-growing computing power, simulation timescales have significantly improved from a few picoseconds to microseconds. The specific-purpose supercomputer, Anton, developed by D.E. Shaw research, accesses the millisecond time regime to study long-time scale processes like protein folding (Lindorff-Larsen, Piana, Dror, & Shaw, 2011). Even with the high-performance computing (HPC) prowess of Anton and several other HPC systems, microseconds and milliseconds of MD simulation are still less common and typically confined to smaller systems. Hence, many highly relevant biochemical processes involving molecular machines, protein–ligand interactions, large-scale conformational motions involved in allostery, and the folding of larger proteins still pose a significant challenge to the MD community. Although specific processes in each system pose different challenges, it is a common theme that these systems suffer from poor sampling in regions of conformational space separated by significant energy barriers. In order to improve sampling of such systems where ergodicity is affected by the systems' energy landscape, we can either keep increasing the simulation time, like the brute-force approach afforded by Anton, or use potential of mean force (Kirkwood, 1935) (PMF) simulations to improve sampling in a targeted region of the system of interest. PMF simulations are applied to examine the variation of the energy landscape along one or more specific variables, known as the reaction coordinate(s) (RC). For example, one can investigate the binding free energy profile of a protein–ligand complex as a function of the distance between the ligand and a protein pocket, where the distance between the ligand and a point in the protein pocket defines the RC. There are several different sampling techniques that have been developed to compute the PMFs of biological processes, including umbrella sampling (Torrie & Valleau, 1977), steered MD (Grubmuller, Heymann, & Tavan, 1996; Isralewitz, Gao, & Schulten, 2001) (SMD), free energy perturbation (Zwanzig, 1954), thermodynamic integration (Tironi & Vangunsteren, 1994), etc. For most of these sampling techniques, a series of biased simulations are conducted followed by analysis using the weighted histogram analysis method (Kumar, Bouzida, Swendsen, Kollman, & Rosenberg, 1992) (WHAM) to construct the unbiased PMF. Herein, we describe examples where we have used umbrella sampling and SMD to compute the PMF of different steps in a catalytic mechanism of an enzyme; including substrate binding, conformational activation, and chemical reaction. This chapter starts with a brief overview of umbrella sampling and SMD, discusses general issues that need to be considered in a PMF study, and gives examples of studies using these methods, as in studies on the protein prenyltransferase FTase, the aromatic prenyltransferases NphB and FtmPT1, and a family of cytochrome P450s (CYPs).

2. METHOD

As discussed earlier, many different sampling techniques have been developed to compute the PMFs of various biological processes. Among them, umbrella sampling and SMD are two of the most commonly used techniques. The basic idea of both umbrella sampling and SMD is to apply biasing forces along given reaction coordinate(s) in order to improve the sampling of high-energy regions of conformational or chemical space. The full details of these methods can be found in many literatures; hence, here we only present a brief introduction to each method.

2.1 Umbrella Sampling

The basic idea of umbrella sampling is to (1) partition a predetermined RC ξ to a number of separate windows, then (2) apply a biasing potential $W(\xi) = \frac{k}{2}(\xi - \xi_0)^2$ (k is a force constant) to enhance the local sampling at a chosen ξ_0 for each sampling window, such that the new potential $V'(\xi) = V(\xi) + W(\xi)$, which is a sum of the true potential $V(\xi)$ and the biasing potential $W(\xi)$, is smoothed out to allow for enhanced sampling in the conformational space along the RC. Finally, (3) the unbiased free energy profile is reconstructed, generally using WHAM. For the interested reader, a thorough review of umbrella sampling technique is given in Kastner (2011).

2.2 Steered MD

Unlike in umbrella sampling where the biasing potential is applied at separate windows along the RC, SMD applies a constant harmonic force to select atoms along a RC to pull them, continuously propagating them at constant velocity. The biasing force has the form of $F = \frac{k}{2}[\xi - (\xi_0 + vt)]$, where v is the pulling velocity, t is time, and k is the force constant, and the work done by the force is $W = \int F \cdot ds$. The Jarzynski's equation (Jarzynski, 1997a, 1997b), $e^{-\Delta G/kT} = \left\langle e^{-W/kT} \right\rangle$, enables the derivation of a PMF from SMD results. For the interested reader, a detailed review of the SMD method can be found in Isralewitz et al. (2001).

Although SMD alone can be applied to study the free energy surface associated with a given RC, it is necessary to apply it multiple times for the same RC in order to obtain accurate results. In fact, we found implementing SMD followed by umbrella sampling was a very effective way to examine the PMF along a RC, especially for studies involving bond forming and breaking where a QM/MM model is needed. In such cases, a fast scan by SMD generates "continuous" snapshots that can be applied as the starting structures for subsequent umbrella sampling simulations, while in the absence of SMD these starting structures are not as straightforward to generate. We have applied this protocol, using SMD first followed by umbrella sampling, to successfully investigate the chemical reaction step of Farnesyltransferase (FTase), NphB, and FtmPT1 catalysis.

3. APPLICATIONS

We have applied PMF simulations, at the MM and/or QM/MM levels, to investigate the structure, dynamics, and function of several different enzymatic processes, including the mechanisms of FTase, NphB, FtmPT1, and a family of CYPs.

3.1 Protein Farnesyltransferase

FTase is a protein prenyltransferase that catalyzes protein farnesylation, which attaches a farnesyl moiety, a 15-carbon isoprenoid to a cysteine residue at or near the C-terminus of protein acceptors, including Ras proteins. This posttranslational modification is critical for many GTP-binding proteins on the receptor tyrosine kinase (RTK) signal transduction pathway. The malfunction of these proteins often results in uncontrolled cell growth, which may lead to cancerous tumors. FTase is a zinc metalloenzyme but also requires a Mg^{2+} ion for optimal reactivity: the binding of Mg^{2+} at the millimolar level boosts the reaction rate by 700-fold (Huang, Hightower, & Fierke, 2000). Unfortunately, to date, a FTase crystal structure, with Mg^{2+} bound, has not been reported. In addition to the Mg^{2+} puzzle, the crystal structure of FTase reactant ternary complex (protein data bank (PDB) ID 1QBQ) revealed a 7.4 Å gap between the reacting centers, the C_1 atom of farnesyldiphosphate (FPP) (the carbon atom connected to the diphosphate moiety) and the S_γ atom of the zinc-bound cysteine residue of the target protein. A hypothesis based on mutagenesis studies proposed that a rotation between the first and second isoprene groups of FPP reduces the gap, resulting in an intermediate complex where the two reacting centers are in much closer proximity (Pickett, Bowers, & Fierke, 2003). Furthermore, the reaction, which involves the diphosphate detaching from FPP and attaching to the zinc-bound cysteine exhibited both S_N1 (Dolence & Poulter, 1995) and S_N2 (Hightower, Huang, Casey, & Fierke, 1998; Mu, Gibbs, Eubanks, & Poulter, 1996) features; hence, a so-called associative mechanism with dissociative features has been proposed (Ho, Vivo, Peraro, & Klein, 2009; Huang et al., 2000). However, for two similar peptide targets, TKCVIF and GCVLS, the measured ^{3}H α-secondary kinetic isotope effects (α-SKIEs) are quite different (1.154 ± 0.006 and 1.00 ± 0.04, respectively), leading to the hypothesis that the rate-limiting step for TKCVIF is the chemical step, while for GCVLS it is the conformational activation step prior to the chemical step (Pais, Bowers, & Fierke, 2006). Later, we demonstrate how PMF simulations can be used to address these questions/hypotheses.

3.1.1 Understanding the Conformational Activation of FPP

FPP has two possible charge states, the deprontonated FPP^{3-} or the monoprotonated $FPP(H)^{2-}$ form. The pH-dependent kinetics experiments revealed a pK_a of 7.4 for the last proton of FPP, and the loss of that proton is key to the Mg^{2+} binding (Saderholm, Hightower, & Fierke, 2000). Therefore, the preferred form for FPP, when Mg^{2+} is present, is FPP^{3-}, while both FPP^{3-} and FPP^{2-} may exist in the absence of Mg^{2+}. With FPP in either form, it is likely that an intermediate complex exists with a smaller gap between two reacting atoms than that observed for the resting state (>7 Å). In addition, the α-SKIE studies revealed that a single change of Ile to Leu at the second to last residue in the target peptide switches the rate-limiting step from the chemical step to the conformational activation step

(Pais et al., 2006). With these in mind, we performed PMF simulations to investigate different scenarios, namely FTase/FPP^{2-}/CVIM, FTase/FPP^{2-}/CVLM, Y300F/FPP^{2-}/CVIM and FTase/FPP^{3-}/CVIM (Cui & Merz, 2007).

By defining the RC as the distance between two potential reacting centers, the C_1 atom of FPP and the S_γ atom of the zinc-bound peptide cysteine, we can implement the umbrella sampling technique to scan the free energy surface associated with the conformational activation prior to the chemical reaction. After the PMF simulations, we applied WHAM to reconstruct the free energy profile as shown in Fig. 1. With the plot, we answered the previous questions regarding the mechanism of FTase activation. First, we learned that in the absence of Mg^{2+}, FPP was likely in its monoprotonated form (FPP^{2-}) because FPP^{3-} cannot undergo the required conformational change that closes the gap between two reacting atoms. Second, an intermediate state was found for all three FTase/FPP^{2-} complexes that feature a much closer distance (5–5.5 Å) between the two reacting centers, with a ~1 to 3 kcal/mol barrier. It is interesting to notice that, by comparing the FTase/FPP^{2-}/CVIM and FTase/FPP^{2-}/CVLM systems, we did discover that it is more difficult for the latter complex to undergo the necessary conformational change; however, the 2.5 kcal/mol barrier to this transition

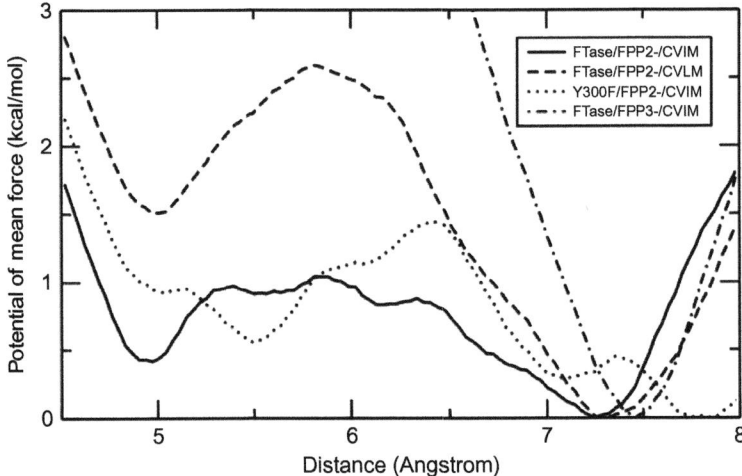

Fig. 1 Free energy profiles of the FTase conformational activation step. *Reprinted with permission from Cui, G., & Merz, K. M., Jr. (2007). Computational studies of the farnesyltransferase ternary complex part II: The conformational activation of farnesyldiphosphate. Biochemistry, 46(43), 12375–12381. doi:10.1021/Bi701324t. Copyright (2007) American Chemical Society.*

does not support the hypothesis that the conformational activation step is the rate-limiting step, as proposed by Fierke and coworkers (Pais et al., 2006). Similarly, we found that the Y300Fβ mutant, which exhibited a 500-fold decrease in reaction rate constant (k_{chem}) (Pickett, Bowers, Hartman, et al., 2003), shifts both the resting state and intermediate state away from the nucleophile by about 0.5 Å, and the transition barrier of Y300Fβ is approximately 0.5 kcal/mol higher than the wild type. However, the barrier change is not sufficient to cause a 500-fold difference. Thus, we hypothesized that the main cause of the significant drop in k_{chem} is most likely the loss of stabilization of the diphosphate leaving group from Y300β.

3.1.2 Identification of the Mg^{2+} Binding Site in FTase

Experiments have been long performed to investigate the role of Mg^{2+} in accelerating FTase catalysis and identifying key residues for Mg^{2+} binding (Pickett, Bowers, & Fierke, 2003; Pickett, Bowers, Hartman, et al., 2003); however, the experimental understanding of Mg^{2+} binding to FTase at the atomistic level was missing. To fill this void, we performed umbrella samplings to investigate Mg^{2+} binding to FTase (Yang, Chakravorty, & Merz, 2010). We first combined knowledge from available experimental studies (Pickett, Bowers, & Fierke, 2003; Pickett, Bowers, Hartman, et al., 2003) and previous computational studies on FTase (Cui & Merz, 2007; Cui, Wang, & Merz, 2005) to generate a number of models for Mg^{2+} binding in the active site pocket of FTase. Subsequently, we conducted regular MD simulations to relax each FTase-Mg^{2+} model followed by applying umbrella sampling to discover the impact of Mg^{2+} binding to the transition from the inactive FTase–FPP complex to the active FTase–FPP complex. The RC is defined as the distance between C_1 atom of the FPP substrate and the S_γ atom of the cysteine in the peptide, the two atom centers that will form a covalent bond in the next step in the catalytic mechanism.

Through multiple nanoseconds of classical MD simulations, two promising Mg^{2+}-binding motifs were identified (see Fig. 2): (1) WT1: the Mg^{2+} ion interacts with two oxygen atoms of the α-phosphate of the FPP, the carboxylate group from D147α and three water molecules; (2) WT2: the Mg^{2+} ion interacts with two oxygen atoms of the α- and β-phosphate of the FPP, the carboxylate group of the catalytically important D352β (Pickett, Bowers, & Fierke, 2003; Pickett, Bowers, Hartman, et al., 2003), and three water molecules. The WT2 showed substantial agreement with the Mg^{2+}-binding motif proposed based on mutagenesis studies which found D352β to

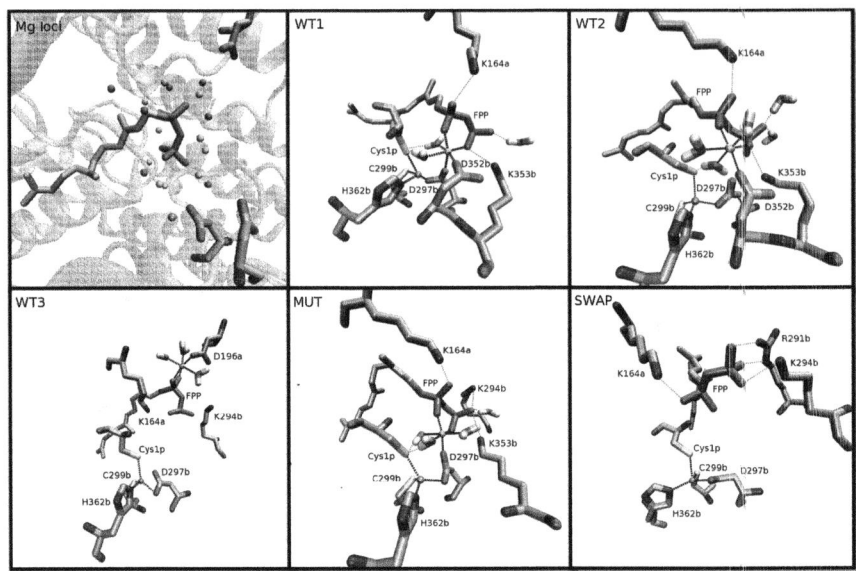

Fig. 2 Snapshots of FTase active site and important hydrogen bond patterns taken from different models. Mg^{2+} (in *green* (*dark gray* in the print version)) and Zn^{2+} (in gray) coordination is shown in *black dashed lines*, while the hydrogen bonds are shown as *red* (*gray* in the print version) *dashed lines*. Reprint with permission from Yang, Y., Chakravorty, D. K., & Merz, K. M. (2010). Finding a needle in the haystack: Computational modeling of Mg^{2+} binding in the active site of protein farnesyltransferase. Biochemistry, 49(44), 9658–9666. doi:10.1021/Bi1008358. Copyright (2010) American Chemical Society.

be a likely Mg^{2+} ligand (Pickett, Bowers, & Fierke, 2003; Pickett, Bowers, Hartman, et al., 2003). In order to validate these models, we subsequently performed umbrella sampling simulations. These PMF simulations revealed that both WT1 and WT2 prefer a shorter distance (d_{RC} is 5.8 and 6.1 Å separately, see Fig. 3) between two reacting centers than that found in the FTase–FPP complex (7.2 Å) (Cui & Merz, 2007) lacking Mg^{2+}. More importantly, from the WT2 PMF we discovered a nearly barrier-less transition ($\sim$0.8 kcal/mol) from the resting/inactive state to the intermediate state, where the two reacting centers are in even closer proximity (d_{RC} is $\sim$5.0 Å). Furthermore, mutation of Dβ352A (MUT system shown in Fig. 3) not only increased the distance between the two reacting centers at the resting state but also made the transition to the intermediate state much more difficult, while replacing the Mg^{2+} in WT2 by a water molecule (named as SWAP, as shown in Fig. 3) reproduced similar salt bridges and hydrogen-bonding patterns around FPP as previously observed in the

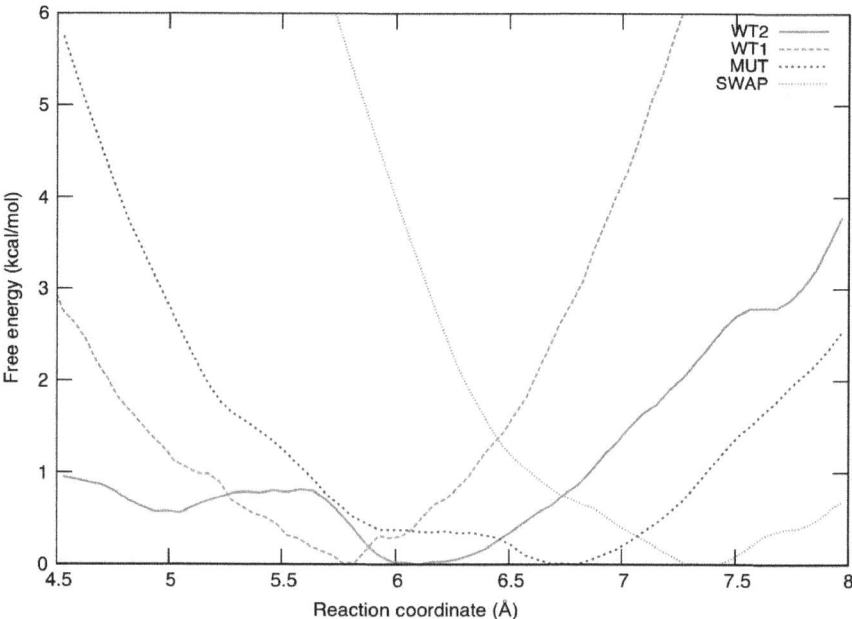

Fig. 3 Computed free energy profiles of the FTase conformational activation step for different FTase/FPP system. *Reprint with permission from Yang, Y., Chakravorty, D. K., & Merz, K. M. (2010). Finding a needle in the haystack: Computational modeling of Mg^{2+} binding in the active site of protein farnesyltransferase. Biochemistry, 49(44), 9658–9666. doi:10.1021/Bi1008358. Copyright (2010) American Chemical Society.*

FTase/FPP system lacking Mg^{2+}. Therefore, our study confirmed that the Mg^{2+} binds to FTase through key Mg–D352β interaction and suggested that Mg^{2+} binding might enhance catalysis by bringing the reaction partners into closer proximity.

3.1.3 Simulating the Farnesylation Reaction

With PMF simulations at the MM level, we gained insight into FTase-Mg^{2+} binding and the FTase activation mechanism; however, in order to fully understand the FTase catalytic mechanism, implementing PMF simulations of the chemical reaction at the QM/MM level was necessary. The reaction involves the breaking of a bond between the C_1 atom of FPP and its connecting phosphorus atom, and the formations of a bond between the C_1 atom of FPP and the S_γ atom of the target cysteine in the peptide. We decided to again use the distance between C_1 and S_γ as the RC because such a choice of RC covers the entire reaction process, and it is consistent with the RC choice for the conformational activation step. As mentioned in the method section, we first applied SMD to scan the RC and generate the initial structures for the subsequent umbrella sampling simulations. We conducted QM/MM PMF simulations for both FTase/FPP^{2-}/CVIM and FTase/FPP^{2-}/CVLS from 8.0 to 1.8 Å along the RC; thus, not only the chemical step but also the activation step were covered in this study using QM/MM methods (Yang, Wang, et al., 2012).

As shown in Fig. 4, the computed transition state free energy barrier for FTase/FPP^{2-}/CVIM is 20.6 kcal/mol, in excellent agreement with experimental measurements of 21.1 kcal/mol for TKCVIF peptide (Pickett, Bowers, Hartman, et al., 2003; Wu et al., 1999). QM/MM PMF simulations confirmed that the chemical step is the rate-limiting step as reported for TKCVIF based on the α-SKIEs (1.154±0.006) experiment (Pais et al., 2006) and was in good agreement with the MM-based PMF simulation results. More importantly, the transition state structure identified through

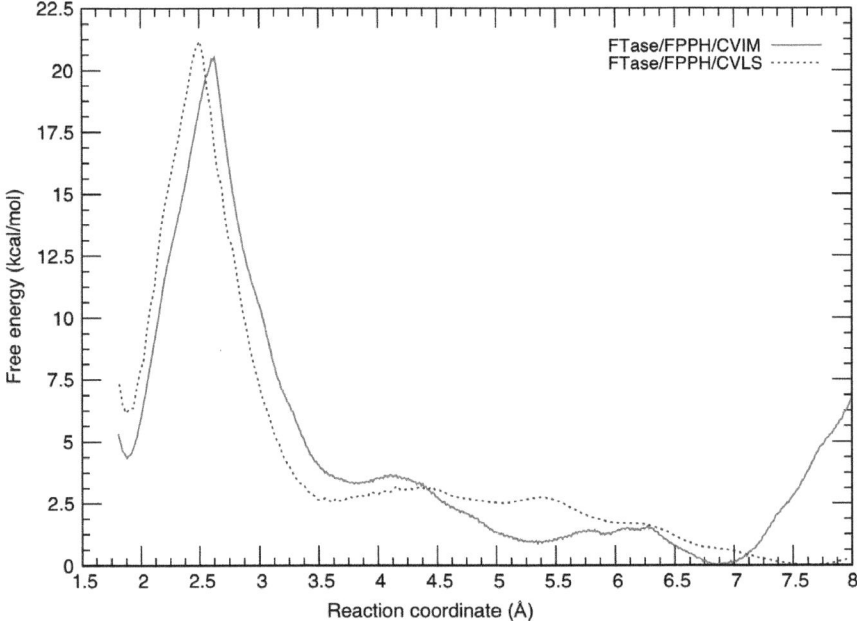

Fig. 4 Computed free energy profile of farnesylation catalyzed by FTase/CVIM (*red (dark gray* in the print version)) and FTase/CVLS (*blue (gray* in the print version)). *Reprint with permission from Yang, Y., Wang, B., Ucisik, M. N., Cui, G. L., Fierke, C. A., & Merz, K. M. (2012). Insights into the mechanistic dichotomy of the protein farnesyltransferase peptide substrates CVIM and CVLS. Journal of the American Chemical Society, 134(2), 820–823. doi:10.1021/ja209650h. Copyright (2012) American Chemical Society.*

PMF simulations confirmed the hypothesis of a so-called associative mechanism with dissociative characteristics. The free energy barrier for the FTase/FPP^{2-}/CVLS system is 21.3 kcal/mol, also in good agreement with experiment (20.0 kcal/mol) (Huang et al., 2000). However, it is clear that the highest barrier is associated with the transition state of the chemical step, contradicting to the hypothesis that the α-SKIE isotope effect of 1.00 ± 0.04 is due to the rate-limiting step being the activation step. We performed further QM calculations to reveal that the reason the two similar peptides have such different α-SKIEs is that FTase/FPP^{2-}/CVLS system features a more symmetric transition state, thus, having a near zero α-SKIE isotope effect. The QM/MM PMF simulations in this case successfully provided additional insights to the experimental observations.

3.2 Aromatic Prenyltransferase NphB

NphB is also a prenyltransferase that catalyzes the transfer of a prenyl group. In contrast to FTase or other protein prenyltransferases, NphB is an aromatic prenyltransferase, and it attaches a 10-carbon geranyl group to a variety of aromatic substrates (Kumano, Richard, Noel, Nishiyama, & Kuzuyama, 2008; Kuzuyama, Noel, & Richard, 2005; Tello, Kuzuyama, Heide, Noel, & Richard, 2008). In addition to its diverse substrate selectivity, NphB also exhibits rich product regioselectivity (Kumano et al., 2008; Kuzuyama et al., 2005; Tello et al., 2008). For example, NphB catalyzes the prenyl transfer between geranyl diphosphate (GPP) and 1,6-dihydroxynaphthalene (1,6-DHN). This prenylation reaction yields three products with the geranyl moiety attaching to different carbon atoms of 1,6-DHN (see Fig. 5). The originally identified major product, 5-geranyl-DHN, and minor product, 2-geranyl-DHN, were characterized with a product ratio of 10:1 (Kuzuyama et al., 2005). Later, another minor product, 4-geranly-DHN was reported with a much lower yield. However, the crystal structure (PDB # IZB6) exhibits a limited access between the C_1 atom of GPP and the C_2 atom of 1,6-DHN. In addition, it was hypothesized that the reaction undergoes a S_N1 type dissociative mechanism that features a carbocation intermediate (Kuzuyama et al., 2005). In order to understand the product regioselectivity and catalytic mechanism, we performed PMF simulations to study not only the dynamics but also the thermodynamics on the binding and reaction of NphB/GPP/1,6-DHN complex (Cui, Li, & Merz, 2007; Yang, Wang, et al., 2012).

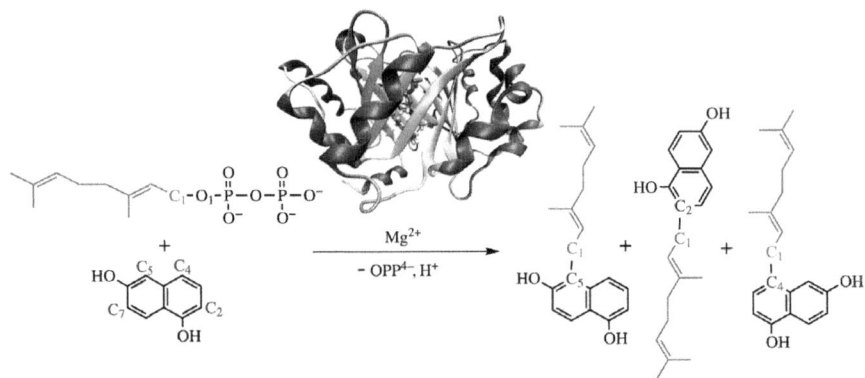

Fig. 5 Schematic representation of the geranylation catalyzed by NphB with GPP and 1,6-DHN. Important atoms, C_1, O_1 from GPP and C_2, C_4, C_5, C_7 from 1,6-DHN are labeled. *Reprint with permission from Yang, Y., Miao, Y. P., Wang, B., Cui, G. L., & Merz, K. M. (2012). Catalytic mechanism of aromatic prenylation by NphB.* Biochemistry, *51(12), 2606–2618. doi:10.1021/Bi201800m. Copyright (2012) American Chemical Society.*

3.2.1 Identifying the Product Regioselectivity Associated with NphB Catalysis

In the crystal structure of NphB/GPP/1,6-DHN complex, the C_1 atom is in close proximity to the C_5 atom of 1,6-DHN, whose subsequent reaction would lead to the major product. However, the C_2 atom of 1,6-DHN is about 7 Å away from the C_1 atom; thus, in order to generate the minor product (2-geranyl-DHN) a conformational rearrangement is needed. Through extensive MD simulations, we observed the rotation of 1,6-DHN, which resulted in three different binding states with two of these states poised to lead to the production of the major product, 5-geranyl-DHN, and the minor product, 2-geranyl-DHN, respectively. In order to quantitatively understand these binding states and the transitions among them, we carried out two-dimensional (2D) PMF simulations, as a function of two variables, the distance between C_1 of GPP and C_5 of 1,6-DHN (D_1) and the distance between C_1 of GPP and C_2 of 1,6-DHN (D_2). Umbrella sampling was used to compute the free energy profile associated with the observed dynamics of 1,6-DHN.

The free energy profile confirmed that there are three binding states, which are named S1, S2, and S3, as shown in Fig. 6. The global minimum, S1, is approximately 2.3 kcal/mol more favorable than S2 and S3 and leads directly to the major product. The S3 state is about 2 kcal/mol less stable

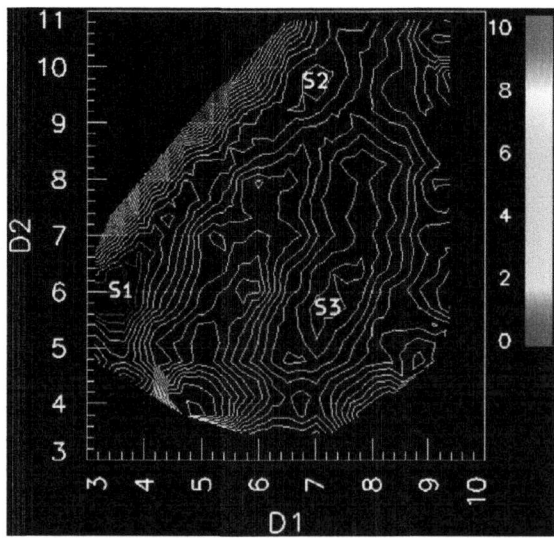

Fig. 6 2D Free energy profile (kcal/mol) of 1,6-DHN binding in NphB as a function of D_1 and D_2 (Å). *Reprint with permission from Cui, G., Li, X., & Merz, K. M., Jr. (2007). Understanding the substrate selectivity and the product regioselectivity of Orf2-catalyzed aromatic prenylations.* Biochemistry, 46(5), 1303–1311. doi:10.1021/bi062076z. *Copyright (2007) American Chemical Society.* (See the color plate.)

than S1 and possesses a shorter distance D2, at approximately 5.5 Å. Therefore, the S3 state allows access to the minor product, 5-geranyl-DHN. However, S1 and S3 are separated by a ~6 kcal/mol barrier; thus, the direct transition between S1 and S3 is somewhat unfavorable. The S2 state, on the other hand, is easily accessible from both the S1 and S3 states and is only around 0.4 kcal/mol less favorable than S3. Hence, we proposed that S2 serves as an intermediate state and provide a low barrier transition pathway between S1 and S3. In addition, through these PMF simulations we identified that both the S2 and S3 states are stabilized by hydrogen bonds to the side chains of Ser214 and Tyr288, similar to the hydrogen bond interaction in the NphB/flaviolin complex found in PDB # 1ZDW. Although we successfully explained the origin of one of the minor product, 2-geranyl-DHN, the 2.3 kcal/mol relative free energy difference between the S1 and S3 states is not in good agreement with the 10:1 product ratio between 5-geranyl-DHN and 2-geranyl-DHN. Thus, we speculated that in the reaction step, 2-geranylation had a lower transition state barrier than 5-geranylation (Fig. 7).

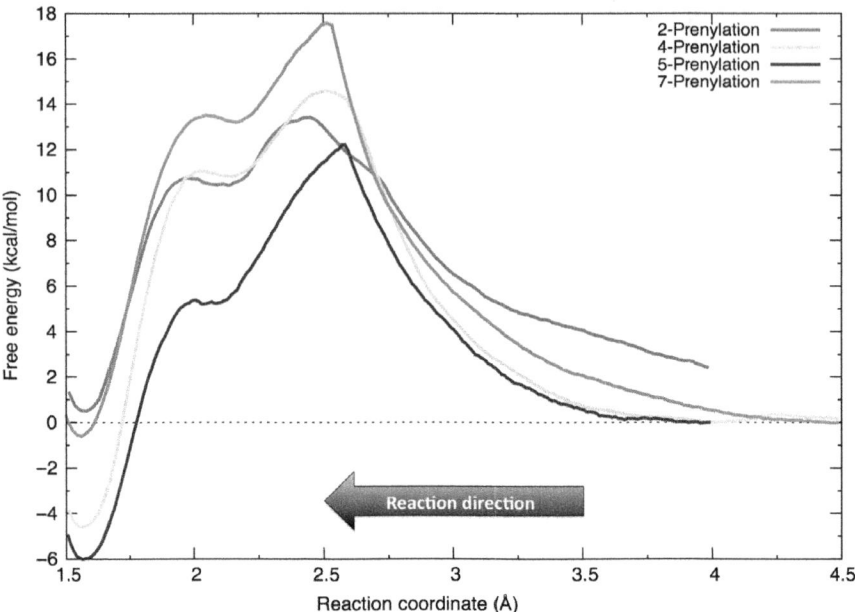

Fig. 7 Free energy profiles for the NphB catalyzed geranylation at different reaction sites in 1,6-DHN. P2 starts at 2.3 kcal/mol, which is the relative binding free energy difference of the P2 state to that of the P5 state. *Reprint with permission from Yang, Y., Miao, Y. P., Wang, B., Cui, G. L., & Merz, K. M. (2012). Catalytic mechanism of aromatic prenylation by NphB. Biochemistry, 51(12), 2606–2618. doi:10.1021/Bi201800m. Copyright (2012) American Chemical Society.*

3.2.2 Elucidating the Reaction Mechanism with QM/MM PMF Simulations

The previous MM PMF study left us with several questions regarding the NphB catalytic mechanism: (1) how does the chemical step compliment the binding step to yield the 10:1 product ratio between 5-geranylation and 2-geranylation? (2) The C_4 atom and C_7 atom can also be placed in close proximity to the C_1 atom of GPP. However, why does 4-geranylation have an even lower yield than 2-geranylation, while 7-geranylation appears inhibited or blocked? Finally, (3) if the reaction is really S_N1-like as proposed by Kuzuyama et al. (2005), how is the carbocation intermediate state stabilized? With these questions in mind, we extended our previous study with another sets of PMF simulations at the QM/MM level (Yang, Miao, Wang, Cui, & Merz, 2012).

Like in the FTase QM/MM PMF study, we carried out SMD simulations, which were followed by umbrella sampling simulations. Four sets of QM/MM PMF simulations were performed, with the RC defined as between the C_1 atom of GPP and the four available geranyl attachment sites, namely the C_2, C_4, C_5, and C_7 atoms of 1,6-DHN. This allowed us to scan the free energy surfaces associated with 2-, 4-, 5-, and 7-prenylation reactions, respectively. The free energy profile illustrates that although the overall barrier of 2-prenylation is higher than 5-prenylation (note the starting points of 2-prenylation and 5-prenylation are placed to represent the 2.3 kcal/mol free energy difference), the latter has a higher transition state barrier which compensates for the relative binding free energy difference to lead to a 0.7 kcal/mol overall free energy difference that matches the 10:1 final product ratio. In addition, 4-prenylation is about 0.7 kcal/mol energetically less favorable than 2-prenylation, thereby, explaining why its yield is even lower. Moreover, 7-prenylation must overcome the highest energy barrier (~17.6 kcal/mol), explaining why 7-geranyl-DHN has not been reported. The PMF profile also reveals an intermediate state for all four prenylation channels, which confirms that NphB catalysis involves a S_N1 mechanism, and presents a novel π-chamber to stabilize the carbocation intermediate. The π-chamber is composed of 1,6-DHN and two aromatic side chains from the protein, Tyr121 and Tyr216 (see Fig. 8), and possibly protects the cation from nearby water molecules. Further QM analysis quantitatively characterizes the energy contribution of the π-chamber to the stabilization of the intermediate of -20.6, -13.4, -6.8, and $+1.0$ kcal/mol for

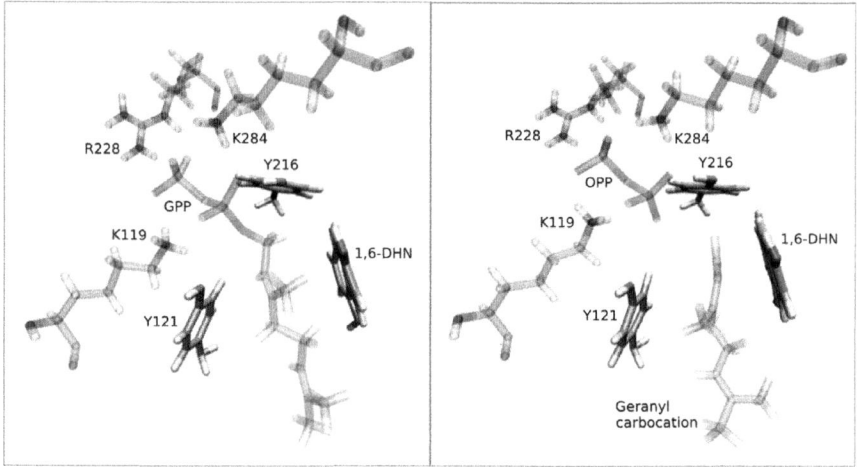

Fig. 8 The novel π-chamber identified in the NphB catalytic channel at the resting state (*left*) and stabilizing the carbocation intermediate state (*right*).

2-, 5-, 4-, and 7-prenylation. Therefore, it is likely that the π-chamber also selectively stabilizes the forming cation toward the favored reaction channel. We also found in the PMF simulations that a water mediated proton transfer facilitates the deprotonation at the prenylation site to lead to the final product in all four channels. However, such a deprotonation is nearly barrier-less with regard to the carbocation intermediate state. The product release was found to take place almost automatically, making carbocation formation the rate-limiting step. With PMF simulations, we were able to reproduce the entire NphB catalytic pathway including substrate binding, the chemical reaction, and product release steps and provide useful insights for future active site engineering.

3.3 Aspergillus Fumigatus Prenyltransferase

Aspergillus Fumigatus Prenyltransferase (FtmPT1) is a fungi indole prenyltransferase that catalyzes the prenyl transfer reactions between dimethyl allyl pyrophosphate (DMAPP) and tryptophan (and its derivatives) in its binding site. Similar to the NphB-catalyzed geranyl transfer, the FtmPT1-catalyzed reactions are composed of a prenyl transfer step and a subsequent proton transfer step. With DMAPP and Brevianamide F (cyclo-L-Trp-L-Pro) bound, FtmPT1 exhibits two competing reaction channels leading to two different products, a cell growth inhibitor Tryprostatin B by bonding between C5@DMAPP and C8@Brevianamide F (channel C5–C8), and a novel five-membered ring compound via bond formation between C2@DMAPP and C9@Brevianamide F (channel C2–C9) (Fig. 9). Experiments revealed that for the wild-type (WT) enzyme the product is Tryprostatin B while for the G115T mutant the product is the novel five-membered ring compound (Jost et al., 2010). For both reaction channels in both the WT FtmPT1 and G115T mutant, the reaction undergoes a S_N1 reaction mechanism. The reaction first occurs with a C–O bond breakage to generate a carbocation, followed by a C–C bond formation between the carbocation and Brevianamide F. The resulting intermediates subsequently deprotonate to form the final products with the extra protons shuttled by water molecules to Glu102, which is the final proton receptor (Fig. 9). The origin of the observed product regioselectivity is poorly understand; hence, we performed PMF simulations at both the MM and QM/MM levels to address the observed product regioselectivity in WT FtmPT1 and the G115T mutant (Pan, Miao, Yang, & Merz, in preparation; Pan, Yang, & Merz, 2014). It was found that PMF simulations

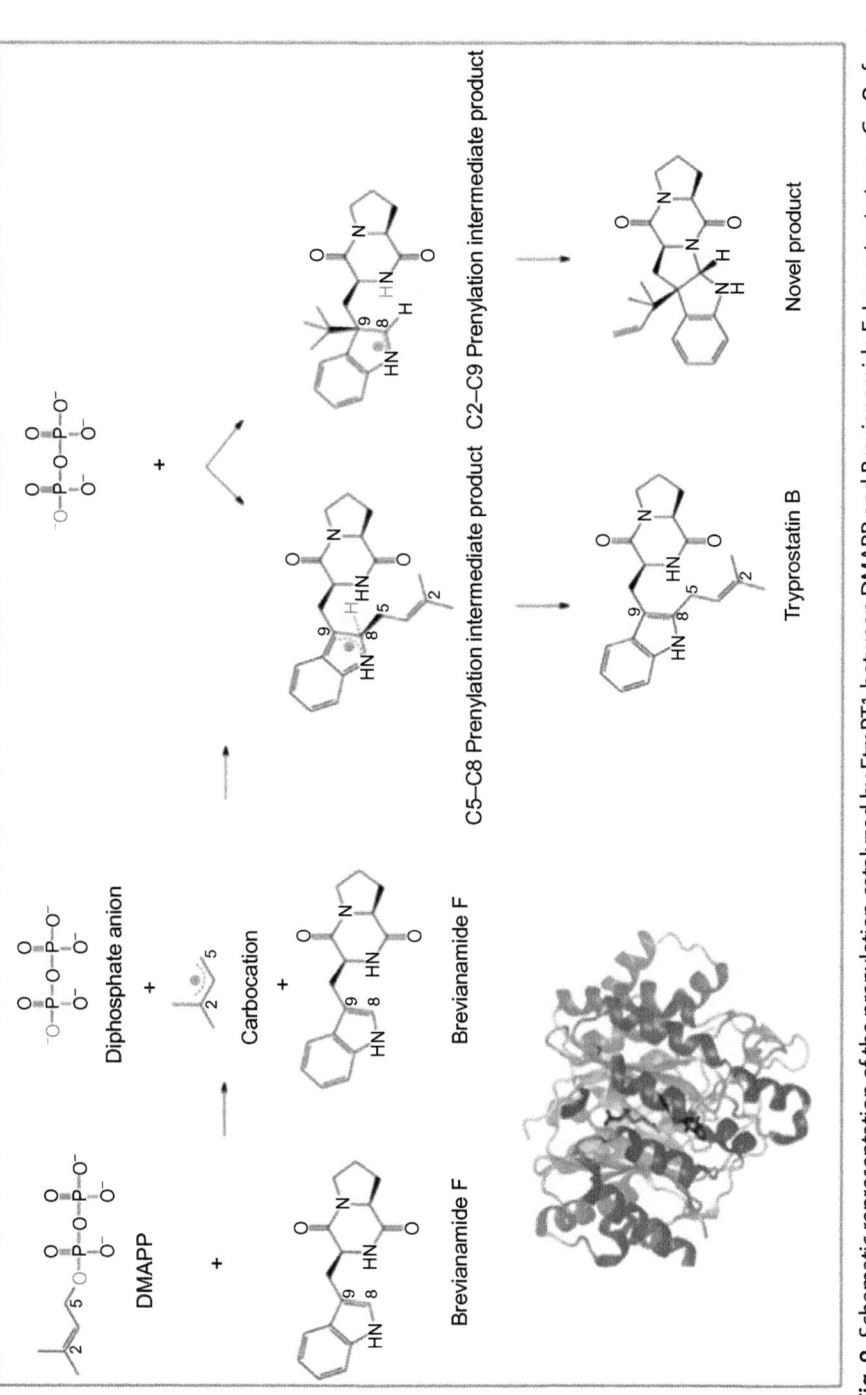

Fig. 9 Schematic representation of the prenylation catalyzed by FtmPT1 between DMAPP and Brevianamide F. Important atoms, C_2, O_5 from DMAPP and C_8, C_9 from Brevianamide F are labeled.

were able to explain why Tryprostatin B is the sole product for WT FtmPT1. However, for the G115T mutant, we introduced dynamic reaction analysis in addition to the PMF simulations to address why the channel C2–C9 is the favored reaction channel allowing for the formation of the novel five-membered ring compound.

3.3.1 Regioselectivity of WT FtmPT1

For the WT FtmPT1, the system selects the C5–C8 channel to produce Tryprostatin B. The highest reaction free energy barrier determined by experiment was ~17.4 kcal/mol (Jost et al., 2010). Our PMF simulations revealed a free energy barrier of 19.3 kcal/mol for the C5–C8 channel and an 18.5 kcal/mol free energy barrier for the C2–C9 channel during the O–C bond breakage (Pan et al., 2014) (Fig. 10) step. Based on PMF simulations, for both reaction channels this step is the rate-limiting step. At this stage, the two channels are competitive as they have similar free energy barriers; however, in the subsequent deprotonation step, the C5–C8 channel has a transition state that is about 15.3 kcal/mol more favorable than that of the C2–C9 channel and a final product which is ~11.6 kcal/mol more stable than the one from C2–C9 channel (Fig. 10). Hence, the deprotonation step makes the C5–C8 channel dominant along the two bifurcating reaction channel, which leads to Tryprostatin B as the sole product for WT FtmPT1 catalysis.

3.3.2 Regioselectivity of G115T Mutant

In the G115T mutant, the novel five-membered ring product is the major product while Tryprostatin B becomes the minor product, found in a final product ratio of 40:1 (Jost et al., 2010). Therefore, in G115T, the C2–C9 channel should be the major channel while the C5–C8 channel should be deprecated. We performed PMF simulations that predicted the O–C bond breakage of DMAPP in the prenylation step to be the rate-limiting step, with a transition state barrier of 24.6 kcal/mol for the C2–C9 channel and of 24.9 kcal/mol for the C5–C8 channel. Therefore, at this stage of the catalysis, the two channels are still competitive. On the other hand, for the deprotonation step, we observed a transition state for the C2–C9 channel of ~3.5 kcal/mol less favorable than the C5–C8 channel, but the final product of the C2–C9 channel was ~3.2 kcal/mol more stable than the C5–C8 channel. Hence, from the QM/MM PMF studies, it was not clear why the novel compound was preferred over the native product in the G115T mutant. Subsequently, we proceeded with further investigations on the reaction dynamics in these systems, which have proved to be an effective

A Complete reaction for C5–C8 reaction pathway to form Tryprostatin B

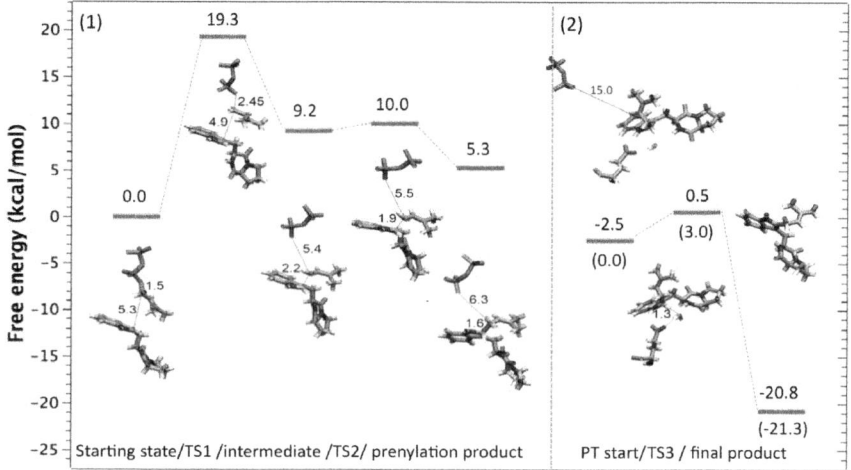

B Complete reaction for C2–C9 reaction pathway to form novel product

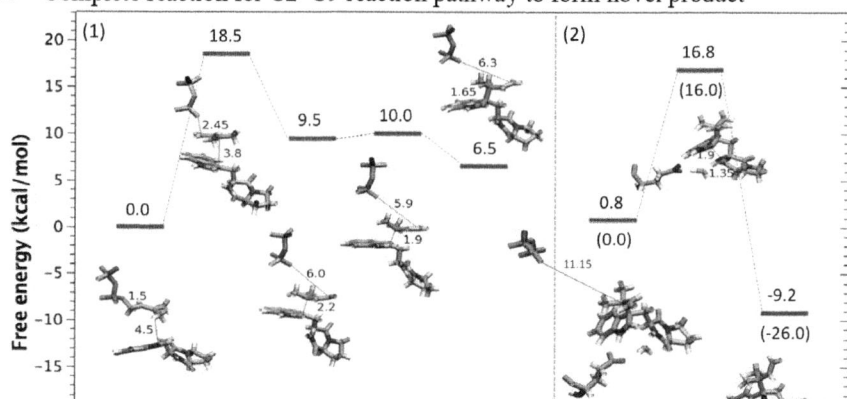

Fig. 10 Complete reaction profile with relative free energy and the associated structure for the C5–C8 and C2–C9 reaction pathways: (1) prenyl transfer reaction and (2) proton transfer reaction. Distances for the relevant reaction coordinates are given on the structures. Free energies compared to the starting state are given above the *green* (*gray* in the print version) and *blue* (*dark gray* in the print version) *lines*. Note the energies in parentheses for the proton transfer step that are compared to their own starting state minima (for the diphosphate and prenylated carbocation). The starting state for reaction 2 (proton transfer) is the relaxed state of the product from reaction 1 (prenylation reaction). *Reprint from Pan, L. L., Yang, Y., & Merz, K. M. (2014). Origin of product selectivity in a prenyl transfer reaction from the same intermediate: Exploration of multiple FtmPT1-catalyzed prenyl transfer pathways. Biochemistry, 53(38), 6126–6138. doi:10.1021/bi500747z. Copyright (2014) American Chemical Society.*

tool to study bifurcated reaction channels (Carpenter, 1985). We started at the rate-limiting step—the transition state for the O–C bond breakage of DMAPP during the prenylation step—and we then carried out "free" (unconstrained) MD simulations. We found that for all C–C bond formation transition states (ie, starting at the C5–C8 or C2–C9 transition states) studied by free MD simulations, the C2–C9 channel is favored over the C5–C8 channel, by a ratio as 61:1, suggesting that reaction dynamics give a preference C2–C9 product formation. Therefore, while the reaction free energies are competitive for the two channels, we observe that the G115T mutant "dynamically" selects the C2–C9 reaction channel. We hypothesize that thermodynamics determines the product formation in WT FtmPT1, while dynamics plays a more important role in the G115T mutant.

3.4 Cytochrome P450s

CYP is a family of enzymes that catalyze the conversion of nonnatural substrates such as drugs to products that can be released by living organisms, which is also referred to as xenobiotic metabolism. CYPs play a vital role in drug metabolism and account for approximately 75% of drug metabolism (Williams et al., 2004). CYPs require an iron heme cofactor for their catalytic power. The iron can possess different charge, spin and coordination states (Shaik et al., 2010), while the reactive species is believed to be compound I (Cpd I) (Rittle & Green, 2010; Sheng et al., 2009), a high valent iron–oxo radical complex. Despite the report that Cpd I exhibits similar catalytic power across different CYPs (Lonsdale, Olah, Mulholland, & Harvey, 2011), different CYP isozymes still exhibit quite different specificity and regioselectivity for any given drug compound. Predicting substrate specificity and product regioselectivity for a set of CYPs is essential to predicting metabolic outcome for drug compounds and is key to drug discovery. Studies have been reported that focus on investigating the role that transition state barriers play in determining product regioselectivity of CYP catalysis (Lonsdale et al., 2013; Olah, Mulholland, & Harvey, 2011). However, as demonstrated in our NphB and FtmPT1 cases, relative binding free energies of different binding poses, in addition to transition state barriers, also play crucial roles in determining product regioselectivity. Indeed, a systemic study focusing on CYP substrate binding had not been reported. For this reason, we performed a set of PMF simulations to study the binding of a well-studied drug, acetaminophen (also known as Tylenol®), to a set of CYPs in order to illustrate the importance of relative binding free energies to product regioselectivity. As known, acetaminophen oxidation catalyzed by CYPs produces two metabolites, the highly reactive and toxic N-acetyl-*p*-benzoquinone imine (NAPQI) and the nontoxic 3-hydroxy-acetaminophen (3-OH-APAP) (Corcoran, Mitchell, Vaishnav, & Horning, 1980; Potter et al., 1973). At high doses, the fast accretion of NAPQI leads to hepatotoxicity (Davidson & Eastham, 1966; Thomson & Prescott, 1966). Several CYP isozymes, including CYP1A2, CYP2A6, CYP2C9, CYP2E1, and CYP3A4, metabolize acetaminophen. However, as for other drugs, different CYPs exhibit different specificity and regioselectivity toward acetaminophen. For instance, CYP2E1 mainly produces NAPQI, while CYP2A6 principally generates 3-OH-APAP (Kalsi, Wood, Waring, & Dargan, 2011).

The study started with docking acetaminophen into crystal structures of different CYPs followed by long MD equilibration simulations. Subsequently, PMF simulations were performed to investigate the thermodynamics of acetaminophen–CYP binding for each of five CYPs included in this study. Similar to the NphB binding study, we decided to perform a set of 2D PMF simulations with the umbrella sampling technique to investigate the relative binding free energies of different acetaminophen-binding states in different CYP isozymes. The two RCs were the distance between the O_1 atom of the iron heme and two possible oxidation sites of acetaminophen, NH and OH, and termed as RC1 (d_{O1-NH}) and RC2 (d_{O1-OH}) (see Fig. 11). With HPC, we were able to perform approximately 3 μs PMF simulations

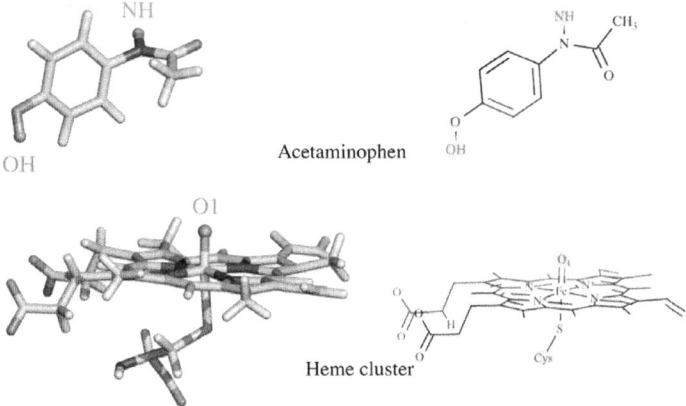

Fig. 11 Labeling of important atoms from acetaminophen and the iron heme cluster. *Reprinted from Yang, Y., Wong, S. E., & Lightstone, F. C. (2014). Understanding a substrate's product regioselectivity in a family of enzymes: A case study of acetaminophen binding in cytochrome P450s. PloS One, 9(2), e87058. doi:10.1371/journal.pone.0087058.*

for each CYP–acetaminophen system and thoroughly examine the free energy surface associated with acetaminophen binding.

In general, the PMF simulations reveal a relatively flat free energy surfaces (Fig. 12) for CYPs with large binding pocket, such as CYP3A4 and CYP2C9, and a preferred binding scheme for CYPs with narrow binding sites, including CYP1A2, CYP2A6, and CYP2E1. For CYP3A4, a binding state (S1) with shorter RC1 but longer RC2 and another binding state (S2) with shorter RC2 but longer RC1 were identified with a rather small energy difference (0.5 kcal/mol, S1 more stable). The S1 and S2 states set two acetaminophen oxidation sites in close proximity to iron heme for the following chemical step that lead to the toxic and nontoxic metabolites, respectively. A third state (S3) with slightly higher free energy (~1.3 kcal/mol) was also identified. S3 could serve as an intermediate for low-energy transitions between S1 and S2. With such a small relative

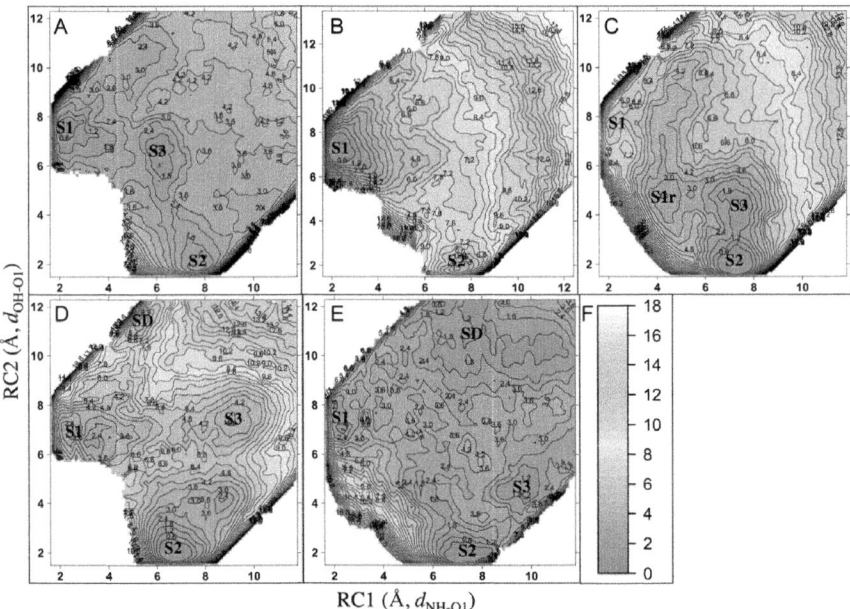

Fig. 12 Free energy profiles for CYP–acetaminophen binding. (A) CYP3A4, (B) CYP2E1, (C) CYP2A6, (D) CYP1A2, (E) CYP2C9. Topography energy legend with energy values (kcal/mol) represented by defined color is given in (F). Binding states, S1, S2, S3, S1r, and SD (not discussed in this chapter) are labeled. *Reprinted from Yang, Y., Wong, S. E., & Lightstone, F. C. (2014). Understanding a substrate's product regioselectivity in a family of enzymes: A case study of acetaminophen binding in cytochrome P450s. PloS One, 9(2), e87058. doi:10.1371/journal.pone.0087058.* (See the color plate.)

binding free energy difference between the productive binding states, it is expected that CYP3A4 does not have a dominant metabolite (Rendic, 2002). Like CYP3A4, CYP2C9 has a relative large pocket, affording acetaminophen a large number of degrees of freedom; therefore, it is not surprising to find that the CYP2C9–acetaminophen complex also exhibits a rather flat free energy surface with a relatively small energy difference between the S1 and S2 states (~2 kcal/mol). Such a small relative binding free difference explains why this CYP isozyme can convert acetaminophen to both metabolites without a significant preference (Rendic, 2002).

The story for CYP2E1 is quite different, as a dominant S1 state was observed. Energetically, the S1 state is about 4.2 kcal/mol more favorable than the S2 state, while the transition barrier between these two states is approximately 7.3 kcal/mol due to the lack of an intermediate state like the S3 state in CYP3A4. Such significant preference for the S1 state over the S2 state would result in an enhanced production of the toxic NAPQI metabolite, which agrees with the fact that CYP2E1 is the major NAPQI producer (Snawder, Roe, Benson, & Roberts, 1994).

In contrast to CYP2E1, CYP2A6 is known to principally convert acetaminophen to the nontoxic metabolite, 3-OH-APAP (Rendic, 2002), with a product ratio of 3:1 (Chen et al., 1998). The PMF simulations discovered a S1 state that does not represent a stable energy basin, but a S1r state was found that can serve as a replacement for the S1 state to position the N-oxidation site for the production of the toxic metabolite. A S2 state was identified to be approximately 2.5 kcal/mol more favorable than the S1r state, while a S3 state was found to provide a energetically favorable pathway for the S1r ↔ S2 transition. The binding free energies agree with the observation that CYPA26 mainly metabolize acetaminophen to 3-OH-APAP. With the known product ratio between 3-OH-APAP and NAPQI being 3:1, we used the relative binding free energy difference coupled with the Curtin–Hammett principle (Carey & Sundberg, 1984) to estimate that the transition state barrier for N-oxidation is about 1.8 kcal/mol (Fig. 13) lower than that for 3-hydroxylation. With such "intrinsic reactivity" identified, we were able to explain why CYP1A2 converts acetaminophen to both metabolites with no obvious preference. In fact, the S2 state of CYP1A2–acetaminophen complex is more than 2 kcal/mol more favorable than the toxic metabolite leading to the S1 state; however, with the relative transition state free energy difference identified in the CYP2A6 case, we would expect that both metabolites could be produced with 3-OH-APAP slightly favored. We also applied the "intrinsic reactivity" concept to make

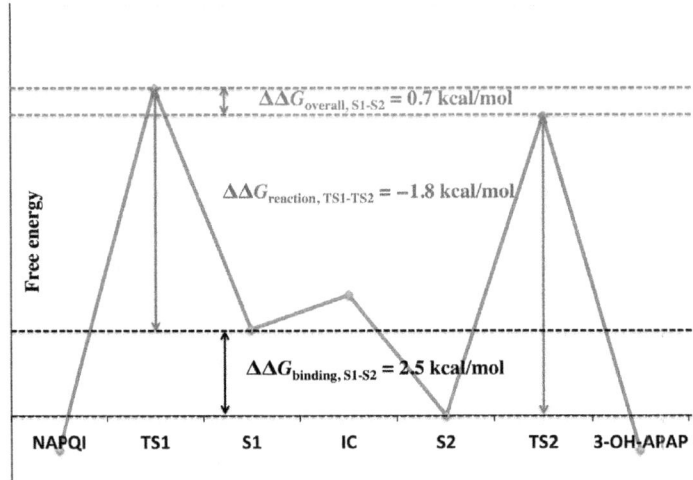

Fig. 13 Cartoon representation of the relative free energy for CYP–acetaminophen metabolism according to the Curtin–Hammett principle. *Reprinted from Yang, Y., Wong, S. E., & Lightstone, F. C. (2014). Understanding a substrate's product regioselectivity in a family of enzymes: A case study of acetaminophen binding in cytochrome P450s. PloS One, 9(2), e87058. doi:10.1371/journal.pone.0087058.*

rational predictions on the product regioselectivity of acetaminophen metabolism by different CYPs (Table 1), and found our predictions to be in good agreement with experimental findings (Rendic, 2002). We suggested that for CYPs that have large binding pockets where a ligand would have many degrees of freedom, the determinant of regioselectivity is more likely due to transition state barriers, while for other CYPs that have relatively small binding sites where the ligand is restrained to a preferred pose, binding free energy plays a more important role in deciding the regioselectivity.

4. SUMMARY

PMF simulation is an important tool to explore the free energy landscape associated with high-energy barrier transitions, such as conformational rearrangements, protein–substrate binding, and chemical reactions. Herein, we presented several studies that applied PMF simulations to study enzyme catalytic mechanisms and addressed several important questions, such as reaction mechanism and product regioselectivity, for FTase, NphB, FtmPT1, and CYPs. Compared to brute-force approaches, PMF simulations are

Table 1 APAP-CYPs Regioselectivity Determined by Experimental Data and Calculation

CYP/Source	Experimental Observation (Rendic, 2002)	Prediction Suggested by Free Energy Profile	Prediction after Including "Intrinsic Reactivity"[a]
CYP1A2	3-OH-APAP and NAPQI (low activity)	3-OH-APAP	3-OH-APAP slightly favored
CYP2A6	3-OH-APAP (major) and NAPQI (minor)	3-OH-APAP	3-OH-APAP
CYP2C9	3-OH-APAP (low rate) and NAPQI (low activity)	3-OH-APAP	3-OH-APAP and NAPQI comparable
CYP2E1	3-OH-APAP (minor) and NAPQI (major)	NAPQI	NAPQI
CYP3A4	3-OH-APAP and NAPQI	3-OH-APAP and NAPQI comparable[b]	3-OH-APAP and NAPQI comparable[b]

[a]Empirical TS barrier different refers to the estimated energy difference between N-oxidation and 3-C-hydroxylation from CYP2A6 study.
[b]Both S2 and S3 could lead to 3-OH-APAP. NAPQI could be slightly favored product.
Reprinted from Yang, Y., Wong, S. E., & Lightstone, F. C. (2014). Understanding a substrate's product regioselectivity in a family of enzymes: A case study of acetaminophen binding in cytochrome P450s. *PloS One, 9*(2), e87058. doi:10.1371/journal.pone.0087058.

parallelizable along a given RC which afford significant speed-ups. However, although the free energy results obtained in these studies are in good agreement with experimental results, we have to be aware that the accuracy of PMF simulations is affected by the accuracy of the chosen MM force field and QM/MM potential. Therefore, improvements made toward more accurate MM force fields and QM/MM implementations will make PMF simulations a more effective and accurate tool to investigate enzyme catalysis.

ACKNOWLEDGMENTS

K.M.M. would like to acknowledge the financial support from the United States National Institutes of Health (RO1's GM044974 and GM066859). Y.Y. and F.C.L. would like to thank Laboratory Directed Research and Development for funding, 12-SI-004, from Lawrence Livermore National Laboratory. Y.Y. and F.C.L. also thank Livermore Computing for the generous computer time. This work was performed under the auspices of the U.S. Department of Energy by Lawrence Livermore National Laboratory under Contract DE-AC52-07NA27344. Release number: LLNL-JRNL-681360.

REFERENCES

Carey, F. A., & Sundberg, R. J. (1984). *Advanced organic chemistry part A: Structure and mechanisms* (2nd ed.). New York, NY: Plenum Press.
Carpenter, B. K. (1985). Trajectories through an intermediate at a fourfold branch point. Implications for the stereochemistry of biradical reactions. *Journal of the American Chemical Society, 107*(20), 5730–5732. http://dx.doi.org/10.1021/Ja00306a021.
Chen, W. Q., Koenigs, L. L., Thompson, S. J., Peter, R. M., Pettie, A. E., Trager, W. F., & Nelson, S. D. (1998). Oxidation of acetaminophen to its toxic quinone imine and nontoxic catechol metabolites by baculovirus-expressed and purified human cytochromes P450 2E1 and 2A6. *Chemical Research in Toxicology, 11*(4), 295–301. http://dx.doi.org/10.1021/Tx9701687.
Corcoran, G. B., Mitchell, J. R., Vaishnav, Y. N., & Horning, E. C. (1980). Evidence that acetaminophen and N-hydroxyacetaminophen form a common arylating intermediate, N-acetyl-para-benzoquinoneimine. *Molecular Pharmacology, 18*(3), 536–542.
Cui, G., Li, X., & Merz, K. M., Jr. (2007). Understanding the substrate selectivity and the product regioselectivity of Orf2-catalyzed aromatic prenylations. *Biochemistry, 46*(5), 1303–1311. http://dx.doi.org/10.1021/bi062076z.
Cui, G., & Merz, K. M., Jr. (2007). Computational studies of the farnesyltransferase ternary complex part II: The conformational activation of farnesyldiphosphate. *Biochemistry, 46*(43), 12375–12381. http://dx.doi.org/10.1021/Bi701324t.
Cui, G., Wang, B., & Merz, K. M., Jr. (2005). Computational studies of the farnesyltransferase ternary complex part I: Substrate binding. *Biochemistry, 44*, 16513.
Davidson, D. G., & Eastham, W. N. (1966). Acute liver necrosis following overdose of paracetamol. *British Medical Journal, 2*(5512), 497–499.
Dolence, J. M., & Poulter, C. D. (1995). A mechanism for posttranslational modifications of proteins by yeast protein farnesyltransferase. *Proceedings of the National Academy of Sciences of the United States of America, 92*(11), 5008–5011.
Grubmuller, H., Heymann, B., & Tavan, P. (1996). Ligand binding: Molecular mechanics calculation of the streptavidin biotin rupture force. *Science, 271*(5251), 997–999. http://dx.doi.org/10.1126/science.271.5251.997.
Hightower, K. E., Huang, C. C., Casey, P. J., & Fierke, C. A. (1998). H-ras peptide and protein substrates bind protein farnesyltransferase as an ionized thiolate. *Biochemistry, 37*(44), 15555–15562.
Ho, M.-H., Vivo, M. D., Peraro, M. D., & Klein, M. L. (2009). Unraveling the catalytic pathway of metalloenzyme farnesyltransferase through QM/MM computation. *Journal of Chemical Theory and Computation, 5*(6), 1657–1666. http://dx.doi.org/10.1021/ct8004722.
Huang, C. C., Hightower, K. E., & Fierke, C. A. (2000). Mechanistic studies of rat protein farnesyltransferase indicate an associative transition state. *Biochemistry, 39*(10), 2593–2602.
Isralewitz, B., Gao, M., & Schulten, K. (2001). Steered molecular dynamics and mechanical functions of proteins. *Current Opinion in Structural Biology, 11*(2), 224–230. http://dx.doi.org/10.1016/S0959-440x(00)00194-9.
Jarzynski, C. (1997a). Equilibrium free-energy differences from nonequilibrium measurements: A master-equation approach. *Physical Review E, 56*(5), 5018–5035. http://dx.doi.org/10.1103/PhysRevE.56.5018.
Jarzynski, C. (1997b). Nonequilibrium equality for free energy differences. *Physical Review Letters, 78*(14), 2690–2693. http://dx.doi.org/10.1103/PhysRevLett.78.2690.
Jost, M., Zocher, G., Tarcz, S., Matuschek, M., Xie, X. L., Li, S. M., & Stehle, T. (2010). Structure-function analysis of an enzymatic prenyl transfer reaction identifies a reaction chamber with modifiable specificity. *Journal of the American Chemical Society, 132*(50), 17849–17858. http://dx.doi.org/10.1021/ja106817c.

Kalsi, S. S., Wood, D. M., Waring, W. S., & Dargan, P. I. (2011). Does cytochrome P450 liver isoenzyme induction increase the risk of liver toxicity after paracetamol overdose? *Open Access Emergency Medicine*, *3*(1), 69–76.

Kastner, J. (2011). Umbrella sampling. *Wiley Interdisciplinary Reviews-Computational Molecular Science*, *1*(6), 932–942. http://dx.doi.org/10.1002/wcms.66.

Kirkwood, J. G. (1935). Statistical mechanics of fluid mixtures. *Journal of Chemical Physics*, *3*(5), 300–313. http://dx.doi.org/10.1063/1.1749657.

Kumano, T., Richard, S. B., Noel, J. P., Nishiyama, M., & Kuzuyama, T. (2008). Chemoenzymatic syntheses of prenylated aromatic small molecules using Streptomyces prenyltransferases with relaxed substrate specificities. *Bioorganic & Medicinal Chemistry*, *16*(17), 8117–8126. http://dx.doi.org/10.1016/J.Bmc.2008.07.052.

Kumar, S., Bouzida, D., Swendsen, R. H., Kollman, P. A., & Rosenberg, J. M. (1992). The weighted histogram analysis method for free-energy calculations on biomolecules. 1. The method. *Journal of Computational Chemistry*, *13*(8), 1011–1021. http://dx.doi.org/10.1002/jcc.540130812.

Kuzuyama, T., Noel, J. P., & Richard, S. B. (2005). Structural basis for the promiscuous biosynthetic prenylation of aromatic natural products. *Nature*, *435*(7044), 983–987. http://dx.doi.org/10.1038/Nature03668.

Lindorff-Larsen, K., Piana, S., Dror, R. O., & Shaw, D. E. (2011). How fast-folding proteins fold. *Science*, *334*(6055), 517–520. http://dx.doi.org/10.1126/science.1208351.

Lonsdale, R., Houghton, K. T., Żurek, J., Bathelt, C. M., Foloppe, N., de Groot, M. J., ... Mulholland, A. J. (2013). Quantum mechanics/molecular mechanics modeling of regioselectivity of drug metabolism in cytochrome P450 2C9. *Journal of the American Chemical Society*, *135*, 8001–8015. http://dx.doi.org/10.1021/ja402016p. ASAP.

Lonsdale, R., Olah, J., Mulholland, A. J., & Harvey, J. N. (2011). Does compound I vary significantly between isoforms of cytochrome P450? *Journal of the American Chemical Society*, *133*(39), 15464–15474. http://dx.doi.org/10.1021/Ja203157u.

Mccammon, J. A., Gelin, B. R., & Karplus, M. (1977). Dynamics of folded proteins. *Nature*, *267*(5612), 585–590. http://dx.doi.org/10.1038/267585a0.

Mu, Y. Q., Gibbs, R. A., Eubanks, L. M., & Poulter, C. D. (1996). Cuprate-mediated synthesis and biological evaluation of cyclopropyl- and tert-butylfarnesyl diphosphate analogs. *Journal of Organic Chemistry*, *61*(23), 8010–8015.

Olah, J., Mulholland, A. J., & Harvey, J. N. (2011). Understanding the determinants of selectivity in drug metabolism through modeling of dextromethorphan oxidation by cytochrome P450. *Proceedings of the National Academy of Sciences of the United States of America*, *108*(15), 6050–6055. http://dx.doi.org/10.1073/Pnas.1010194108.

Pais, J. E., Bowers, K. E., & Fierke, C. A. (2006). Measurement of the alpha-secondary kinetic isotope effect for the reaction catalyzed by mammalian protein farnesyltransferase. *Journal of the American Chemical Society*, *128*(47), 15086–15087. http://dx.doi.org/10.1021/Ja065838m.

Pan, L. L., Miao, Y., Yang, Y., & Merz, K. M., Jr. (in preparation). Mechanism of formation of the non-standard product in the prenyltransfease reaction of the G115T mutant of FtmPT1: A case of reaction dynamics calling the shots?

Pan, L. L., Yang, Y., & Merz, K. M. (2014). Origin of product selectivity in a prenyl transfer reaction from the same intermediate: Exploration of multiple FtmPT1-catalyzed prenyl transfer pathways. *Biochemistry*, *53*(38), 6126–6138. http://dx.doi.org/10.1021/bi500747z.

Pickett, J. S., Bowers, K. E., & Fierke, C. A. (2003). Mutagenesis studies of protein farnesyltransferase implicate aspartate beta 352 as a magnesium ligand. *Journal of Biological Chemistry*, *278*(51), 51243–51250. http://dx.doi.org/10.1074/Jbc.M309226200.

Pickett, J. S., Bowers, K. E., Hartman, H. L., Fu, H. W., Embry, A. C., Casey, P. J., & Fierke, C. A. (2003). Kinetic studies of protein farnesyltransferase mutants establish active substrate conformation. *Biochemistry, 42*(32), 9741–9748. http://dx.doi.org/10.1021/Bi0346852.

Potter, W. Z., Davis, D. C., Mitchell, J. R., Jollow, D. J., Gillette, J. R., & Brodie, B. B. (1973). Acetaminophen-induced hepatic necrosis. 3. Cytochrome P-450-mediated covalent binding in-vitro. *Journal of Pharmacology and Experimental Therapeutics, 187*(1), 203–210.

Rendic, S. (2002). Summary of information on human CYP enzymes: Human P450 metabolism data. *Drug Metabolism Reviews, 34*(1–2), 83–448.

Rittle, J., & Green, M. T. (2010). Cytochrome P450 compound I: Capture, characterization, and C–H bond activation kinetics. *Science, 330*(6006), 933–937. http://dx.doi.org/10.1126/Science.1193478.

Saderholm, M. J., Hightower, K. E., & Fierke, C. A. (2000). Role of metals in the reaction catalyzed by protein farnesyltransferase. *Biochemistry, 39*(40), 12398–12405. http://dx.doi.org/10.1021/Bi0011781.

Shaik, S., Cohen, S., Wang, Y., Chen, H., Kumar, D., & Thiel, W. (2010). P450 enzymes: Their structure, reactivity, and selectivity-modeled by QM/MM calculations. *Chemical Reviews (Washington, DC, United States), 110*(2), 949–1017. http://dx.doi.org/10.1021/Cr900121s.

Sheng, X., Zhang, H. M., Im, S. C., Horner, J. H., Waskell, L., Hollenberg, P. F., & Newcomb, M. (2009). Kinetics of oxidation of benzphetamine by compounds I of cytochrome P450 2B4 and its mutants. *Journal of the American Chemical Society, 131*(8), 2971–2976. http://dx.doi.org/10.1021/Ja808982g.

Snawder, J. E., Roe, A. L., Benson, R. W., & Roberts, D. W. (1994). Loss of Cyp2e1 and Cyp1a2 activity as a function of acetaminophen dose: Relation to toxicity. *Biochemical and Biophysical Research Communications, 203*(1), 532–539. http://dx.doi.org/10.1006/Bbrc.1994.2215.

Tello, M., Kuzuyama, T., Heide, L., Noel, J. P., & Richard, S. B. (2008). The ABBA family of aromatic prenyltransferases: Broadening natural product diversity. *Cellular and Molecular Life Sciences, 65*(10), 1459–1463. http://dx.doi.org/10.1007/S00018-008-7579-3.

Thomson, J. S., & Prescott, L. F. (1966). Liver damage and impaired glucose tolerance after paracetamol overdosage. *British Medical Journal, 2*(5512), 506–507.

Tironi, I. G., & Vangunsteren, W. F. (1994). A molecular-dynamics simulation study of chloroform. *Molecular Physics, 83*(2), 381–403. http://dx.doi.org/10.1080/00268979400101331.

Torrie, G. M., & Valleau, J. P. (1977). Non-physical sampling distributions in Monte-Carlo free-energy estimation – Umbrella sampling. *Journal of Computational Physics, 23*(2), 187–199. http://dx.doi.org/10.1016/0021-9991(77)90121-8.

Williams, J. A., Hyland, R., Jones, B. C., Smith, D. A., Hurst, S., Goosen, T. C., … Ball, S. E. (2004). Drug-drug interactions for UDP-glucuronosyltransferase substrates: A pharmacokinetic explanation for typically observed low exposure (AUC(i)/AUC) ratios. *Drug Metabolism and Disposition, 32*(11), 1201–1208. http://dx.doi.org/10.1124/Dmd.104.000794.

Wu, Z., Demma, M., Strickland, C. L., Radisky, E. S., Poulter, C. D., Le, H. V., & Windsor, W. T. (1999). Farnesyl protein transferase: Identification of K164 alpha and Y300 beta as catalytic residues by mutagenesis and kinetic studies. *Biochemistry, 38*(35), 11239–11249.

Yang, Y., Chakravorty, D. K., & Merz, K. M. (2010). Finding a needle in the haystack: Computational modeling of Mg^{2+} binding in the active site of protein farnesyltransferase. *Biochemistry, 49*(44), 9658–9666. http://dx.doi.org/10.1021/Bi1008358.

Yang, Y., Miao, Y. P., Wang, B., Cui, G. L., & Merz, K. M. (2012). Catalytic mechanism of aromatic prenylation by NphB. *Biochemistry*, *51*(12), 2606–2618. http://dx.doi.org/10.1021/Bi201800m.

Yang, Y., Wang, B., Ucisik, M. N., Cui, G. L., Fierke, C. A., & Merz, K. M. (2012). Insights into the mechanistic dichotomy of the protein farnesyltransferase peptide substrates CVIM and CVLS. *Journal of the American Chemical Society*, *134*(2), 820–823. http://dx.doi.org/10.1021/ja209650h.

Zwanzig, R. W. (1954). High-temperature equation of state by a perturbation method. 1. Nonpolar gases. *Journal of Chemical Physics*, *22*(8), 1420–1426.

CHAPTER TWO

Empirical Force Fields for Mechanistic Studies of Chemical Reactions in Proteins

A.K. Das, M. Meuwly[1]
University of Basel, Basel, Switzerland
[1]Corresponding author: e-mail address: m.meuwly@unibas.ch

Contents

1. Introduction	31
2. Computational Approaches	35
2.1 Adiabatic Reactive Molecular Dynamics	35
2.2 Multisurface ARMD	38
2.3 Empirical Valence Bond	40
3. Applications	41
3.1 Rebinding Dynamics in MbNO	41
3.2 NO Detoxification Reaction in trHbN	42
3.3 Competitive Ligand Binding in trHbN	43
4. Outlook	49
Acknowledgments	51
References	51

Abstract

Following chemical reactions in atomistic detail is one of the most challenging aspects of current computational approaches to chemistry. In this chapter the application of adiabatic reactive MD (ARMD) and its multistate version (MS-ARMD) are discussed. Both methods allow to study bond-breaking and bond-forming processes in chemical and biological processes. Particular emphasis is put on practical aspects for applying the methods to investigate the dynamics of chemical reactions. The chapter closes with an outlook of possible generalizations of the methods discussed.

1. INTRODUCTION

Chemical reactions involve bond-breaking and bond-forming processes and are fundamental in chemistry and the life sciences in general. In many cases, mechanistic aspects of the reactions ("which reaction partners interact at which time with each other") are of interest. However, experimentally most atomistic aspects in bond-breaking and bond-forming remain elusive because the reactive step itself is a transient process, and the transition state is unstable and short-lived (Brooks, 1988; Neumark, 1992). Thus, the most interesting regions along a reaction path cannot be investigated experimentally in a direct fashion. A more direct approach is afforded by theoretical and computational work which has become invaluable to experimental efforts in understanding particular reaction schemes.

The computational investigation of a chemical or biological system requires means to compute the total energy of the system. There are two fundamentally different concepts to do that: either by solving the electronic Schrödinger equation or by working with a suitably defined empirical potential energy surface (PES) (Hehre, 2003). The most advanced quantum chemistry approaches applicable to long-time and/or large-system applications (tens to hundreds of heavy atoms) have been refined to a degree that allows one to carry out calculations with "chemical accuracy"—ie, accuracies for relative energies within 1 kcal/mol for the chemically bonded region and less accurately for transition states. As a practical aside, a quantum chemical calculation makes no assumption on the bonding pattern in the molecule, contrary to empirical force fields which are based on the concept of atom connectivities. To obtain realistic reaction profiles, electronic structure calculations at a sufficiently high level of theory are required, particularly in the region of the transition. Through statistical mechanics and assuming idealized models such as rigid rotor or harmonic oscillators, average internal energies, enthalpies, and by including entropic effects, also free energies can be calculated (Kuczera, Gao, Tidor, & Karplus, 1990; McQuarrie, 2000; Simonson, Archontis, & Karplus, 2002). However, although such computations are by now standard, they can realistically and routinely only be carried out for systems including several tens of heavy atoms in the quantum region. This is due to the N^3 scaling of the secular determinants that need to be diagonalized, where N is the number of basis functions. An exception to this are semiempirical quantum methods, such as density functional tight binding (DFTB) (Gaus, Cui, & Elstner, 2011).

Alternative approaches to solving the electronic Schrödinger equation are available and used in dynamics studies. For small molecules, London's work on the H + H$_2$ reaction uses a 2 × 2 valence bond treatment (London, 1929). Further refined and extended approaches led to the London–Eyring–Polanyi (Eyring & Polanyi, 1931) and to the London–Eyring–Polanyi–Sato surfaces (Sato, 1955a, 1955b). Further along the lines of valence bond theory, the diatomics-in-molecules (DIM) method (Ellison, 1963) was developed. Following a slightly different perspective, Pauling profoundly influenced the theoretical description of chemical reactivity through his work on molecular structure and the nature of the chemical bond (Pauling, 1932, 1960). Empirical relationships such as the one between bond length and bond order later became foundations to empirical models for treating chemical reactivity (Johnston & Parr, 1963; van Duin, Dasgupta, Lorant, & Goddard, 2001).

Excluding all electronic effects finally leads to empirical force fields. They were developed by focusing on the chemical structure and dynamics of macromolecules, including peptides and proteins (Brooks et al., 1983; Hermans, Berendsen, van Gunsteren, & Postma, 1984; Hwang, Stockfisch, & Hagler, 1994; Jorgensen & Tirado-Rives, 1988; Levitt & Lifson, 1969; Lifson & Warshel, 1968; Maple et al., 1994; Weiner et al., 1984). Thus, originally their primary applications concerned sampling and characterizing conformations of large molecules without considering the reorganization of chemical bonds. The typical mathematical form of empirical force fields

$$V_{\text{bond}} = \sum K_b (r - r_e)^2$$
$$V_{\text{angle}} = \sum K_\theta (\theta - \theta_e)^2 \qquad (1)$$
$$V_{\text{dihe}} = \sum K_\phi (1 + \cos(n\phi - \delta))$$

is thus not geared toward describing chemical reactions where bonds are broken and formed. In Eq. (1) the K are the force constants associated with the particular type of interaction, r_e and θ_e are equilibrium values, n is the periodicity of the dihedral and δ is the phase which determines the location of the maximum. The sums run over all respective terms. Nonbonded interactions include electrostatic and van der Waals terms

$$V_{\text{elstat}} = \frac{1}{4\pi\epsilon_0} \sum \frac{q_i q_j}{r_{ij}}$$
$$V_{\text{vdW}} = \sum \epsilon_{ij} \left[\left(\frac{R_{\min,ij}}{r_{ij}} \right)^{12} - 2 \left(\frac{R_{\min,ij}}{r_{ij}} \right)^6 \right] \qquad (2)$$

where the sums run over all nonbonded atom pairs. q_i and q_j are the partial charges of the atoms i and j, and ϵ_0 is the vacuum dielectric constant. For the van der Waals terms, the potential energy is expressed as a Lennard-Jones potential with well depth $\epsilon_{ij} = \sqrt{\epsilon_i \epsilon_j}$ and range $R_{\min,ij} = (R_{\min,i} + R_{\min,j})/2$ at the Lennard-Jones minimum. This interaction captures long range dispersion ($\propto -r^{-6}$) and exchange repulsion ($\propto r^{-12}$) where the power of the latter is chosen for convenience. The combination of Eqs. (1) and (2) constitutes a minimal model for a force field.

An important step to investigate reactions by simulation methods has been the introduction of mixed quantum mechanical/classical mechanics methods (QM/MM) (Alagona, Ghio, & Kollman, 1986; Bash, Field, & Karplus, 1987; Warshel & Levitt, 1976). In QM/MM the total system is divided into a (typically small) reaction region for which the energy is calculated quantum mechanically—ie, from electronic structure methods - and a (bulk) environment which is treated with a conventional force field. The majority of applications of QM/MM methods to date use semiempirical (such as AM1, PM3 (Claeyssens, Ranaghan, Manby, Harvey, & Mulholland, 2005), SCC-DFTB (Konig et al., 2006; Zhou, Tajkhorshid, Frauenheim, Suhai, & Elstner, 2002)) or DFT methods. Typically, the QM part contains several tens of atoms. It should also be noted that studies of reactive processes in the condensed phase often employ energy evaluations along predefined progression coordinates (Claeyssens et al., 2005; Cui, Elstner, & Karplus, 2003); ie, the system is forced to move along a set of more or less well-suited coordinates. One of the main reasons why ab initio QM/MM calculations are not yet used routinely in fully quantitative studies is related to the fact that the energy and force evaluations for the QM region are computationally too expensive to allow meaningful configurational sampling which is required for reliably estimating essential quantities such as free energy changes.

For these reasons, alternatives to QM/MM methods have been developed whereby empirical force fields are used to investigate chemical reactions by combining them in suitable ways. They include RMD (reactive molecular dynamics) (Meuwly, Becker, Stote, & Karplus, 2002; Nutt & Meuwly, 2006; Dayal, Weyand, McNeish, & Mosey, 2011), EVB (empirical valence bond) (Warshel & Weiss, 1980), and its variants AVB (approximate valence bond) (Grochowski, Lesyng, Bala, & McCammon, 1996) and MCMM (multiconfiguration molecular mechanics) (Kim, Corchado, Villa, Xing, & Truhlar, 2000).

Force field-based treatments of chemical reactivity start from conventional force fields (FFs) and employ the diabatic picture of electronic states to define reactant and product states (Van Voorhis et al., 2010). From an FF perspective, in a diabatic state the connectivity of the atoms does not change. Low-amplitude vibrations and conformational motion in these states can be efficiently described by conventional FFs, which on the other hand yield very high potential energies for geometries far from the equilibrium configuration. For example, force-field evaluation of a chemical bond at its equilibrium geometry with a force field for the unbound state (bonded term is replaced by electrostatic and van der Waals interactions) yields a very high energy for the unbound state due to van der Waals repulsion. This is because typically the sum of van der Waals radii for the two atoms is considerably larger than their equilibrium separation. This large energy difference can be exploited to define a dominant force field which has the lowest energy for almost all accessible configurations and makes the energy difference a useful coordinate. Other methods use geometric formulas to switch on and off interactions individually (eg, ReaxFF). The various methods differ mainly in the choice of switching method and parameter.

This chapter describes adiabatic reactive molecular dynamics (ARMD) (Danielsson & Meuwly, 2008; Nutt & Meuwly, 2006) and its multisurface variant (MS-ARMD) (Nagy, Yosa Reyes, & Meuwly, 2014; Tong et al., 2012; Yosa Reyes, Nagy, & Meuwly, 2014). Both methods have been developed with the aim to combine the accuracy of quantum methods and the speed of FF simulations such that the processes of interest can be sampled in a statistically meaningful manner. This then allows to determine suitable averages which can be compared with experimental data.

2. COMPUTATIONAL APPROACHES

In the following the formal aspects of ARMD and MS-ARMD are discussed. Also, for comparison, other methods are briefly mentioned to highlight similarities and differences between the approaches.

2.1 Adiabatic Reactive Molecular Dynamics

In ARMD (Danielsson & Meuwly, 2008; Nutt & Meuwly, 2006) at least two parametrized PESs, V_1 for the reactant (educt) and V_2 for the product states, are considered. The adiabatic dynamics of the nuclei takes place on the lowest PES, while the energy of the higher states is also determined at each configuration. Whenever the energy of the current state equals that of a higher state, the simulation is restarted from a few time steps $t_s/2$ (where t_s is the switching time) prior to the detected crossing and during time interval t_s. Over this time window the PESs are mixed in different proportions by multiplying them with a suitable time-dependent smooth switching function $f(t)$ (eg, a tanh function) (Danielsson & Meuwly, 2008; Nutt & Meuwly, 2006)

$$V_{\text{ARMD}}(x,t) = (1-f(t))\,V_1(x) + f(t)V_2(x) \quad (3)$$

At the beginning of the mixing the system is fully in state 1 ($f(0) = 0$), while at the end it is fully in state 2 ($f(t_s) = 1$). The algorithm of ARMD is schematically shown for a collinear atom transfer reaction in Fig. 1A.

As during surface crossing $V_{\text{ARMD}}(x,t)$ is explicitly time dependent, the total energy of the system cannot be conserved in a strict sense. For large systems (eg, proteins in solution) total energy was found to be conserved to within ≈ 1 kcal/mol which is sufficient for most applications. This allowed successful application of ARMD simulation method to the investigation of rebinding dynamics of NO molecule in myoglobin (Danielsson & Meuwly, 2008; Nutt & Meuwly, 2006) or the dioxygenation of NO into NO_3^- by oxygen-bound truncated hemoglobin (trHbN) (Mishra & Meuwly, 2010).

However, for highly energetic reactions of small molecules in the gas phase considerable violations of energy conservation in constant number of atom (N), volume (V), and energy (E) (NVE) simulations was found, for example, for vibrationally induced photodissociation of H_2SO_4 (Yosa & Meuwly, 2011; Yosa Reyes et al., 2014). If, however, several crossings between the states involved can take place or the course of the dynamics after the reaction is of interest—eg, for a final state analysis—energy conservation becomes crucial. The magnitude of energy violation ΔE for a simple 1D system (see Fig. 1B) with effective mass m crossing between two linear potentials $V_1(x) = \alpha x$ and $V_2(x) = \beta x$ using a linear switching function $f(t) = t/t_s$ is (Nagy et al., 2014):

$$\Delta E = \frac{\beta(\alpha-\beta)t_s^2}{24m} \quad (4)$$

Hence, exact or nearly exact energy conservation, $\Delta E \approx 0$, can be achieved with ARMD (a) if the steepness of the two PESs along the trajectory during crossing are the same ($\alpha = \beta$), (b) if the second surface has a small slope ($\beta \approx 0$) in the crossing region thus accidental cancellation of violations can occur, (c) if the system has a large effective mass m, which is often true for biomolecular systems, where both partners are heavy or the reaction is accompanied by the rearrangement of a solvation shell involving many solvent molecules, (d) if the switching time is short, however, for $t_s \to 0$ the connection between the PESs will be unphysically sharp and thus fixed-step size integrators fail to conserve energy.

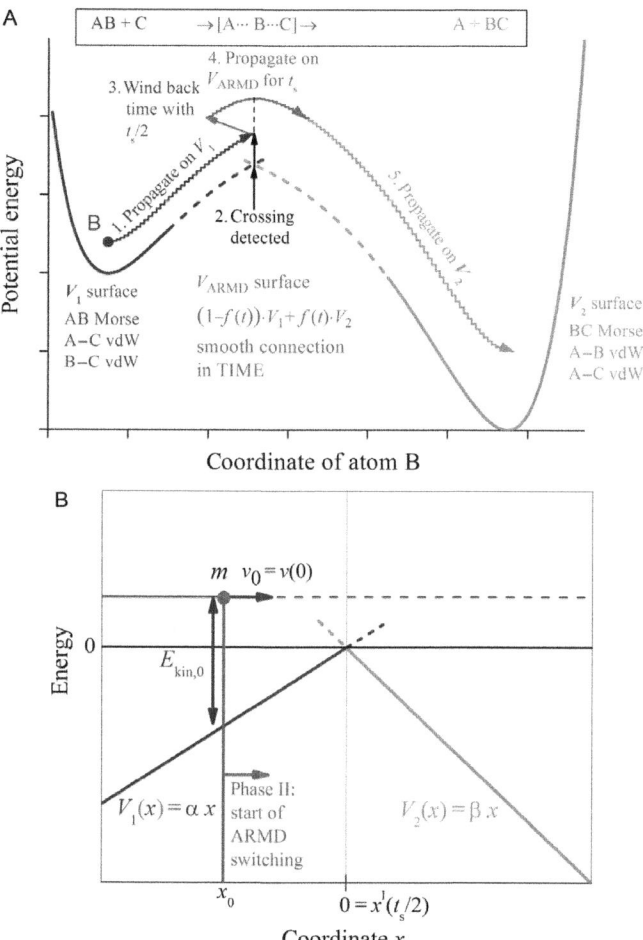

Fig. 1 (A) Schematic figure of the ARMD simulation method for a collinear reaction, where atom B is transferred from donor atom A to acceptor atom C. During crossing the surfaces are switched in time and the Morse bond is replaced by van der Waals (vdW) interactions and vice versa. (B) Simple model for estimating energy violation in ARMD simulations. The system with mass m is approaching from the left on PES $V_1(x)$ (phase I). At $t = 0$ time it is at x_0 with velocity v_0 and kinetic energy $E_{kin,0}$. After crossing is detected at $x = 0$ the time is rewound by $t_s/2$ and the dynamics is resimulated, while $V_1(x)$ is being switched to $V_2(x)$ in t_s (phase II). *Figure partially adapted and reprinted with permission from Nagy, T., Yosa Reyes, J., & Meuwly, M. (2014). Multi-surface adiabatic reactive molecular dynamics.* Journal of Chemical Theory and Computation, 10(4), 1366–1375. doi: 10.1021/ct400953f. *Copyright 2014 American Chemical Society.*

ARMD involves two or multiple PESs defined by individual sets of force-field parameters. For macromolecular systems the number of energy terms by which the PESs differ is much smaller compared to the total number of energy terms. Thus, by providing only a smaller number of additional parameters compared to a standard MD simulation, it is possible to describe the difference between the states of interest with limited computational overhead (Danielsson & Meuwly, 2008). Because the FFs for the individual states are separately parametrized, they need to be related to each other by an offset Δ which puts the asymptotic energy differences between the states in the correct order (Danielsson & Meuwly, 2008).

2.2 Multisurface ARMD

To alleviate the problem with violating the conservation of total energy, an alternative approach was developed. In the multisurface (MS) variant of ARMD, the effective potential energy is a linear combination of n PESs with coordinate-dependent weights $w_i(\mathbf{x})$. With this, total energy is conserved during crossing:

$$V_{\text{MS-ARMD}}(\mathbf{x}) = \sum_{i=1}^{n} w_i(\mathbf{x}) V_i(\mathbf{x}) \qquad (5)$$

The $w_i(\mathbf{x})$ are obtained by renormalizing the raw weights $w_{i,0}(\mathbf{x})$, which were calculated by using a simple exponential decay function of the energy difference between surface i and the minimum energy surface over a characteristic energy scale ΔV (switching parameter):

$$w_i(\mathbf{x}) = \frac{w_{i,0}(\mathbf{x})}{\sum_{i=1}^{n} w_{i,0}(\mathbf{x})}, \qquad \text{where} \quad w_{i,0}(\mathbf{x}) = \exp\left(-\frac{V_i(\mathbf{x})}{\Delta V}\right) \qquad (6)$$

Only those surfaces will have significant weights, whose energy is within a few times of ΔV from the lowest-energy surface. A 1D and 2D example for MS-ARMD is reported in Fig. 2.

The CHARMM (Brooks et al., 2009) implementation (available from v39a2) of MS-ARMD allows adding/removal and reparametrization of terms in any conventional force field. Morse potentials and generalized Lennard–Jones potentials (MIE potential; Kramer, Gedeck, & Meuwly, 2013; Mie, 1903) are also available in the implementation in order to improve the simultaneous description of PES regions close to the equilibrium and the crossing zone. Furthermore, as the energy of each force field

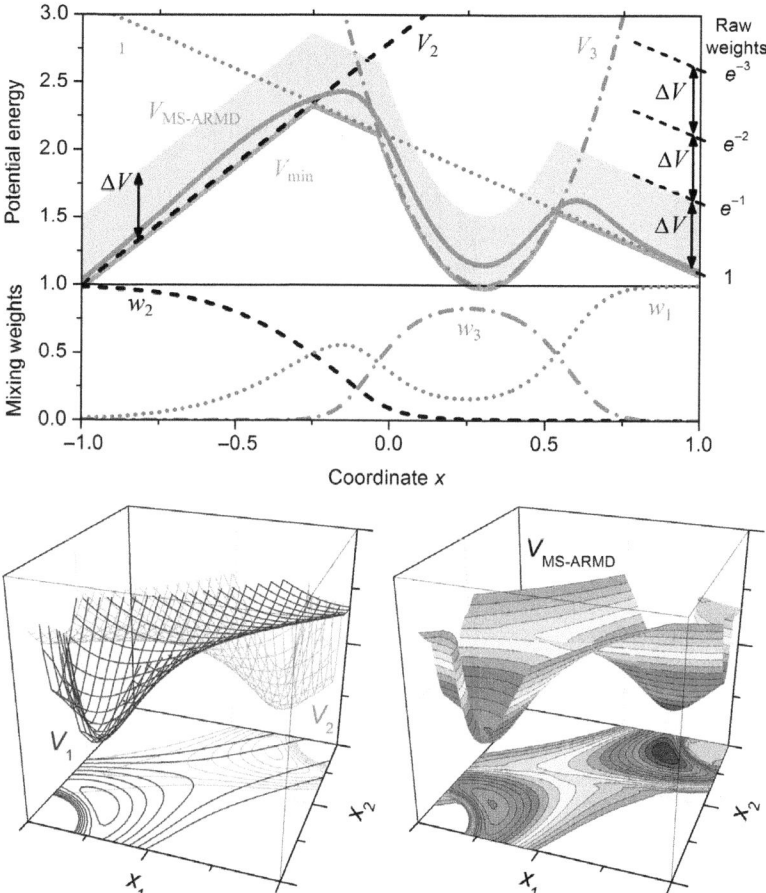

Fig. 2 The MS-ARMD switching method applied in one and two dimensions to three and two surfaces ($V_{1,2,3}$). The effective surface is ($V_{MS-ARMD}$) always close to the lowest-energy surface (V_{min}), except for regions where other surfaces are within a few times ΔV (here = 0.5) in energy. Here, the algorithm switches smoothly among them by varying their weights ($w_{1,2,3}$; *lower left panel*). *Figure partially adapted and reprinted with permission from Nagy, T., Yosa Reyes, J., & Meuwly, M. (2014). Multisurface adiabatic reactive molecular dynamics. Journal of Chemical Theory and Computation, 10(4), 1366–1375. doi: 10.1021/ct400953f. Copyright 2014 American Chemical Society.*

is measured from its own global minimum, an additive constant has to be defined for bringing each force field to a common energy scale to reproduce reaction energies.

Force fields separately optimized for reactant and product states sometimes predict an unrealistic, high-energy crossing point. According to MS-ARMD the transition point between two PESs has a weight of 0.5 from both contributing states. In order to adjust and reshape the barrier region to match energies obtained from electronic structure calculations, products of Gaussian and polynomial functions (GAPOs) ($k = 1,..,n_{ij}$) of the energy difference $\Delta V_{ij}(\mathbf{x}) = V_j(\mathbf{x}) - V_i(\mathbf{x})$ can be applied acting between any two surfaces (i and j):

$$\Delta V_{\text{GAPO},k}^{ij}(\mathbf{x}) = \exp\left(-\frac{\left(\Delta V_{ij}(\mathbf{x}) - V_{ij,k}^0\right)^2}{2\sigma_{ij,k}^2}\right) \cdot \sum_{l=0}^{m_{ij,k}} a_{ij,kl}\left(\Delta V_{ij}(\mathbf{x}) - V_{ij,k}^0\right)^l \quad (7)$$

Here, $V_{ij,k}^0$ and $\sigma_{ij,k}$ denote the center and the standard deviation of the Gaussian function, respectively. Whenever the energy difference between the two PESs deviates from $V_{ij,k}^0$ more than a few times of $\sigma_{ij,k}$, the corresponding GAPO functions will be negligible provided that $V_{ij,k}^0$ and $\sigma_{ij,k}$ are small. The global MS-ARMD PES with this extension is a weighted sum of PESs and GAPO functions scaled with the sum of the weights of the two corresponding surfaces:

$$V_{\text{MS-ARMD}}(\mathbf{x}) = \sum_{i=1}^{n} w_i(\mathbf{x}) V_i(\mathbf{x}) + \sum_{i=1}^{n-1}\sum_{j=i+1}^{n} [w_i(\mathbf{x}) + w_j(\mathbf{x})] \sum_{k=1}^{n_{ij}} \Delta V_{\text{GAPO},k}^{ij}(\mathbf{x}) \quad (8)$$

MS-ARMD as implemented in CHARMM is a general tool for constructing global PESs from empirical force fields for modeling chemical reactions in gas, surface, and condensed phases, relevant to homogeneous, heterogeneous, and enzymatic catalysis.

2.3 Empirical Valence Bond

One of the established methods to investigate chemical reactions based on empirical force fields is the EVB method (Hong, Rosta, & Warshel, 2006; Warshel & Weiss, 1980). EVB starts from the fact that valence bond states are suitable to distinguish between ionic and covalent resonance forms of a chemical bond which reflects chemical intuition. Since the environment of a chemical reaction primarily interacts through electrostatics with the reactive species, empirical force fields can be used to describe the resonant forms of the reactant and product states. For a bond-breaking reaction AB → A + B, three resonance forms are introduced: $\psi_1 = \text{AB}$, $\psi_2 = \text{A}^-\text{A}^+$, and

$\psi_3 = A^+ A^-$. If A is more electronegative than B, resonance structure ψ_3 is largely irrelevant and the process can be described by ψ_1 and ψ_2.

For a collection of covalent and ionic states, matrix elements for the EVB Hamiltonian have to be determined. They include diagonal elements for the covalent and ionic states, and off-diagonal elements that couple configurations (bonding patterns) that differ by the location of an electron pair. All other off-diagonal matrix elements $H_{ij} = 0$. The justification for this is that such matrix elements are proportional to the square or higher powers of the overlap between atomic orbitals, but they may also be retained (Coulson & Danielsson, 1954; Warshel & Weiss, 1980). The diagonal matrix elements H_{ii} correspond essentially to an empirical force field, whereas for the ionic diagonal matrix elements the bonded terms are replaced by electrostatic interactions between the charged fragments and the formation energy of $A^- B^+$ from AB has also to be added. For the two-fragment system AB the matrix elements are $H_{11} = D_e(1 - \exp[-\beta(r - r_e)])^2$ and $H_{22} = \Delta - \frac{e^2}{r} + V_{nb}$, where Δ is the gas-phase formation energy of $A^- B^+$ from AB at infinite separation, and V_{nb} is the nonbonded interaction potential such that the minimum of $(\frac{-e^2}{r} + V_{nb})$ is given by the sums of the ionic radii of A^+ and B^-. In the original version of EVB the off-diagonal element $H_{12} = H_{21}$ is determined through the requirement that the eigenvalues of the Hamiltonian E satisfy the relation $H_{12} = \sqrt{(H_{11} - E)(H_{22} - E)}$ and E is the experimentally determined ground-state bond energy. In a later and slightly more general approach, the off-diagonal elements are parametrized functions $H_{ij} = A \exp(-\mu(r - r_0))$ which depend on a predefined reaction coordinate r (Hong et al., 2006). The definition and parametrization of the off-diagonal terms have been a source of considerable discussion in the field, in particular the assumption that upon transfer of the reaction from the gas phase to the solution phase these elements do not change significantly.

3. APPLICATIONS

3.1 Rebinding Dynamics in MbNO

Myoglobin, besides being an important model system for understanding the relation between structure and function of proteins, has also been of interest due to its ligand binding properties. In particular, the migration pathways and rebinding dynamics of diatomic ligands such as O_2, NO, and CO inside the protein matrix have been studied by both experimental and computational methods. While rebinding of CO is nonexponential at low temperature but becomes exponential at high temperature (timescale of 100 ns), rebinding of NO remains nonexponential at all temperature with time constants of the order of tens of picoseconds which is ideal for reactive MD simulations (Austin, Beeson, Eisenstein, Frauenfelder, & Gunsalus, 1975; Kim, Jin, & Lim, 2004).

The rebinding dynamics of MbNO was studied employing the ARMD method. To this end, two force fields corresponding to the bound and dissociated states were employed. They differ in a number of energy terms. The dissociating Fe–N bond was a Morse potential to describe the anharmonic nature of the bond. Multiple trajectory simulations were carried out for asymptotic separations of $\Delta = 60, 65$, and 70 kcal/mol (Danielsson & Meuwly, 2008). The crossing seam including all observed crossing geometries is rather wide in both radial and angular geometries with a maximum in the iron–ligand distance around 3 Å (Danielsson & Meuwly, 2008) and rather insensitive toward the value of Δ.

The fraction of trajectories without crossing provides information about the kinetics of rebinding. The choice of Δ is found to substantially affect the time constant associated with the rebinding reaction, although for all values of Δ the rebinding dynamics remains nonexponential. For $\Delta = 65$ kcal/mol, the time constants are 3.6 and 373 ps (Danielsson & Meuwly, 2008) compared to the experimental (from ultrafast IR spectroscopy) value of 5.3 and 133 ps (Kim et al., 2004; Petrich et al., 1991). While the fast rebinding component is well reproduced by ARMD, the agreement for the slower component is poor which is due to insufficient sampling of the slow timescale by ARMD.

3.2 NO Detoxification Reaction in trHbN

trHbN is a recently discovered heme protein found in plants, bacteria, and lower eukaryotes. The trHbN of *Mycobacterium tuberculosis* has been proposed to play an important role in the survival of the bacteria causing tuberculosis in host cells by converting toxic NO to harmless NO_3^-. The large second-order rate constant of 7.5×10^8 M^{-1} s^{-1} has been attributed to the existence of a continuous tunnel inside the protein which assists ligand migration (Milani et al., 2004; Ouellet et al., 2002). However, an atomistic understanding about the mechanism of the detoxification reaction had remained elusive.

ARMD was used to shed light on the reaction by dividing the overall reaction into following four steps (Bourassa, Ives, Marqueling, Shimanovich, & Groves, 2001):

$$\text{Fe(II)} - \text{O}_2 + \text{NO} \rightarrow \text{Fe(III)}[-\text{OONO}] \quad \text{O2} - \text{N distance} \quad \text{(I)}$$
$$\text{Fe(III)}[-\text{OONO}] \rightarrow \text{Fe(IV)}=\text{O} + \text{NO}_2 \quad \text{O1} - \text{O2 distance} \quad \text{(II)}$$
$$\text{Fe(IV)}=\text{O} + \text{NO}_2 \rightarrow \text{Fe(III)}[-\text{ONO}_2] \quad \text{O1} - \text{N distance} \quad \text{(III)}$$
$$\text{Fe(III)}[-\text{ONO}_2] \rightarrow \text{Fe}^+(\text{III}) + [\text{NO}_3]^- \quad \text{Fe} - \text{O1 distance} \quad \text{(IV)}$$

The right-hand side column of the above reaction steps indicates the bond broken/formed in the reaction step. First, oxy-trHbN reacts with free NO and forms a peroxynitrite intermediate which then undergoes homolytic fission followed by the rebinding of free NO_2 to the oxo-ferryl species to form the heme-bound nitrato complex which then undergoes heme-ligand dissociation resulting in free NO_3^- and penta-coordinated heme.

The force-field parameters associated with the reactants and products of each of the reaction steps are obtained from ab initio calculations. Each of the reaction steps was then studied by running multiple ARMD trajectories with a range of Δ values (Mishra & Meuwly, 2010). For reaction steps I, III, and IV, the ARMD simulations yielded rate constants on the picosecond timescale. The choice of the free parameter Δ had only limited effects on the reaction rate.

For step II, however, no reactive events even on the nanosecond timescale were found (Mishra & Meuwly, 2010). From DFT calculations it is known that this step involves a barrier of 6.7 kcal/mol (Blomberg, Blomberg, & Siegbahn, 2004). Umbrella sampling simulations with ARMD yielded a barrier of 12–15 kcal/mol which corresponds to timescales on the order of micro- to milliseconds. Since experimentally, the overall reaction is on the picosecond timescale, it is unlikely that the reaction occurs via step II. This proposition is in line with the lack of experimental detection of free NO_2 radical in several studies which propose an alternative mechanism where peroxynitrite intermediate rearranges to nitrato complex (Herold, 1999; Herold, Exner, & Nauser, 2001). To further corroborate this, ARMD simulations for the rearrangement reaction were carried out and found this process to occur within picoseconds, explaining the fast overall detoxification reaction (Mishra & Meuwly, 2010).

3.3 Competitive Ligand Binding in trHbN

Physiologically, the reaction sequence for denitrification starts at the FeNO-bound state due to the larger binding energy of NO compared to O_2

(Rovira, Kunc, Hutter, Ballone, & Parrinello, 1997). Hence, the first step in the reaction leading to NO_3^- is the displacement (or ligand exchange) reaction

$$\text{Fe(II)} - \text{NO} + O_2 \rightarrow \text{Fe(II)} - O_2 + \text{NO} \qquad (V)$$

This is an example for competitive ligand binding at a heme center. One possibility for a computational investigation of this process is to precompute a two- or three-dimensional PES for each ligand interacting with the heme as has been done for CO and NO in Mb (Soloviov & Meuwly, 2014, 2015). However, the electronic structure of O_2 interacting with heme-Fe turns out to be complicated due to closely spaced electronic states which eventually point toward a multireference treatment which is, however, computationally very demanding (Siegbahn, Blomberg, & Chen, 2010). Several attempts have been made to characterize the PES along the Fe–O_2 distance, with variable success. Earlier work (Jensen & Ryde, 2004) using the BP86 functional and 6-31G(d) basis set for all atoms except iron for which double-ζ basis set augmented with two p, one d and one f function was used, gives completely different triplet (^{3}A) and singlet (^{1}A) Fe–O_2 binding curves with very low dissociation energy of ~5 kcal/mol. Rigid scans along the Fe-O_2 distance for three different geometries (with the Fe atom, in or out of the porphyrin plane) (Franzen, 2002) using the BLYP functional and double-ζ plus extra polarization (DNPP) basis set with implicit solvent provides potential energy curves that neither reflect the results from Kepp (2013) and Jensen and Ryde (2004). A slightly modified B3LYP (15% exact exchange) functional together with the cc-pvtz(-f) basis set with and without dispersion correction reproduces the (Siegbahn et al., 2010) experimental Fe–O_2 dissociation energy with dispersion correction. Further work using the B3LYP functional and the 6-31G(d,p) basis set for all atoms except iron (LANL2DZ used for iron) obtained a double crossing PES (^{1}A $\rightarrow$ ^{3}A $\rightarrow$ ^{7}A) with very low dissociation energy of 7 kcal/mol (Ali, Sanyal, & Oppeneer, 2012). Finally, recent work using TPSSh-D3/def2-TZVPP/Cosmo ($\epsilon = 10$) with implicit solvent and dispersion corrections found that both triplet (^{3}A) and singlet (^{1}A) states are parallel to each other with a dissociation energy of ~ 17 kcal/mol which agrees well with experiment (Kepp, 2013). However, using the same method overestimates the ΔH of O_2 binding by 4 kcal/mol. On the other hand, more standard B3LYP-D3/def2-TZVPP/Cosmo calculations ($\epsilon = 10$) yield a very low dissociation energy of 5 kcal/mol (Kepp, 2013). Since the calculation of dissociation energy and the Fe–O_2

binding curves strongly depends on the method used due to both multireference and van der Waals effects and a multireference character of this interaction cannot be excluded, an alternative path to parametrize a force field is illustrated in the following.

As precomputing two- or three-dimensional PESs for O_2 interacting with heme yields conflicting results and appears to sensitively depend on slight adjustments in the methods employed, a viable and meaningful alternative is to base the parametrization on experimentally available data. They include structural data and stabilization energies of the heme–O_2 complex in the protein. The experimental binding energy of O_2 ligand with *Aplysia* myoglobin is 14.6 kcal/mol (Antonini & Brunori, 1971) and the binding energy of NO ligand to doubly protonated iron tetrapyridyl porphyrin is 24.8 ± 0.7 kcal/mol (Chen, Grob, Liechty, & Ridge, 1999). Using these data for NO- and O_2-bound heme simulations, ligand exchange dynamics were carried out. To study ligand exchange four different reactive surfaces were considered: Fe–O1O2 + NO, (isomerization) Fe–O2O1 + NO, Fe–NO +O_2, and Fe + NO +O_2. Different force-field parameters were used for the five- and six-coordinated heme (Meuwly et al., 2002). The O_2, NO, Fe–O_2, and Fe–NO bonds were all represented by Morse potentials. The Morse parameters are given in Table 1 (Konowalow & Hirschfelder, 1961). The van der Waals parameters R_{min} used for O and N are 2.0 and 2.05 Å for the bound state, whereas for the unbound state $R_{min} = 1.2$ Å is used for the O atom.

The simulation system consists of a solvated trHbN (trHbN) (protein data bank entry 1IDR) (Milani et al., 2001) in an orthorhombic, periodic TIP3P (Jorgensen, Chandrasekhar, Madura, Impey, & Klein, 1983) water box of size $a = 78.25$, $b = 53.10$, and $c = 53.10$ Å (see Fig. 3). This X-ray structure contains a bound dioxygen molecule and a free NO ligand. All MD simulations were carried out with the CHARMM (Brooks et al., 1983) suite of programs with a time step of 1 fs. The nonbonded interactions (electrostatic and Lennard–Jones) were truncated at a distance of 14 Å and switched between 10 and 12 Å. The CHARMM22 (MacKerell et al., 1998) force field was used for this simulation together with the necessary parameters to describe bond breaking an bond formation (see above). After preparation, the system was heated to 300 K. Then the system was equilibrated for 1 ns.

Because the experimentally determined bond strengths of the Fe–O_2 (14–16 kcal/mol (Collman, Brauman, Doxsee, Halbert, & Suslick, 1978)) and the Fe–NO (20–24 kcal/mol (Chen et al., 1999)) are much larger than

Table 1 Morse Parameters for the Respective Bonds

Bonds	D_e (kcal/mol)	β (Å^{-1})	r_e (Å)
Fe–O$_2$	15.0	2.215	1.80
Fe–NO	24.0	2.775	1.73
O–O (bound)	215.0	1.500	1.23
N–O (bound)	229.3	2.415	1.16
O–O (unbound)	290.0	1.500	1.21
N–O (unbound)	298.2	1.953	1.15

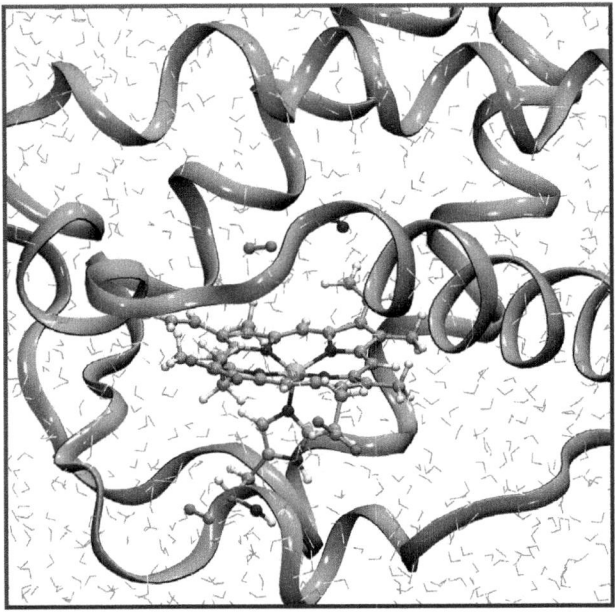

Fig. 3 Truncated hemoglobin solvated in water is shown here. protein shown in ribbon and the active cite with two ligands are shown in *ball* and *sticks*. Water molecules shown by *lines*. Atom color code used, Fe (*green; light gray* in the print version), C (*cyan; gray* in the print version), N (*blue; dark gray* in the print version), O (*red; gray* in the print version), and H (*white*).

the thermal energy, spontaneous ligand exchange on typical MD timescales (tens to hundreds of nanoseconds) will not occur and biased simulations will be required. For direct investigation of the ligand exchange reaction on conventionally accessible MD timescales the dissociation energies D_e of the Fe–NO and Fe–O$_2$ bonds were first lowered to 2 kcal/mol. This allows to illustrate the working principle of MS-ARMD. Two simulations were performed using different starting structures. One in which the O_2 is bound and the other with the NO bound to the heme iron. In both cases ligand exchange takes place within 15 ps. The distance variation between the heme iron and the ligands is shown in Fig. 4. During the entire dynamics the total energy is well conserved (see Fig. 5).

In a next step, umbrella sampling simulations with the correct parametrizations for the ligand dissociation energies (see above) were carried out. The ratio between the Fe and the two ligands $r_c = \dfrac{d_{Fe-NO}}{d_{Fe-O_2}}$ was the reaction coordinate and each umbrella was simulated for 25 ps with a force constant of 100 kcal/mol. Fig. 6 shows the transition state ensemble separating the Fe–O_2 and Fe–NO bound states. As can be seen, the transition state is very broad. This provides now a basis for more detailed investigations of the dynamics of the ligand exchange reaction which are currently under way. These representative structures can be compared with a transition state calculation for the Fe–O_2/NO system at the UB3LYP level of theory with the 6-31G(d) basis set reported in Fig. 7. As can be seen, the two structures compare favorably but the MS-ARMD simulations yield an entire ensemble representative of the transition state from quantum chemical calculations.

One of the clear advantages of MS-ARMD (and empirical force fields in general) is the possibility to better control the states included in the calculations. This is particularly relevant for systems involving complicated electronic structure, such as Fe–O_2. Hence, the simulations can be carried out

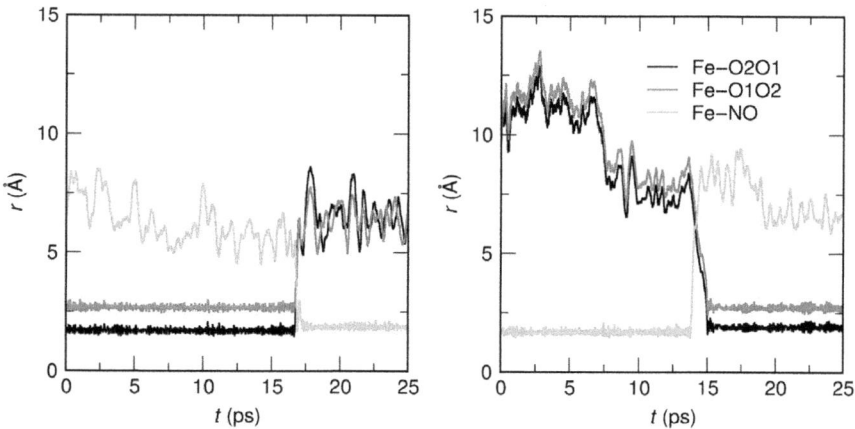

Fig. 4 Variation of distance between ligands and the metal during the ligand exchange reaction. (See the color plate.)

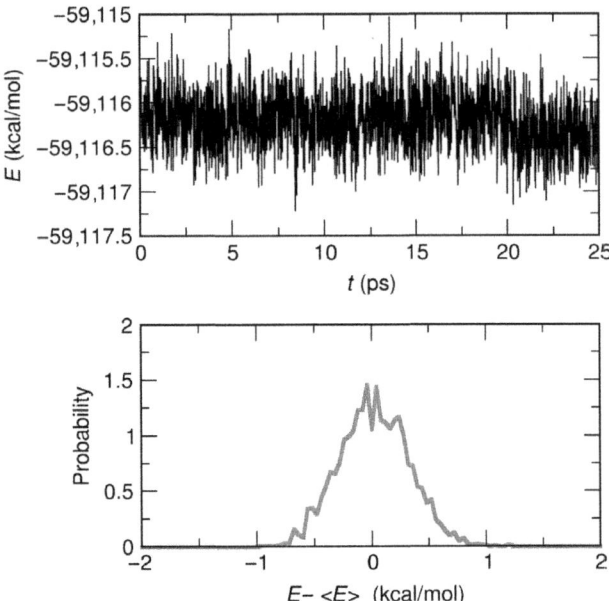

Fig. 5 Variation of total energy along the trajectory (*upper panel*) together with the distribution around the mean (*lower panel*). (See the color plate.)

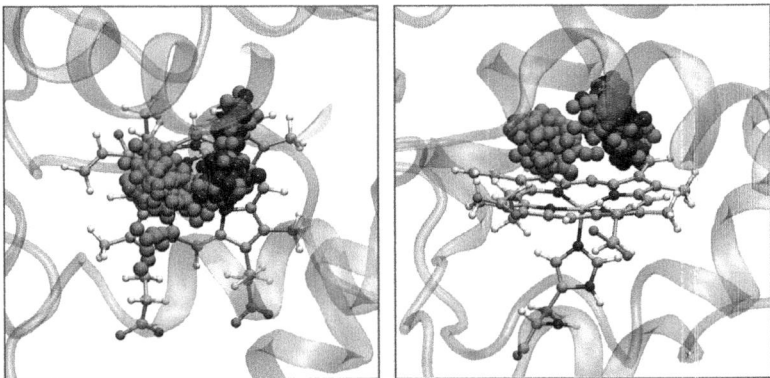

Fig. 6 Transition state ensemble during the ligand exchange reaction in trHbN.

for the electronic states relevant to the particular experiment. On the other hand, electronic structure calculations at the (single reference) density functional theory level—which are customarily used for mixed QM/MM simulations (see above)—are plagued with difficulties of convergence to the wrong electronic state or convergence at all. Therefore, routine

Fig. 7 Transition state structure for the Fe–NO + O_2 ⇔ Fe–O_2 + NO exchange reaction at the UB3LYP/6-31G(d) level of theory. The transition state ensemble shown in Fig. 6 is representative of this structure calculated with quantum chemistry.

application of QM/MM to such problems is typically difficult on the multinanosecond timescale.

4. OUTLOOK

This chapter discusses the background, parametrization, and application of the (MS-)ARMD method which allows to follow chemical reactions in atomistic simulations. Smooth switching between multiple PESs in ARMD and MS-ARMD is carried out by time- and energy-dependent switching functions controlled by the switching time t_s (in ARMD) or a switching parameter ΔV (in MS-ARMD), respectively. These switching variables are one dimensional (time or energy) and serve as a natural reaction coordinate, whereas in the case of geometrical switching variables the determination of reaction coordinate is neither straightforward nor unambiguous. Furthermore, the potential energy and its gradient are analytic functions of the individual force-field energies.

Due to the explicit time dependence of the ARMD Hamiltonian, the ARMD crossing is inherently a dynamical method and this is also the reason for violations of energy conservation observed during crossing in gas-phase reactions of highly excited small molecules. On the other hand, in MS-ARMD the PES is stationary and thus can be used in NVE simulations. The ARMD method cannot describe crossings in regions properly where more than two surfaces are close in energy, whereas the MS-ARMD technique can be applied to the simultaneous switching among multiple PESs. Finally, in MS-ARMD, the height and the shape of the crossing region can be adjusted in a flexible manner, which makes it very similar in functionality to EVB, but without the need of referring to any geometrical reaction coordinate. This extension makes the CHARMM implementation of MS-ARMD method (Brooks et al., 2009) comparable to QM/MM methods in capability for modeling complex chemical reaction in gas, surface, and condensed phases albeit at the accelerated speed of a conventional force-field simulation and with the only added effort of parametrization.

Future improvements of the ARMD and MS-ARMD methods include the more routine development of accurate force-field parametrizations which still poses an obstacle to a more routine use of such approaches in all areas of chemistry and biophysics. For ARMD, which requires an asymptotic offset Δ between the states, an improvement could be achieved by making this parameter coordinate dependent. On the other hand, this would introduce one or several geometric progression coordinates whose definition may be difficult. Nevertheless, a recent study of nitric oxide rebinding to Mb found that indeed, Δ is not constant but depends on the iron-out-of-plane position in this particular situation (Soloviov & Meuwly, 2015). Hence, a better reaction energetics and a more realistic modeling of the ligand-(re)binding dynamics is obtained from such improvements.

The current implementation of MS-ARMD requires the separate definition and evaluation of force fields for each possible product connectivity. This means several possibilities in solvent-phase or surface-phase simulations where the reaction can take place with any of the solvent molecules or at any of the surface sites. Though the common part of all force fields are evaluated only once, the calculation and even the enumeration of all cases are unnecessary. In future developments, this will be constrained to a small number of momentary physically sensible connectivities by an automatic geometrical preselection.

Force field-based approaches to study chemical reactions are becoming more widespread as the functionalities are made available in commonly used atomistic simulation programs. They allow to study chemical reactivity on timescales relevant to the real processes and provide insight complementary to experiment provided that the underlying force fields are accurate. The fitting of reactive force fields remains a challenge but generalizing and simplifying this step will make such approaches valuable additions to the toolbox of computational and experimental chemists interested in chemical reactivity.

ACKNOWLEDGMENTS
This work was supported by the Swiss National Science Foundation through Grant 200021-117810, the NCCR MUST, and the University of Basel, which is gratefully acknowledged. We thank all coworkers mentioned in the references for their contributions to the methods and their applications discussed in this work.

REFERENCES
Alagona, G., Ghio, C., & Kollman, P. A. (1986). Simple-model for the effect of GLU165-]ASP165 mutation on the rate of catalysis in triose phosphate isomerase. *Journal of Molecular Biology, 191*, 23–27.

Ali, M. E., Sanyal, B., & Oppeneer, P. M. (2012). Electronic structure, spin-states, and spin-crossover reaction of heme-related Fe-porphyrins: A theoretical perspective. *Journal of Physical Chemistry B, 116*(20), 5849–5859. http://dx.doi.org/10.1021/jp3021563.

Antonini, E., & Brunori, M. (1971). *Hemoglobin and myoglobin in their reactions with ligands. In Frontiers of biology.* Amsterdam, London: North-Holland Publishing Co. (Vol. 21).

Austin, R. H., Beeson, K. W., Eisenstein, L., Frauenfelder, H., & Gunsalus, I. C. (1975). Dynamics of ligand-binding to myoglobin. *Biochemistry, 14*, 5355–5373.

Bash, P. A., Field, M. J., & Karplus, M. (1987). Free-energy perturbation method for chemical-reactions in the condensed phase—A dynamical-approach based on a combined quantum and molecular mechanics potential. *Journal of the American Chemical Society, 109*, 8092–8094.

Blomberg, L. M., Blomberg, M. R. A., & Siegbahn, P. E. M. (2004). A theoretical study of myoglobin working as a nitric oxide scavenger. *Journal of Biological Inorganic Chemistry, 9*, 923–935.

Bourassa, J. L., Ives, E. L., Marqueling, A. L., Shimanovich, R., & Groves, J. T. (2001). Myoglobin catalyzes its own nitration. *Journal of the American Chemical Society, 123*, 5142–5143.

Brooks, P. R. (1988). Spectroscopy of transition region species. *Chemical Reviews, 88*(2), 407–428. http://dx.doi.org/10.1021/cr00084a004.

Brooks, B. R., Brooks, C. L., III, Mackerell, A. D., Jr., Nilsson, L., Petrella, R. J., Roux, B., ... Karplus, M. (2009). CHARMM: The biomolecular simulation program. *Journal of Computational Chemistry, 30*(10), 1545–1614.

Brooks, B., Bruccoleri, R., Olafson, B., States, D., Swaminathan, S., & Karplus, M. (1983). CHARMM: A program for macromolecular energy, minimization, and dynamics calculations. *Journal of Computational Chemistry, 4*, 187–217.

Chen, O., Grob, S., Liechty, A., & Ridge, D. P. (1999). *Journal of the American Chemical Society, 121*, 11910.

Claeyssens, F., Ranaghan, K. E., Manby, F. R., Harvey, J. N., & Mulholland, A. J. (2005). Multiple high-level QM/MM reaction paths demonstrate transition-state stabilization in chorismate mutase: Correlation of barrier height with transition-state stabilization. *Chemical Communications, 40*, 5068–5070.

Collman, J., Brauman, J., Doxsee, K., Halbert, T., & Suslick, K. (1978). Model compounds for T-state of hemoglobin. *Proceedings of the National Academy of Sciences of the United States of America*, *75*(2), 564–568.

Coulson, C. A., & Danielsson, U. (1954). Ionic and covalent contributions to the hydrogen bond. Part 1. *Arkiv fr Fysik*, *8*, 239–244.

Cui, Q., Elstner, T., & Karplus, M. (2003). A theoretical analysis of the proton and hydride transfer in liver alcohol dehydrogenase (LADH). *Journal of Physical Chemistry B*, *106*, 2721–2740.

Danielsson, J., & Meuwly, M. (2008). Atomistic simulation of adiabatic reactive processes based on multi-state potential energy surfaces. *Journal of Chemical Theory and Computation*, *4*, 1083–1093.

Dayal, P., Weyand, S. A., McNeish, J., & Mosey, N. J. (2011). Temporal quantum mechanics/molecular mechanics: Extending the time scales accessible in molecular dynamics simulations of reactions. *Chemical Physics Letters*, *516*, 263–267.

Ellison, F. O. (1963). A method of diatomics in molecules. 1. General theory and application to H2O. *Journal of the American Chemical Society*, *85*, 3540–3544.

Eyring, H., & Polanyi, M. (1931). Concerning simple gas reactions. *Zeitschrift fr Physikalische Chemie. Abteilung B*, *12*, 279–311.

Franzen, S. (2002). Spin-dependent mechanism for diatomic ligand binding to heme. *Proceedings of the National Academy of Sciences of the United States of America*, *99*(26), 16754–16759. http://dx.doi.org/10.1073/pnas.252590999.

Gaus, M., Cui, Q., & Elstner, M. (2011). DFTB-3rd: Extension of the self-consistent-charge density-functional tight-binding method SCC-DFTB. *Journal of Chemical Theory and Computation*, *7*, 931–948.

Grochowski, P., Lesyng, B., Bala, P., & McCammon, J. A. (1996). Density functional based parametrization of a valence bond method and its applications in quantum-classical molecular dynamics simulations of enzymatic reactions. *International Journal of Quantum Chemistry*, *60*, 1143–1164.

Hehre, W. (2003). *A guide to molecular mechanics and quantum chemical calculations*. Irvine, CA: Wavefunction Press.

Hermans, J., Berendsen, H. J. C., van Gunsteren, W. F., & Postma, J. P. (1984). A consistent empirical potential for water-protein interactions. *Biopolymers*, *23*, 1.

Herold, S. (1999). Kinetic and spectroscopic characterization of an intermediate peroxynitrite complex in the nitrogen monoxide induced oxidation of oxyhemoglobin. *FEBS Letters*, *443*, 81–84.

Herold, S., Exner, M., & Nauser, T. (2001). Kinetic and mechanistic studies of the NO-mediated oxidation of oxymyoglobin and oxyhemoglobin. *Biochemistry*, *40*, 3385–3395.

Hong, G., Rosta, E., & Warshel, A. (2006). Using the constrained DFT approach in generating diabatic surfaces and off diagonal empirical valence bond terms for modeling reactions in condensed phases. *Journal of Physical Chemistry B*, *110*, 19570–19574.

Hwang, M. J., Stockfisch, T. P., & Hagler, A. T. (1994). Derivation of class II force fields: 2. Derivation and characterization of a class II force field, CFF93, for the alkyl functional group and alkane molecules. *Journal of the American Chemical Society*, *116*, 2515–2525.

Jensen, K. P., & Ryde, U. (2004). How O2 binds to heme: Reasons for rapid binding and spin inversion. *Journal of Biological Chemistry*, *279*(15), 14561–14569. http://dx.doi.org/10.1074/jbc.M314007200.

Johnston, H. S., & Parr, C. (1963). Activation energies from bond energies. I. Hydrogen transfer reactions. *Journal of the American Chemical Society*, *85*, 2544–2551.

Jorgensen, W. L., Chandrasekhar, J., Madura, J., Impey, R., & Klein, M. (1983). Comparison of simple potential functions for simulating liquid water. *Journal of Chemical Physics*, *79*, 926–935.

Jorgensen, W. L., & Tirado-Rives, J. (1988). The OPLS potential functions for proteins—Energy minimizations for crystals of cyclic-peptides and crambin. *Journal of the American Chemical Society, 110,* 1657–1666.

Kepp, K. P. (2013). O2 binding to heme is strongly facilitated by near-degeneracy of electronic states. *Chemphyschem, 14*(15), 3551–3558. http://dx.doi.org/10.1002/cphc.201300658.

Kim, Y., Corchado, J. C., Villa, J., Xing, J., & Truhlar, D. G. (2000). Multiconfiguration molecular mechanics algorithm for potential energy surfaces of chemical reactions. *Journal of Chemical Physics, 112,* 2718–2735.

Kim, S., Jin, G., & Lim, M. (2004). Dynamics of geminate recombination of NO with myoglobin in aqueous solution probed by femtosecond mid-IR spectroscopy. *Journal of Physical Chemistry B, 108,* 20366–20375.

Konig, P. H., Ghosh, N., Hoffmann, M., Elstner, M., Tajkhorshid, E., Frauenheim, T., & Cui, Q. (2006). Toward theoretical analysis of long-range proton transfer kinetics in biomolecular pumps. *Journal of Physical Chemistry A, 110,* 548–563.

Konowalow, D., & Hirschfelder, J. (1961). Morse potential parameters for O-O, N-N, and N-O interactions. *Physics of Fluids, 4*(5), 637–642. http://dx.doi.org/10.1063/1.1706374.

Kramer, C., Gedeck, P., & Meuwly, M. (2013). Multipole-based force fields from ab initio interaction energies and the need for jointly refitting all intermolecular parameters. *Journal of Chemical Theory and Computation, 9*(3), 1499–1511.

Kuczera, K., Gao, J., Tidor, B., & Karplus, M. (1990). Free energy of sickling: A simulation analysis. *Proceedings of the National Academy of Sciences of the United States of America, 87*(21), 8481–8485. http://dx.doi.org/10.1073/pnas.87.21.8481.

Levitt, M., & Lifson, S. (1969). Refinement of protein conformations using a macromolecular energy minimization procedure. *Journal of Molecular Biology, 46,* 269–279.

Lifson, S., & Warshel, A. (1968). Consistent force field for calculations of conformations vibrational spectra and enthalpies of cycloalkane and n-alkane molecules. *Journal of Chemical Physics, 49,* 5116–5129.

London, F. (1929). Quantum mechanical interpretation of the process of activation. *Zeitschrift fr Elektrochemie, 35,* 552–555.

MacKerell, A. D., Jr., Bashford, D., Bellott, M., Dunbrack, R. L., Jr., Evanseck, J. D., Field, M. J., … Karplus, M. (1998). All-atom empirical potential for molecular modeling and dynamics studies of proteins. *Journal of Physical Chemistry B, 102,* 3586.

Maple, J. R., Hwang, M. J., Stockfisch, T. P., Dinur, U., Waldman, M., Ewig, C. S., & Hagler, A. T. (1994). Derivation of class-II force-fields. 1. Methodology and quantum force-field for the alkyl functional-group and alkane molecules. *Journal of Computational Chemistry, 15,* 162–182.

McQuarrie, D. (2000). *Statistical mechanics.* Sausalito, CA: University Science Books.

Meuwly, M., Becker, O. M., Stote, R., & Karplus, M. (2002). NO rebinding to myoglobin: A reactive molecular dynamics study. *Biophysical Chemistry, 98,* 183–207.

Mie, G. (1903). Zur kinetischen theorie der einatomigen körper. *Annals of Physics, 11,* 657–697.

Milani, M., Pesce, A., Ouellet, Y., Ascenzi, P., Guertin, M., & Bolognesi, M. (2001). Mycobacterium tuberculosis hemoglobin N displays a protein tunnel suited for O2 diffusion to the heme. *The EMBO Journal, 20*(15), 3902–3909. http://dx.doi.org/10.1093/emboj/20.15.3902.

Milani, M., Pesce, A., Ouellet, Y., Dewilde, S., Friedman, J., Ascenzi, P., … Bolognesi, M. (2004). Heme-ligand tunneling in group I truncated hemoglobins. *Journal of Biological Chemistry, 279,* 21520–21525.

Mishra, S., & Meuwly, M. (2010). Atomistic simulation of NO dioxygenation in group I truncated hemoglobin. *Journal of the American Chemical Society, 132,* 2968.

Nagy, T., Yosa Reyes, J., & Meuwly, M. (2014). Multisurface adiabatic reactive molecular dynamics. *Journal of Chemical Theory and Computation, 10*(4), 1366–1375. http://dx.doi.org/10.1021/ct400953f.

Neumark, D. M. (1992). Transition state spectroscopy of bimolecular chemical reactions. *Annual Review of Physical Chemistry*, *43*(1), 153–176. http://dx.doi.org/10.1146/annurev.pc.43.100192.001101.

Nutt, D. R., & Meuwly, M. (2006). Studying reactive processes with classical dynamics: Rebinding dynamics in MbNO. *Biophysical Journal*, *90*, 1191–1201.

Ouellet, H., Ouellet, Y., Richard, C., Labarre, M., Wittenberg, B., Wittenberg, J., & Guertin, M. (2002). Truncated hemoglobin HbN protects Mycobacterium bovis from nitric oxide. *Proceedings of the National Academy of Sciences of the United States of America*, *99*, 5902–5907.

Pauling, L. (1932). The nature of the chemical bond. IV. The energy of single bonds and the relative electronegativity of atoms. *Journal of the American Chemical Society*, *54*, 3570–3582.

Pauling, L. (1960). *The nature of the chemical bond*. Ithaca, NY: Cornell University Press.

Petrich, J. W., Lambry, J. C., Kuczera, K., Karplus, M., Poyart, C., & Martin, J. L. (1991). Ligand binding and protein relaxation in heme proteins: A room temperature analysis of NO geminate recombination. *Biochemistry*, *30*, 3975–3987.

Rovira, C., Kunc, K., Hutter, J., Ballone, P., & Parrinello, M. (1997). Equilibrium geometries and electronic structure of iron–porphyrin complexes: A density functional study. *Journal of Physical Chemistry A*, *101*, 8914–8925.

Sato, S. (1955a). On a new method of drawing the potential energy surface. *Journal of Chemical Physics*, *23*, 592–593.

Sato, S. (1955b). Potential energy surface of the system of three atoms. *Journal of Chemical Physics*, *23*, 2465–2466.

Siegbahn, P. E. M., Blomberg, M. R. A., & Chen, S.-L. (2010). Significant van der Waals effects in transition metal complexes. *Journal of Chemical Theory and Computation*, *6*(7), 2040–2044. http://dx.doi.org/10.1021/ct100213e.

Simonson, T., Archontis, G., & Karplus, M. (2002). Free energy simulations come of age: Protein–ligand recognition. *Accounts of Chemical Research*, *35*(6), 430–437.

Soloviov, M., & Meuwly, M. (2014). CO-dynamics in the active site of cytochrome c oxidase. *Journal of Chemical Physics*, *140*(14), 145101.

Soloviov, M., & Meuwly, M. (2015). Reproducing kernel potential energy surfaces in biomolecular simulations: Nitric oxide binding to myoglobin. *Journal of Chemical Physics*, *143*(10), 105103.

Tong, X., Nagy, T., Reyes, J. Y., Germann, M., Meuwly, M., & Willitsch, S. (2012). State-selected ion-molecule reactions with Coulomb-crystallized molecular ions in traps. *Chemical Physics Letters*, *547*, 1–8. http://dx.doi.org/10.1016/j.cplett.2012.06.042.

van Duin, A. C. T., Dasgupta, S., Lorant, F., & Goddard, W. A., III. (2001). ReaxFF: A reactive force field for hydrocarbons. *Journal of Physical Chemistry A*, *105*, 9396–9409.

Van Voorhis, T., Kowalczyk, T., Kaduk, B., Wang, L.-P., Cheng, C.-L., & Wu, Q. (2010). The diabatic picture of electron transfer, reaction barriers, and molecular dynamics. *Annual Review of Physical Chemistry*, *61*(1), 149–170. http://dx.doi.org/10.1146/annurev.physchem.012809.103324. PMID: 20055670.

Warshel, A., & Levitt, M. (1976). Theoretical studies of enzymic reactions: Dielectric, electrostatic and steric stabilization of the carbonium ion in the reaction of lysozyme. *Journal of Molecular Biology*, *103*, 227–249.

Warshel, A., & Weiss, R. M. (1980). An empirical valence bond approach for comparing reactions in solutions and in enzymes. *Journal of the American Chemical Society*, *102*, 6218–6226.

Weiner, S. J., Kollman, P. A., Case, D. A., Singh, U., Ghio, C., Alagona, G., … Weiner, P. (1984). A new force-field for molecular mechanical simulation of nucleic-acids and proteins. *Journal of the American Chemical Society*, *106*, 765–784.

Yosa, J., & Meuwly, M. (2011). Vibrationally induced dissociation of sulfuric acid (H2SO4). *Journal of Physical Chemistry A*, *115*(50), 14350–14360.

Yosa Reyes, J., Nagy, T., & Meuwly, M. (2014). Competitive reaction pathways in vibrationally induced photodissociation of H2SO4. *Physical Chemistry Chemical Physics*, *16*, 18533–18544. http://dx.doi.org/10.1039/C4CP01832J.

Zhou, H. Y., Tajkhorshid, E., Frauenheim, T., Suhai, S., & Elstner, M. E. (2002). Performance of the AM1, PM3, and SCC-DFTB methods in the study of conjugated Schiff base molecules. *Chemical Physics*, *277*, 91–103.

CHAPTER THREE

Generalized Ensemble Sampling of Enzyme Reaction Free Energy Pathways

D. Wu*,[1], M.I. Fajer[†,‡,1], L. Cao[‡], X. Cheng[†,§,2], W. Yang*,[‡,2]
*Institute of Molecular Biophysics, Florida State University, Tallahassee, FL, United States
[†]UT-ORNL Center for Molecular Biophysics, Oak Ridge National Laboratory, Oak Ridge, TN, United States
[‡]Florida State University, Tallahassee, FL, United States
[§]The University of Tennessee, Knoxville, TN, United States
[2]Corresponding authors: e-mail address: chengx@ornl.gov; yyang2@fsu.edu

Contents

1. Introduction	58
2. Collective Variable and Reaction Order Parameter	59
3. Traditional Importance Sampling vs GE Sampling	62
4. Dimensionality Limit	64
5. One of the First Metadynamics-Based Enzyme Reaction Studies	65
6. GE-Based String Optimization: The OTPRW Method	67
7. OTPRW Study of a Substrate-Assisted Glycosylation Reaction	69
8. Final Remarks	70
Acknowledgments	71
References	71

Abstract

Free energy path sampling plays an essential role in computational understanding of chemical reactions, particularly those occurring in enzymatic environments. Among a variety of molecular dynamics simulation approaches, the generalized ensemble sampling strategy is uniquely attractive for the fact that it not only can enhance the sampling of rare chemical events but also can naturally ensure consistent exploration of environmental degrees of freedom. In this review, we plan to provide a tutorial-like tour on an emerging topic: generalized ensemble sampling of enzyme reaction free energy path. The discussion is largely focused on our own studies, particularly ones based on the metadynamics free energy sampling method and the on-the-path random walk path sampling method. We hope that this minipresentation will provide interested practitioners some meaningful guidance for future algorithm formulation and application study.

---

[1] Equal contribution to this work.

1. INTRODUCTION

Powered by thermal reservoir, chemical reaction systems aimlessly fluctuate in their surrounding media. Through molecular interactions, energy is channeled to the reaction center and ultimately leads to large amplitude of fluctuations that cause chemical bond breaking and formation. Because the overall probability for a system to be adequately activated and successfully form reactive configurations is low, a chemical reaction is usually in orders of magnitudes slower than the elementary vibration that is directly responsible for the reactive event. Based on the transition state theory (Chandler, 1978; Wigner, 1938), among all the possible regions, from which the system can barrierlessly proceed to the product basin, there is a characteristic transition state (TS) region, the reaching of which from the reactant basin requires the least activation input, and therefore the TS region is expected to attract a majority of reaction fluxes to pass through. Based on such a simplified picture, understanding the mechanism of a chemical reaction largely means (a) elucidating how a chemical system proceeds along the most probable pathway, in particular from its reactant basin to the TS region and (b) quantifying the free energy change along such a pathway. Correspondingly, free energy path calculation constitutes a major task in computational analysis of chemical reaction mechanisms, such as those occurring in enzymatic environment.

Enzyme reaction represents a special class of chemical reactions, the rates of which are usually significantly higher than their counterparts in aqueous solution; for instance, enzymatic rate enhancement (Miller & Wolfenden, 2002) can be as high as $\sim 10^{17}$. Applying molecular dynamics (MD) simulation methods to elucidate enzyme reaction mechanisms and understand the corresponding catalytic strategies has been a classical topic in computational chemistry and biophysics (Gao et al., 2006; Gao & Truhlar, 2002; Hu & Yang, 2008; Orlando & Jorgensen, 2010; Senn & Walter, 2007; Warshel et al., 2006). In the recent years, method development efforts for this topic, in particular toward quantitative depiction of free energy pathways, have been reviving, partly due to ever-increasing interest in enzyme designs and partly due to an increasing demand of predicting atomistic level details that can serve as meaningful hypothesis for experimental test. Generally speaking, the quality of an enzyme reaction free energy path calculation relies on both potential energy function accuracy and sampling adequacy. Constrained by ever-limited computing power, tremendous algorithm developments have focused on (a) how to reduce energy and force evaluation cost while maintaining minimum accuracy loss and (b) how to more efficiently perform reaction free energy path sampling. Development efforts in the former aspect have been widely acknowledged, for instance, as represented by the combined quantum mechanical and molecular mechanical (QM/MM) scheme (Field, Bash, & Karplus, 1990; Warshel & Levitt, 1976) and recent multiscale reactive force field models (Schmitt & Voth, 1998; Swanson et al., 2007; Voth, 2006). In contrast, advancing sampling methods for free energy path calculation only became flourishing about a decade ago. Notably, recent advancements have been mainly catalyzed by the development of generalized ensemble (GE) sampling techniques (Okamoto, 2004; Yang, Nymeryer, Zhou, Berg, & Brüschweiler, 2008; Zuckerman, 2011) and path optimization algorithms (Chen & Yang, 2009; Dickson, Huang, & Post, 2012; Elber & Karplus, 1987; Henkelman, Uberuaga, & Jonsson, 2000; Maragliano, Fischer, Vanden-Eijnden, & Ciccotti, 2006; Pan, Sezer, & Roux, 2008; E, Ren, & Vanden-Eijnden, 2002, 2005). In this review, we focus our discussion on an emerging topic: GE sampling of enzyme reaction free energy path. Our review is largely based on our earlier and recent studies; these studies, respectively, represent early applications of the metadynamics free energy sampling method (Laio & Parrinello, 2002) and the on-the-path random walk (OTPRW) path sampling method (Cao, Lv, & Yang, 2013; Chen & Yang, 2009) on enzyme reaction systems. We hope that our presentation will provide interested practitioners some meaningful guidance for future algorithm formulation and application study.

2. COLLECTIVE VARIABLE AND REACTION ORDER PARAMETER

Although enzymes reaction rates are accelerated, in comparison with timescales commonly accessible to MD simulations, these processes are still far too slow. Estimated based on the diffusion limit, commonly, enzyme reaction free energy barriers are higher than 11–12 kcal/mol (Alberty & Hammes, 1958). Together with the fact that costly electronic calculation has to be included for chemical transformation treatment, there has never been a hope that a reactive trajectory can be obtained by a canonical ensemble QM/MM MD simulation. Therefore, activation biases need to be introduced in order to sufficiently sample high free energy regions, particularly regions that are >2 kT above the reactant basin; for instance, a simplest way to realize such activation is to restrain the system around to-be-activated regions. Following this requirement, a key technical question arises: which degrees of freedom need to be chosen for sampling enhancement.

In correspondence, often the first step in an enzyme reaction simulation study is to choose a set of essential collective variables (CVs; commonly geometric variables) that is hopefully sufficient to describe the target reaction. In the context of free energy path sampling, being an "essential" collective variable indicates that free energy flattening along this CV can lead to random walk dynamics in certain portions of the reaction pathway, which is otherwise unachievable along another candidate CV. In an application study, collective variable selection is commonly dictated by chemical intuition, eg, simulators' educated guess or priori knowledge of the reaction mechanism. In practice, simulators often choose "reaction bond order parameters" (RBOPs), which can distinguish the reactant basin, the product basin, and plausible metastable intermediates, as candidate CVs. Here is an example. Inosine monophosphate dehydrogenase (IMPDH) catalyzes two sequential chemical transformations: (1) a dehydrogenase reaction between IMP and NAD^+ that produces a Cys319-linked intermediate E-XMP* and NADH, and (2) a hydrolysis reaction that releases XMP. Before our study (Min et al., 2008) on the hydrolysis step catalyzed by the IMPDH Arg418Gln variant, it was known that Tyr419 needs to be deprotonated; and thereby it can act as a general base to activate a water molecule so that it can become a better nucleophile to replace Cys319 on the XMP ring (Fig. 1A). Two reaction processes, proton transfer between water and Tyr419 and nucleophilic substitution between the activated water (if completely activated, a hydroxyl group) and Cys319, may occur either sequentially or concertedly; even within each chemical process, two subevents, for instance, proton deattachment from water and protonation of Tyr419 during proton transfer, may also occur either sequentially or concertedly. Thus four pairs of distances, corresponding to four chemical bonds [O(water)–H(water), O(Tyr419)–H(water), O(water)–C(XMP), and S(Cys319)–C(XMP)] that form either in the reactant basin, the product basin, or at possible metastable intermediate states, were initially chosen as candidate CVs for this study. It should be noted that by definition, "order parameters" are defined to distinguish stable basins. As is generally known, order parameters are often insufficient to describe protein conformational changes, because between two conformational basins, slow reorganization changes are likely to occur along

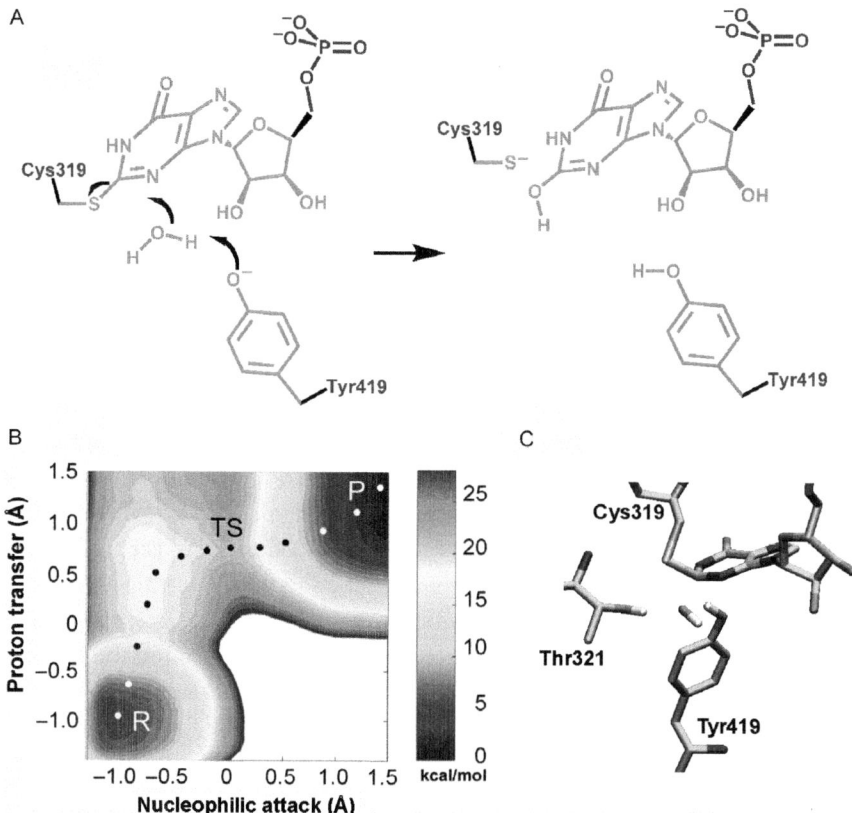

Fig. 1 Wang–Landau (flat-histogram) metadynamics simulation of the hydrolysis step in the IMPDH Arg418Gln variant. Figure was originally published in PLoS Biology (open-access, doi:10.1371/journal.pbio.0060206.g003). (A) The proposed mechanism on the hydrolysis of E-XMP* with Tyr419 acting as the general base. (B) The free energy landscape of the Tyr419 pathway in the Ar418Gln variant. *P*, product; *R*, reactant; and *TS*, transition state. (C) The corresponding transition state structure. (See the color plate.)

orthogonal degrees of freedom. In contrast, enzyme active sites are generally preorganized for chemical steps and reactive events usually closely follow high-frequency fluctuations that directly involve chemical bond vibrations. Therefore unless chemical transformation occurs through a conformation that differs from that of the starting structure or the reaction involves slow conformational changes, the RBOP-based CV identification strategy is generally effective; considering possibly complex electronic structural effects involved in enzyme reactions, sometimes bond formation/breaking angles can be essential CVs (Chen & Yang, 2009).

3. TRADITIONAL IMPORTANCE SAMPLING VS GE SAMPLING

Upon the identification of CV candidates, the next question is how to use them to sample enzyme reaction free energy pathways. Till about a decade ago, most-related studies had been carried out based on the traditional importance sampling methods: either the umbrella sampling method (Patey & Valleau, 1975) or the blue moon ensemble method (Sprik & Ciccotti, 1998), where independent MD simulations are performed with the CV either restrained around or constrained at a series of values that can cover the entire reaction span. Due to the computing power limitation, a majority of these sampling efforts were along a single CV. Despite that these studies had played a significant role in advancing the field of computational enzymology and deepening our understanding of enzymatic catalysis, their sampling deficiency is obvious. When a single CV is employed, it is very likely that it cannot sufficiently describe the whole reaction pathway. As shown in Fig. 2, due to the missing of certain essential CVs, the hidden CV issue may lead the sampling CV to proceed traversely along the physical reaction path. Due to the restriction or the loss of dynamics along the sampling CV, accurately exploring the traverse region can be forbiddingly challenging to the umbrella sampling and the blue moon ensemble methods; for instance under the restraint treatment, transitions between A and B or between B and C (Fig. 2) have to be through a region kinetically inaccessible

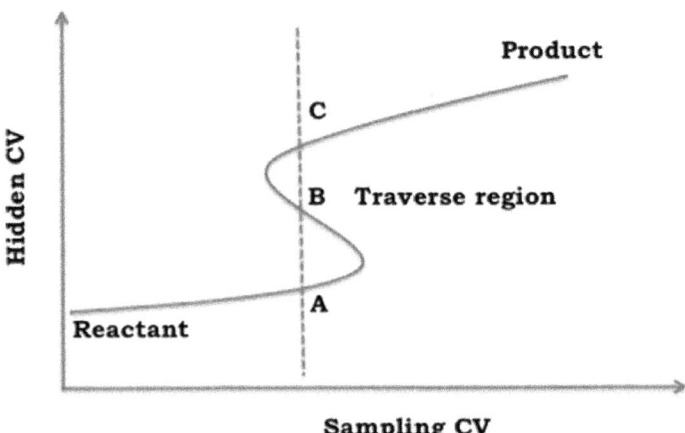

Fig. 2 The schematic illustration of the hidden CV issue.

to the physical process. It should be specially noted that unless missing essential CVs can be specifically guessed, the hidden CV issue is practically undetectable from such simulations themselves because there is no phase space connectivity information on samples obtained from independently restrained/constrained windows.

Different from the traditional importance sampling strategy, GE methods (Okamoto, 2004; Yang et al., 2008; Zuckerman, 2011) do not require any restriction of dynamics along the sampling CV. Instead, activation along the sampling CVs is enabled through the modification of the Hamiltonian, as follows:

$$H_m = H_o + f_m\left[\vec{\theta}\,(X)\right], \tag{1}$$

where H_o represented the original Hamiltonian and $f_m\left[\vec{\theta}\,(X)\right]$ stands for the biasing potential along the prechosen CV set $\vec{\theta}\,(X)$. Commonly, the target function of $f_m\left[\vec{\theta}\,(X)\right]$ is set to be $-G_o\left[\vec{\theta}\,(X)\right]$, the negative of the free energy surface (FES) mapped along $\vec{\theta}\,(X)$. $G_o\left[\vec{\theta}\,(X)\right]$ is the sampling target, which is unknown a priori. To adaptively obtain $G_o\left[\vec{\theta}\,(X)\right]$, three major recursion approaches have been developed, including the adaptive umbrella sampling method (Bartels & Karplus, 1997), in which free energy estimations are based on probability distributions, the adaptive biasing force (ABF) method (Darve & Pohorille, 2001), in which free energy estimations are based on the thermodynamic integration formula (Carter, Ciccotti, Hynes, & Kapral, 1989; Kirkwood, 1935), and the metadynamics (Laio & Parrinello, 2002)/local elevation (Huber, Torda, & van Gunsteren, 1994) method, which is realized through continuous deposition of repulsive basis functions. Through GE sampling, target FESs are explored through single continuous trajectories. It allows moderate hidden CV problems to be possibly bypassed and severe hidden CV problems to be detectable. In addition, through random walks following a single trajectory, the remaining environment degrees of freedom that are not subject to biased activation can be consistently sampled along the reaction pathway.

Among the above recursion methods, metadynamics has attracted the most attention. In the past years, tremendous efforts have been made to improve its robustness and convergence behavior. The original metadynamics is as simple as generating $f_m\left[\vec{\theta}(X)\right]$ by continuously depositing Gaussian functions:

$$f_m\left[\vec{\theta}(X)\right] = \sum_t h \prod_{i=1}^m exp\left(-\frac{(\theta_i - \theta_i^t)^2}{2w_i^2}\right), \quad (2)$$

where θ_i stands for the ith collective variable and θ_i^t stands for the value of the ith collective variable at the scheduled time t, h is the height of the basis Gaussian function, and w_i is the width of the ith component of the basis Gaussian function. Realizing the fact that the error of free energy estimation

$$\left[G_o\left[\vec{\theta}(X)\right] = -\sum_t h \prod_{i=1}^m exp\left(-\frac{(\theta_i - \theta_i^t)^2}{2w_i^2}\right)\right]$$ strongly depends on the size

of the basis Gaussian function, our group introduced the first systematic improvement (Min, Liu, Carbone, & Yang, 2007) for metadynamics by strategically reducing its size through the Wang–Landau flat-histogram procedure. Since this beginning, there have been several ingenious and rigorous improvements, such as well-tempered metadynamics (Barducci, Bussi, & Parrinello, 2008; Dama, Parrinello, & Voth, 2014), transition-tempered metadynamics (Dama, Rotskoff, Parrinello, & Voth, 2014), and recently very promising metabasin metadynamics (Dama, Hocky, Sun, & Voth, 2015), etc., formulated. Among them, well-tempered metadynamics has become a widely applied method. As one can expect, the recent more elegant metadynamics methods (Dama et al., 2015; Dama, Rotskoff, et al., 2014) will soon prove their unique advantages for sampling enzyme reaction free energy paths.

4. DIMENSIONALITY LIMIT

In theory, reaction free energy sampling can be performed in any number of dimensions. In practice, it is challenging to simultaneously sample more than three CVs. Such dimensionality limit is commonly considered being the result of the sampling manifold issue. As discussed earlier, an essential CV should play its sampling role in certain portions of the reaction pathway; in the other portions of the pathway, ideally it should not be activated so as to confine sampling within an one-dimension reaction channel. In high-dimension GE sampling, CVs are unselectively activated even when they are not around their individual working regions. Therefore the sampling manifold is much larger than the size of the reaction channel.

With the increase of the sampling dimensionality, the diffusion time in regions unrelated to the physical process is expected to grow drastically.

Indeed, the origin of dimensionality limit can be physical. For instance, gas phase reactions can be readily studied via three-dimensional metadynamics (Ensing & Klein, 2005). On contrary, based on our observation and experience, for enzyme reactions, it is likely that two dimensions are a common practical limit, while with a careful choice of CVs, higher-dimension GE sampling might be marginally possible. When multiple CVs are applied for GE sampling, lower-frequency collective motions inaccessible to the physical process are likely to be promoted. As one can imagine, a boost of 11–12 kcal/mol or above on these lower-frequency motions, which involve collective interplays of these sampling CVs, can be detrimental to the overall structural integrity and the system stability.

5. ONE OF THE FIRST METADYNAMICS-BASED ENZYME REACTION STUDIES

In 2008, we reported one of the first metadynamics-based enzyme reaction studies (Min et al., 2008), which is on the IMPDH-catalyzed hydrolysis step. As discussed earlier, for the reaction catalyzed by the IMPDH Arg418Gln variant, we initially identified four candidate distance CVs [$d1$: O(water)–H(water); $d2$: O(Tyr419)–H(water); $d3$: O(water)–C (XMP); and $d4$: S(Cys319)–C(XMP)]. To reduce the CV dimensionality at least to two, we redefined $d4-d3$ as CV1 (θ_1) to sample the nucleophilic substitution process and $d2-d1$ as CV2 (θ_2) to sample the proton transfer process. It is noted that taking CV difference is a common means to reduce CV dimensionality; however, it should be applied with caution because possible high degeneracy may introduce large diffusion sampling overhead.

As shown in Fig. 1B, the Wang–Landau metadynamics (flat-histogram metadynamics) simulation led to a detailed and nicely converged FES, on which besides the reactant and product basins, there is a metastable intermediate occurring between the proton transfer step and the nucleophilic substitution step. From the FES, we could generate a string of CV(θ)-space points to describe the minimum free energy path (MFEP) between the centers of the reactant and product basins. These points satisfy the following string condition (E et al., 2002): $\left[\frac{\partial G}{\partial \theta}\right]^{\perp} = 0$, in which $\frac{\partial G}{\partial \theta}\left(\frac{\partial G}{\partial \theta_1}, \frac{\partial G}{\partial \theta_2}\right)$ stands for the free energy gradient vector and $\perp$ denotes the projection perpendicular to the minimum free energy curve. Along the MFEP, the two chemical events proceed in a stepwise manner. The free energy barrier of the second step is higher (about 17 kcal/mol); therefore the nucleophilic substitution process is the rate-limiting step. Taking into the account the free energy penalty for Tyr deprotonation, the overall free energy barrier is about 21–22 kcal/mol, which is in good accord with the barrier observed for the reactions of the IMPDH Arg418Gln and Arg418Ala variants (about 20–21 kcal/mol). The location of the TS reveals that both the proton transfer step and the nucleophilic displacement step are concerted. As shown in Fig. 3C, at the TS of the rate-limiting step, the S(Cys319)–C(XMP) bond partially breaks and the O(water)–C(XMP) bond partially forms.

Back in 2008, obtaining an enzyme reaction FES with the above quality was rare; it was impossible without our own implementation of the metadynamics method in the CHARMM program (Brooks et al., 2009). Interestingly, this early implementation has many worth-noting features. For instance, Gaussian functions are deposited to grids with their heights determined by the second-order spline function; in addition, we introduced a mechanism to uniformly delete Gaussian functions to prevent Gaussian functions from flooding outside predefined boundaries.

For this review, we did a careful literature search and found that one metadynamics-based enzyme reaction study (Stanton, Kuo, Mundy, Laino, & Houk, 2007) was published before our above study; in this work, reported by the Houk group in 2007, 1D metadynamics sampling was employed to exam the direct decarboxylation mechanism catalyzed by the most proficient enzyme: Orotidine-5′-monophosphate decarboxylase (ODCase). Since then, there have been only ~30 metadynamics-based

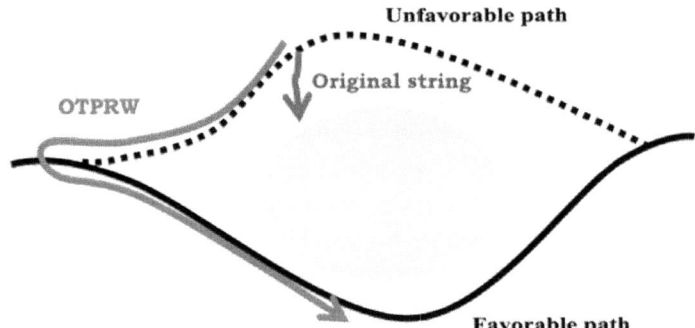

Fig. 3 The schematic illustration of the path switching mechanism in OTPRW.

enzyme reaction studies reported. Considering the popularity of metadynamics, this small number likely reflects the practical challenge in applying metadynamics to explore enzyme reaction pathways. As mentioned earlier, recent advancement of the metadynamics method (Dama et al., 2015; Dama, Parrinello, et al., 2014; Dama, Rotskoff, et al., 2014) will certainly lead to more successful applications. Nevertheless dimensionality limit still requires simulators to creatively design low-dimension CVs (Okamoto, 2004) and often apply them in a trial-and-error manner (McCullagh, Saunders, & Voth, 2014).

6. GE-BASED STRING OPTIMIZATION: THE OTPRW METHOD

The above case study based on FES sampling demonstrates the indirect reaction free energy path calculation strategy (Ensing, Laio, Parrinello, & Klein, 2005). As an alternative, the chain-of-states (COS) path optimization strategy can be employed to directly obtain reaction pathways. In comparison with the FES sampling-based strategy, the path optimization strategy only requires one-dimension sampling and thus has no dimensionality limit issue; eg, multiple candidate CVs $\theta(X) = (\theta_1(X), \ldots, \theta_m(X))$ can be employed to represent the path space. Among various COS algorithms, the string (FTS) method has attracted a great deal of attention, in particular recently for enzyme reaction mechanism studies (Aranda, Zinovjev, Roca, & Tunon, 2014; Khavrutskii, Legler, Friedlander, & Wallqvist, 2014; Kumari et al., 2015; Lans et al., 2012; Rosta, Nowotyny, Yang, & Hummer, 2011; Sanchez-Martinez, Field, & Crehuet, 2015; Zhang et al., 2015; Zinovijev, Ruiz-Pernia, & Tunon, 2013). Based on string method, MFEP can be obtained according to the minimization criterion $\left[M\frac{\partial G}{\partial \theta}\right]^\perp = 0$ (M is the diffusion tensor matrix), which, in comparison with $\left[\frac{\partial G}{\partial \theta}\right]^\perp = 0$, can more accurately reflect the curvilinear nature of CVs.

In common string method applications, sampling is performed on a series of noncommunicating images between two prechosen end points $Z^A = (z_1^A, \ldots, z_m^A)$ and $Z^B = (z_1^B, \ldots, z_m^B)$. Regarding the original string method, two sampling issues are worth noting: (a) because images are independently explored, the on-the-path continuity of the environmental degrees of freedom cannot be guaranteed; (b) the CV degrees of freedom are restricted from regular MD sampling and thus it is challenging for a being-optimized path that represents an unfavorable mechanism to be switched into the correct reaction channel. To overcome these issues, we developed a GE sampling-based string path optimization method: the OTPRW algorithm (Cao et al., 2013; Chen & Yang, 2009). In OTPRW, the CV-space pathway is represented by a set of λ-dependent functions $Z(\lambda) = (z_1(\lambda), \ldots, z_m(\lambda))$, in which $Z(0) = Z^A$ and $Z(1) = Z^B$, where the progressing parameter λ is set equal to the percentage of the on-the-path distance of the corresponding state from the starting point Z^A. Dynamic propagation in OTPRW is based on the following extended-dynamics Hamiltonian,

$$H_\lambda = H_o + \frac{p_\lambda^2}{2m_\lambda} + \sum_{i=1}^{m} \frac{1}{2} K_i (\theta_i(X) - z_i(\lambda))^2 + f_m(\lambda), \qquad (4)$$

where λ is treated as a one-dimension dynamic particle with a mass of m_λ and its momentum of p_λ and is propagated based on Langevin dynamics; via the $\sum_{i=1}^{m} \frac{1}{2} K_i (\theta_i(X) - z_i(\lambda))^2$ term, the system is restrained on the path from the latest optimization update. Through the biasing function $f_m(\lambda)$, which can be adaptively obtained via either the metadynamics (Chen & Yang, 2009) or the ABF method (Cao et al., 2013), the target system, instead of being constrained on noncommunicating images, can randomly walk along the instantaneous path to collect samples for the following path optimization. Thereby, the structural continuity of the environmental degrees of freedom can be naturally ensured; notably at a joint image state between two pathways, the system can switch from an unfavorable pathway to a better reaction channel (Fig. 3).

The OTPRW method has been successfully applied to the studies of the transformation between Chorismate and Prephenate (Chen & Yang, 2009), where 8 CVs were used, and the DNA base extrusion process (Cao et al., 2013), where 10 CVs were employed. In these studies, single-trajectory OTPRW simulations led to nicely converged MFEPs, which could be convincingly validated via the committor analysis. Interestingly in both studies, unexpected essential CVs were identified despite the fact that these systems had been immensely investigated. For instance, it was shown that besides RBOPs, bond breaking angles are intimately involved in the formation of the TS between Chorismate and Prephenate (Chen & Yang, 2009); and in the DNA base extrusion process, rather than commonly assumed base

7. OTPRW STUDY OF A SUBSTRATE-ASSISTED GLYCOSYLATION REACTION

Recently, we applied the OTPRW method to study the substrate-assisted glycosylation (SAG) reaction (Fig. 4A) that is catalyzed by a β-hexosaminidase protein, OfHex1. The SAG reaction involves two key chemical processes: "proton transfer" between the general acid (GluH) and glycosidic oxygen atoms and "nucleophilic substitution" around the central anomeric carbon. In this study, eight distance CVs were selected to describe the reaction path. Four of these CVs are RBOPs, corresponding to the chemical bonds directly involved in the bond forming and breaking events.

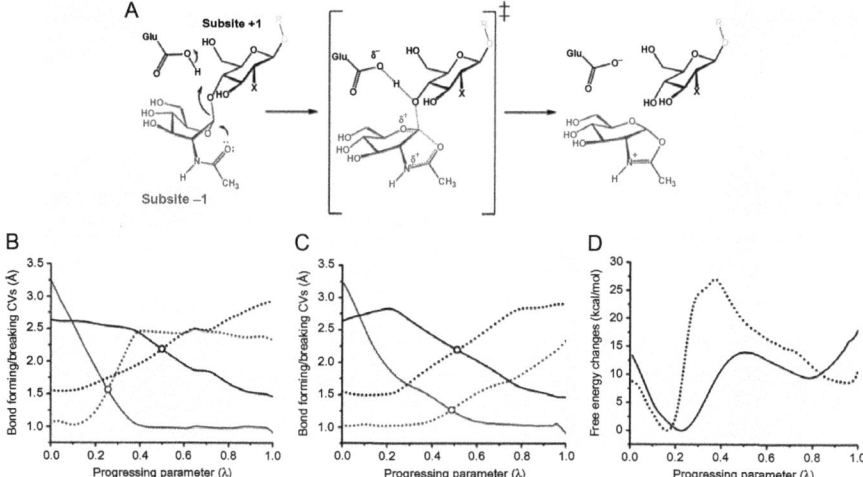

Fig. 4 The OTPRW simulation of the substrate-assisted glycosylation reaction in OfHex1. (A) The proposed mechanism on the substrate-assisted glycosylation reaction in OfHex1. (B) The chemical order parameter changes along the initial minimum energy path. *Red*: The CVs for the proton transfer process. *Blue*: The CVs for the nucleophilic substitution process. (C) The chemical order parameter changes along the OTPRW optimized minimum free energy path. *Red*: The CVs for the proton transfer process. *Blue*: The CVs for the nucleophilic substitution process. (D) The free energy changes along the initial minimum energy path (the *dotted line*) and along the OTPRW optimized minimum free energy path (the *solid line*). (See the color plate.)

To accurately describe geometrical constraints on the proton transfer process, the distance between the proton transfer donor and acceptor oxygen atoms and the distance between the to-be-transferred proton and the nonproton-donor oxygen of GluH were included in the CV set. In addition, two extra distance CVs around the nucleophilic substitution center were defined to describe possible bond formation and breaking angle changes during the reaction.

The OTPRW simulation began with a minimum energy path (MEP), along which the two chemical events are largely desynchronized. As shown in Fig. 4B, along this MEP, the proton transfer event proceeds earlier than the nucleophilic substitution event. Using the on-the-path ABF method (Cao et al., 2013), we calculated the free energy profile along this initial-guess reaction path. As shown by the dotted line in Fig. 4D, the "apparent" TS corresponds to the state of $\lambda = 0.371$, which is right between the midpoints of the two chemical events (Fig. 4B); and the overall free energy barrier is about 27.0 kcal/mol. Within 7 ns, the OTPRW simulation converged. As shown in Fig. 4C, along the MFEP, the two chemical processes are precisely synchronous and the subevents in each of the processes are highly concerted. As shown by the solid line in Fig. 4D, the TS corresponds to the state of around $\lambda = 0.49$, which is also the midpoints of the two chemical processes. The corresponding free energy barrier is about 13.1 kcal/mol, which is in excellent agreement with the experimental value. This study clearly demonstrates the importance of the MFEP sampling over the MEP calculation; obviously an MEP obtained based on a nondynamic enzyme environment can be drastically different from the target MFEP. Based on this result, we strongly discourage any future attempt to perform free energy calculation along a MEP to estimate the reaction free energy barrier.

8. FINAL REMARKS

Free energy path sampling plays an essential role in computational understanding of chemical reactions, particularly ones occurring in enzymatic environments. Among a variety of MD simulation approaches, the GE sampling strategy is uniquely attractive for the fact that it not only can enhance the sampling of rare chemical events but also can naturally ensure consistent exploration of environmental degrees of freedom. In this review, we plan to provide a tutorial-like tour on an emerging topic: GE sampling of enzyme reaction free energy path. The discussion is largely focused on our own studies, particularly ones based on the metadynamics free energy sampling method and the OTPRW path sampling method. We hope that this minipresentation will provide interested practitioners some meaningful guidance for future algorithm formulation and application study.

We would also like to point out that the GE sampling strategy is still far from being adequate. Necessary human input on the preselection of essential CVs is still greatly hindering enzyme reaction studies from reaching the predictive stage. In addition, there is still scarce of convincingly successful case study on enzyme reactions that couple with slow conformational transitions. Currently, we are actively working on further enriching the orthogonal space sampling scheme (Lv et al., 2016; Zheng, Chen, & Yang, 2008, 2009; Zheng & Yang, 2012) to overcome these challenges.

ACKNOWLEDGMENTS

Funding support from the National Science Foundation (MCB1158284) and National Institute of Health (R01GM054403 and R01GM111886) is acknowledged. This research used resources of the Oak Ridge Leadership Computing Facility at the Oak Ridge National Laboratory, which is supported by the Office of Science of the U.S. Department of Energy under Contract No. DE-AC05-00OR22725.

REFERENCES

Alberty, R. A., & Hammes, G. G. (1958). Application of the theory of diffusion-controlled reactions to enzyme kinetics. *The Journal of Physical Chemistry, 62*, 154–159.

Aranda, J., Zinovjev, K., Roca, M., & Tunon, I. (2014). Dynamics and reactivity in thermos aquaticus N6-adenine methyltransferases. *Journal of the American Chemical Society, 136*, 16227–16239.

Barducci, A., Bussi, G., & Parrinello, M. (2008). Well-tempered metadynamics: A smoothly converging and tunable free-energy method. *Physical Review Letters, 100*, 020603.

Bartels, C., & Karplus, M. (1997). Multidimensional adaptive umbrella sampling: Applications to main chain and side chain people. *Journal of Computational Chemistry, 18*, 1450–1462.

Brooks, B. R., Brooks, C. L., Mackerell, A. D., Nilsson, L., Petrella, R. J., Roux, B., ... Karplus, M. (2009). CHARMM: The biomolecular simulation program. *Journal of Computational Chemistry, 30*, 1545–1614.

Cao, L., Lv, C., & Yang, W. (2013). Hidden conformation events in DNA base extrusions: A generalized-ensemble path optimization and equilibrium simulation study. *Journal of Chemical Theory and Computation, 9*, 3756–3768.

Carter, E. A., Ciccotti, G., Hynes, J. T., & Kapral, R. (1989). Constrained reaction coordinate dynamics for the simulation of rare events. *Chemical Physics Letters, 156*, 472–477.

Chandler, D. (1978). Statistical mechanics of isomerization dynamics in liquids and the transition state approximation. *The Journal of Chemical Physics, 68*, 2959–2970.

Chen, M. E., & Yang, W. (2009). On-the-path random walk sampling for efficient optimization of minimum free-energy path. *Journal of Computational Chemistry, 30*, 1649–1653.

Dama, J. F., Hocky, G. M., Sun, R., & Voth, G. A. (2015). Exploring valleys without climbing every peak: More efficient and forgiving metabasin metadynamics via robust on-the-fly bias domain restriction. *Journal of Chemical Theory and Computation, 11*, 5638–5650.

Dama, J. F., Parrinello, M., & Voth, G. A. (2014). Well-tempered metadynamics converges asymptotically. *Physical Review Letters, 112*, 240602.

Dama, J. F., Rotskoff, J., Parrinello, M., & Voth, G. A. (2014). Transition-tempered metadynamics: Robust, convergent metadynamics via on-the-fly transition barrier estimation. *Journal of Chemical Theory and Computation, 10*, 3626–3633.

Darve, E., & Pohorille, A. (2001). Calculating free energies using average force. *The Journal of Chemical Physics, 115*, 9169–9183.

Dickson, B. M., Huang, H., & Post, C. B. (2012). Unrestrained computation of free energy along a path. *The Journal of Physical Chemistry. B, 116*, 11046–11055.

E, W. N., Ren, W., & Vanden-Eijnden, E. (2002). String method for the study of rare events. *Physical Review B, 66*, 052301.

E, W. N., Ren, W., & Vanden-Eijnden, E. (2005). Finite temperature string method for the study of rare events. *The Journal of Physical Chemistry. B, 109*, 6688–6693.

Elber, R., & Karplus, M. (1987). A method for determining reaction paths in large molecules: Application to myoglobin. *Chemical Physics Letters, 139*, 375–380.

Ensing, B., & Klein, M. L. (2005). Perspective on the reactions between F^- and CH_3CH_2F: The free energy landscapes of the E2 and SN_2 reaction channels. *Proceedings of the National Academy of Sciences of the United States of America, 102*, 6755–6759.

Ensing, B., Laio, A., Parrinello, M., & Klein, M. L. (2005). A recipe for the computation of the free energy barrier and the lowest free energy path of concerted reactions. *The Journal of Physical Chemistry. B, 109*, 6676–6687.

Field, M. J., Bash, P. A., & Karplus, M. (1990). A combine quantum-mechanical and molecular mechanical potential for molecular-dynamics simulations. *Journal of Computational Chemistry, 11*, 700–733.

Gao, J. L., Ma, S. H., Major, D. T., Nam, K., Pu, J. Z., & Truhlar, D. (2006). Mechanisms and free energies of enzymatic reactions. *Chemical Reviews, 106*, 3188–3209.

Gao, J. L., & Truhlar, D. G. (2002). Quantum mechanical methods for enzyme kinetics. *Annual Review of Physical Chemistry, 53*, 467–505.

Henkelman, G., Uberuaga, B. P., & Jonsson, H. (2000). A climbing image nudged elastic band method for finding saddle points and minimum energy paths. *The Journal of Chemical Physics, 113*, 9901–9904.

Hu, H., & Yang, W. T. (2008). Free energies of chemical reactions in solution and in enzymes with ab initio quantum mechanics/molecular mechanics methods. *Annual Review of Physical Chemistry, 59*, 573–601.

Huber, T., Torda, A. E., & van Gunsteren, W. F. (1994). Local elevation: A method for improving the searching properties of molecular dynamics simulation. *Journal of Computer-Aided Molecular Design, 8*, 695–708.

Khavrutskii, I. V., Legler, P. M., Friedlander, A. M., & Wallqvist, A. (2014). A reaction path study of the catalysis and inhibition of the Bacillus anthracis CapD gamma-glutamyl transpeptidase. *Biochemistry, 53*, 6954–6967.

Kirkwood, J. G. (1935). Statistical mechanics of fluid mixtures. *The Journal of Chemical Physics, 3*, 300–313.

Kumari, M., Kozmon, S., Kuhanek, P., Stepan, J., Tvaroska, I., & Koca, J. (2015). Exploring reaction pathways for O-GlcNAc transferase catalysis. *The Journal of Physical Chemistry. B, 119*, 4371–4381.

Laio, A., & Parrinello, M. (2002). Escaping free-energy minima. *Proceedings of the National Academy of Sciences of the United States of America, 99*, 12562–12566.

Lans, I., Medina, M., Rosta, E., Hummer, G., Garcia-Viloca, M., Lluch, J. M., & Gonzalez-Lafont, A. (2012). Theoretical study of the mechanism of the hydride transfer between ferredoxin-NADP(+) reductase and NADP(+): The role of Tyr303. *Journal of the American Chemical Society, 134*, 20544–20553.

Lv, C., Aitchison, E. W., Wu, D., Zheng, L. Q., Cheng, X. L., & Yang, W. (2016). Comparative exploration of hydrogen sulfide and water transmembrane free energy surfaces via orthogonal space tempering free energy sampling. *Journal of Computational Chemistry, 37*, 567–574.

Maragliano, L., Fischer, A., Vanden-Eijnden, E., & Ciccotti, G. (2006). String method in collective variables: Minimum free energy paths and siocommittor surfaces. *The Journal of Chemical Physics, 125*, 024106.

McCullagh, M., Saunders, M. G., & Voth, G. A. (2014). Unraveling the mystery of ATP hydrolysis in actin filaments. *Journal of the American Chemical Society, 136*, 13053–13058.

Miller, B. G., & Wolfenden, R. (2002). Catalytic proficiency: The unusual case of OMP decarboxylase. *Annual Review of Biochemistry, 71*, 847–885.

Min, D., Josephine, H. R., Li, H. Z., Lakner, C., MacPherson, I. S., Naylor, G. J., ... Yang, W. (2008). An enzymatic atavist revealed in dual pathways for water activation. *PLoS Biology, 6*, e206.

Min, D., Liu, Y., Carbone, I., & Yang, W. (2007). On the convergence improvement in the metadynamics simulations: A Wang-Landau recursion approach. *The Journal of Chemical Physics, 126*, 194104.

Okamoto, Y. (2004). Generalized-ensemble algorithms: Enhanced sampling techniques for Monte Carlo and molecular dynamics simulations. *Journal of Molecular Graphics & Modelling, 22*, 425–439.

Orlando, A., & Jorgensen, W. L. (2010). Advances in quantum and molecular mechanical (QM/MM) simulations for organic and enzymatic reactions. *Accounts of Chemical Research, 43*, 142–151.

Pan, A. C., Sezer, D., & Roux, B. (2008). Finding transition pathways using the string method with swarms of trajectories. *The Journal of Physical Chemistry. B, 112*, 3432–3440.

Patey, G. N., & Valleau, J. P. (1975). A Monte Carlo method for obtaining the interionic potential of mean force in ionic solution. *The Journal of Chemical Physics, 63*, 2334–2339.

Rosta, E., Nowotyny, M., Yang, W., & Hummer, G. (2011). Catalytic mechanism of RNA backbone cleavage by ribonuclease H from quantum mechanics/molecular mechanics simulations. *Journal of the American Chemical Society, 133*, 8934–8941.

Sanchez-Martinez, M., Field, M., & Crehuet, R. (2015). Enzymatic minimum free energy path calculations using swarms of trajectories. *The Journal of Physical Chemistry. B, 119*, 1103–1113.

Schmitt, U. W., & Voth, G. A. (1998). Multistate empirical valence bond model for proton transport in water. *The Journal of Physical Chemistry. B, 102*, 5547–5551.

Senn, H. M., & Walter, T. (2007). QM/MM studies of enzymes. *Current Opinion in Chemical Biology, 11*, 182–187.

Sprik, M., & Ciccotti, G. (1998). Free energy from constrained molecular dynamics. *The Journal of Chemical Physics, 109*, 7737–7744.

Stanton, C. L., Kuo, I. F. W., Mundy, C. J., Laino, T., & Houk, K. N. (2007). QM/MM metadynamics study of the direct decarboxylation mechanism for orotidine-5′-monophosphate decarboxylase using two different QM regions: Acceleration too small to explain rate of enzyme catalysis. *The Journal of Physical Chemistry. B, 111*, 12573–12581.

Swanson, J. M. J., Maupin, C. M., Chen, H. N., Petersen, M. K., Xu, J. C., Wu, Y. J., & Voth, G. A. (2007). Proton solvation and transport in aqueous and biomolecular systems: Insights from computer simulations. *The Journal of Physical Chemistry. B, 111*, 4300–4314.

Voth, G. A. (2006). Computer simulation of proton solvation and transport in aqueous and biomolecular systems. *Accounts of Chemical Research, 39*, 143–150.

Warshel, A., & Levitt, M. (1976). Theoretical studies of enzymic reactions—Dielectric, electrostatic and steric stabilization of carbonium-ion in reaction of lysozyme. *Journal of Molecular Biology, 103*, 227–249.

Warshel, A., Sharma, P. K., Kato, M., Xiang, Y., Liu, H. B., & Olsson, M. H. M. (2006). Electrostatic basis for enzyme catalysis. *Chemical Reviews, 106*, 3210–3235.

Wigner, E. (1938). The transition state method. *Transactions of the Faraday Society, 34*, 29–41.

Yang, W., Nymeryer, H., Zhou, H. X., Berg, B. A., & Brüschweiler, R. (2008). Quantitative computer simulations of biomolecules: A snapshot. *Journal of Computational Chemistry, 29*, 668–672.

Zhang, S. X., Ganguly, A., Goyal, P., Bingaman, J. L., Bevilacqua, P. C., & Hammes-Schiffer, S. (2015). Role of the active site guanine in the glmS ribozyme self-cleavage mechanism: Quantum mechanical/molecular mechanical free energy simulations. *Journal of the American Chemical Society, 137*, 784–798.

Zheng, L. Q., Chen, M. E., & Yang, W. (2008). Random walk in orthogonal space to achieve efficient free-energy simulation of complex systems. *Proceedings of the National Academy of Sciences of the United States of America, 105*, 20227–20232.

Zheng, L. Q., Chen, M. E., & Yang, W. (2009). Simultaneous escaping of explicit and hidden free energy barriers: Application of the orthogonal space random walk strategy in generalized ensemble based conformational sampling. *The Journal of Chemical Physics, 130*, 234105.

Zheng, L. Q., & Yang, W. (2012). Practically efficient and robust free energy calculations: Double-integration orthogonal space tempering. *Journal of Chemical Theory and Computation, 8*, 810–823.

Zinovijev, K., Ruiz-Pernia, J. J., & Tunon, I. (2013). Toward an automatic determination of enzymatic reaction mechanisms and their activation free energies. *Journal of the American Chemical Society, 9*, 3740–3749.

Zuckerman, D. M. (2011). Equilibrium sampling in biomolecular simulation. *Annual Review of Biophysics, 40*, 41–62.

CHAPTER FOUR

Methods for Efficiently and Accurately Computing Quantum Mechanical Free Energies for Enzyme Catalysis

F.L. Kearns*, P.S. Hudson*, S. Boresch[†,1], H.L. Woodcock*[,1]
*University of South Florida, Tampa, FL, United States
[†]Faculty of Chemistry, University of Vienna, Vienna, Austria
[1]Corresponding authors: e-mail address: stefan@mdy.univie.ac.at; hlw@usf.edu

Contents

1. Introduction 76
   1.1 Background on FES 77
   1.2 Direct vs Indirect Free Energy Simulations 79
   1.3 Efficiently Calculating $\Delta A(0_{low} \rightarrow 0_{high})$ 81
2. Alchemical FES 86
   2.1 Examples 88
3. Reaction Profiles 92
   3.1 Examples 95
Appendix 97
   A.1 Relevant Nomenclature and Usage 97
Acknowledgment 98
References 98

Abstract

Enzyme activity is inherently linked to free energies of transition states, ligand binding, protonation/deprotonation, etc.; these free energies, and thus enzyme function, can be affected by residue mutations, allosterically induced conformational changes, and much more. Therefore, being able to predict free energies associated with enzymatic processes is critical to understanding and predicting their function. Free energy simulation (FES) has historically been a computational challenge as it requires both the accurate description of inter- and intramolecular interactions and adequate sampling of all relevant conformational degrees of freedom. The hybrid quantum mechanical molecular mechanical (QM/MM) framework is the current tool of choice when accurate computations of macromolecular systems are essential. Unfortunately, robust and efficient approaches that employ the high levels of computational theory needed to accurately describe many reactive processes (ie, ab initio, DFT), while also including explicit solvation effects and accounting for extensive conformational sampling are *essentially nonexistent.* In this chapter, we will give a brief overview of two recently developed methods that mitigate several major challenges associated with QM/MM FES: the QM non-Boltzmann Bennett's acceptance ratio method and the QM nonequilibrium work method. We will also describe usage of these methods to calculate free energies associated with (1) relative properties and (2) along reaction paths, using simple test cases with relevance to enzymes examples.

1. INTRODUCTION

Understanding enzyme structure and function requires accurately accounting for free energy changes associated with processes such as (de)protonation, ligand binding, transition state stabilization, and reactive pathways. Experimentalists investigate these properties with mutation, titration, kinetic, and crystallographic studies. From a computational perspective, free energy simulation (FES) is the critical methodology. The standard approach to FES is to conduct a series of simulations—each consisting of 10^5–10^7 energy and force calculations—and estimate $\Delta A(0 \rightarrow 1)$ using free energy perturbation (FEP) (Zwanzig, 1954), thermodynamic integration (TI) (Kirkwood, 1935), or Bennett's acceptance ratio (BAR) (Bennett, 1976) (we will consider only FEP and BAR here).

Although molecular mechanical (MM) simulations are often the best choice for generating sufficient sampling in FES, in the case of many enzyme properties an MM description of interactions is likely insufficient. Enzymes provide a range of subtle yet complex interactions that often cannot be accurately modeled with MM force fields; eg, covalent bond formation, electrostatic stabilization, etc. In many cases this leads to modulation of conformational degrees of freedom. Additionally, solvent molecules very often play critical roles and thus need to be accounted for explicitly. This begs the question: how can we incorporate *accurate* descriptions of inter- and intramolecular interactions (ie, quantum mechanical—QM) while still *adequately* sampling configurational space (ie, conducting sufficiently long simulations)? (König & Boresch, 2011; König, Hudson, Boresch, & Woodcock, 2014).

The hybrid quantum mechanical/molecular mechanical (QM/MM) (Warshel & Levitt, 1976) framework is the current tool of choice for tackling such daunting problems. However, robust and efficient approaches that employ the high levels of computational theory needed to accurately describe many processes in enzymes (eg, ab initio, DFT), while also including explicit solvation effects and accounting for extensive conformational sampling, are essentially nonexistent. Application of MM FES techniques would require us to conduct long QM/MM simulations and estimate free energy differences using FEP or BAR, but this solution is intractably expensive (de Ruiter, Boresch, & Oostenbrink, 2013; Leitgeb, Schröder, & Boresch, 2005). Therefore, past attempts to do QM/MM FES have mostly employed semiempirical QM/MM (SQM/MM) or empirical valence bond (EVB) models, where the generated ensembles are reasonably accurate and conducting long simulations is feasible. However, despite the fact that recent developments have improved the accuracy of SQM methods while maintaining computational efficiency (Cui, Elstner, Kaxiras, Frauenheim, & Karplus, 2001; Elstner et al., 1998; Repasky, Chandrasekhar, & Jorgensen, 2002; Thiel & Voityuk, 1996; Tuttle & Thiel, 2008; Yang, Yu, York, Elstner, & Cui, 2008), weaknesses still remain among them (Bowman, Grant, & Mulholland, 2008; Claeyssens et al., 2006; Lonsdale, Harvey, & Mulholland, 2010; Lonsdale, Hoyle, Grey, Ridder, & Mulholland, 2012) and it is evident that ab initio or DFT methods are still required for the most accurate FES. Herein we describe our recently developed techniques for *efficiently* incorporating high-level energetic detail in FES.

1.1 Background on FES

We will begin with a brief discussion of free energy estimators; however an in-depth description of FES is outside the scope of this chapter so some familiarity will be assumed. Those interested in reviewing FES in more detail should refer to Pohorille, Jarzynski, and Chipot (2010), Yang, Cui, Min, and Li (2010), and Shirts, Mobley, and Chodera (2007). Additionally, we will use nomenclature to succinctly represent thermodynamic states and mathematical expressions. In general "0"/"1" will refer to two distinct chemical/conformational states (ie, solvated vs gas phase, protonated vs deprotonated, etc.) with subscripts indicating the level of theory describing these states. Energy evaluations will be denoted as U_A^X, where "A" represents the chemical/conformational state (eg, "0"/"1"), and X represents the level of theory at which the energy was evaluated. See the Appendix for the complete set of terminology and definitions.

FEP is defined by Zwanzig's exponential formula (Zwanzig, 1954),

$$\Delta A(0 \to 1) = -k_B T \ln \langle \exp[-(U_1 - U_0)/k_B T] \rangle_0 \tag{1}$$

where k_B is Boltzmann's constant and T the temperature. FEP is a one-sided method, meaning that if $\Delta A(0 \to 1)$ is the free energy of interest, then a simulation at state 0 is conducted with potential energies of the trajectory snapshots evaluated at both endstates, ie, U_0 and U_1, and the ensemble average is calculated as shown in Eq. (1). BAR (Eq. 2) is widely recognized as one of the most efficient free energy estimators, often converging easily for calculations where FEP does not (Bennett, 1976; Boresch & Bruckner, 2011; König & Boresch, 2011; Lu, Kofke, & Woolf, 2004; Pohorille et al., 2010; Shirts & Pande, 2005):

$$\Delta A(0 \rightarrow 1) = k_B T \left(\ln \frac{\langle f(U_0 - U_1 + C) \rangle_1}{\langle f(U_1 - U_0 - C) \rangle_0} \right) + C \qquad (2)$$

where $f(x)$ denotes the Fermi function $f(x) = \left(1 + \exp\left(\frac{x}{k_B T}\right)\right)^{-1}$ and

$$C = k_B T \ln \frac{Q_0 N_1}{Q_1 N_0} \qquad (3)$$

Q_0 and Q_1 are the canonical partition functions, and N_0 and N_1 are the number of data points used to compute the ensemble averages. Eq. (2) is solved iteratively until the condition

$$\langle f(U_0 - U_1 + C) \rangle_1 = \langle f(U_1 - U_0 - C) \rangle_0 \qquad (4)$$

is fulfilled. With C determined in this manner, one obtains

$$\Delta A(0 \rightarrow 1) = -k_B T \ln \frac{N_1}{N_0} + C \qquad (5)$$

As opposed to FEP, BAR is a two-sided method: if $\Delta A(0 \rightarrow 1)$ is the free energy of interest, two simulations must be conducted at states 0 and 1, and the snapshots of each simulation are to be evaluated at both states of interest, ie, $\langle ... \rangle_0$ requires U_0 and U_1 for every snapshot sampled at state 0, and $\langle ... \rangle_1$ requires U_0 and U_1 for every snapshot sampled at state 1 (Bennett, 1976; König et al., 2014).

As discussed in Pohorille et al. (2010), for one-sided methods (eg, FEP) to yield converged results a sufficient number of configurations sampled in state 0 must be representative of state 1 (ie, low energy conformations). This can be checked by plotting the forward and backward potential energy difference probability distributions, $p(\Delta U_{0 \rightarrow 1})$ and $p(-\Delta U_{1 \rightarrow 0})$, respectively.[a] For two-sided methods (eg, BAR), even limited overlap between $p(\Delta U_{0 \rightarrow 1})$ and $p(-\Delta U_{1 \rightarrow 0})$ is sufficient. In Fig. 1A, a sizable number of configurations

---

[a] Where the "state" on the left side of the arrow is used to generate the ensemble. For example, $0 \rightarrow 1$ indicates that the ensemble was generated at state 0.

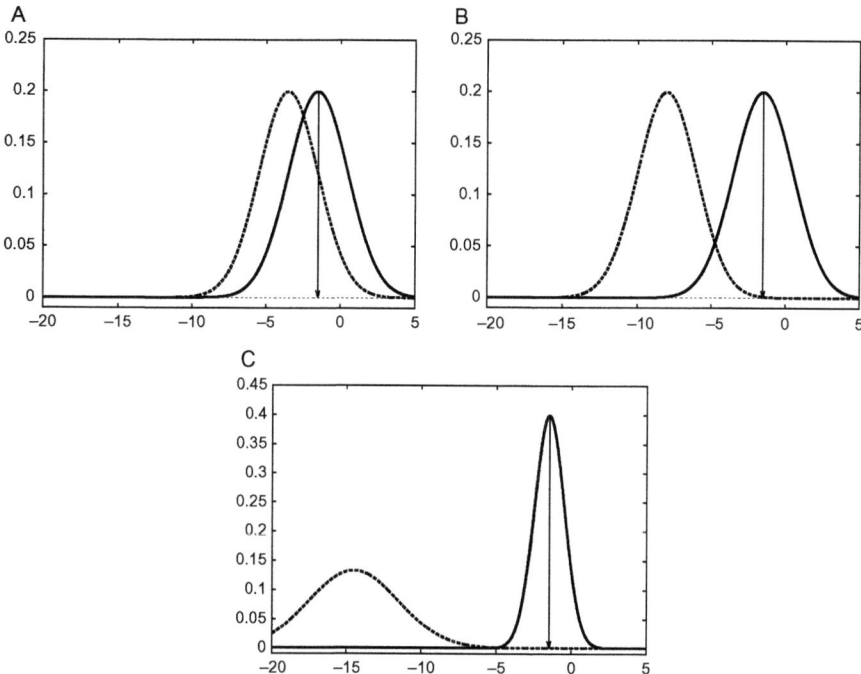

Fig. 1 Illustration of overlap between distributions of forward ($p(\Delta U_{0\to 1})$, *black dashed*) and backward ($p(-\Delta U_{1\to 0})$, *black solid*) energy differences. (A) Excellent overlap: a sufficient number of samples from the forward distribution is in the peak region of the backward distribution; FEP likely to work. (B) Weak overlap; BAR likely to work, FEP likely to fail. (C) No overlap, even BAR is likely to fail. *Note*: Normal distributions used for illustrative purposes; real distributions are likely to have marked deviations from the Gaussian case.

of the forward distribution fall into the peak region of the backward distribution; under these circumstances, FEP is likely to converge and result in correct free energy differences. By contrast, in Fig. 1B forward and backward energy distributions overlap, but only very few—if any within a finite simulation—configurations of $p(\Delta U_{0\to 1})$ are near the peak of $p(-\Delta U_{1\to 0})$. While BAR is likely to work in this situation, FEP will most likely fail. Finally, Fig. 1C depicts a situation where both FEP and BAR will not converge.

1.2 Direct vs Indirect Free Energy Simulations

As previously stated, many enzyme processes may not be modeled accurately enough with an MM description. For this reason we now turn our attention to calculating *high*-level or QM/MM free energy differences. There are two accepted approaches for calculating $\Delta A(0_{high} \to 1_{high})$: "direct" and "indirect." The direct approach requires one to conduct QM/MM simulations at state 0 and use FEP to evaluate $\Delta A(0_{high} \to 1_{high})$, or to conduct QM/MM simulations at states 0 and 1 and use BAR to evaluate $\Delta A(0_{high} \to 1_{high})$.

Alternatively, one could take an "indirect" approach to calculating $\Delta A(0_{high} \to 1_{high})$ by devising an advantageous thermodynamic cycle around the ΔA of interest, Fig. 2, and summing over the other legs of the cycle: $\Delta A(0_{high} \to 1_{high}) = \Delta A(0_{low} \to 1_{low}) + \Delta A(1_{low} \to 1_{high}) - \Delta A(0_{low} \to 0_{high})$ (Chandrasekhar, Smith, & Jorgensen, 1984, 1985; Cisneros, Liu, Lu, & Yang, 2005; Gao & Freindorf, 1997; Gao, Luque, & Orozco, 1993; Gao & Xia, 1992; Heimdal & Ryde, 2012; Hu, Lu, Parks, Burger, & Yang, 2008; Hu, Lu, & Yang, 2007; Kollman, 1993; Li & Yang, 2007; Luzhkov & Warshel, 1992; Rod & Ryde, 2005a, 2005b; Valiev, Bylaska, Dupuis, & Tratnyek, 2008; Valiev et al., 2007; Wesolowski & Warshel, 1994; Woods, Manby, & Mulholland, 2008; Zheng, Chen, & Yang, 2008; Zheng & Merz, 1992).

In Fig. 2, $\Delta A(0_{low} \to 1_{low})$ can be calculated using any common free energy estimator (eg, BAR), while (ii) and (iv) are typically calculated with FEP as it does not require *high*-level simulations:

$$\Delta A(0_{low} \to 0_{high}) = -k_B T \ln \left\langle \exp\left[-(U_0^{high} - U_0^{low})/k_B T\right] \right\rangle_{0,low} \quad (6)$$

It is important to emphasize the computational savings when choosing the indirect approach, as all simulations can be done with a low level of theory and only simulation snapshots have to be evaluated with high-level

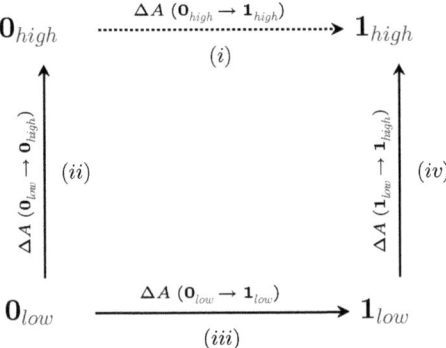

Fig. 2 Scheme illustrating the thermodynamic cycle used in the indirect approach. $\Delta A(0_{high} \to 1_{high}) = \Delta A(0_{low} \to 1_{low}) + \Delta A(1_{low} \to 1_{high}) - \Delta A(0_{low} \to 0_{high})$.

energetics. Additionally, when the "low" level is chosen to be MM, then tools such as soft-core potentials can be used to avoid end-point problems.

Although FEP is the most popular choice for calculating $\Delta A(0_{low} \rightarrow 0_{high})$, recall that it has demonstrable convergence and accuracy issues. As performance problems with FEP have become more recognized, attempts have been made to alleviate errors by "freezing" QM region atoms (ie, atoms described quantum mechanically within the QM/MM framework) during MM simulations (Rod & Ryde, 2005a, 2005b). Unfortunately, excluding QM region flexibility ignores entropic contributions. Warshel and coworkers have worked to maintain QM region flexibility when using FEP (Bentzien, Muller, Florián, & Warshel, 1998; Frushicheva & Warshel, 2012), namely by replacing MM simulations with EVB Hamiltonians parameterized to specifically reproduce QM endstates; still many of these calculations suffered the same poor convergence and inaccuracies typical of FEP (Bentzien et al., 1998; Plotnikov, Kamerlin, & Warshel, 2011). A more robust method for computing $\Delta A(0_{low} \rightarrow 0_{high})$ is clearly needed. This challenge is not new for FES (Hudson, Lee Woodcock, & Boresch, 2015; Hudson, White, et al., 2015; König et al., 2014), as we and many others (Chandrasekhar et al., 1984, 1985; Cisneros et al., 2005; Gao & Freindorf, 1997; Gao et al., 1993; Gao & Xia, 1992; Heimdal & Ryde, 2012; Hu et al., 2008, 2007; Kollman, 1993; Li & Yang, 2007; Luzhkov & Warshel, 1992; Rod & Ryde, 2005a, 2005b; Valiev et al., 2008, 2007; Wesolowski & Warshel, 1994; Woods et al., 2008; Zheng et al., 2008; Zheng & Merz, 1992) have been working to reconcile these conflicting requirements, but the subtle and complex nature of enzyme catalysis only stands to amplify this challenge.

1.3 Efficiently Calculating $\Delta A(0_{low} \rightarrow 0_{high})$

One of the main reasons FEP fails in "vertical" (ie, $\Delta A(0_{low} \rightarrow 0_{high})$ Fig. 2) calculations is poor overlap between the *low*-level and *high*-level potential energy surfaces. In principle, using BAR to calculate $\Delta A(0_{low} \rightarrow 0_{high})$ would be ideal; however, the requirement of simulations at both endstates (specifically 0_{high}) makes this approach untenable. So, how can one improve the convergence of FES between *low* and *high* levels of theory? We have recently developed two new methods for doing exactly this: the QM non-Boltzmann Bennett method (QM-NBB) and the QM nonequilibrium work (QM-NEW) method. In the next section we will describe the derivation of these methods and their basic usage.

QM-NBB: Recent work by our groups described a practical and conceptually simple means to utilize *high*-level energies efficiently in BAR without *high*-level simulations. The QM-NBB[b] method employs biasing potentials to extract QM quality free energies from *low*-level simulations. Torrie and Valleau described how to extract unbiased ensemble averages from biased simulations (Torrie & Valleau, 1977):

$$\langle X \rangle = \frac{\langle X \exp(\beta V^b) \rangle_{bias}}{\langle \exp(\beta V^b) \rangle_{bias}} \quad (7)$$

where β is $1/k_B T$. The notation $\langle \ldots \rangle_{bias}$ denotes these ensemble averages are evaluated from simulations operating under a biasing potential, while $\langle X \rangle$ is an unbiased ensemble average (König et al., 2014). The biasing potential V^b is evaluated in general as $V^b = U^b - U'$, where U^b is the energy in the presence of the bias and U' is the energy in the absence of the bias (Torrie & Valleau, 1977). To extract *high*-level ensemble averages from *low*-level simulations, one uses an *unusual* biasing potential, $V^b = U^{low} - U^{high}$ (König et al., 2014). For example, consider $\Delta A(0_{QM} \rightarrow 1_{QM})$: to extract QM quality results from MM simulations, the MM simulations are regarded as QM simulations operating under this *unusual* biasing potential (König et al., 2014):

$$V^b = U^b - U' = U^{low} - U^{high} = U^{MM} - U^{QM} \quad (8)$$

In practice, V^b essentially reweights conformations from the MM ensemble based on overlap with the QM ensemble (König et al., 2014). Substituting Eqs. (7) and (8) into Eq. (2) we arrive at the *general* NBB method (König et al., 2014),

$$\Delta A(0 \rightarrow 1) = k_B T \ln \left(\frac{\langle f(U_0 - U_1 + C) \exp(\beta V_1^b) \rangle_{1,b} \langle \exp(\beta V_0^b) \rangle_{0,b}}{\langle f(U_1 - U_0 - C) \exp(\beta V_0^b) \rangle_{0,b} \langle \exp(\beta V_1^b) \rangle_{1,b}} \right) + C \quad (9)$$

The notation follows Eq. (2), with an additional subscript b, indicating that ensemble averages were obtained in the presence of a biasing potential (Eq. 8) (König et al., 2014). As Eq. (9) is written, one could calculate $\Delta A(0_{high} \rightarrow 1_{high})$ in one step. However, many practical cases have

---

[b] The term QM-NBB refers to the case in which NBB is applied to calculate free energies at a QM level of theory. The term "NBB" will be used to mean switching between arbitrary levels of theory, eg, MM → (S)QM.

insufficient configurational overlap between endstates of interest (even when using BAR), requiring many intermediate states.[c] Thus, we focus on the use of NBB to calculate $\Delta A(0_{low} \rightarrow 0_{high})$ via the "indirect" approach.

Using two-sided methods such as BAR and NBB requires performing two simulations. For example, to compute $\Delta A(0_{low} \rightarrow 0_{high})$ via BAR means that the 0_{low} ensemble would likely be generated via a computationally affordable level of theory (eg, MM). On the other hand, the target *high* level of theory is often QM, which quickly makes BAR untenable due to the need of the 0_{high} ensemble. In contrast, QM-NBB circumvents the need for a 0_{high} ensemble by introducing a "middle" level of theory; this 0_{mid} is ideally more computationally affordable (than QM) to generate, while also yielding QM-like ensembles. This 0_{mid} ensemble can then be reweighted with a biasing potential (ie, $V^b = U_0^{mid} - U_0^{high}$) to compute $\Delta A(0_{low} \rightarrow 0_{high})$ via QM-NBB.

Fig. 3 summarizes how QM-NBB can be applied to (*ii*) and (*iv*) of Fig. 2. For example, to execute $\Delta A(0_{MM} \rightarrow 0_{QM})$ via QM-NBB, one can chose a *low* MM simulation and a *mid* SQM simulation (Hudson, White, et al., 2015). For the simulation at state 0_{MM}, U_0^{MM}, and U_0^{QM} need to be evaluated for every snapshot. For the simulation at state 0_{SQM}, U_0^{SQM}, U_0^{QM}, U_0^{MM}, and $V_{0,SQM}^b$ need to be evaluated for every snapshot of the trajectory. Where $V_{0,SQM}^b = U_0^{SQM} - U_0^{QM}$. The other state, 0_{MM}, is not to be biased; therefore $V_{0,MM}^b = U_0^{MM} - U_0^{MM} = 0.0$. These energies are then each substituted into their appropriate spots in Eq. (9) to give Eq. (10) which describes the particular use of QM-NBB to calculate $\Delta A(0_{MM} \rightarrow 0_{QM})$ using an SQM middle state:

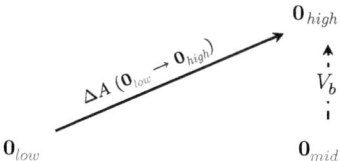

Fig. 3 Illustration of NBB use in calculating $\Delta A(0_{low} \rightarrow 0_{high})$ in an indirect scheme. The same scheme can be constructed to calculate $\Delta A(1_{low} \rightarrow 1_{high})$ to calculate to other "vertical" leg of an indirect cycle.

---

[c] One notable example of this is the computation of solvation free energies using MM- and QM-based implicit solvation models (König et al., 2014).

$$\Delta A(0_{MM} \to 0_{QM})$$

$$= k_B T \ln \left(\frac{\left\langle f(U_0^{MM} - U_0^{QM} + C) \exp\left(\beta V_{0,SQM}^b\right) \right\rangle_{0,SQM}}{\left\langle f(U_0^{QM} - U_0^{MM} - C) \right\rangle_{0,MM}} \cdot \frac{1}{\left\langle \exp\left(\beta V_{0,SQM}^b\right) \right\rangle_{0,SQM}} \right) + C$$

(10)

Brief Note: For the particular situation of computing free energy differences between an MM and QM representation of a system, one should keep in mind that it may not be possible to obtain $p(-\Delta U^{QM \to MM})$ at all. Fig. 1C illustrates a case where 0 and 1 endstates are very distinct, and this can be applied to the MM $\to$ QM case if 0 is replaced with 0_{MM} and 1 is replaced with 0_{QM}. This illustrates how far apart *low-* and *high-*level descriptions of a system can be. An additional complication arises if the distribution $p(\Delta U^{MM \to QM})$ is broad, as in Fig. 1C. In this case, the exponential average $1/N \sum_i^N \exp(-\beta \Delta U_i)$ will be very sensitive to the negative tail of the distribution. Dellago and Hummer quantify "broad" as $\sigma^2(\Delta U_i) \gg (kT)^2$ (Dellago & Hummer, 2014). If the distribution were Gaussian, as assumed for simplicity in Fig. 1, the exponential averaging can be avoided by the use of the cumulant expansion (Zwanzig, 1954), which in the case of normal distributions converges faster and more reliably (Dellago & Hummer, 2014). Unfortunately, in practice $p(\Delta U)$ often is not distributed normally, and use of the cumulant approximation may lead to systematic errors (Dellago & Hummer, 2014).

QM-NEW: Although the use of QM-NBB is appealing from both a conceptual and technical perspective, unfortunately there are inherent limitations to using equilibrium simulations for calculating $\Delta A(0_{low} \to 0_{high})$ (vide supra). These limitations have primarily been attributed to mismatches between the so-called "stiff" degrees of freedom (eg, bond-stretching terms) observed in *low* and *high*-level descriptions (eg, MM and QM, respectively) (Cave-Ayland, Skylaris, & Essex, 2015; Genheden, Cabedo Martinez, Criddle, & Essex, 2014; Genheden, Ryde, & Söderhjelm, 2015; Heimdal & Ryde, 2012; Hudson, White, et al., 2015; König & Brooks, 2012, 2015; König et al., 2014; Sampson, Fox, Tautermann, Woods, & Skylaris, 2015). Typically, these problems have been addressed by modifying how the low-level ensemble is generated (Bentzien et al., 1998; Plotnikov et al., 2011) or by excluding the energy terms responsible for poor overlap (Cave-Ayland et al., 2015; Genheden et al., 2015; Sampson et al., 2015). However, we have recently demonstrated that "stiff" degrees of freedom mismatches can easily be overcome by using more robust FES techniques, namely, nonequilibrium work (NEW) methods such as Jarzynski's equation (JAR; Eq. 11) (Hudson, Lee Woodcock et al., 2015; Jarzynski, 1997).[d]

$$\Delta A(0_{MM} \to 0_{QM}) = -k_B T \ln \left\langle \exp\left[\frac{-W_0^{MM \to QM}}{k_B T}\right] \right\rangle_0 \qquad (11)$$

JAR is the NEW equivalent to FEP and is calculated by replacing the potential energy differences of states 0_{MM} and 0_{QM} in FEP with the NEW (ie, $W_0^{MM \to QM}$). These work values are obtained from relatively short switching simulations between the 0_{MM} and 0_{QM}, effectively bridging the conformational gap by slowly connecting the ensembles generated at the *low* and *high* levels of theory.

In practice, this entails a number of steps. First (1), one must carry out an extended equilibrium simulation at the *low* level where both coordinate and velocity information are saved for a sufficient number of snapshots. Next (2), this saved information is then used to start switching simulations from $0_{low} \to 0_{high}$ (eg, $0_{MM} \to 0_{QM}$). Then (3), the potential energies of the first and last frames of each switching simulation are used to calculate a NEW value. Finally (4), all NEW data are used as input into Eq. (11) and $\Delta A(0_{low} \to 0_{high})$ is obtained.

Extending Fig. 1, we illustrate why fast switching improves convergence (Fig. 4). Specifically, the use of JAR allows us to fulfill a fundamental requirement of any one-sided method that at least some configurations sampled at the *low* level should be low energy structures at the *high* level (Pohorille et al., 2010). In the case of FEP, and most likely BAR, the ΔU distributions in Fig. 4 would not lead to converged results.

Although NEW simulations are effective at circumventing mismatches in "stiff" degrees of freedom, a more problematic issue arises, differences in "soft" degrees of freedom (eg, conformational and/or environmental rearrangements). In such cases, both one- or two-sided methods (eg, FEP, BAR) would be very likely to fail; however, NEW methods are able to overcome this poor overlap albeit, at a potentially high computational cost. For example, to overcome overlap deficiencies in "soft" degrees of freedom one would have to significantly extend the length of the switching simulation to accommodate for slower conformational changes.

---

[d] A NEW equivalent to BAR, Crook's equation (Crooks, 2000, eq. 19), can also be realized by replacing forward and backward energy differences in BAR with forward and backward NEW values, but use of this will not be covered in the scope of this chapter (Crooks, 2000; Hudson, Lee Woodcock et al., 2015).

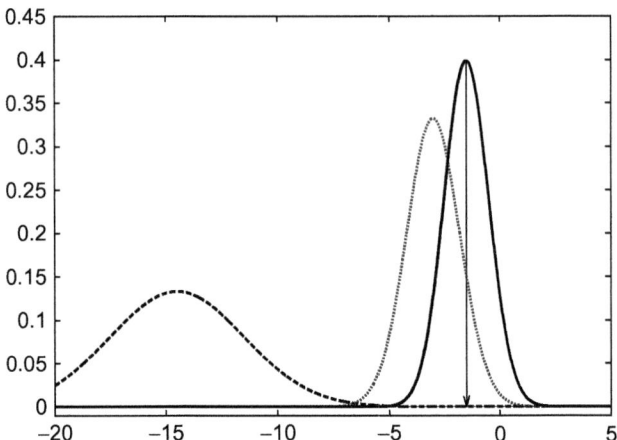

Fig. 4 Illustration of overlap between distributions of forward ($p(\Delta U_{low \to high})$, black dashed) and of (negative) backward energy ($p(\Delta U_{high \to low})$, black solid) differences compared to the distribution obtained from NEW simulations ($p(W_{low \to high})$, gray dashed).

2. ALCHEMICAL FES

Calculating free energy differences, ΔA, between two chemically distinct (eg, different small molecules) states, 0 and 1, requires the introduction of "nonphysical" intermediate states due to insufficient overlap. These intermediate λ or "hybrid" states represent the degree of perturbation from $0 \to 1$ and are governed by a potential energy function which mixes the endstate energies (typically linearly), eg, $E(\lambda) = (1 - \lambda)E_0 + \lambda E_1$, $0.0 \leq \lambda \leq 1.0$. After establishing the hybrid states, the total free energy difference can be found as the sum of free energy differences between each state: $\Delta A(\lambda^{0.0} \to \lambda^{1.0}) = \Delta A(\lambda^{0.0} \to \lambda^{0.1}) + \Delta A(\lambda^{0.1} \to \lambda^{0.2}) + \cdots + \Delta A(\lambda^{0.8} \to \lambda^{0.9}) + \Delta A(\lambda^{0.9} \to \lambda^{1.0})$. These individual ΔA values can then computed in a number of different ways,[e] eg, FEP (Zwanzig, 1954), TI (Kirkwood, 1935), or BAR (Bennett, 1976). In MM force field-based FES all three methods are widely used; however, several recent studies have shown that BAR is the most efficient and robust (Bruckner & Boresch, 2011a, 2011b; Pohorille et al., 2010; Shirts & Pande, 2005).

In practice, most alchemical FES in enzymes require λ-states (eg, residue mutations, protonation/deprotonation, etc.). Thus it is again clear that the use of the "indirect" FES scheme is preferred over the "direct" approach

---

[e] We, as in most alchemical FES, follow the "dual-topology" paradigm in which a hybrid molecule is constructed with characteristic groups of both chemical states 0 and 1 included in the topology file.

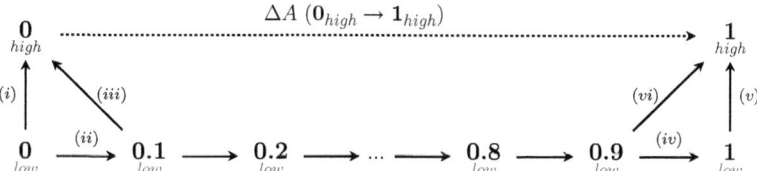

Fig. 5 General scheme for conducting alchemical FES with λ-states. Here $\Delta A(0_{high} \to 1_{high})$ is tackled by simulating λ-states at a lower level of theory, ie, $\Delta A(0_{low} \to 1_{low})$. Legs (iii) and (vi) illustrate "corner cutting."

(see Section 1.2) (Yang, Cui, Min, & Li, 2010). For example, the direct calculation of $\Delta A(\lambda^{0.0}_{(S)QM/MM} \to \lambda^{1.0}_{(S)QM/MM})$ would stipulate each λ state be simulated at the (S)QM/MM level of theory (Hu & Yang, 2005; Li & Yang, 2007; Min et al., 2010; Reddy, Singh, & Erion, 2004, 2007, 2011; Riccardi & Cui, 2007; Riccardi, Schaefer, & Cui, 2005; Riccardi et al., 2006; Woodcock, Moran, Pastor, MacKerell, & Brooks, 2007; Zheng et al., 2008), very quickly becoming cost prohibitive. Fortunately, since free energy is a state function, we can avoid treating λ-states with the more expensive computational methods. In Fig. 5, λ-state simulations are all conducted at the *low* level of theory, such as MM, while QM-NBB or QM-NEW can be used to connect these simulations to the desired *high* level of theory, thus minimizing the total number of QM evaluations needed. Also, one can easily incorporate a more gradual shift from $0 \to 1$, by using more λ-states, at the *low* level of theory without adding *high*-level overhead.

QM-NBB in Alchemical FES: We have already described how QM-NBB can be used to calculate the "vertical" legs of an indirect cycle (see Section 1.3), and a similar calculation can be used in the case of λ-states with one additional note: when connecting levels of theory (ie, low $\to$ high), the chemical/conformational states *do not* have to remain constant. One can see this in Fig. 5 where leg (iii) connects $\Delta A(\lambda^{0.1}_{low} \to \lambda^{0.0}_{high})$ and leg (vi) connects $\Delta A(\lambda^{0.9}_{low} \to \lambda^{1.0}_{high})$; thus we have chemical/conformational changes along with level of theory changes; we refer to this as "corner cutting." Because there is a chemical/conformational change involved with "corner cutting," a *mid*-level of theory is not required. Using QM-NBB to calculate $\Delta A(\lambda^{0.1}_{MM} \to \lambda^{0.0}_{QM})$ without a *mid*-level of theory would look like the following:

$$\Delta A(\lambda^{0.1}_{MM} \to \lambda^{0.0}_{QM})$$
$$= k_B T \ln \left(\frac{\langle f(U^{MM}_{0.1} - U^{QM}_{0.0} + C) \exp(\beta V^b_{0.0,MM}) \rangle_{0.0,MM}}{\langle f(U^{QM}_{0.0} - U^{MM}_{0.1} - C) \rangle_{0.1,MM}} \frac{1}{\langle \exp(\beta V^b_{0.0,MM}) \rangle_{0.0,MM}} \right) + C$$

(12)

where

$$V^b_{0.1,MM} = U^{MM}_{0.1} - U^{MM}_{0.1} = 0.0, \quad V^b_{0.0,MM} = U^{MM}_{0.0} - U^{QM}_{0.0} \quad (13)$$

An alternative, and perhaps more robust, use of QM-NBB avoids "corner cutting" entirely. Instead, legs (*ii*) and (*iv*) in Fig. 5 would be computed at the *low*-level (here MM) and the use of QM-NBB would follow according to procedure laid out in Eq. (10) and Fig. 3. In our experience, a good SQM method (eg, SCC-DFTB) serves as an appropriate *mid* level of theory for connecting MM → QM and has several benefits over the "corner cutting," with the most obvious being that overlap between states $\lambda^{0.0}_{MM}$ and $\lambda^{0.0}_{QM}$ should be improved over that of $\lambda^{0.1}_{MM}$ and $\lambda^{0.0}_{QM}$.

QM-NEW in Alchemical FES: The application of QM-NEW techniques in alchemical FES follows the latter usage of QM-NBB; ie, legs (*ii*) and (*iv*) in Fig. 5 would be computed using the *low*-level method to yield a converged $\Delta A(\lambda^{0.0}_{MM} \to \lambda^{1.0}_{MM})$. Following this procedure, extended equilibrium simulations are needed at both $\lambda^{0.0}_{MM}$ and $\lambda^{1.0}_{MM}$ states. Note that QM-NEW simulations require coordinate and velocity information to be saved at regular intervals during the $\lambda^{0.0}_{MM}$ and $\lambda^{1.0}_{MM}$ simulations. Next, starting from each coordinate/velocity set saved, "switching" simulations are initiated such that the $\lambda^{0.0}_{MM}$ state is smoothly converted to $\lambda^{0.0}_{QM}$ (and likewise with $\lambda^{1.0}_{MM} \to \lambda^{1.0}_{QM}$). Subsequently, the indirect thermodynamic cycle (Fig. 2) is completed to yield $\Delta A(\lambda^{0.0}_{QM} \to \lambda^{1.0}_{QM})$.

2.1 Examples

Alchemical FES are exceptionally useful for calculating relative properties in enzymes such as free energies as a function of mutating residues, relative binding free energies of ligands, free energies of protonation state changes (ie, pK_a), and more. We have thus selected two cases that exemplify how to use QM-NBB and QM-NEW to efficiently calculate QM/MM free energy differences. Additionally, we will describe special considerations for calculating ΔAs that involve a change in net charge (eg, pK_as). The Appendix contains a section with general notes and caveats about the examples presented herein.

QM-NBB $\Delta A(Etha_{QM} \to MeOH_{QM})$: In König et al. (2014), we calculated relative solvation free energies between small molecules of similar size using alchemical FES; eg, ethane (Etha) → methanol (MeOH). Such an alchemical change absolutely requires the use of nonphysical intermediates or λ-states. In order to calculate $\Delta A(\text{Etha}_{QM} \rightarrow \text{MeOH}_{QM})$, we used the indirect cycle presented in Fig. 5 where: $\lambda^{0.0} = \text{Etha}$, $\lambda^{1.0} = \text{MeOH}$, $low = \text{MM}$, and $high = \text{QM/MM}$. As in Fig. 5, 11 λ-states were used to move $0 \rightarrow 1$, each at increments of $\Delta\lambda = 0.1$.

Conducting simulations at each λ-state requires constructing a "dual-topology" hybrid molecule from the two endstates and using a simulation module capable of parsing hybrid molecule energies (we used CHARMM's multiscale modeling module, MSCALE) (Brooks et al., 2009; Woodcock et al., 2011). A "dual-topology" molecule is constructed by identifying what atom types the two molecules have in common, then connecting distinctive groups to this "core" group in such a way that interactions between the distinctive groups are not considered (ie, angles and dihedrals between the distinctive groups are deleted). This means that atoms from both endstates are present in the molecular topology file during all simulations, but the two halves do not interact and energetic contributions from either state are scaled by $(1 - \lambda)$ and λ. Etha and MeOH both share a common CH_3 and the distinctive groups are the additional CH_3 and the OH group respectively; thus we defined a hybrid molecule from Etha and MeOH as seen in Fig. 6. It is best to choose a "core" group (Fig. 6) in the hybrid molecule from atoms in the two endstate molecules that have identical atom types, so that van der Waals parameters are appropriately represented at either endstate. This is somewhat straightforward for many enzyme problems as in an enzyme, a clear choice for "core" atoms would be the C_α and backbone atoms as these are the same for most residues.

In the case of $\Delta A(\text{Etha}_{QM} \rightarrow \text{MeOH}_{QM})$ simulations were done at each λ-state using the CHARMM generalized force field (Vanommeslaeghe et al., 2010); QM calculations were done with Q-Chem (Shao et al., 2015, 2006), and QM/MM calculations were done using the Q-Chem/CHARMM interface (Woodcock et al., 2007). Please refer to König et al. (2014) for more specifics on how λ-state simulations were conducted.

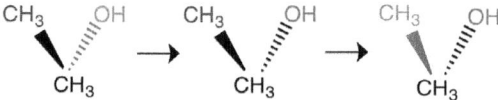

Fig. 6 Illustration of hybrid molecule constructed in the Etha → MeOH alchemical mutation. All atoms are present in the hybrid's topology at all times. *Black atoms* represent the "core" group, *blue (dark gray* in the print version) *atoms* represent those distinctive of ethane, *red (gray* in the print version) *atoms* represent those distinctive of methanol atoms, and *gray atoms* are those that are still present but are not contributing to the energy function at the endstates.

BAR was used to evaluate ΔAs between $\lambda^{0.1}$ and $\lambda^{0.9}$ (as seen in Fig. 5). QM-NBB was used to "corner-cut" and evaluate $\Delta A(\lambda_{MM}^{0.1} \to \lambda_{QM}^{0.0})$ and $\Delta A(\lambda_{MM}^{0.9} \to \lambda_{QM}^{1.0})$, and all ΔAs were summed to give $\Delta A(\lambda_{QM}^{0.0} \to \lambda_{QM}^{1.0})$. Our calculations resulted in free energy differences that strongly agreed with experiment (error of 0.03 kcal mol^{-1}), illustrating QM-NBB's successful application to alchemical FES (see König et al., 2014, p. 1412, Table 2 for Etha $\to$ MeOH data).

QM-NEW $\Delta A(Ala_{QM} \to Ser_{QM})$: Although QM-NBB can be ideal for many problems of interest in computational enzymology, there are also cases where QM-NBB can result in poorly converged FES. One such case that is directly related to enzymatic systems, and where we have previously observed FES convergence problems, is the free energy difference between the blocked N-acetyl-methylamide amino acids alanine (Ala) and serine (Ser) ("blocked" amino acids, often referred to as alanine and serine "dipeptides"), $\Delta A(Ala_{QM} \to Ser_{QM})$. In an effort to overcome these convergence problems, which we attributed to poor overlap stemming from mismatches in "stiff" degrees of freedom, we sought to apply our recently developed QM-NEW technique (Hudson, Lee Woodcock et al., 2015).

As previously highlighted, QM-NEW simulations require only simulations at the "physical" endstates of the alchemical thermodynamic cycle (Fig. 5). Here, the procedure for obtaining the necessary NEW data for Ala and Ser is enumerated. Using the MSCALE module in CHARMM (Brooks et al., 2009; Woodcock et al., 2011), we are able to linearly combine multiple potential energy functions. For this example we used CHARMM22 (C22) as our *low* level and SCC-DFTB as our *high* level; however, MSCALE is extremely flexible and works with more accurate methods in CHARMM (eg, Q-Chem and GAMESS QM/MM interfaces) and easily interfaces to external MM and QM packages (eg, AMBER, Tinker, Psi4, etc.) (Case et al., 2005; Ponder, 2012; Turney et al., 2012). Further, MSCALE supports the PERT free energy facility of CHARMM; thus, the degree of mixing between MM and QM is not only controllable but can also be modified continually through the course of a MD simulation.

As stated previously, to begin the QM-NEW procedure one must obtain a set of coordinate/velocity information (ie, restart files) from an equilibrium simulation in the canonical ensemble. Using this information, "fast-switching" simulations were carried out to obtain a set of work values (W). Switches were made linearly from the initial to final state using the MSCALE and PERT facilities. Fast switches over 100 steps of Langevin

Dynamics were carried out for all restart files saved. Next, the ensemble of work values ($W_0^{MM \to SQM}$) was used to compute $\Delta A(0_{MM} \to 0_{SQM})$ via Eq. (11).

Currently, we are continuing to explore the utility of QM-NEW methods for computing $\Delta A(MM \to QM)$ started in Hudson, Lee Woodcock et al. (2015). Results so far give some additional insights concerning the difficulty of obtaining converged numbers for this quantity. First, if the two levels of theory lead to large differences in conformational preferences (as, eg, was the case for alanine and serine in Hudson, Lee Woodcock et al., 2015), then the computation of the free energy difference becomes much more difficult; ie, convergence is poor. This confirms the concluding remark/considerations of Hudson, Lee Woodcock et al. (2015).[f] Second, we have started to apply QM-NEW fast-switching simulations to solutes/solvent systems, ie, switching from a pure MM system to an SQM/MM system, with SQM (specifically, SCC-DFTB) used to describe the solute and MM for the solvent. In several cases the protocols (switching lengths of 0.25–1 ps) which were verified to work well in the gas phase, failed in aqueous solution. Switching lengths of 5–10 ps were required to regain convergence. It turns out that in these cases the charge distribution of the solute, ie, the regular partial charges of the force field, and the (average) Mulliken charges of the SCC-DFTB method, is rather different. Consequently, the water configurations sampled during the MM simulations are mostly incorrect when switching to an SQM description of the solute. If switching of interactions is carried out over 5–10 ps, the waters near the solute can adapt to the new charge distribution, whereas switching times <1 ps are insufficient for water to rearrange. We would like to point out that this is a complication that would be difficult, if not impossible, to detect by equilibrium methods in which the "switch" is instantaneous.

Brief Note: Many enzyme systems of interest may require a change in charge, for example, calculating free energies associated with mutating charged residues such as glutamate or lysine to neutral residues like leucine, and calculating the free energy between protonation states (ie, pK_a). As briefly mentioned, simulation of λ-states with explicit water is very

---

[f] "Finally, our results show that differences between *low* (eg, MM) and *high* (eg, QM) levels of theory are not restricted to the stiff degrees of freedom (eg, bond stretching, angle bending); in fact, differences in conformational preferences may well prove to be the more daunting challenge when connecting levels of theory."

important as hydrogen bonds and explicit electrostatic interactions are likely to provide large amounts of stabilization to either endstate. Periodic boundary conditions (PBCs) are typically used in MM simulations to mimic the effects of infinite solvent, while particle mesh Ewald (PME) is used to compute the long-range electrostatic interactions in such periodic conditions. However, PME summation requires that the overall system charge be neutral. As λ-states deviate from this neutral charge, errors from electrostatic treatment are expected. Thus, one should avoid using PBCs and PME in these cases and instead employ solvent boundary potential methods. Such methods model water explicitly in a user-defined volume of interest and then model beyond the explicit solvent with a dielectric constant appropriate for the rest of the system (eg, water or protein). We have begun applying efficient QM-FES to predicting residue pK_as, and for this task we model infinite solvation with the generalized solvent boundary potential module in CHARMM. The interested reader should also review Li and Cui (2003)'s scheme 2 and associated equations to understand how to convert a free energy involved in protonation/deprotonation into a pK_a.

Additionally, we should mention that when calculating $\Delta A(0_{MM} \rightarrow 0_{QM/MM})$, if 0_{MM} simulations are conducted using PBC and PME, $U_0^{QM/MM}$ should be calculated in such a way to ensure that QM region atoms are included in the Ewald summation. In the past this was not possible and required creative workarounds (König et al., 2014), but QM/MM Ewald functionality has recently been developed and implemented in Q-Chem and the Q-Chem/CHARMM QM/MM interface (Holden, Richard, & Herbert, 2013; Woodcock et al., 2007).

3. REACTION PROFILES

Enzyme active sites are tuned to carry out a wide range of biochemical processes. To date, most *high*-level computations on enzymes have focused on modeling reactive processes, the hallmark of (S)QM/MM methodology. A plethora of methods, to this end, exist and differ predominantly based on the details of how the reaction coordinate, ξ, is defined and propagated. For example, discrete points (ξ^i) along the path are typically enforced by either suitable restraints (umbrella potentials) or constraints.

Further, various methods can be used to compute the free energy along the path investigated, ie, the potential of mean force (PMF) along ξ. Self-penalty walk-based methods, the finite temperature string method and its derivatives, QM/MM-MFEP, QTCP, and nudged elastic band variants make up a diverse class of methods that differ in both how ξ is treated and how PMFs are computed (eg, via free energy differences with FEP, WHAM, etc.) (Acevedo & Jorgensen, 2010; Burger & Ayers, 2010; Burger & Yang, 2006, 2007; Czerminski & Elber, 1990a, 1990b; Dey & Ayers, 2006; Elber & Karplus, 1987; Ghasemi & Goedecker, 2011; Goodrow, Bell, & Head-Gordon, 2008, 2009, 2010; Hu et al., 2008, 2007; Hu & Yang, 2009; Klimes, Bowler, & Michaelides, 2010; Koslover & Wales, 2007; Maeda & Morokuma, 2011; Maeda, Ohno, & Morokuma, 2009; Maragliano, Fischer, Vanden-Eijnden, & Ciccotti, 2006; Parks, Hu, Rudolph, & Yang, 2009; Peters, Heyden, Bell, & Chakraborty, 2004; Quapp, 2007, 2009; Ranaghan et al., 2003; Rosta, Woodcock, Brooks, & Hummer, 2009; Sheppard, Terrell, & Henkelman, 2008; Trygubenko & Wales, 2004; Weinan, Ren, & Vanden-Eijnden, 2005a, 2005b, 2007; Woodcock, Hodoscek, & Brooks, 2007; Woodcock et al., 2008; Zeng, Hu, Hu, & Yang, 2009). Alternatively, one can use the average force in the direction of reaction coordinate $-\langle \partial U/\partial \xi \rangle_{\xi^i}$ and compute the PMF in analogy to TI.

As highlighted later, connection schemes between levels of theory are starting to become essential when developing robust reaction path methods. However, unlike alchemical FES, where one only has to care about connecting *low* and *high*-levels of theory at the endstates, PMF generation requires accurate free energy differences at *every* step; ie, all intermediate states/reaction path steps have to be modeled accurately. Of course, accurate PMFs could be generated "directly," by simulating every reaction intermediate at a *high* level of theory, but as we have discussed, this is resource intensive and can easily lead to poor results. Instead, in the interest of *efficiency*, one should incorporate *high*-level energy evaluations through either QM-NBB or QM-NEW to construct a *high*-level PMF from an extensively sampled *low*-level PMF. It is worth mentioning that for reaction path mapping, a *low* level of theory will typically have to be chosen that can account for bond breaking/forming. Therefore, *low* will likely be either a reactive MM force field or an SQM/MM simulation, while *high* level of theory will likely be QM/MM.

Here we will describe possible means for calculating free energies along a path using QM-NBB and QM-NEW and then illustrate the process using a simple example. We will denote an arbitrary reaction coordinate as ξ_X^i, where i denotes particular reaction intermediates and X denotes the level of theory used. We will denote the change in free energy between two reaction coordinates as $\Delta A(\xi^{i-1} \to \xi^i)$. To use NBB or NEW to extract a *high*-level PMF from *low*-level simulations at each ξ^i, one must first generate a *low*-level PMF and then calculate $\Delta A(\xi^i_{low} \to \xi^i_{high})$ which represents a "correction" from the *low*-level PMF to the *high*-level PMF at each reaction intermediate:

$$f_{high}(\xi^i) = \left[\sum_{k=1}^{i} \Delta A(\xi^{k-1}_{low} \to \xi^k_{low})\right] + \Delta A(\xi^i_{low} \to \xi^i_{high}) \qquad (14)$$

QM-NBB: Fig. 7 illustrates how this calculation would be done where *high*, *mid*, and *low* descriptors are denoted. Recall that QM-NBB can be used to calculate $\Delta A(\xi^i_{low} \to \xi^i_{high})$ so long as there are two simulations, and we can advantageously choose one simulation to be at a *mid*-level of theory to avoid overlap problems between the *low* and *high* levels of theory. Additionally, there is no requirement that the number of snapshots used at the *low* and *mid* levels be equal. This facilitates the use of more accurate (ie, QM/MM) methods as the *mid* level without becoming cost prohibitive. As Fig. 7 illustrates, BAR is used to calculate $\Delta A(\xi^{i-1}_{low} \to \xi^i_{low})$ and then QM-NBB is used to calculate corrections to the high-level $\Delta A(\xi^i_{low} \to \xi^i_{high})$ at every intermediate via a *mid*-level simulation to provide the *high*-level ensemble. The value of the free energy at each step ξ is then found by evaluating Eq. (14).

QM-NEW: As noted previously, if overlap problems occur when using QM-NBB to connect *low* → *high* PMFs then QM-NEW methods can be used as a more robust alternative. To accomplish this, one need to simply follow the procedure laid out for using QM-NEW methods (eg, JAR) to compute $\Delta A(\text{Ala}_{MM} \to \text{Ala}_{QM})$.

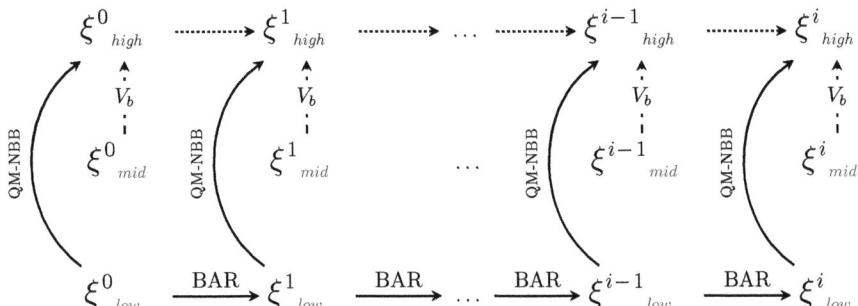

Fig. 7 QM-NBB applied in an indirect approach along a general reaction path with *i* steps.

3.1 Examples

Although (S)QM/MM is becoming more popular for alchemical FES, its traditional application has been for mapping reaction pathways in the condensed phase (ie, water, protein, etc.). To this end, a significant number of methods have been developed for calculating the free energies of these reactive processes. However, most of these dictate that fairly substantial simulations be carried out at the (S)QM/MM level of theory. Here, we will present a simple example (the torsional potential of gas phase butane) to illustrate the use of both QM-NBB and QM-NEW for accurately and efficiently computing pathway free energies. Our interest in such a simple example stems from the fact that FEP has significant problems yielding an accurate and converged free energy surface when using MM and Hartree–Fock as the *low* and *high* levels, respectively (see Fig. 6, Hudson, White, et al., 2015).

Butane's profile is characterized by the central C–C–C–C dihedral taking on values from $-180°$ to $+180°$, over which four distinct conformations are observed, the anti, eclipsed-Gauche, Gauche, and eclipsed conformations. The path does not require bond breakage/formation, and thus MM is sufficient for the *low*-level simulation; however, this will not always be the case. In such cases, SQM/MM methods serve as attractive alternatives for the *low* level of theory. Because the reaction coordinate in this case is butane's central dihedral, we will denote each reaction coordinate as ϕ rather than the more general ξ, where ϕ can take on values between $-180° \leq \phi \leq +180°$.

Butane Torsional Rotation, QM-NBB: In Hudson, White, et al. (2015), we were able to show that the use of "indirect" approach was both efficient and accurate for calculating butane's torsional PMF. Of course, this approach first requires the PMF to be generated at the *low* level of theory. Subsequently, the QM-NBB method was used, via a *mid*-level ensemble, to calculate QM corrections to the *low*-level PMF as shown in Fig. 7.

The reaction path was discretized into 37 states, one for each 10° rotation around butane's central torsion (ie, from $-180° \rightarrow 180°$). A harmonic dihedral restraint was applied to ensure butane sampled conformations around the ϕ of interest. Two simulations were conducted for every discrete step along the reaction path: ϕ_{MM} and ϕ_{SQM}, where SCC-DFTB was chosen as the SQM/*mid* level of theory. BAR (Eq. 2) was then used to calculate the *low*-level PMF (ie, $\Delta A(\phi_{MM}^{-180°} \rightarrow \phi_{MM}^{-170°})$, $\Delta A(\phi_{MM}^{-170°} \rightarrow \phi_{MM}^{-160°})$, etc.). Next, QM energy evaluations, at the HF/6-31G* level of theory, were performed on snapshots saved from both MM and SQM simulations for every

ϕ along the reaction path. The value of the PMF, $f_{QM}(\phi)$, at each ϕ is then found by evaluating Eq. (14) and plotting. Complete details of this procedure can be found in Hudson, White, et al. (2015).

From Fig. 7 in Hudson, White, et al. (2015), we were able to illustrate that QM-NBB produced a converged PMF with only 10K (ie, 10,000) MM snapshots and only 1K SQM snapshots (at each ϕ). When FEP was used to evaluate $\Delta A(\phi_{MM} \to \phi_{QM})$ (without a *mid* level of theory) 200K MM points and 200K QM evaluations were required at each ϕ to achieve a converged *high*-level PMF; however, even this PMF still had irregularities. Two important points should be gleaned from this: a *mid* or intermediate level of theory can be used to better approximate the *high* level of theory, and QM-NBB calculating the "vertical" legs of an indirect cycle does *not* require the same number of snapshots from the *low*- and *mid*-level simulations. These points are of particular interest when considering the application of QM-NBB to enzyme systems, where SQM/MM methods will likely be needed to account for bond breaking/making. Thus, even if QM/MM simulations are needed as the *mid* level of theory, they can be run for a much shorter period of time.

Butane Torsional Rotation, QM-NEW

As noted above for QM-NBB usage, if SQM/MM is used for *low*-level simulations, then an "appropriate" *mid* level of theory must also be used. Unfortunately, the choice of this *mid* level may not always be clear. Thus, avoiding this choice via the use of QM-NEW methods is very attractive; however, it will likely come with a significant increase in computational cost when compared to QM-NBB. To demonstrate the advantages, we will again use the generation of butane's *high*-level PMF (ie, $\Delta A(\phi_{MM} \to \phi_{QM})$), as was done in Hudson, Lee Woodcock et al. (2015), as our example.

The application of QM-NEW begins much the way QM-NBB does, ie, by computing an underlying PMF at the *low* level of theory (MM in this case). We used BAR to calculate the MM PMF in analogy to the procedure listed above (see Hudson, Lee Woodcock et al., 2015 for full details). To evaluate the QM-NEW methodology, we only considered half of butane's PMF (ie, from 0° to 180°); since it is symmetric, this was deemed to be sufficient for proof of concept. The most significant difference between QM-NBB and QM-NEW *low*-level simulations is the necessity to save both coordinate and velocity information, every picosecond in this example.

Using each of these "restart" files, short switching simulations were used to smoothly "switch" from MM $\to$ QM (HF/6-31G*). Again, the MSCALE module of CHARMM (*vide infra*) was used to carry out these switching simulations, 50 steps each. Upon completion of all simulations, the potential energy difference between steps 1 and 50 of each simulation was computed and used to determine $\Delta A(\phi_{MM} \rightarrow \phi_{QM})$ via Eq. (11). To evaluate performance of QM-NEW simulations, the "gold standard" from Hudson, White, et al. (2015) was used (Fig. 1, Hudson, Lee Woodcock et al., 2015). QM-NEW results were obtained using 1K switching simulations per discrete reaction coordinate (ie, ϕ value) and yielded a near perfect match to the reference PMF (root mean squared error of 0.03 kcal mol^{-1}).

One final consideration is the total computational cost of these procedures. Comparing PMF generation via FEP and QM-NEW yields a surprising conclusion, QM-NEW is actually more efficient. For example, FEP results used 100K snapshots per MM simulation per ϕ value; these were saved and used for QM single point energy evaluations. In contrast, QM-NEW required 50K QM energy and force evaluations per ϕ value. Taking into account the cost of single point energy vs energy + force calculations, QM-NEW simulations were roughly three-fourths the cost of FEP calculations!

APPENDIX

A.1 Relevant Nomenclature and Usage

We will use shorthand notations to succinctly describe thermodynamic endstates and mathematical expressions. The chemical/conformational state descriptor will be denoted by A_X, where A represents the chemical/conformational state taking values 0 or 1, the optional subscript, X, denotes the level of theory at which to simulate of A. The λ-state descriptor will be denoted by λ_X^L, where L is the value of λ and X is the level of theory at which the λ-state was simulated. The energy descriptor will be denoted by U_A^X, where the subscript and superscript indicate that the energy should be evaluated according to chemical/conformational state A and level of theory X, respectively. Ensemble averages will be denoted with angular brackets as $\langle \ldots \rangle_{A,X}$ where A and X represent the chemical/conformational state and level of theory in which the ensemble was generated, respectively. It is important to differentiate chemical/conformational changes from the level of theory changes. A chemical/conformational change will be denoted $\Delta A(0_X \rightarrow 1_X)$, where X is the constant level of theory, if X is not given then any level of theory could be chosen. A level of theory change will be denoted as $\Delta A(A_{low} \to A_{high})$, where A is the constant chemical/conformational state having value "0"/"1."

We will discuss "levels of theory" such as MM, SQM, QM, QM/MM, and SQM/MM methodologies; (S)QM will refer to either SQM or QM. We will also more generally use the terms "*low*" and "*high*" to illustrate how levels of theory can be selected based on system demands. The term "*low*" will to refer to levels of theory that can be simulated easily, "*high*" will refer to more rigorous/expensive levels of theory, and "*mid*" will refer to some level of theory between "*low*" and "*high*." Depending on system, "*low*" may be MM, SQM, or SQM/MM, and "*high*" will most often be (S)QM/MM calculations; "*mid*" will likely be SQM/MM or polarizable MM force fields. The saved coordinates from a simulation will be called "snapshots."

Finally, we describe methods in context of their use via the CHARMM and/or Q-Chem packages (Brooks et al., 2009; Shao et al., 2015, 2006). Although all major simulation or quantum packages be used for the QM-NBB data generation, it is unclear how many packages support MSCALE-like functionality for carrying out MM $\to$ QM switching simulations.

ACKNOWLEDGMENT

The authors would like to highlight that this material is based upon work supported by the National Science Foundation under CHE-1464946. Further, HLW and PSH thank USF Research Computing (Circe) and the NSF for support via their Major Research Instrumentation Program (CHE-1531590).

REFERENCES

Acevedo, O., & Jorgensen, W. L. (2010). Advances in quantum and molecular mechanical (QM/MM) simulations for organic and enzymatic reactions. *Accounts of Chemical Research, 43*, 142–151.

Bennett, C. H. (1976). Efficient estimation of free energy differences from Monte Carlo data. *Journal of Computational Physics, 22*, 245–268.

Bentzien, J., Muller, R. P., Florián, J., & Warshel, A. (1998). Hybrid ab initio quantum mechanics/molecular mechanics calculations of free energy surfaces for enzymatic reactions: The nucleophilic attack in subtilisin. *Journal of Physical Chemistry A, 102*, 2293–2301.

Boresch, S., & Bruckner, S. (2011). Avoiding the van der Waals endpoint problem using serial atomic insertion. *Journal of Computational Chemistry, 32*, 2449–2458.

Bowman, A. L., Grant, I. M., & Mulholland, A. J. (2008). QM/MM simulations predict a covalent intermediate in the hen egg white lysozyme reaction with its natural substrate. *Chemical Communications, 37*, 4425–4427.

Brooks, B. R., Brooks, C. L., III, Mackerell, A. D., Jr., Nilsson, L., Petrella, R. J., Roux, B., … Karplus, M. (2009). CHARMM: The biomolecular simulation program. *Journal of Computational Chemistry, 30*, 1545.

Bruckner, S., & Boresch, S. (2011a). Efficiency of alchemical free energy simulations. I. A practical comparison of the exponential formula, thermodynamic integration, and Bennett's acceptance ratio method. *Journal of Computational Chemistry, 32*, 1303–1319.

Bruckner, S., & Boresch, S. (2011b). Efficiency of alchemical free energy simulations. II. Improvements for thermodynamic integration. *Journal of Computational Chemistry, 32*, 1320–1333.

Burger, S. K., & Ayers, P. W. (2010). Methods for finding transition states on reduced potential energy surfaces. *Journal of Chemical Physics, 132*, 234110.

Burger, S. K., & Yang, W. T. (2006). Quadratic string method for determining the minimum-energy path based on multiobjective optimization. *Journal of Chemical Physics, 124*, 054109.

Burger, S. K., & Yang, W. (2007). Sequential quadratic programming method for determining the minimum energy path. *Journal of Chemical Physics, 127*, 164107.

Case, D. A., Cheatham, T. E., Darden, T., Gohlke, H., Luo, R., Merz, K. M., … Woods, R. J. (2005). The Amber biomolecular simulation programs. *Journal of Computational Chemistry, 26*, 1668–1688.

Cave-Ayland, C., Skylaris, C.-K., & Essex, J. W. (2015). Direct validation of the single step classical to quantum free energy perturbation. *Journal of Physical Chemistry B, 119*, 1017–1025.

Chandrasekhar, J., Smith, S. F., & Jorgensen, W. L. (1984). SN2 reaction profiles in the gas phase and aqueous solution. *Journal of the American Chemical Society, 106*, 3049–3050.

Chandrasekhar, J., Smith, S. F., & Jorgensen, W. L. (1985). Theoretical examination of the SN2 reaction involving chloride ion and methyl chloride in the gas phase and aqueous solution. *Journal of the American Chemical Society, 107*, 154–163.

Cisneros, G. A., Liu, H., Lu, Z., & Yang, W. (2005). Reaction path determination for quantum mechanical/molecular mechanical modeling of enzyme reactions by combining first order and second order "chain-of-replicas" methods. *Journal of Chemical Physics, 122*, 114502.

Claeyssens, F., Harvey, J. N., Manby, F. R., Mata, R. A., Mulholland, A. J., Ranaghan, K. E., … Werner, H.-J. (2006). High-accuracy computation of reaction barriers in enzymes. *Angewandte Chemie International Edition, 45*, 6856–6859.

Crooks, G. E. (2000). Path-ensemble averages in systems driven far from equilibrium. *Physical Review E, 61*, 2361–2366.

Cui, Q., Elstner, M., Kaxiras, E., Frauenheim, T., & Karplus, M. (2001). A QM/MM implementation of the self-consistent charge density functional tight binding (SCC-DFTB) method. *Journal of Physical Chemistry B, 105*, 569–585.

Czerminski, R., & Elber, R. (1990a). Reaction path study of conformational transitions in flexible systems: Applications to peptides. *Journal of Chemical Physics, 92*, 5580–5601.

Czerminski, R., & Elber, R. (1990b). Self-avoiding walk between 2 fixed-points as a tool to calculate reaction paths in large molecular systems. *International Journal of Quantum Chemistry, 24*, 167–186.

de Ruiter, A., Boresch, S., & Oostenbrink, C. (2013). Comparison of thermodynamic integration and Bennett's acceptance ratio for calculating relative protein–ligand binding free energies. *Journal of Computational Chemistry, 34*, 1024–1034.

Dellago, C., & Hummer, G. (2014). Computing equilibrium free energies using nonequilibrium molecular dynamics. *Entropy, 16*, 41–61.

Dey, B. K., & Ayers, P. W. (2006). A Hamilton-Jacobi type equation for computing minimum potential energy paths. *Molecular Physics, 104*, 541–558.

Elber, R., & Karplus, M. (1987). A method for determining reaction paths in large molecules: Application to myoglobin. *Chemical Physics Letters, 139*, 375–380.

Elstner, M., Porezag, D., Jungnickel, G., Elsner, J., Haugk, M., Frauenheim, T., ... Seifert, G. (1998). Self-consistent-charge density-functional tight-binding method for simulations of complex materials properties. *Phys. Rev. B, 58*, 7260–7268.

Frushicheva, M. P., & Warshel, A. (2012). Towards quantitative computer-aided studies of enzymatic enantioselectivity: The case of Candida antarctica lipase A. *ChemBioChem, 13*, 215–223.

Gao, J., & Freindorf, M. (1997). Hybrid ab initio QM/MM simulation of N-methylacetamide in aqueous solution. *Journal of Physical Chemistry A, 101*, 3182–3188.

Gao, J., Luque, F. J., & Orozco, M. (1993). Induced dipole moment and atomic charges based on average electrostatic potentials in aqueous solution. *Journal of Chemical Physics, 98*, 2975.

Gao, J., & Xia, X. (1992). A priori evaluation of aqueous polarization effects through Monte Carlo QM-MM simulations. *Science, 258*, 631–635.

Genheden, S., Cabedo Martinez, A. I..Criddle, M. P., & Essex, J. W. (2014). Extensive all-atom Monte Carlo sampling and QM/MM corrections in the SAMPL4 hydration free energy challenge. *Journal of Computer-Aided Molecular Design, 28*, 187–200.

Genheden, S., Ryde, U., & Söderhjelm, P. (2015). Binding affinities by alchemical perturbation using QM/MM with a large QM system and polarizable MM model. *Journal of Computational Chemistry, 36*, 2114–2124.

Ghasemi, S. A., & Goedecker, S. (2011). An enhanced splined saddle method. *Journal of Chemical Physics, 135*, 014108.

Goodrow, A., Bell, A. T., & Head-Gordon, M. (2008). Development and application of a hybrid method involving interpolation and ab initio calculations for the determination of transition states. *Journal of Chemical Physics, 129*, 174109.

Goodrow, A., Bell, A. T., & Head-Gordon, M. (2009). Transition state-finding strategies for use with the growing string method. *Journal of Chemical Physics, 130*, 244108.

Goodrow, A., Bell, A. T., & Head-Gordon, M. (2010). A strategy for obtaining a more accurate transition state estimate using the growing string method. *Chemical Physics Letters, 484*, 392–398.

Heimdal, J., & Ryde, U. (2012). Convergence of QM/MM free-energy perturbations based on molecular-mechanics or semiempirical simulations. *Physical Chemistry Chemical Physics, 14*, 12592–12604.

Holden, Z. C., Richard, R. M., & Herbert, J. M. (2013). Periodic boundary conditions for QM/MM calculations: Ewald summation for extended Gaussian basis sets. *Journal of Chemical Physics, 139*, 244108.

Hu, H., Lu, Z., Parks, J. M., Burger, S. K., & Yang, W. (2008). Quantum mechanics/molecular mechanics minimum free-energy path for accurate reaction energetics in solution and enzymes: Sequential sampling and optimization on the potential of mean force surface. *Journal of Chemical Physics, 128*, 034105.

Hu, H., Lu, Z., & Yang, W. (2007). QM/MM minimum free energy path: Methodology and application to triosephosphate isomerase. *Journal of Chemical Theory and Computation, 3*, 390–406.

Hu, H., & Yang, W. (2005). Dual-topology/dual-coordinate free-energy simulation using QM/MM force field. *Journal of Chemical Physics, 123*, 041102.

Hu, H., & Yang, W. T. (2009). Development and application of ab initio QM/MM methods for mechanistic simulation of reactions in solution and in enzymes. *Journal of Molecular Structure: THEOCHEM, 898*, 17–30.

Hudson, P. S., Lee Woodcock, H., & Boresch, S. (2015). Use of nonequilibrium work methods to compute free energy differences between molecular mechanical and quantum mechanical representations of molecular systems. *Journal of Physical Chemistry Letters, 6*, 4850–4856.

Hudson, P. S., White, J. K., Kearns, F. L., Hodoscek, M., Boresch, S., & Lee Woodcock, H. (2015). Efficiently computing pathway free energies: New approaches based on chain-of-replica and non-Boltzmann Bennett reweighting schemes. *Biochimica et Biophysica Acta, 1850*, 944–953.

Jarzynski, C. (1997). Nonequilibrium equality for free energy differences. *Physical Review Letters, 78*, 2690–2693.

Kirkwood, J. G. (1935). Statistical mechanics of fluid mixtures. *Journal of Chemical Physics, 3*, 300–313.

Klimes, J., Bowler, D. R., & Michaelides, A. (2010). A critical assessment of theoretical methods for finding reaction pathways and transition states of surface processes. *Journal of Physics: Condensed Matter, 22*, 074203.

Kollman, P. (1993). Free energy calculations: Applications to chemical and biochemical phenomena. *Chemical Reviews, 93*, 2395–2417.

König, G., & Boresch, S. (2011). Non-Boltzmann sampling and Bennett's acceptance ratio method: How to profit from bending the rules. *Journal of Computational Chemistry, 32*, 1082–1090.

König, G., & Brooks, B. R. (2012). Predicting binding affinities of host-guest systems in the SAMPL3 blind challenge: The performance of relative free energy calculations. *Journal of Computer-Aided Molecular Design, 26*, 543–550.

König, G., & Brooks, B. R. (2015). Correcting for the free energy costs of bond or angle constraints in molecular dynamics simulations. *Biochimica et Biophysica Acta, 1850*, 932–943.

König, G., Hudson, P. S., Boresch, S., & Woodcock, H. L. (2014). Multiscale free energy simulations: An efficient method for connecting classical MD simulations to QM or QM/MM free energies using non-Boltzmann Bennett reweighting schemes. *Journal of Chemical Theory and Computation, 10*, 1406–1419.

Koslover, E. F., & Wales, D. J. (2007). Comparison of double-ended transition state search methods. *Journal of Chemical Physics, 127*, 134102.

Leitgeb, M., Schröder, C., & Boresch, S. (2005). Alchemical free energy calculations and multiple conformational substates. *Journal of Chemical Physics, 122*, 084109.

Li, G., & Cui, Q. (2003). pKa calculations with QM/MM free energy perturbations. *Journal of Physical Chemistry B, 107*, 14521–14528.

Li, H., & Yang, W. (2007). Sampling enhancement for the quantum mechanical potential based molecular dynamics simulations: A general algorithm and its extension for free energy calculation on rugged energy surface. *Journal of Chemical Physics, 126*, 114104.

Lonsdale, R., Harvey, J. N., & Mulholland, A. J. (2010). Inclusion of dispersion effects significantly improves accuracy of calculated reaction barriers for cytochrome P450 catalyzed reactions. *Physical Chemistry Letters, 1*, 3232–3237.

Lonsdale, R., Hoyle, S., Grey, D. T., Ridder, L., & Mulholland, A. J. (2012). Determinants of reactivity and selectivity in soluble epoxide hydrolase from quantum mechanics/molecular mechanics modeling. *Biochemistry, 51*, 1774–1786.

Lu, N., Kofke, D. A., & Woolf, T. B. (2004). Improving the efficiency and reliability of free energy perturbation calculations using overlap sampling methods. *Journal of Computational Chemistry, 25*, 28–39.

Luzhkov, V., & Warshel, A. (1992). Microscopic models for quantum mechanical calculations of chemical processes in solutions: LD/AMPAC and SCAAS/AMPAC calculations of solvation energies. *Journal of Computational Chemistry, 13*, 199–213.

Maeda, S., & Morokuma, K. (2011). Finding reaction pathways of type $A + B \to X$: Toward systematic prediction of reaction mechanisms. *Journal of Chemical Theory and Computation, 7*, 2335–2345.

Maeda, S., Ohno, K., & Morokuma, K. (2009). An automated and systematic transition structure explorer in large flexible molecular systems based on combined global reaction route mapping and microiteration methods. *Journal of Chemical Theory and Computation, 5*, 2734–2743.

Maragliano, L., Fischer, A., Vanden-Eijnden, E., & Ciccotti, G. (2006). String method in collective variables: Minimum free energy paths and isocommittor surfaces. *Journal of Chemical Physics, 125*, 024106.

Min, D., Zheng, L., Harris, W., Chen, M., Lv, C., & Yang, W. (2010). Practically efficient QM/MM alchemical free energy simulations: The orthogonal space random walk strategy. *Journal of Chemical Theory and Computation, 6*, 2253–2266.

Parks, J. M., Hu, H., Rudolph, J., & Yang, W. (2009). Mechanism of Cdc25B phosphatase with the small molecule substrate p-nitrophenyl phosphate from QM/MM-MFEP calculations. *Journal of Physical Chemistry B, 113*, 5217–5224.

Peters, B., Heyden, A., Bell, A. T., & Chakraborty, A. (2004). A growing string method for determining transition states: Comparison to the nudged elastic band and string methods. *Journal of Chemical Physics, 120*, 7877.

Plotnikov, N. V., & Kamerlin, S. C. L..Warshel, A. (2011). Paradynamics: An effective and reliable model for ab initio QM/MM free-energy calculations and related tasks. *Journal of Physical Chemistry B, 115*, 7950–7962.

Pohorille, A., Jarzynski, C., & Chipot, C. (2010). Good practices in free-energy calculations. *Journal of Physical Chemistry B, 114*, 10235–10253.

Ponder, J. W.. (2012). In *TINKER molecular modeling package, v6.1.*

Quapp, W. (2007). Finding the transition state without initial guess: The growing string method for Newton trajectory to isomerization and enantiomerization reaction of alanine dipeptide and poly(15)alanine. *Journal of Computational Chemistry, 28*, 1834–1847.

Quapp, W. (2009). The growing string method for flows of Newton trajectories by a second-order method. *Journal of Chemical Theory and Computation, 8*, 101–117.

Ranaghan, K. E., Ridder, L., Szefczyk, B., Sokalski, W. A., Hermann, J. C., & Mulholland, A. J. (2003). Insights into enzyme catalysis from QM/MM modelling: Transition state stabilization in chorismate mutase. *Molecular Physics, 101*, 2695–2714.

Reddy, M. R., Singh, U. C., & Erion, M. D. (2004). Development of a quantum mechanics-based free-energy perturbation method: Use in the calculation of relative solvation free energies. *Journal of the American Chemical Society, 126*, 6224–6225.

Reddy, M. R., Singh, U. C., & Erion, M. D. (2007). Ab initio quantum mechanics-based free energy perturbation method for calculating relative solvation free energies. *Journal of Computational Chemistry, 28*, 491–494.

Reddy, M. R., Singh, U. C., & Erion, M. D. (2011). Use of a QM/MM-based FEP method to evaluate the anomalous hydration behavior of simple alkyl amines and amides: Application to the design of FBPase inhibitors for the treatment of type-2 diabetes. *Journal of the American Chemical Society, 133*, 8059–8061.

Repasky, M. P., Chandrasekhar, J., & Jorgensen, W. L. (2002). PDDG/PM3 and PDDG/MNDO: Improved semiempirical methods. *Journal of Computational Chemistry, 23*, 1601–1622.

Riccardi, D., & Cui, Q. (2007). pKa analysis for the zinc-bound water in human carbonic anhydrase II: Benchmark for "multiscale" QM/MM simulations and mechanistic implications. *Journal of Physical Chemistry A, 111*, 5703–5711.

Riccardi, D., Schaefer, P., & Cui, Q. (2005). pKa calculations in solution and proteins with QM/MM free energy perturbation simulations: A quantitative test of QM/MM protocols. *Journal of Physical Chemistry B, 109*, 17715–17733.

Riccardi, D., Schaefer, P., Yang, Y., Yu, H., Ghosh, N., Prat-Resina, X., … Cui, Q. (2006). Development of effective quantum mechanical/molecular mechanical (QM/MM) methods for complex biological processes. *Journal of Physical Chemistry B, 110*, 6458–6469.

Rod, T. H., & Ryde, U. (2005a). Quantum mechanical free energy barrier for an enzymatic reaction. *Physical Review Letters*, *94*(13), 138302.

Rod, T. H., & Ryde, U. (2005b). Accurate QM/MM free energy calculations of enzyme reactions: Methylation by catechol O-methyltransferase. *Journal of Chemical Theory and Computation*, *1*, 1240–1251.

Rosta, E., Woodcock, H. L., Brooks, B. R., & Hummer, G. (2009). Artificial reaction coordinate "tunneling" in free-energy calculations: The catalytic reaction of RNase H. *Journal of Computational Chemistry*, *30*, 1634–1641.

Sampson, C., Fox, T., Tautermann, C. S., Woods, C., & Skylaris, C.-K. (2015). A "stepping stone" approach for obtaining quantum free energies of hydration. *Journal of Physical Chemistry B*, *119*, 7030–7040.

Shao, Y., Gan, Z., Epifanovsky, E., Gilbert, A. T. B..Wormit, M., Kussmann, J., ... Head-Gordon, M. (2015). Advances in molecular quantum chemistry contained in the Q-Chem 4 program package. *Molecular Physics*, *113*, 184–215.

Shao, Y., Molnar, L. F., Jung, Y., Kussmann, J., Ochsenfeld, C., Brown, S. T., ... Head-Gordon, M. (2006). Advances in methods and algorithms in a modern quantum chemistry program package. *Physical Chemistry Chemical Physics*, *8*, 3172–3191.

Sheppard, D., Terrell, R., & Henkelman, G. (2008). Optimization methods for finding minimum energy paths. *Journal of Chemical Physics*, *128*, 134106.

Shirts, M. R., Mobley, D. L., & Chodera, J. D. (2007). Alchemical free energy calculations: Ready for prime time? *Annual Reports in Computational Chemistry*, *3*, 41–59.

Shirts, M. R., & Pande, V. S. (2005). Comparison of efficiency and bias of free energies computed by exponential averaging, the Bennett acceptance ratio, and thermodynamic integration. *Journal of Chemical Physics*, *122*, 144107.

Thiel, W., & Voityuk, A. A. (1996). Extension of MNDO to d orbitals: Parameters and results for the second-row elements and for the zinc group. *Journal of Physical Chemistry*, *100*, 616–626.

Torrie, G. M., & Valleau, J. P. (1977). Non-physical sampling distributions in Monte-Carlo free energy estimation—Umbrella sampling. *Journal of Computational Physics*, *23*, 187–199.

Trygubenko, S. A., & Wales, D. J. (2004). Analysis of cooperativity and localization for atomic rearrangements. *Journal of Chemical Physics*, *121*, 6689–6697.

Turney, J. M., Simmonett, A. C., Parrish, R. M., Hohenstein, E. G., Evangelista, F. A., Fermann, J. T., ... Crawford, T. D. (2012). Psi4: An open-source ab initio electronic structure program. *WIREs Computational Molecular Science*, *2*, 556–565.

Tuttle, T., & Thiel, W. (2008). OMx-D: Semiempirical methods with orthogonalization and dispersion corrections. Implementation and biochemical application. *Physical Chemistry Chemical Physics*, *10*, 2159–2166.

Valiev, M., Bylaska, E. J., Dupuis, M., & Tratnyek, P. G. (2008). Combined quantum mechanical and molecular mechanics studies of the electron-transfer reactions involving carbon tetrachloride in solution. *Journal of Physical Chemistry A*, *112*(12), 2713–2720.

Valiev, M., Garrett, B. C., Tsai, M.-K., Kowalski, K., Kathmann, S. M., Schenter, G. K., & Dupuis, M. (2007). Hybrid approach for free energy calculations with high-level methods: Application to the SN2 reaction of CHCl3 and OH$^-$ in water. *Journal of Chemical Physics*, *127*, 051102.

Vanommeslaeghe, K., Hatcher, E., Acharya, C., Kundu, S., Zhong, S., Shim, J., ... Mackerell, A. D. (2010). CHARMM general force field: A force field for drug-like molecules compatible with the CHARMM all-atom additive biological force fields. *Journal of Computational Chemistry*, *31*, 671–690.

Warshel, A., & Levitt, M. (1976). Theoretical studies of enzymic reactions: Dielectric, electrostatic and steric stabilization of the carbonium ion in the reaction of lysozyme. *Journal of Molecular Biology*, *103*, 227–249.

Weinan, E., Ren, W. Q., & Vanden-Eijnden, E. (2005a). Finite temperature string method for the study of rare events. *Journal of Physical Chemistry B, 109,* 6688–6693.

Weinan, E., Ren, W. Q., & Vanden-Eijnden, E. (2005b). Transition pathways in complex systems: Reaction coordinates, isocommittor surfaces, and transition tubes. *Chemical Physics Letters, 413,* 242–247.

Weinan, E., Ren, W., & Vanden-Eijnden, E. (2007). Simplified and improved string method for computing the minimum energy paths in barrier-crossing events. *Journal of Chemical Physics, 126,* 164103.

Wesolowski, T., & Warshel, A. (1994). Ab initio free energy perturbation calculations of solvation free energy using the frozen density functional approach. *Journal of Physical Chemistry, 98,* 5183–5187.

Woodcock, H. L., Hodoscek, M., & Brooks, B. R. (2007). Exploring SCC-DFTB paths for mapping QM/MM reaction mechanisms. *Journal of Physical Chemistry A, 111,* 5720–5728.

Woodcock, H. L., III, Hodoscek, M., Gilbert, A. T. B., Gill, P. M. W., Schaefer, I., Henry, F., & Brooks, B. R. (2007). Interfacing Q-Chem and CHARMM to perform QM/MM reaction path calculations. *Journal of Computational Chemistry, 28,* 1485–1502.

Woodcock, H. L., Miller, B. T., Hodoscek, M., Okur, A., Larkin, J. D., Ponder, J. W., & Brooks, B. R. (2011). MSCALE: A General Utility for Multiscale Modeling. *Journal of Chemical Theory and Computation, 7,* 1208–1219.

Woodcock, H. L., Moran, D., Pastor, R. W., MacKerell, A. D., & Brooks, B. R. (2007). Ab initio modeling of glycosyl torsions and anomeric effects in a model carbohydrate: 2-Ethoxy tetrahydropyran. *Biophysical Journal, 93,* 1–10.

Woodcock, H. L., Zheng, W., Ghysels, A., Shao, Y., Kong, J., & Brooks, B. R. (2008). Vibrational subsystem analysis: A method for probing free energies and correlations in the harmonic limit. *Journal of Chemical Physics, 129,* 214109.

Woods, C. J., Manby, F. R., & Mulholland, A. J. (2008). An efficient method for the calculation of quantum mechanics/molecular mechanics free energies. *Journal of Chemical Physics, 128,* 014109.

Yang, W., Cui, Q., Min, D., & Li, H. . (2010). Chapter 4—QM/MM alchemical free energy simulations: Challenges and recent developments. In R. A. Wheeler (Ed.), *Vol. 6. Annual reports in computational chemistry* (pp. 51–62): Elsevier.

Yang, Y., Yu, H., York, D., Elstner, M., & Cui, Q. (2008). Description of phosphate hydrolysis reactions with the self-consistent-charge density-functional-tight-binding (SCC-DFTB) theory. 1. Parameterization. *Journal of Chemical Theory and Computation, 4,* 2067–2084.

Zeng, X., Hu, H., Hu, X., & Yang, W. (2009). Calculating solution redox free energies with ab initio quantum mechanical/molecular mechanical minimum free energy path method. *Journal of Chemical Physics, 130,* 164111.

Zheng, L., Chen, M., & Yang, W. (2008). Random walk in orthogonal space to achieve efficient free-energy simulation of complex systems. *Proceedings of the National Academy of Sciences of the United States of America, 105,* 20227–20232.

Zheng, Y. J., & Merz, K. M. (1992). Mechanism of the human carbonic anhydrase II-catalyzed hydration of carbon dioxide. *Journal of the American Chemical Society, 114,* 10498–10507.

Zwanzig, R. W. (1954). High-temperature equation of state by a perturbation method. 1. nonpolar gases. *Journal of Chemical Physics, 22,* 1420–1426.

CHAPTER FIVE

Born–Oppenheimer Ab Initio QM/MM Molecular Dynamics Simulations of Enzyme Reactions

Y. Zhou*, S. Wang[†], Y. Li[‡], Y. Zhang[†,§,1]

*Laboratory of Mesoscopic Chemistry, Collaborative Innovation Center of Chemistry for Life Sciences, Institute of Theoretical and Computational Chemistry, School of Chemistry and Chemical Engineering, Nanjing University, Nanjing, China
[†]New York University, New York, NY, United States
[‡]International Center of Quantum and Molecular Structures, and Shanghai Key Laboratory of High Temperature Superconductors, Shanghai University, Shanghai, China
[§]NYU-ECNU Center for Computational Chemistry at NYU Shanghai, Shanghai, China
[1]Corresponding author: e-mail address: yingkai.zhang@nyu.edu

Contents

1. Introduction 106
2. Methods 107
   2.1 The Pseudobond Approach to Describe the QM/MM Boundary Across Covalent Bonds 107
   2.2 A Dual Focal aiQM/MM-PME Potential with the Periodic Boundary Condition 108
   2.3 Microiterative Optimization and Reaction Coordinate Driving 110
   2.4 Born–Oppenheimer aiQM/MM-MD with Umbrella Sampling 111
   2.5 Implementation 111
3. Examples 112
   3.1 Dissociation of *tert*-Butyl Chloride in Water 112
   3.2 First Step of Acylation Reaction for a Serine Protease 112
4. Enzyme Simulation Protocol 114
5. Conclusion 116
Acknowledgments 116
References 116

Abstract

There are two key requirements for reliably simulating enzyme reactions: one is a reasonably accurate potential energy surface to describe the bond-forming/breaking process as well as to adequately model the heterogeneous enzyme environment; the other is to perform extensive sampling since an enzyme system consists of at least thousands of atoms and its energy landscape is very complex. One attractive approach to meet both daunting tasks is Born–Oppenheimer ab initio QM/MM molecular dynamics (aiQM/MM-MD) simulation with umbrella sampling. In this chapter, we describe our recently developed pseudobond Q-Chem–Amber interface, which employs a combined electrostatic–mechanical embedding scheme with periodic boundary condition and the particle mesh Ewald method for long-range electrostatics interactions. In our implementation, Q-Chem and the sander module of Amber are combined at the source code level without using system calls, and all necessary data communications between QM and MM calculations are achieved via computer memory. We demonstrate the applicability of this pseudobond Q-Chem–Amber interface by presenting two examples, one reaction in aqueous solution and one enzyme reaction. Finally, we describe our established aiQM/MM-MD enzyme simulation protocol, which has been successfully applied to study more than a dozen enzymes.

1. INTRODUCTION

The idea of combining quantum mechanical (QM) and molecular mechanical (MM) methods (Warshel & Levitt, 1976) aims to take advantage of the applicability of QM methods for modeling chemical reactions and the computational efficiency of MM calculation. Such a combined QM/MM approach allows to study systems significant larger than a pure QM method can do, and has been widely used in modeling chemical reactions in complex system, from the solid and surface catalysis to solution and enzyme reactions (Gao & Truhlar, 2002; Hu & Yang, 2008; Senn & Thiel, 2009; Zhang, 2006). Meanwhile, due to the complexity of the energy landscape of an enzyme system in aqueous solution, which consists of at least thousands of degrees of freedom and is highly flexible, extensive sampling with a reasonably accurate potential energy surface is needed. The free energy changes associated with an enzyme reaction process are not only better defined but also can be directly compared with corresponding experimental results.

One most attractive approach to meet both above daunting tasks—a reasonably accurate potential energy surface for describing enzyme reactions and extensive sampling to calculate free energy change—is Born–Oppenheimer ab initio QM/MM molecular dynamics (aiQM/MM-MD) simulation: at each time step, the atomic forces as well as the total energy of the enzyme system are calculated with the aiQM/MM method on-the-fly, and Newton equations of motion are integrated. In spite of the conventional wisdom that such simulations would be computationally too expensive to be feasible for simulating enzyme reactions, in last several years we have advanced the aiQM/MM-MD approach by interfacing Q-Chem (Shao et al., 2006) with Tinker (Ponder, 2004) and Amber (Case et al., 2012) programs at the source code level and substantially enhancing its efficiency. We have demonstrated this state-of-the-art approach to be feasible and successful in elucidating the catalytic power of enzymes (Wang, Hu, & Zhang, 2007) and characterizing complicated multistep reaction mechanisms (Ke, Smith, Zhang, & Guo, 2011; Lior-Hoffmann et al., 2012; Rooklin, Lu, & Zhang, 2012; Shi, Zhou, Wang, & Zhang, 2013; Wu, Wang, Zhou, Cao, & Zhang, 2010; Zhou, Wang, & Zhang, 2010). Very recently, we have successfully applied this state-of-the-art approach to simulate covalent inhibition (Lei, Zhou, Xie, & Zhang, 2015) as well as to design a time-dependent HDAC2-selective inhibitor (Zhou et al., 2015).

In the below, first we describe our recently developed pseudobond Q-Chem–Amber interface with a combined electrostatic–mechanical embedding scheme, periodic boundary condition, and the particle mesh Ewald method for long-range electrostatics interactions; then we present two examples to demonstrate its applicability to simulate chemical reactions in aqueous solution and enzymes; and finally we describe our established aiQM/MM-MD enzyme simulation protocol and discuss several important computational details regarding aiQM/MM-MD simulations of enzyme reactions.

2. METHODS

2.1 The Pseudobond Approach to Describe the QM/MM Boundary Across Covalent Bonds

The treatment of the QM/MM boundary across covalent bonds is a critical problem concerning the accuracy and applicability of the combined QM/MM methods to study enzyme reactions (Field, 2002; Field, Bash, & Karplus, 1990). In the pseudobond approach (Zhang, 2005, 2006; Zhang, Lee, & Yang, 1999), as illustrated in Fig. 1, the boundary atom of the QM subsystem Y is replaced by a one-free-valence boundary C_{ps} atom to form a pseudobond C_{ps}–X. The C_{ps} atom has seven valence electrons,

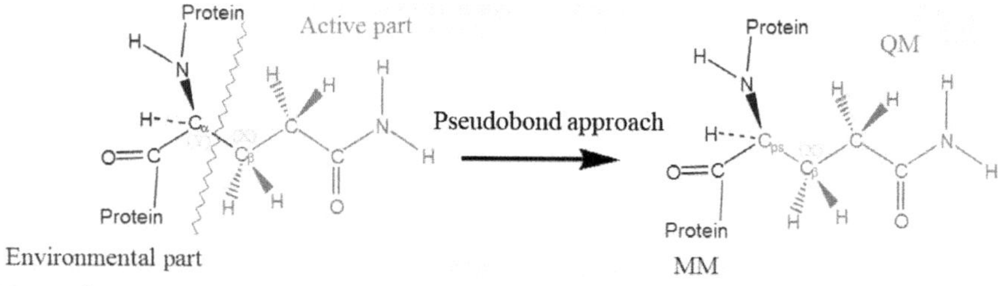

Fig. 1 Illustration of the pseudobond approach in the treatment of the QM/MM boundary problem. (See the color plate.)

nuclear charge as +7, an angular-momentum-independent effective core potential, and a parameterized STO-2G basis set. The C_{ps} atom is parameterized to make the C_{ps}–X pseudobond mimic the original Y–X bond with similar bond length and strength, and also similar effects on the rest of the active part. The C_{ps} atoms and all atoms in the active part form a well-defined (often closed-shell) QM subsystem, which can be treated by quantum mechanical methods. The rest atoms form the MM subsystem, which will be represented by a molecular mechanical force field. The developed pseudobonds are independent of the molecular mechanical force field and offer a smooth connection at the QM/MM interface without introducing additional degrees of freedom into the system. Up to now, the $C_{ps}(sp^3)$–$C(sp^3)$, $C_{ps}(sp^3)$–$C(sp^2$, carbonyl), and $C_{ps}(sp^3)$–$N(sp^3)$ pseudobonds have been developed for the cutting of protein backbones and nucleic acid bases with the 6-31G* basis set. The parameterization is performed with density functional calculations using hybrid B3LYP exchange–correlation functional, but the same set of parameters is applicable to Hartree–Fock and MP2 methods, as well as DFT calculations with other exchange–correlation functionals (Zhang, 2006).

2.2 A Dual Focal aiQM/MM-PME Potential with the Periodic Boundary Condition

In our implementation of aiQM/MM-MD with Q-Chem (Shao et al., 2006) and Tinker (Ponder, 2004), the spherical boundary condition is employed. Although this description is sufficient for many applications, it would not be appropriate to study chemical processes that are strongly influenced by long-range interactions and dynamics. A more reliable and robust description is the periodic boundary condition with the Ewald method for long-range electrostatics interactions (Nam, Gao, & York, 2005; Walker, Crowley, & Case, 2008). As shown in Fig. 2, the total potential energy of the QM/MM (PBC) system can be written as

$$E^{PBC}[\rho+q,\rho+q] = E^{RS}[\rho,\rho] + E^{RS}[\rho,q] + \Delta E^{PBC} + E^{PBC}[q,q],$$
$$\Delta E^{PBC} = \left(E^{PBC}[\rho,\rho] - E^{RS}[\rho,\rho]\right) + \left(E^{PBC}[\rho,q] - E^{RS}[\rho,q]\right),$$

where $E^{RS}[\rho,\rho]$ and $E^{RS}[\rho,q]$ only include interactions within the same cell and are treated by an effective Hamiltonian H_{eff} as done in nonperiodic boundary systems. The $E^{PBC}[q,q]$ is an MM term and can be treated by the particle mesh Ewald method. One key idea of the QM/MM-Ewald

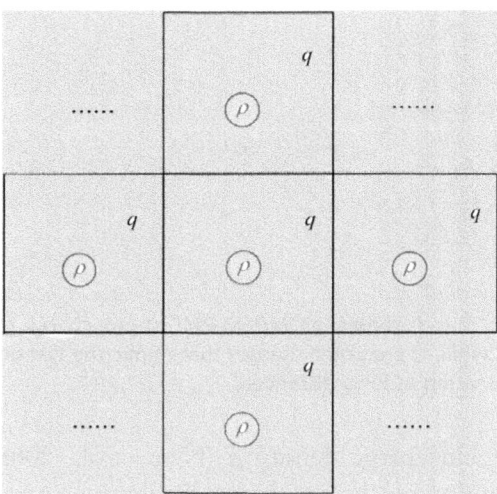

Fig. 2 Illustration of a conventional QM/MM periodical boundary system: *yellow* (*gray* in the print version) *circles* indicate the QM subsystems, while the shaded area q is the MM subsystem.

method (Nam et al., 2005; Walker et al., 2008) is to approximate the charge density ρ with atomic charges, Q, in calculating the PBC correction term:

$$\Delta E^{\text{PBC}} \approx \left(E^{\text{PBC}}[Q, Q] - E^{\text{RS}}[Q, Q]\right) + \left(E^{\text{PBC}}[Q, q] - E^{\text{RS}}[Q, q]\right)$$
$$= \Delta E^{\text{PBC}}[Q, Q] + \Delta E^{\text{PBC}}[Q, q]$$

Although such an implementation of aiQM/MM-Ewald method (Okamoto et al., 2011; Torras, Seabra, Deumens, Trickey, & Roitberg, 2008) is feasible, its computational cost is higher than that with the spherical boundary condition since a significant more number of point charges are included in the effective Hamiltonian. In addition, the imaging of the QM subsystem is conceptually not appealing, nor is it necessary since the interactions are approximated as charge–charge interactions in calculations anyway. Here we have developed a dual-focal aiQM/MM-PME approach: there is only one QM subsystem, whose electron density is polarized by nearby MM atoms, while interactions with all remote atoms are represented by classical coulomb interactions, as illustrated in Fig. 3.

q_{eff} represents the point charges in the effective Hamiltonian which are chosen by residue, and q represents the charges of surroundings. For a periodic boundary system, only the MM atoms in the central box are considered in the effective Hamiltonian. At distances on the order of unit cell size away, the differences are very small between the potential calculated from fitted

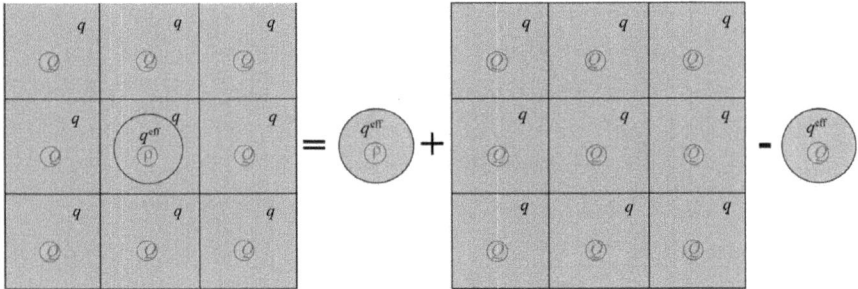

Fig. 3 Illustration of the dual-focal aiQM/MM-PME approach, q_{eff} refers to the nearby MM atom charges, while Q are point charges that mimic the QM density which would be a good approximation at long distance.

point charge Q and charge density ρ (Nam et al., 2005); therefore, an approximate quantum mechanics charge distribution Q is used to represent ρ in image boxes. The electrostatic potential fitted (ESP) charges (Cox & Williams, 1981; Momany, 1978) are employed in our interface. This can also be considered as a combined electrostatic–mechanical embedding scheme. The implementation is quite straightforward: the first term would be calculated with the aiQM/MM method as implemented in Q-Chem using the spherical boundary condition; with the determined charges, the second term would be calculated with the particle mesh Ewald (PME) (Darden, York, & Pedersen, 1993) method implemented in Amber, and the third term can also be calculated easily with the MM method.

2.3 Microiterative Optimization and Reaction Coordinate Driving

To optimize an enzyme system with an aiQM/MM method, we employ the microiterative minimization approach, in which QM and MM subsystems are minimized separately and iteratively (Zhang, Liu, & Yang, 2000). In each circle, the QM subsystem is optimized one step using aiQM/MM calculations with frozen MM environment, and then the MM system is fully optimized using MM calculations with a fixed QM subsystem in which the QM charge density represented by ESP charges calculated from the QM/MM method. The iterations will not stop until convergence criterion is reached for both minimizations. To determine the reaction path, we use the reaction coordinate driving (RCD) method: by choosing a proposed reaction coordinate (RC), we carry out a series of restrained minimizations for different points of RC from the initial state to final state to determine a minimal energy path (Zhang et al., 2000). The choice of a proper reaction coordinate is crucial in this approach, which is generally a linear combination of internal coordinates. Sometimes, two-dimensional minimal energy surfaces would be needed for studying complicated reactions.

2.4 Born–Oppenheimer aiQM/MM-MD with Umbrella Sampling

In Born–Oppenheimer aiQM/MM-MD simulations, in which the atomic forces as well as the total energy of the whole system are calculated at each time step on-the-fly with the hybrid QM/MM potential, and Newton equations of motion are integrated. Thus, it takes accounts of dynamics of reaction center and its environment on an equal footing. In order to make aiQM/MM-MD to be time reversible, besides using the time-reversible integrator Verlet for atom positions, the lossless time-reversible density matrix propagation (Niklasson, Tymczak, & Challacombe, 2006) has been implemented for the electronic degrees of freedom. For a new time step, the initial guessed density matrix for SCF is constructed from the converged SCF density matrix of current time step P_n and the initial guessed density matrix for the previous time step $\tilde{P}_{n-1}$ in a time-reversible way, $\tilde{P}_{n+1} = 2P_n - \tilde{P}_{n-1}$. In this way, aiQM/MM-MD would be time reversible and would achieve much better energy conservation than employing $\tilde{P}_{n+1} = P_n$. To determine the free energy profile for a given chemical reaction, umbrella sampling (Patey & Valleau, 1975) would be carried out by employing a series of harmonic potentials centered along the chosen reaction coordinate.

2.5 Implementation

We have implemented a pseudobond Q-Chem–Amber interface with a combined electrostatic–mechanical embedding scheme, periodic boundary condition, and the PME method for long-range electrostatics interactions. In our implementation, the quantum chemistry package Q-Chem4.0.1 (Krylov & Gill, 2013) and the sander module of molecular mechanics package Amber12 (Case et al., 2012) are combined at the source code level without using system calls. There is only one executable file for Q-Chem–Amber, and all the necessary data communications between Q-Chem and Amber are via Q-Chem internal files and in memory that makes the user interface clear and friendly. In this way, QM/MM iterative optimization and QM/MM-MD can be run continuously without restarting Q-Chem or Amber executable programs. Apart from the Q-Chem and Amber standard input files, only a few parameters need to be added to specify QM/MM interface. Besides the output of Q-Chem restart files, the interface generates Amber restart and trajectory files by default which can be visualized and analyzed by MD analysis tools. The microiterative approach has been implemented for QM/MM geometry minimizations and reaction path explorations. In Born–Oppenheimer molecular dynamics simulations, a lossless time-reversible density matrix propagation scheme has been used for the electronic degrees of freedom. Accuracy, efficiency, and applicability of this pseudobond Q-Chem–Amber interface have been extensively examined by simulating chemical reactions in aqueous solution and enzymes.

3. EXAMPLES

3.1 Dissociation of *tert*-Butyl Chloride in Water

To examine the applicability of the Q-Chem–Amber interface, we simulated the SN1 fragmentation of *tert*-butyl chloride in water (Hartsough & Merz, 1995; Peters, 2007), $t-Bu-Cl \rightarrow t-Bu^+ + Cl^-$ in which water plays an indispensable role. For such a charge separation reaction, the proper treatment of long-range electrostatic interactions and boundary condition is expected to be very important and spherical boundary condition would not be sufficient in this case. Here the reactant was solvated in a box with a 35 Å buffer of water molecules and equilibrated with 1 ns NPT classical MD simulations. By choosing the C–Cl distance as the reaction coordinate, aiQM/MM-MD simulations with umbrella sampling have been used to determine free energy profiles with different *qmcut* values. *qmcut* is the distance cutoff for MM charges to be treated as q_{eff}. For each window, 110 ps B3LYP/6-31+G(d) QM/MM-MD simulation was performed, and the last 60 ps data were used for analysis. As shown in Fig. 4, all free energy profiles have converged reasonably well in spite of different *qmcut* values, and the calculated free energy barriers for dissociation are between 18.7 ± 0.2 and $19.0 \pm .2$ kcal/mol, in good agreement with the experimental derived value of 19.5 kcal/mol (Grunwald & Effio, 1974; Winstein & Fainberg, 1957).

3.2 First Step of Acylation Reaction for a Serine Protease

We have employed the Q-Chem–Amber interface to simulate the first step of acylation reaction catalyzed by porcine-trypsin using periodic boundary condition, which has been previously simulated with the Q-Chem–Tinker

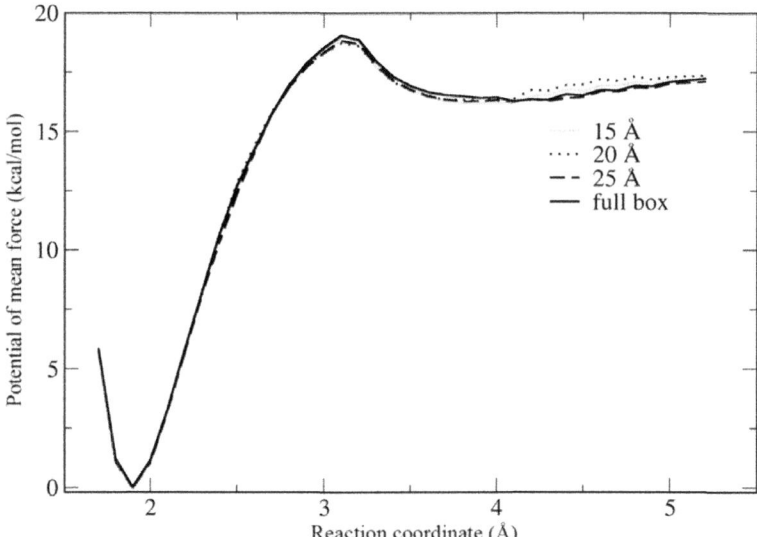

Fig. 4 Free energy profile for dissociation of *tert*-butyl chloride in water determined by B3LYP/6-31+G(d) QM/MM molecular dynamics simulations and umbrella sampling with different *qmcut* values, which indicates the distance for MM charges to be included in electrostatic embedding. *Full box* indicates that all MM charges in the primary box are treated as q_{eff} in Fig. 3.

interface under spherical boundary condition (Zhou & Zhang, 2011). The simulations were carried out at the B3LYP/6-31+G* QM/MM level with the pseudobond approach, which is the same as in the previous work (Zhou & Zhang, 2011). We tested 20 Å *qmcut* and full box for q_{eff}. The computational time for full box q_{eff} is about 1.5 times of that for the 20 Å *qmcut*, which is mainly caused by the increasing number of one-electron terms in the effective Hamiltonian. The free energy profiles, as shown in Fig. 5, are very similar between two *qmcut* values. By considering both efficiency and accuracy, a reasonable default *qmcut* value would be 20 Å. Thus we recommend using 20 Å *qmcut* as the default value to minimize the computational cost. The free energy barrier for the first step is about 15 kcal/mol, in good agreement with the experimental value for activation energy of 15–20 kcal/mol (Fersht, 1999). The free energy barrier is 16.4 kcal/mol from our previous study with spherical boundary condition, and the shape of two free energy curves is almost the same for the two boundary conditions. This indicates that the spherical boundary condition is reasonably accurate for enzyme reactions that are not strongly influenced by long-range electrostatic interactions and dynamics.

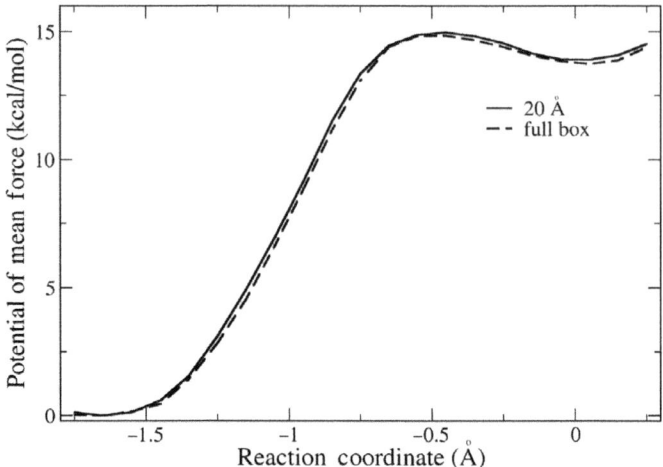

Fig. 5 Free energy profile for the first step of acylation reaction for porcine-trypsin determined by B3LYP/6-31+G(d) QM/MM molecular dynamics simulations and umbrella sampling.

4. ENZYME SIMULATION PROTOCOL

In this section, we present our established aiQM/MM-MD enzyme simulation protocol, as shown in Fig. 6, which has been successfully applied to study more than a dozen enzymes. For initial preparation and classical MD simulations, the protocol is the same as typical classical simulations of biomolecular systems. Thus here we would focus on the QM/MM modeling part and mainly address a few essential details that are important for aiQM/MM-MD simulations.

- *QM/MM system setup.* Given a structural snapshot from MM modeling, the QM/MM partition needs to be carried out, and the covalent boundary would be treated with the improved pseudobond approach. Since pseudobonds are basis set dependent, it is important that the QM atoms directly bonded to pseudoatoms are treated with 6-31G* basis set. Meanwhile, for hydrogen atoms of water molecules and hydroxyl hydrogen that have been partitioned into the QM subsystem, their van der Waals (vdW) parameters need to be changed from zero to a tiny value, such as 0.001. Otherwise, in later aiQM/MM-MD simulations, sometimes a hydrogen atom with zero vdW parameters can jump to collide with an MM atom, which is not physical and would lead to the simulation failure.

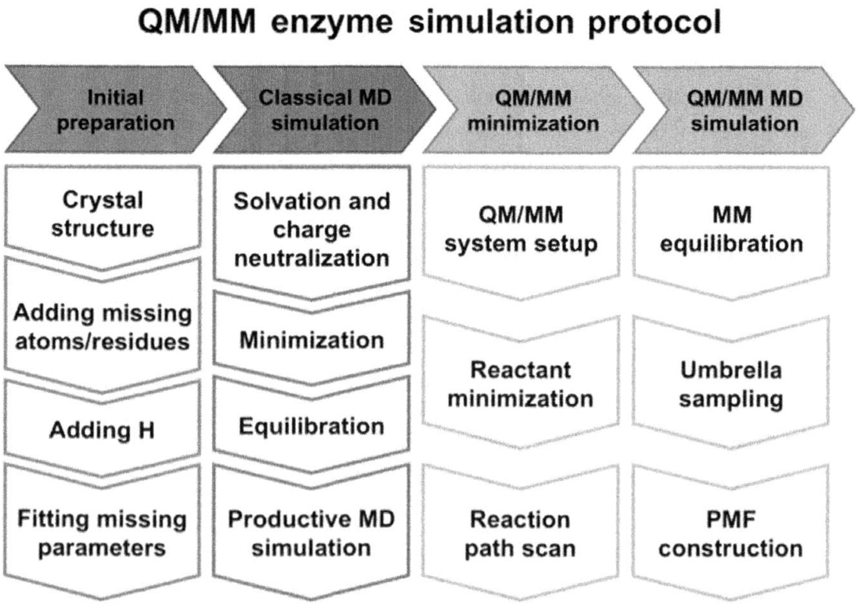

QM/MM enzyme simulation protocol

Initial preparation	Classical MD simulation	QM/MM minimization	QM/MM MD simulation
Crystal structure	Solvation and charge neutralization	QM/MM system setup	MM equilibration
Adding missing atoms/residues	Minimization	Reactant minimization	Umbrella sampling
Adding H	Equilibration		
Fitting missing parameters	Productive MD simulation	Reaction path scan	PMF construction

Fig. 6 Illustration of our enzyme simulation protocol. (See the color plate.)

- *Reactant minimization.* This step is carried out by the iterative microiterative optimization without employing any restraints. Typically, this step would be straightforward if all previous steps have been done appropriately.
- *Reaction path scan.* In this step, it is important to check whether there are significant discontinuities in some geometry elements along the reaction path. This information would be useful to help you select more appropriate reaction coordinates. In some cases, two-dimensional minimum energy surfaces would need to be mapped out.
- *MM equilibration.* This step is to employ nanosecond classical molecular dynamics simulations to equilibrate the MM subsystem with QM subsystem fixed at previously QM/MM-minimized structures. Since the length of aiQM/MM-MD simulations is quite short, this step is important that the MM subsystem can be relaxed given the change of QM subsystem.
- *Umbrella sampling.* In this step, aiQM/MM-MD simulations with restraint will be carried out. For each window along the reaction coordinate, an initial force constant is chosen based on the estimated slope of the determined minimum energy curve; ie, larger restraint is needed for regions of a steeper slope or for allowing direct sampling of the transition

state and a smaller restraint is added at flatter regions. Histogram between neighboring windows will be checked to make sure that there are sufficient overlaps. In some challenging cases, two-dimensional umbrella sampling needs to be carried out.
- *PMF construction*. After adequate sampling by umbrella sampling, the free energy profile or free energy surface can be obtained with the weighted histogram analysis method method (Ferrenberg & Swendsen, 1988; Kumar, Bouzida, Swendsen, Kollman, & Rosenberg, 1992; Souaille & Roux, 2001). One prerequisite is that the individual-biased simulations in each window must be sampled along the proper reaction coordinate and overlapped well with each other. The convergence of umbrella sampling can be tested by the PMF calculated from different time spans.

5. CONCLUSION

In this chapter, we have described a Born–Oppenheimer aiQM/MM-MD approach to simulate enzyme reactions. In comparison with many other computational methods that can be used to simulate enzyme reactions, one unique strength of the aiQM/MM-MD approach is to study biological reactions whose mechanism has not been fully elucidated, such as for novel enzymes and new covalent modifiers. With the continuing advances in computer technology and algorithm development, the aiQM/MM-MD approach is expected to become the method of choice to simulate novel biochemical reactions in the very near future.

ACKNOWLEDGMENTS
We would like to acknowledge the support by NIH (R01-GM079223 to Zhang) and the National Natural Science Foundation of China (Grant No. 21203090 to Zhou, Grant No. 21503130 to Li). And Li is also partly supported by Shanghai Key Laboratory of High Temperature Superconductors (No. 14DZ2260700). We thank NYU-ITS and NYUAD for providing computational resources.

REFERENCES
Case, D. A., Darden, T. A., Cheatham, T. E., III, Simmerling, C. L., Wang, J., Duke, R. E., et al. (2012). *AMBER 12*. San Francisco, CA: University of California.
Cox, S. R., & Williams, D. E. (1981). Representation of the molecular electrostatic potential by a net atomic charge model. *Journal of Computational Chemistry, 2*, 304–323.
Darden, T., York, D., & Pedersen, L. (1993). Particle mesh Ewald: An N log(N) method for Ewald sums in large systems. *Journal of Chemical Physics, 98*, 10089–10092.

Ferrenberg, A. M., & Swendsen, R. H. (1988). New Monte Carlo technique for studying phase transitions. *Physical Review Letters, 61*, 2635–2638.
Fersht, A. (1999). *Structure and mechanism in protein science. A guide to enzyme catalysis and protein folding* (2nd ed.). New York, NY: W. H. Freeman and Company.
Field, M. J. (2002). Simulating enzyme reactions: Challenges and perspectives. *Journal of Computational Chemistry, 23*, 48–58.
Field, M. J., Bash, P. A., & Karplus, M. (1990). A combined quantum mechanical and molecular mechanical potential for molecular dynamics simulation. *Journal of Computational Chemistry, 11*, 700–733.
Gao, J. L., & Truhlar, D. G. (2002). Quantum mechanical methods for enzyme kinetics. *Annual Review of Physical Chemistry, 53*, 467–505.
Grunwald, E., & Effio, A. (1974). Solution thermodynamics in nonideal mixed solvents under endostatic conditions. *Journal of the American Chemical Society, 96*, 423–430.
Hartsough, D. S., & Merz, K. M. (1995). Potential of mean force calculations on the SN1 fragmentation of tert-butyl chloride. *Journal of Physical Chemistry, 99*, 384–390.
Hu, H., & Yang, W. T. (2008). Free energies of chemical reactions in solution and in enzymes with ab initio quantum mechanics/molecular mechanics methods. *Annual Review of Physical Chemistry, 59*, 573–601.
Ke, Z., Smith, G. K., Zhang, Y., & Guo, H. (2011). Molecular mechanism for eliminylation, a newly discovered post-translational modification. *Journal of the American Chemical Society, 133*, 11103–11105.
Krylov, A. I., & Gill, P. M. W. (2013). Q-Chem: An engine for innovation. *Wiley Interdisciplinary Reviews: Computational Molecular Science, 3*, 317–326.
Kumar, S., Bouzida, D., Swendsen, R. H., Kollman, P. A., & Rosenberg, J. M. (1992). The weighted histogram analysis method for free-energy calculations on biomolecules. I: The method. *Journal of Computational Chemistry, 13*, 1011–1021.
Lei, J., Zhou, Y., Xie, D., & Zhang, Y. (2015). Mechanistic insights into a classic wonder drug—Aspirin. *Journal of American Chemical Society, 137*, 70–73.
Lior-Hoffmann, L., Wang, L., Wang, S., Geacintov, N. E., Broyde, S., & Zhang, Y. (2012). Preferred WMSA catalytic mechanism of the nucleotidyl transfer reaction in human DNA polymerase kappa elucidates error-free bypass of a bulky DNA lesion. *Nucleic Acids Research, 40*, 9193–9205.
Momany, F. A. (1978). Determination of partial atomic charges from ab initio electrostatic potentials. Application to formamide, methanol and formic acid. *Journal of Physical Chemistry, 82*, 592–601.
Nam, K., Gao, J. L., & York, D. M. (2005). An efficient linear-scaling Ewald method for long-range electrostatic interactions in combined QM/MM calculations. *Journal of Chemical Theory and Computation, 1*, 2–13.
Niklasson, A. M. N., Tymczak, C. J., & Challacombe, M. (2006). Time-reversible Born-Oppenheimer molecular dynamics. *Physical Review Letters, 97*, 123001.
Okamoto, T., Yamada, K., Koyano, Y., Asada, T., Koga, N., & Nagaoka, M. (2011). A minimal implementation of the AMBER-GAUSSIAN interface for ab initio QM/MM-MD simulation. *Journal of Computational Chemistry, 32*, 932–942.
Patey, G. N., & Valleau, J. P. (1975). A Monte Carlo method for obtaining the interionic potential of mean force in ionic solution. *Journal of Chemical Physics, 63*, 2334–2339.
Peters, K. S. (2007). Nature of dynamic processes associated with the S(N)1 reaction mechanism. *Chemical Reviews, 107*, 859–873.
Ponder, J. W. (2004). *Software tools for molecular design, TINKER version 4.2.* June.
Rooklin, D. W., Lu, M., & Zhang, Y. (2012). Revelation of a catalytic calcium-binding site elucidates unusual metal dependence of a human apyrase. *Journal of the American Chemical Society, 134*, 15595–15603.

Senn, H. M., & Thiel, W. (2009). QM/MM methods for biomolecular systems. *Angewandte Chemie, International Edition, 48*, 1198–1229.

Shao, Y., Molnar, L. F., Jung, Y., Kussmann, J., Ochsenfeld, C., Brown, S. T., et al. (2006). Advances in methods and algorithms in a modern quantum chemistry program package. *Physical Chemistry Chemical Physics, 8*, 3172–3191.

Shi, Y., Zhou, Y., Wang, S., & Zhang, Y. (2013). Sirtuin deacetylation mechanism and catalytic role of the dynamic cofactor binding loop. *Journal of Physical Chemistry Letters, 4*, 491–495.

Souaille, M., & Roux, B. (2001). Extension to the weighted histogram analysis method: Combining umbrella sampling with free energy calculations. *Computer Physics Communications, 135*, 40–57.

Torras, J., Seabra, G. D. M., Deumens, E., Trickey, S. B., & Roitberg, A. E. (2008). A versatile AMBER-Gaussian QM/MM interface through PUPIL. *Journal of Computational Chemistry, 29*, 1564–1573.

Walker, R. C., Crowley, M. F., & Case, D. A. (2008). The implementation of a fast and accurate QM/MM potential method in Amber. *Journal of Computational Chemistry, 29*, 1019–1031.

Wang, S., Hu, P., & Zhang, Y. (2007). Ab initio quantum mechanical/molecular mechanical molecular dynamics simulation of enzyme catalysis: The case of histone lysine methyltransferase SET7/9. *The Journal of Physical Chemistry. B, 111*, 3758–3764.

Warshel, A., & Levitt, M. (1976). Theoretical studies of enzymic reactions: dielectric, electrostatic and steric stabilization of the carbonium ion in the reaction of lysozyme. *Journal of Molecular Biology, 103*, 227–249.

Winstein, S., & Fainberg, A. H. (1957). Correlation of solvolysis rates. IV. Solvent effects on enthalpy and entropy of activation for solvolysis of t-butyl chloride. *Journal of the American Chemical Society, 79*, 5937–5950.

Wu, R., Wang, S., Zhou, N., Cao, Z., & Zhang, Y. (2010). A proton-shuttle reaction mechanism for histone deacetylase 8 and the catalytic role of metal ions. *Journal of the American Chemical Society, 132*, 9471–9479.

Zhang, Y. K. (2005). Improved pseudobonds for combined ab initio quantum mechanical/molecular mechanical methods. *Journal of Chemical Physics, 122*, 024114.

Zhang, Y. K. (2006). Pseudobond ab initio QM/MM approach and its applications to enzyme reactions. *Theoretical Chemistry Accounts, 116*, 43–50.

Zhang, Y. K., Lee, T. S., & Yang, W. T. (1999). A pseudobond approach to combining quantum mechanical and molecular mechanical methods. *Journal of Chemical Physics, 110*, 46–54.

Zhang, Y. K., Liu, H. Y., & Yang, W. T. (2000). Free energy calculation on enzyme reactions with an efficient iterative procedure to determine minimum energy paths on a combined ab initio QM/MM potential energy surface. *Journal of Chemical Physics, 112*, 3483–3492.

Zhou, J., Li, M., Chen, N., Wang, S., Luo, H.-B., Zhang, Y., et al. (2015). Computational design of a time-dependent histone deacetylase 2 selective inhibitor. *ACS Chemical Biology, 10*, 687–692.

Zhou, Y., Wang, S., & Zhang, Y. (2010). Catalytic reaction mechanism of acetylcholinesterase determined by Born-Oppenheimer ab initio QM/MM molecular dynamics simulations. *The Journal of Physical Chemistry. B, 114*, 8817–8825.

Zhou, Y., & Zhang, Y. (2011). Serine protease acylation proceeds with a subtle re-orientation of the histidine ring at the tetrahedral intermediate. *Chemical Communications, 47*, 1577–1579.

CHAPTER SIX

QM/MM Calculations on Proteins

U. Ryde[1]
Chemical Centre, Lund University, Lund, Sweden
[1]Corresponding author: e-mail address: ulf.ryde@teokem.lu.se

Contents

1. Introduction 120
2. Methods 122
   2.1 QM Methods 123
   2.2 MM Methods 124
   2.3 QM/MM Calculations 125
   2.4 The Big-QM Approach 128
   2.5 QM/MM Free-Energy Calculations 130
   2.6 Combination with Experiments 133
3. Applications 135
   3.1 Energy Components in Glutamate Mutase 135
   3.2 Mo Oxo-Transfer Enzymes 140
   3.3 Blue Copper Proteins and Multicopper Oxidases 143
4. Suggested Approach for QM/MM Investigations 147
Acknowledgments 150
References 150

Abstract

In this chapter, I discuss combined quantum mechanics (QM) and molecular mechanics (MM; QM/MM) calculations for proteins. In QM/MM, a small but interesting part of the protein is treated by accurate QM methods, whereas the remainder is treated by faster MM methods. The prime problems with QM/MM calculations are bonds between the QM and MM systems, the selection of the QM system, and the local-minima problem. The two first problems can be solved by the big-QM approach, including in the QM calculation all groups within 4.5–6 Å of the active site and all buried charges in the protein. The third problem can be solved by calculating free energies. It is important to study QM/MM energy components to ensure that the results are stable and reliable. They can also be used to understand the reaction and the effect of the surroundings, eg, by dividing the catalytic effect into bonded, van der Waals, electrostatic, and geometric components and to deduce which parts of the protein contribute most to the catalysis. It should be ensured that the QM calculations are reliable and converged by extending the basis set to quadruple-zeta quality, including a proper treatment of dispersion, as well as relativistic, zero-point, thermal, and entropy effects, and performing calculations with both pure and hybrid density functional theory methods. If the latter give differing results, calibration with high-level QM methods is needed. Reactions that change the net charge should be avoided. QM/MM calculations can be combined with experimental methods.

1. INTRODUCTION

During the latest two decades, quantum mechanical (QM) calculations have been established as a powerful complement to experiments for the study of structural and energetic properties of proteins and enzyme reactions. QM calculations have the advantage of directly providing energetic information, which govern all chemical processes. Moreover, they are not subject to experimental limitations, such as invisible, short-lived, expensive, or hazardous compounds. In particular, QM calculations have been used to characterize and study the transition states and activation energies of enzyme reaction, allowing for the discrimination between alternative suggested reaction mechanisms. With QM calculations, reactions are studied by computers, avoiding all use of chemicals, which make them the ultimate type of environment-friendly green chemistry. On the other hand, QM calculations are approximate, so the results should be calibrated against experimental data and you should ensure that the calculations are converged.

There are two approaches to treat proteins with QM methods. In the QM-cluster approach, a small number of residues (typically 30–200 atoms; cf. Fig. 1A) are cut out from the protein and are studied by QM in a continuum solvent (Blomberg, Borowski, Himo, Liao, & Siegbahn, 2014; Himo & Siegbahn, 2009). This allows for a full control of the structures of the studied system. On the other hand, the results may be biased by the choice of included residues and the surrounding protein is studied in a very crude manner. Alternatively, the active site is treated by a similar QM model, but the surrounding protein and solvent are modeled by molecular mechanics (MM), giving the QM/MM approach (Lin & Truhlar, 2007; Senn & Thiel, 2009; Sousa, Fernandes, & Ramos, 2012; Fig. 1B). This provides a more detailed description of the surroundings, but the calculations become more complicated, and involve thousands of atoms, which cannot be easily controlled.

In this chapter, I provide my view of how QM/MM studies of proteins should be performed to give useful and reliable results, based on my more than 20 years experience and method development (Delcey et al., 2014;

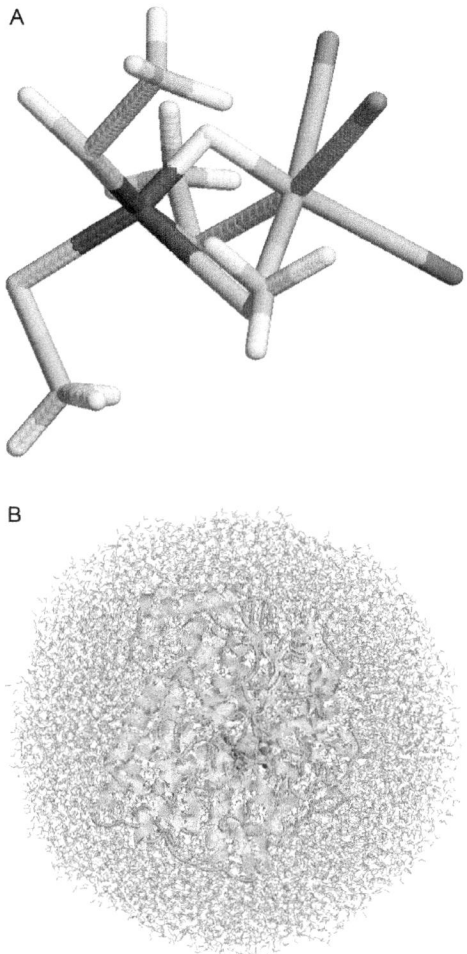

Fig. 1 Examples of (A) QM-cluster, (B) QM/MM (the QM system is shown as *balls*), and

Dong & Ryde, 2016; Hedegård, Kongsted, & Ryde, 2015; Heimdal, Kaukonen, Srnec, Rulíšek, & Ryde, 2011; Hu, Söderhjelm, & Ryde, 2011, 2013; Kaukonen, Söderhjelm, Heimdal, & Ryde, 2008a; Li, Farrokhnia, Rulíšek, & Ryde, 2015; Rod & Ryde, 2005a; Ryde, 1995, 1996; Ryde & Olsson, 2001; Ryde, Olsen, & Nilsson, 2002; Söderhjelm & Ryde, 2006; Sumner, Söderhjelm, & Ryde, 2013). I start with a short description of the methods involved and the special approaches

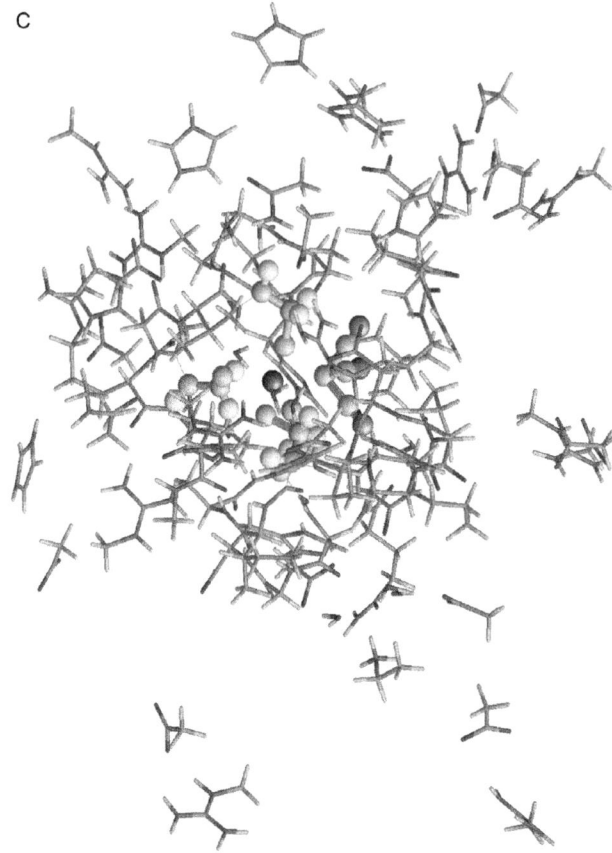

Fig. 1—Cont'd (C) big-QM models for [NiFe] hydrogenase (Dong & Ryde, 2016). (See the color plate.)

developed in our group. Then, I give some typical applications, illustrating various aspects of these methods. In the final section, I provide a detailed scheme of the recommended QM/MM approach, together with some concluding remarks.

2. METHODS

In this section, an introduction to the methods employed is given. For QM and MM methods, only the very basics are given and the interested reader is referred to textbooks in computational chemistry (eg, Cramer, 2006; Jensen, 2007). For the more specific QM/MM approaches, a somewhat deeper introduction is given, but for technical details, the reader is referred to review articles (Lin & Truhlar, 2007; Senn & Thiel, 2009; Sousa et al., 2012).

2.1 QM Methods

In QM calculations, the Schrödinger equation is solved for a molecular system (Cramer, 2006; Jensen, 2007). The input is the coordinates, the net charge, and the total spin. The result is the total energy and the wavefunction, from which all measureable properties of the system can be calculated. Unfortunately, the Schrödinger equation can be analytically solved only for a few systems with a single electron—for all large systems, only approximate numerical solutions can be obtained. The most basic QM approach is the Hartree–Fock (HF) method (Fock, 1930; Hartree, 1928). Unfortunately, it gives quite approximate results and it is therefore mainly used as a starting point of more accurate methods, based on perturbation theory or series expansions, eg, Møller–Plesset second-order perturbation theory (MP2; Møller & Plesset, 1934) or coupled cluster calculations with single, double, and perturbatively treated triple excitations (CCSD(T); Raghavachari, Tucks, Pople, & Head-Gordon, 1989). The latter is currently considered as the gold standard QM method. Unfortunately, it is computationally a very demanding method so that it in practice can be used only for molecules with up to $\sim$30 atoms. However, recently methods have been developed that take advantage of the local nature of the wavefunction, eg, LCCSD(T0), which can be used for systems with more than 100 atoms (Werner & Schütz, 2011).

Today, the great majority of QM calculations are performed with density functional theory (DFT). It is not based on the wavefunction, which is a function of the coordinates of all involved particles, but instead on the electron density, which is a function of only the three coordinates of Cartesian space. Still, there is a one-to-one correspondence between a wavefunction and the electron density (Hohenberg & Kohn, 1964). DFT also involves approximations and there are a great number of variants, eg, BP86 (Becke, 1988; Perdew, 1986), PBE (Perdew, Burke, & Ernzerhof, 1996), and TPSS (Tao, Perdew, Staroverov, & Scuseria, 2003). So-called hybrid DFT methods involve a fraction of exchange from the HF method, eg, B3LYP (Becke, 1993; Lee, Yang, & Parr, 1988). In general, DFT methods give results of MP2 quality or better at a lower computational effort.

All mentioned QM methods expand the wavefunction in a set of known functions, the basis set. In general, the larger the basis set, the more accurate will the results be. Reasonable structures can be obtained with one basis function for the core electrons, two for the valence electrons, and one set of basis function with an angular momentum one step higher than for the valence electrons for nonhydrogen atoms. This is called a polarized split-valence basis set (SVP), eg, def2-SV(P) (Weigend & Ahlrichs, 2005). Energies normally require at least three basis functions for the valence electrons (valence triple-zeta basis sets, VTZ), eg, def2-TZVP and the convergence should be checked with four basis functions for the valence electrons (valence quadruple-zeta basis sets), eg, def2-QZVP (Weigend & Ahlrichs, 2005). Anions require the use of more diffuse basis functions, eg, def2-QZVPD (Rappoport & Furche, 2010).

QM calculations can be sped up by a factor of $\sim$1000 by using a minimal basis set and replacing all integrals by empirical parameters, the semiempirical QM methods (SEQM; Thiel, 2014). Many such methods have been suggested, eg, AM1 (Dewar, Zoebisch, Healy, & Stewart, 1985), PM3 (Stewart, 1989), and PM6 (Stewart, 2007), but the accuracy is typically lower than for QM and DFT methods and somewhat unpredictable, especially for unusual electronic structures, such as transition states.

Neither HF nor DFT calculations provide a proper description of the London dispersion interactions. Several approaches have been suggested to correct this problem (Grimme, 2011), but Grimme's DFT-D2 and D3 methods are most widely used (Grimme, 2006; Grimme, Antony, Ehrlich, & Krieg, 2010).

QM calculations are normally performed on isolated molecules in gas phase. However, most experiments are performed in condensed phases, eg, water solution. Methods have been developed to model a surrounding homogeneous solvent in QM calculations, continuum solvation methods (Tomasi, Mennucci, & Cammi, 2005), eg, the polarizable continuum model (PCM) (Cossi, Tomasi, & Cammi, 1995), or the conductor-like solvent model (COSMO) (Klamt & Schüürmann, 1994). For protein-sized systems, methods based on the Poisson–Boltzmann (PB) equation or the generalized Born (GB) approach are more common (Bashford & Case, 2000; Honig, Sharp, & Yang, 1997).

2.2 MM Methods

In MM methods, no attempt is made to solve the Schrödinger equation and electrons are ignored. Instead, molecules are considered as a collection of balls, connected by springs. The energy of the system as a function of the coordinates is described by an empirical function, a force field. For proteins, it typically contains terms for the distortion of bonds, angles, and dihedrals, as well as the nonbonded exchange-repulsion, dispersion (van der Waals), and electrostatic interaction energies (Mackerell, 2004). Such an energy can be calculated for a whole protein in seconds, allowing for extensive sampling of the accessible phase space by molecular dynamics (MD) or Monte Carlo methods. The protein is typically solvated with several thousands of explicit water molecules and possibly counterions to provide a more realistic account of the surroundings. To mimic infinite systems, periodic boundary conditions are employed and long-range electrostatic interactions are often treated by Ewald summation (Darden, York, & Pedersen, 1993).

2.3 QM/MM Calculations

The philosophy behind the QM/MM approach is that a QM method is used for a small, but interesting, part of the protein, eg, the active site, whereas the remainder of the protein as well as the surrounding solvent is treated by a MM method (Fig. 1B). This is intended to combine the accuracy of the QM method with the speed of the MM method. This is accomplished by adding QM and MM energies and forces, avoiding double-counting any interactions.

There are many ways to obtain a valid QM/MM energy function (Senn & Thiel, 2009). The simplest approach is:

$$E_{ME}^{QM/MM} = E_1^{QM} + E_{12}^{MM} - E_1^{MM} \tag{1}$$

where E_1^{QM} is the QM energy of the QM system (subsystem 1), E_1^{MM} is the MM energy of the same system, and E_{12}^{MM} is the MM energy of all atoms. Such an energy can be obtained with any combination of QM and MM programs without any change in the code and it is employed in the ONIOM approach (Svensson et al., 1996).

Eq. (1) indicates that all interactions between the QM and MM systems are treated at the MM level. This is appropriate for van der Waals interactions, which are hard to describe accurately with QM, but it is more questionable for electrostatic interactions, which often provide the dominating catalytic effect. Alternatively, the electrostatic interactions between the QM and MM systems can be treated with QM by including the MM point charges in the QM calculation (E_{1,q_2}^{QM}) and turning off the corresponding interactions in the MM calculations by zeroing the charges in the QM system ($E_{12,q_1=0}^{MM}$ and $E_{1,q=0}^{MM}$) (Ryde, 1996):

$$E_{\text{EE}}^{\text{QM/MM}} = E_{1,q_2}^{\text{QM}} + E_{12,q_1=0}^{\text{MM}} - E_{1,q=0}^{\text{MM}} \tag{2}$$

This is called electrostatic (electronic) embedding (EE), in contrast to the mechanical embedding (ME) in Eq. (1). It allows the QM system to be polarized by the MM charges, but the MM system is not polarized by the MM charges. The most accurate approach is to allow both the QM and MM systems to be simultaneously polarized (polarized embedding) (Poulsen, Kongsted, Osted, Ogilby, & Mikkelsen, 2001; Söderhjelm, Husberg, Strambi, Olivucci, & Ryde, 2009), but this requires a polarizable MM potential and a QM program that allows for the inclusion of MM polarizabilities.

If there are covalent bonds between the QM and MM systems, so-called junctions, special action is needed. First, the QM system needs to be truncated in a proper way. This can be done in two different ways. One is to truncate the QM system with hydrogen atoms (as in the QM-cluster approach; the H-link-atom approach) or with parametrized boundary atoms (Senn & Thiel, 2009; Von Lilienfeld, Tavernelli, & Rothlisberger, 2005; Zhang, 2006). The alternative is to use frozen localized orbitals at the boundary (Warshel & Levitt, 1976). The localized self-consistent field, suggested by Rivail and coworkers (Ferré, Assfeld, & Rivail, 2002; Théry, Rinaldi, Rivail, Maigret, & Ferenczy, 1994) and later extended and parametrized by Friesner and coworkers (Philipp & Friesner, 1999), places the frozen orbital on the last QM atom, whereas the generalized hybrid orbital method by Gao and coworkers instead introduces a boundary atom with a localized orbital (Gao, Amara, Alhambra, & Field, 1998; Pu, Gao, & Truhlar, 2005). In theory, the localized orbitals should give the more accurate results when properly parametrized, but in practice the two approaches typically give similar results (Nicoll, Hindle, MacKenzie, Hillier, & Burton, 2001). The localized orbital approaches require specialized QM software, whereas the H-link-atom approach can use any QM software. Therefore, most QM/MM studies use the latter approach (Ryde, 2003).

It should be recognized that junctions provide a severe problem for the QM/MM calculations. The H-link-atom approach introduces an incorrect atom at an erroneous position. These problems are further reinforced in EE approaches by the presence of nearby point charges, which give rise to strong artificial interactions (Hu, Söderhjelm, et al., 2011). Therefore, many schemes have been suggested to remove or redistribute charges around the junctions (König, Hoffman, Trauenheim, & Cui, 2005; Lin & Truhlar,

2005; Sherwood et al., 1997). However, we did not find any consistent improvement by any of these approaches in a large-scale test, except that the charge on the atom that is converted to the link atom should be excluded (Hu, Söderhjelm, et al., 2011). It has also been noted that all problems with the link atoms can be corrected at the MM level (Rod & Ryde, 2005b; Vreven et al., 2006). However, in practice, such corrections do more harm than good if the junctions are close to the reactive center (Hu, Söderhjelm, et al., 2011).

ME approaches are not affected by these problems because the electrostatic interactions are evaluated at the MM level in the full system without any junctions. However, they typically employ some sort of QM charges, which have been obtained with link atoms and therefore have to be adapted to the real system in some more or less arbitrary way (Hu, Söderhjelm, et al., 2011). Alternative ME approaches have been developed, eg, with QM charges from a polarized wavefunction, with a proper treatment of polarization, and avoiding instabilities in the determination of QM charges, which give significantly better results than standard ME and EE approaches, especially when the junctions are close to the reactive system (Hu, Söderhjelm, et al., 2011).

Finally, it has often been noted that EE approaches give rise to an overpolarization of the QM system. This has been attributed to the charges close to the link atoms (Senn & Thiel, 2009), but it is also caused by the inconsistent treatment of polarization (the QM system is polarizable, but not the MM system; Hu, Söderhjelm, et al., 2011). Several groups have suggested that the van der Waals parameters around the junctions should be modified (Freindorf, Shao, Furlani, & Kong, 2005; Murphy, Philipp, & Friesner, 2000; Riccardi, Li, & Cui, 2004). However, it seems dangerous to correct a problem in the electrostatics by modifying van der Waals parameters. In practice, the problems are best solved by minimizing the number of junctions and moving them as far away as possible from the reactive site (Hu, Söderhjelm, et al., 2011).

One of the greatest advantages with the QM/MM calculations is that they allow for a thorough interpretation and understanding of the results. Once it has been shown that the QM/MM calculations give reasonable results that reproduce key experimental findings, an analysis of the QM/MM results and components can show how the surroundings affect the structures and reaction energies. Moreover, such an analysis also gives an indication whether the results are reliable and therefore should always be performed. This is thoroughly illustrated by the application in Section 3.1.

The QM/MM approach was introduced by Warshel and Levitt (1976) for energies and by Singh and Kollman (1986) for geometry optimization. QM/MM programs are now available in many QM or MM software, eg, Gaussian (ONIOM; Svensson et al., 1996), Q-site (Murphy et al., 2000), AMBER (Götz, Clark, & Walker, 2014), and CHARMM (Riahi & Rowley, 2014), and also in some independent QM/MM interfaces that combine various QM and MM software, eg, ChemShell (Sherwood et al., 2003) and ComQum (Ryde, 1996; Ryde & Olsson, 2001). In QM/MM applications, it is important to have a versatile MM software with the opportunity of doing MD simulations for structure and solvent equilibration and also with utilities to parameterize nonstandard molecules in the protein—typically the MM calculations are much harder to set up than the QM calculations because they rely on parameterized empirical potentials.

2.4 The Big-QM Approach

One of the prime problems of both the QM-cluster and QM/MM approaches is that the selection of the QM system may bias the results, ie, that if the QM system is extended with more groups, the energies may change significantly. Himo and coworkers have shown that QM-cluster results may depend on the size of the QM system and also the value of the dielectric constant used for the continuum solvation model. However, when 150–200 atoms have been included, the results typically become essentially independent of the dielectric constant, which has been used as a convergence criterion (Hopmann & Himo, 2008; Sevastik & Himo, 2007). On the other hand, we have shown that even with QM systems of 400 atoms, the energies may be unstable: We obtained a difference of over 50 kJ/mol if the QM residues were selected according to their distance to the active site or by QM/MM energy components (Hu, Eliasson, Heimdal, & Ryde, 2009; Sumner et al., 2013). Similar differences have also been found in other studies, eg, a 45 kJ/mol difference in QM-cluster calculations with 300 and 1800 atoms (Sumowski & Ochsenfeld, 2009) and 55 or 27 kJ/mol differences between QM systems of 27 or 135 and 220 atoms, respectively (Liao, Yu, & Himo, 2011).

Several studies have indicated that QM/MM calculations converge faster than QM-cluster calculations with respect to the size of the QM system (Flaig, Beer, & Ochsenfeld, 2012; Sumowski & Ochsenfeld, 2009). On the other hand, other investigations have shown that also QM/MM energies strongly depend on the size of the QM system (Tian, Strid, &

Eriksson, 2011). In fact, we showed that the convergence of QM/MM calculations for a model of [NiFe] hydrogenase was not much better than QM-cluster calculations unless specialized ME approaches were used (Hu, Söderhjelm, et al., 2011).

Our convergence studies of QM-cluster and QM/MM calculations gave clear indications of what groups need to be included in the QM system to give converged results (Hu et al., 2009; Hu, Söderhjelm, et al., 2011): Neutral groups contribute significantly to reaction energies only when within 4.5 Å of the reactive system. Charged groups, on the other hand, influence the energies even at distances of 16 Å. However, it is known from experiments that solvent-exposed charges do not affect pK_a values and other energies (André, Kesvatera, Jönsson, Åkerfeldt, & Linse, 2004). Moreover, we have already emphasized that junctions should be moved away from the reactive site. Based on these results, we suggested the big-QM approach to obtain stable energies for protein reactions (Hu et al., 2013). In this approach, all chemical groups within 4.5–6 Å of the minimal QM system, all buried charged groups in the protein, and two capped amino acids around each residue in the minimal QM system are included in the QM calculations (Fig. 1C). This typically amounts to 600–1000 atoms, for which single-point energies easily can be calculated at the DFT level with SVP or even VTZ basis sets. Liao and Thiel (2013) have also argued for large QM systems (408–657 atoms), selected by a charge deletion approach (ie, single-point QM/MM calculations with the charges of one MM residue deleted).

The big-QM energies can be calculated in vacuum, with a point-charge model, or in a continuum solvent. The former can then be corrected by an ME-QM/MM term, according to Eq. (1), whereas the EE approach (Eq. 2) can be used for the energy obtained with the point-charge model. In practice, all three approaches give similar results, within 14 kJ/mol (Hu et al., 2013). A point-charge model typically gives the fastest convergence in the QM calculations and is therefore recommended, especially as all junctions are far from the reactive system.

The effect of the size of the QM system in the geometry optimizations has also been examined (Sumner et al., 2013). It was shown that QM/MM optimizations give much more stable and reliable structures than QM-cluster optimizations with various sets of fixed atoms. The raw QM/MM energies showed a quite large variation with the size of the QM system (up to 70 kJ/mol), but converged after the inclusion of 6–13 residues. The big-QM energies, on the other hand, were very stable, with variations of less than 15 kJ/mol, mainly depending on residues involving junctions. Therefore, QM/MM optimizations can employ quite small QM systems if energies are evaluated by the big-QM approach.

2.5 QM/MM Free-Energy Calculations

The third serious issue with QM/MM calculations comes from the local-minima problem. A minimization typically converges to the closest local minimum, which is not necessarily the global minimum of the system. There are no optimization methods that always find the global minimum (except an exhaustive systematic search). In practice, it is normally not necessary that all groups reside in their global minimum, but it is essential that all distant groups remain in the same local minimum throughout a reaction sequence; otherwise the energy will be blurred by irrelevant energy components. For example, the total energy may change by $\sim$20 kJ/mol if a single water molecule at the periphery changes its hydrogen-bond pattern. For small QM-cluster models, you can see by the eye whether all states in a reaction mechanism belong to the same local minimum. However, for a solvated protein, this is impossible—there are thousands of atoms and essentially an infinite number of local minima.

This is a very serious problem for QM/MM calculations and several approaches have been suggested to detect and avoid it (Senn & Thiel, 2009). One way is to run calculations forth and back between each pair of states in a reaction mechanism until the energy differences are constant. This is tedious and time consuming, especially if the mechanism contains many states. Another way is to keep the relaxed MM system as small as possible. However, this introduces the risk that important changes in the surroundings are missed, in particular the dielectric relaxation of the surroundings in effect of changes in the charge distribution of the active site. Moreover, even a rather small part of the protein may have many local minima.

A more common approach is to perform a MD simulation and then calculate QM/MM structures for several (3–10) snapshots from the simulations (Hu et al., 2013; Lonsdale, Harvey, & Mulholland, 2010; Metz & Thiel, 2009). This shows possible variations of the energies and indicates how certain the results are. However, it is not obvious how the results should be averaged to give reliable activation barriers (Cooper & Kästner, 2014).

The local-minima problem is caused by the fact that you consider minimized structures. The proper way to cure it is to calculate free energies, which involves sampling and averaging over all relevant thermally accessible structures. Several strict methods are available to obtain valid free energies, in particular free-energy simulations (FES), based on MD or Monte Carlo simulations and calculating the free energies by exponential averaging, thermodynamic integration, or Bennett acceptance ratio (Bennett, 1976; Kirkwood, 1935; Zwanzig, 1954). Such calculations are today routinely performed at the MM level, often giving quite accurate results (Hansen & van Gunsteren, 2014; Mikulskis, Genheden, & Ryde, 2014).

FES calculations can be performed also with QM/MM if an SEQM method is employed (Reddy & Erion, 2007; Senn & Thiel, 2009; Senn, Thiel, & Thiel, 2005). However, the rather poor accuracy of the SEQM methods makes such an approach risky, especially for transition states with their complicated electronic structure (Heimdal & Ryde, 2012).

QM/MM FES with DFT or high-level QM methods are rare and restricted to short simulation times owing to the large computational effort (Senn & Thiel, 2009; VandeVondele & Rothlisberger, 2002). Therefore, methods have been developed to avoid the costly QM/MM simulations. They are typically based on the thermodynamic cycle in Fig. 2. The aim is to estimate the free-energy difference between two states, A and B, at the QM/MM level, but this is computationally prohibitive. However, at the MM level this free energy can easily be obtained by FES methods. With these MM simulations, we can also estimate the free-energy difference of going from MM to QM/MM for the A and B states, completing the thermodynamic cycle. Such reference-potential methods were first suggested by Luzhkov and Warshel (1992) and they have been reinvented several times (Iftimie, Salahub, Wei, & Schofield, 2000; Wood, Yezdimer, Sakane, Barriocanal, & Doren, 1999), eg, the QM/MM thermodynamic cycle perturbation approach (QTCP; Rod & Ryde, 2005a, 2005b).

The problem with such an approach is that the MM → QM/MM perturbations need to converge in a single step if QM/MM simulations should

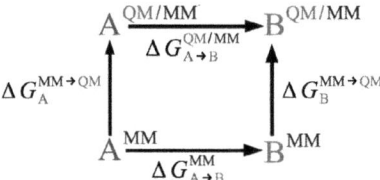

Fig. 2 The thermodynamic cycle that is the basis of the reference-potential methods and QTCP, showing that $\Delta G_{A \to B}^{QM/MM} = \Delta G_{A \to B}^{MM} - \Delta G_{A}^{MM \to QM} + \Delta G_{B}^{MM \to QM}$ (Luzhkov & Warshel, 1992; Rod & Ryde, 2005a).

be avoided, although FES studies (eg, the A→B perturbation at the MM level) typically need to be divided into 10–40 small steps to ensure proper convergence. This is normally solved by keeping the QM system fixed, which ignores differences in the QM and MM energy function for the internal degrees of freedom of the QM system (Rod & Ryde, 2005b). This means that the entropy of the QM system is omitted, but it can be estimated from QM frequencies of the isolated QM system. This is often a valid approximation for enzyme reactions, but it is more questionable when studying the binding of a ligand to a protein. Therefore, there has recently been quite some interest to obtain full QM/MM free energies with a flexible QM system, eg, by employing MD simulations at the SEQM/MM level or with a tailored MM potential (Heimdal & Ryde, 2012), or by employing the non-Boltzmann Bennett acceptance ratio (NBB) approach (König & Boresch, 2011). Recently, we have shown that converged QM/MM binding free energies require ~700,000 QM calculations if interaction energies are used and that exponential averaging with the cumulant expansion gives both a faster convergence and requires fewer QM calculations than NBB (Olsson, Söderhjelm, & Ryde, 2016).

Alternative approaches have been suggested, involving some restricted QM/MM MD simulations. In the paradynamics approach, one short QM/MM MD simulation is performed for each state and the free energy is estimated by the linear response approximation (Plotnikov, Kamerlin, & Warshel, 2011). Woods, Shaw, and Mulholland (2015) instead used the Metropolis–Hastings Monte Carlo approach to reduce the number of QM/MM calculations needed.

Several more approximate methods have also been developed to obtain QM/MM free energies. In the QM/MM-FE (free energy) approach (Hu & Yang, 2008; Zhang, Liu, & Yang, 2000) the MM→QM/MM perturbations are replaced by single-point MM→QM/MM extrapolations (ie, $\Delta E_{QM/MM} - \Delta E_{MM}$ for the QM/MM structures of the two states A and B). Even if this is a quite severe approximation, it has been shown to reproduce QTCP free energies within 5–9 kJ/mol for two sets of proton-transfer reactions (Kaukonen et al., 2008a). However, this approach is still quite time consuming as it involves FES at the MM level with extensive MD sampling.

The MM/PBSA (molecular mechanics combined with Poisson–Boltzmann and surface area solvation) is a popular approach to estimate binding free energies (Genheden & Ryde, 2015; Kollman et al., 2000). In this approach, free energies are estimated from internal, electrostatic, and van der Waals energies, calculated at the MM level, combined with PB or GB electrostatic continuum solvation energies, nonpolar solvation energies from the solvent-accessible surface area, as well as entropies from vibrational frequencies obtained at the MM level. Several groups have developed QM/MM-PBSA approaches, in which the MM energies are replaced by QM/MM calculations, keeping the solvation-free energies and entropies (Gräter, Schwarzl, Dejaegere, Fischer, & Smith, 2005; Kaukonen, Söderhjelm, Heimdal, & Ryde, 2008b; Retegan, Milet, & Jamet, 2009; Ryde & Söderhjelm, 2016). We have shown that such an approach reproduces QTCP proton-transfer reaction free energies within 7–13 kJ/mol even if only minimized QM/MM structures are used, which is appreciably better than the raw QM/MM energies (Kaukonen et al., 2008b). Such energies can be obtained from QM/MM structures with a minimal effort.

2.6 Combination with Experiments

QM/MM investigations are typically based on protein crystal structures. Unfortunately, crystal structures cannot be used directly to extract QM energies, owing to systematic errors in both the QM and experimental methods (Ryde et al., 2002). Therefore, most QM/MM projects involve reoptimization of crystal structures. Outer atoms are then typically fixed at the original crystal structure, whereas the active site and surrounding residues are freely optimized. Thereby, the latter atoms may move significantly away from the true structure, whereas the positions of the outer atoms may be biased by uncertainties in the crystal structure. A more satisfactory approach would be to restrain all atoms toward the experimental raw data, ie, the crystallographic structure factors.

We developed such a quantum-refinement approach in 2002 (Ryde et al., 2002). Standard crystallographic refinement is already in the form of an optimization in which the deviation between the experimental structure factors and those calculated from the current coordinates should be minimized (measured by the crystallographic R factor or more sophisticated statistical measures, E_{12}^{cryst}; Kleywegt & Jones, 1997). Moreover, the crystallographic data are typically supplemented by a MM-like energy function to ensure that bond lengths and angles are chemically reasonable:

$$E^{X\text{-refine}} = w_A E_{12}^{cryst} + E_{12}^{MM} \qquad (3)$$

The w_A weight factor is needed to because E_{12}^{cryst} typically is unit-less, whereas the MM term is in energy units. w_A gives the relative weight of the crystallographic and MM data and it is normally determined by requiring that the crystallographic and MM forces during a short MD simulation are of a similar magnitude (showing that standard crystallographic structures actually are 50% theoretical).

QM can be introduced into this function, by replacing the MM potential for a small, but interesting, part of the protein by QM calculations, giving the energy function (Ryde et al., 2002):

$$E^{\text{QM-refine}} = w_A E_{12}^{\text{cryst}} + E_1^{\text{QM}} + E_{12}^{\text{MM}} - E_1^{\text{MM}} \qquad (4)$$

We have shown that such an approach works well and that it can improve crystal structures locally, because DFT calculations give structures with a better accuracy than low- and medium-resolution crystal structures (Ryde, 2007; Ryde & Nilsson, 2003). Moreover, it can be used to deduce the protonation state of metal-bound ligands in crystal structures, although the protons are not explicitly seen (Nilsson & Ryde, 2004), or to discriminate between alternate structural interpretations of intermediates in the reaction mechanism (Heimdal, Rydberg, & Ryde, 2008; Hersleth, Ryde, Rydberg, Görbitz, & Andersson, 2006; Ryde, Hsiao, Rulíšek, & Solomon, 2007; Söderhjelm & Ryde, 2006). However, the application of the method to other metalloproteins has been hampered by the fact that metal sites are often partly reduced during the data collection, so that the raw data describes a mixture of structures (Rulíšek & Ryde, 2006; Söderhjelm & Ryde, 2006). Quantum refinement can detect such photoreduction, but no accurate models can be obtained. The method has recently been extended also to neutron crystallographic data (Manzoni, Oksanen, Logan, & Ryde, 2016).

A similar approach can be used also for NMR data: The raw data are different, but standard refinement still involves optimization of an energy function containing one term for the agreement between experimental data and the coordinates and a standard MM energy function (Cavanagh, Fairbrother, Palmer, & Skelton, 1996). Therefore, an energy function of the same type as the one in Eq. (4) can be used to introduce QM/MM calculations for an interesting part of the protein (Hsiao, Drakenberg, & Ryde, 2005). This was tested for two calcium-binding sites in protein S, with strongly improved structures compared to standard NMR refinement.

Finally, we have also developed methods to combine QM-cluster or QM/MM calculations with EXAFS (extended X-ray absorption fine structure) structures (Hsiao, Tao, Shokes, Scott, & Ryde, 2006). This was more complicated because the standard EXAFS approach gives only information of a few metal–atom distances. This approach has been used to study sitting-atom intermediates in the metallation of porphyrins, to decide the structure of the peroxide adduct in multicopper oxidases and of the oxygen-evolving complex in photosystem I (Hsiao & Ryde, 2006; Li, Siegbahn, & Ryde, 2015; Li, Sproviero, Ryde, Batista, & Chen, 2013; Ryde et al., 2007).

3. APPLICATIONS

In this section, I give a few examples from our QM/MM studies, which have led to our current QM/MM approach or illustrate some of the important aspects of QM/MM investigations.

3.1 Energy Components in Glutamate Mutase

In the first example, I want to illustrate how QM/MM calculations can be interpreted using energy components to give a complete understanding of an enzyme reaction and the catalytic effect (Jensen & Ryde, 2005). Glutamate mutase is a bacterial enzyme that catalyzes the conversion of glutamate to L-*threo*-methylaspartate (Marsh, 2000). It is a radical reaction, initiated by the coenzyme adenosyl cobalamine (AdoCbl), a vitamin B_{12} derivative. AdoCbl contains a Co ion in the center of a corrin ring. In the resting state, Co is in the +III state, forming a Co–C bond to the C5′ methylene group of ribose moiety of adenosine. This bond is relatively weak and can be homolytically cleaved to Co(II) and an adenosyl radical.

In gas phase, the calculated homolytic bond dissociation energies (BDE) of AdoCbl and the related coenzymes RibCbl and MeCbl (with ribose or a methyl group bound to Co) at the BP86/SVP level of theory are 143–156 kJ/mol (Jensen & Ryde, 2005), in reasonable agreement with experimental estimates of 126–155 kJ/mol (Hay & Finke, 1986; Martin & Finke, 1992). However, QM/MM calculations indicated that in the protein, the BDE is only 13 kJ/mol (Jensen & Ryde, 2005), in accordance with experimental estimates of an equilibrium constant close to unity (Padmakumar, Padmakumar, & Banerjee, 1997). Our aim is to understand this difference in BDE.

The BDE is calculated with respect to the infinitely separated products. However, in the protein, the two moieties cannot dissociate to a Co–C distance of more than ~3.5 Å. At this distance, there are still significant interactions between the two moieties and the BDE is only 122 kJ/mol. This has been termed the cage effect (Dölker, Maseras, & Siegbahn, 2004) and it explains 20 kJ/mol of the catalytic effect. The QM/MM energy difference between the Co(III) and Co(II) states in the protein is 8 kJ/mol.

The latter energy can be divided into the QM and MM energy components, according to Eq. (2), which amount to 33 and 25 kJ/mol, respectively. E_{1,q_2}^{QM} is the QM energy, including a point-charge model of the surroundings. The corresponding energy without the point-charge model is 67 kJ/mol and the difference (34 kJ/mol) represents the direct electrostatic effect of the surroundings. All these energies have been calculated for QM/MM structures. Using instead geometries optimized in vacuum, we get the gas-phase BDE, but still with a Co–C distances of 3.5 Å for the dissociated state, 122 kJ/mol. The difference (56 kJ/mol) represents the geometric effect of the protein. Consequently, we have shown that the catalytic effect of the protein (143−8=135 kJ/mol) can be divided into MM, electrostatic, geometric, and cage effects of 25, 34, 56, and 20 kJ/mol, respectively (Fig. 3).

Each of these effects can be further understood: All MM terms are pairwise decomposable, so they can be completely divided into contributions from the bonded, van der Waals, and electrostatic interactions between

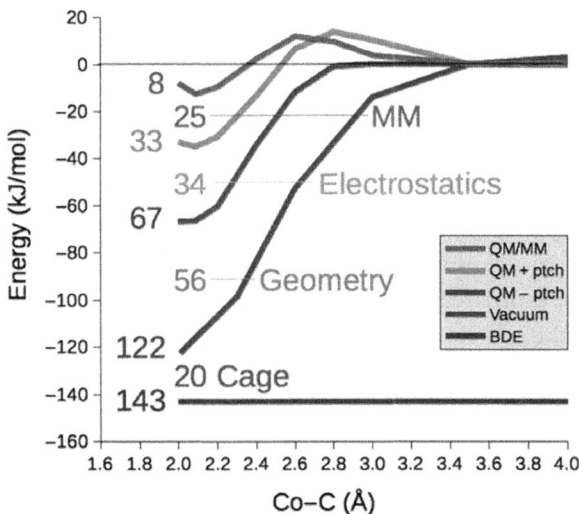

Fig. 3 Energy components for glutamate mutase (Jensen & Ryde, 2005). The *colored numbers* to the *left* are the energies for the five curves at Co–C=2.0 Å (using the energy at Co–C=3.5 Å as the reference). The *numbers* to the *right* are the pairwise differences between these numbers. All energies are in kJ/mol. The components are further explained in the main text. (See the color plate.)

two atoms (three or four atoms for angles and dihedrals). However, the number of terms rapidly becomes overwhelmingly large, so it is normally best to group them. For example, the MM term can be divided into contributions from the QM system (subsystem 1), the optimized part of the MM system (subsystem 2), and the fixed part of the MM system (subsystem 3). The pairwise contributions from these three subsystems are shown in Fig. 4. All terms are dominated by the electrostatic and van der Waals contributions. It can be seen that three of these contributions are negligible: 1–3 (owing to the large distance between these two subsystems), 2–3 (by chance), and 3–3 (because subsystem 3 is fixed). The other three terms are rather small and of a similar magnitude, 5–11 kJ/mol. This is good because it is in these terms (especially 2–2) the local-minima problem is expected. Therefore, the MM term should always be investigated and if it is large, its components should be thoroughly checked.

The direct electrostatic term is normally the dominating effect of the protein. It comes from QM calculations, making it somewhat harder to interpret. It can be residue-wise decomposed by repeating the QM calculation with the point charges of one residue deleted (Liao & Thiel, 2013). However, this is somewhat tedious and time consuming, and it still ignores the polarization effects (ie, all residue components will not add up to the total effect). Normally, the electrostatic effects are similar if instead ME-QM/MM is used, ie, if the electrostatic effects are calculated with MM (Eq. 1) with a QM charge model of the QM system (Ryde, 1996). Then, the electrostatic components can be obtained in a single (cheap)

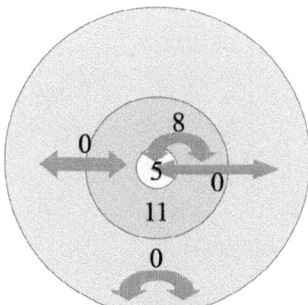

Fig. 4 MM energy components in kJ/mol (Jensen & Ryde, 2005). The *yellow* (*white* in the print version), *green* (*light gray* in the print version), and *cyan* (*light gray* in the print version) *circles* represent the QM, relaxed MM, and fixed MM systems, respectively, and the energy components are the energies within or between the various systems, as illustrated by the *arrows*.

calculation. Fig. 5 shows the major residue contributions from such an analysis of glutamate mutase, indicating that it is mainly charged residues around the adenosyl moiety that affect the reaction energies.

The geometric term is unusually large in this enzyme. It can be interpreted by calculating strain energies in different parts of the QM system, ie, the energy difference between the molecule optimized in vacuum and in the protein. As can be seen in Fig. 6, the strain energy is dominated by

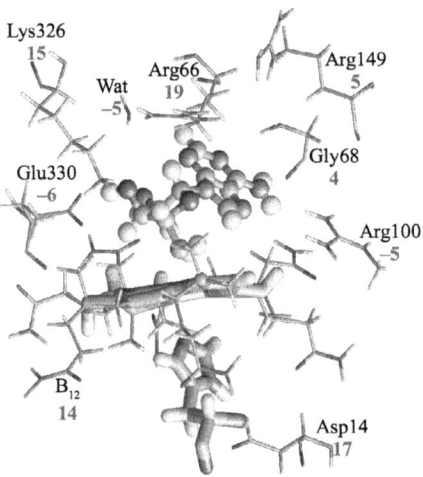

Fig. 5 Residues with large electrostatic energy components (*numbers* in *green* (*gray* in the print version); kJ/mol) for the catalytic effect of glutamate mutase (Jensen & Ryde, 2005). The axial His ligand and the corrin ring are shown in *thick sticks*, whereas the adenosine moiety is shown with *balls and sticks*.

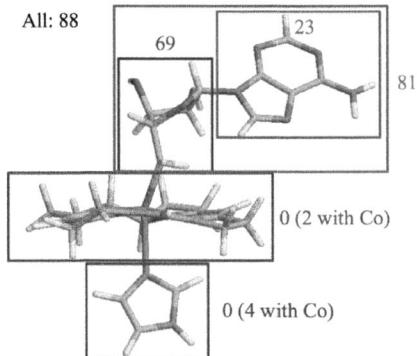

Fig. 6 Strain energy in various moieties of the AdoCbl coenzyme in kJ/mol (Jensen & Ryde, 2005).

contributions from the adenosyl moiety and especially the ribose ring. Further calculations with systematically removed constraints to the QM/MM structure show that it is especially the Co–C–C angle that is constrained, by ~40 kJ/mol.

Next, we can turn the attention to the surrounding protein and ask how it constrains the geometry. By repeating the QM/MM optimization with zeroed charges of the protein, we showed that the electrostatic effect on the geometry is rather small, ~24 kJ/mol (Jensen & Ryde, 2005). On the other hand, if also the van der Waals contributions were turned off, the catalytic effect of the protein almost disappeared. Finally, the residues that are most important for the geometric effect could be determined by repeating the QM/MM optimizations with some residues around the active site mutated to Gly. The results in Fig. 7 show that the effect comes mainly from two residues (Lys326 and Glu330) and one of the side chains of the cobalamine coenzyme, although the effects are strongly cooperative. All important groups form hydrogen bonds to the ribose moiety of the coenzyme, showing that this part is needed as a handle. This was also confirmed by repeating the QM/MM calculations with the MeCbl and RibCbl coenzymes—the latter still gave a catalytic effect of 109 kJ/mol, whereas it was reduced to only 42 kJ/mol for MeCbl.

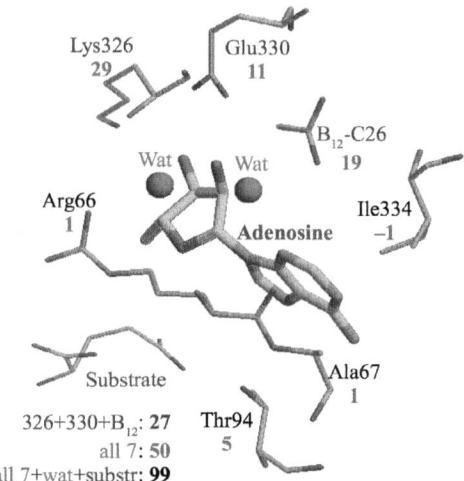

Fig. 7 Decrease in catalytic effect when various residues are mutated to Gly in kJ/mol (Jensen & Ryde, 2005). The cooperative effect of simultaneously mutating three or the seven shown residues is given in the *lower left corner*, together with the effect of mutating the same seven residues and deleting two water molecules and the substrate (also shown in the figure).

Finally, it should be noted that several aspects of this reaction are controversial, both regarding what DFT functional should be used (Jensen & Ryde, 2003; Kozlowski et al., 2012; Siegbahn, Blomberg, & Chen, 2010) and the importance of geometric vs electrostatic effects (Jensen & Ryde, 2009; Kwiecien, Khavrutskii, Musaev, Morokuma, Banerjee, & Paneth, 2006; Román-Meléndez, von Glehn, Harvey, Mulholland, & Marsh, 2014; Sharma, Chu, Olsson, & Warshel, 2007). The prime problem is that there are no reliable crystal structures with an intact Co(III)–C bond, owing to photoreduction of the structures. Still, this example nicely shows that QM/MM results are not restricted to geometries and energies; much more information can be obtained from the QM/MM energy components and additional test calculations.

3.2 Mo Oxo-Transfer Enzymes

The second application emphasizes the accuracy of the QM calculations and how it depends on the size of the QM system. A large number of oxo-transfer enzymes employ molybdenum in the active site. For example, DMSO reductase converts dimethyl sulfoxide (DMSO) to dimethyl sulfide and the active-site Mo ion is concomitantly oxidized from Mo(IV) to Mo(VI)=O (Hille, Hall, & Basu, 2014). At least six research groups have performed QM-cluster calculations on this enzyme and all have agreed on a two-step reaction mechanism with an activation enthalpy of 68–80 kJ/mol (Ryde, Dong, Li, Andrejic, & Mata, 2016), ie, close to the experimental estimate of 63 kJ/mol (Cobb, Conrads, & Hille, 2005).

However, we found that the reaction is unusually sensitive to the QM method (Li, Andrejic, Mata, & Ryde, 2015; Li, Mata, et al., 2013). The activation enthalpy increased by 74 kJ/mol if calculated with TZP instead of SVP basis sets, whereas a further increase to QZP changed it by 9 kJ/mol. The effect of dispersion for the QM system (estimated with DFT-D2) was large, eg, 33 kJ/mol for the activation energy. Moreover, this energy needs to be balanced by the nonpolar solvation energies (which also contain a dispersion component), especially when a molecule binds or dissociates from the active site (Ryde, Mata, & Grimme, 2011). Unfortunately, such terms are problematic for enzyme active sites, for which it has to be decided whether there is a cavity before the ligand binds and if it is filled with solvent or not. These terms also contribute by ~38 kJ/mol to the activation enthalpy.

Furthermore, different DFT functionals gave widely different activation and reaction energies, owing to the change in the oxidation states of Mo and the substrate: The activation enthalpy was low (23–34 kJ/mol) with pure functionals, but much higher with hybrid functionals, 46–139 kJ/mol, increasing with the amount of HF exchange. To know what results to trust, we performed LCCSD(T0) calculations, extrapolated to the complete-basis-set limit (CBS). These showed that no single functional gave reliable results—instead B3LYP gave the best activation enthalpy, whereas the pure functionals gave better results for the other states. The DFT results were also much closer to the LCCSD(T0) reference if a DFT-D correction was included, showing that dispersion corrections clearly improve DFT results. Our final results, based on the LCCSD(T0)/CBS energies and including relativistic, zero-point-energy, thermal, and solvation corrections (both polar and nonpolar terms), reproduced experimental results both for the enzyme reaction and for an inorganic model (Li, Mata, et al., 2013; Li, Sproviero, et al., 2013).

Interestingly, our calculations gave nearly the same activation enthalpy as the previous computational studies, although our estimate included dispersion and nonpolar solvation corrections of 71 kJ/mol that were not considered before. The reason is that the previous studies either used a method (BP86) or a basis set (SVP) that gave a too low barrier. Thus, all previous studies reproduced the experimental results for the wrong reason. On the other hand, they gave widely different reaction energies (−50 to −131 kJ/mol) because no experimental results were available for this energy.

A similar strong method dependence was obtained also for the sulfite oxidase reaction $(SO_3^{2-} + Mo^{6+}=O \rightarrow SO_4^{2-} + Mo^{4+})$, which we solved with LCCSD(T0)/CBS calculations and the same type of energy corrections (Van Severen et al., 2014). However, in this case QM-cluster calculations gave a too high activation energy even in a water-like continuum solvent, owing to extensive Coulombic repulsion between the negatively charged active site and substrate. In the protein, this is compensated by three Arg residues. This is best studied by QM/MM methods enzyme (Caldararu, Andrejic, Ciloboc, Mata, & Ryde, 2016). It turned out that a large QM system was needed to obtain stable structures and energies, including the three Arg residues, as well as five additional groups and two water molecules (164 atoms; Fig. 8A). With such a model, a reasonable activation barrier could be obtained (∼60 kJ/mol). Moreover, this large and carefully selected QM system gave almost identical energies (within 4 kJ/mol) to those of an 805-atom big-QM system (Fig. 8B).

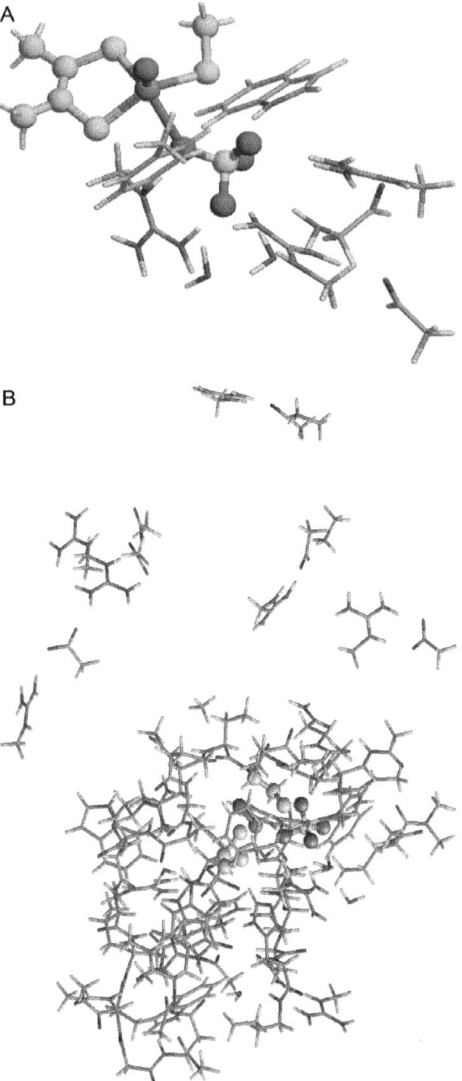

Fig. 8 The QM system needed for a proper QM/MM study of sulfite oxidase (A) and the corresponding big-QM system (B; Caldararu et al., 2016).

These studies show several important aspects of QM studies of proteins:
- Always test both pure and hybrid DFT functionals for your reaction of interest and use LCCSD(T0)/CBS or similar methods as a reference if the results differ.
- Basis sets of QZP quality should be used to ensure convergence of the energies.

- DFT energies should always be supplemented by dispersion corrections.
- Corrections for relativistic, zero-point-energy, thermal, and solvation effects should always be included.
- Solvation energies should include also the nonpolar terms, especially when molecules bind or dissociate. These terms are quite problematic in QM-cluster calculations of an enzyme active site.
- The QM system in QM/MM studies should be selected to give stable structures and energies, including all significant movements of the surroundings. Such a properly chosen QM system often gives energies close to those obtained with the big-QM approach.

3.3 Blue Copper Proteins and Multicopper Oxidases

Blue copper proteins are small electron carriers. They contain a Cu site, typically with one Cys, two His, and possibly one or two additional weaker ligands, Met or a carbonyl group (cf. Fig. 9A). The structure is normally trigonal with the three strong ligands in an approximate plane and 0–2 axial ligands. The Cu ion alternates between the +I and +II oxidation states, but both states give similar structures, mainly owing to the properties of the Cu–Cys bond (Ryde, Olsson, & Pierloot, 2001). On the other hand, the redox potential can vary quite extensively, from 0.18 to over 1 V (Liu et al., 2014). Many computations studies of the redox potentials of blue copper proteins have been published (Hadt et al., 2012; Olsson, Hong, & Warshel, 2003; Van den Bosch et al., 2005).

It is extremely demanding to obtain stable QM/MM redox potentials for metal sites in proteins, as is illustrated by the results in Table 1 (M.-C. Van Severen, M. Kaukonen, & U. Ryde, unpublished results): QM/MM potentials depend strongly on the oxidation state for which the MM system is optimized with differences of 1.5 V, even when averaged over five independent structures. Fortunately, this variation is reduced to only 0.1 V with QTCP. On the other hand, the results then depend on the treatment of long-range solvation effects, outside the simulated system (40 Å radius): Born/Onsager, Ewald, PB, and GB give results that vary from -2.1 to $+0.7$ V. This variation can be reduced to 0.2–0.8 V by neutralizing all charged groups on the protein surface. However, the largest problem is that the results also depend on the size of the QM systems: For example, we got average redox potentials at the QM/MM level of 1.3, -1.0, 1.2, 0.5, and 0.0 V using increasingly larger QM systems of 33, 124, 170, 197, and 577 atoms (shown in Fig. 9). Apparently, it is very hard to get converged QM/MM absolute redox potentials. Another problem is that different DFT methods (in particular pure and

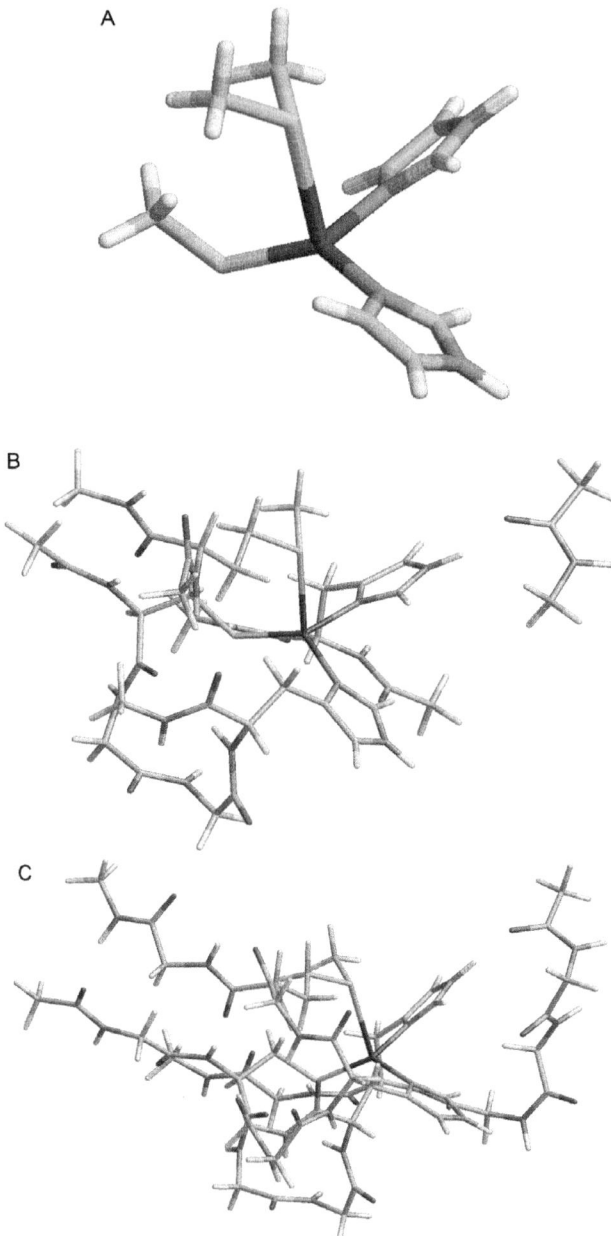

Fig. 9 QM systems of five sizes for plastocyanin: (A) minimal model, (B) junctions moved one residue away and including hydrogen bonds to the first-sphere ligands, (C) junctions moved two residues away, (D) junctions moved three residues away, and (E) a big-QM model.

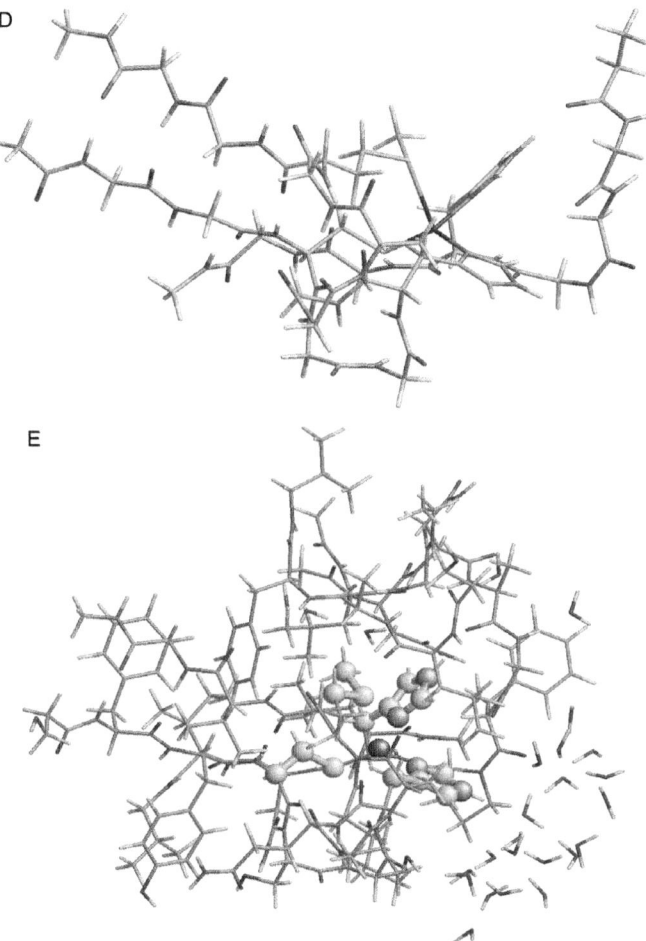

Fig. 9—Cont'd

hybrid functionals) often give redox potentials that differ by 50–100 kJ/mol (Chen, Liu, VandeVondele, Sulpizi, & Sprik, 2014; Heimdal et al., 2011; Li, Mata, et al., 2013).

The former problem can partly be reduced by not calculating absolute potentials, but instead study the internal electron transfer within the protein. Such calculations require the extension of QM/MM to the use of two separate QM systems (Blumberger, 2010; Hu, Farrokhnia, et al., 2011). Such an approach was tested for the multicopper oxidases, which contain both a blue copper site and a trinuclear Cu cluster, in which O_2 is reduced to water. We studied the transfer of electrons between these two sites (~14 Å distance) for

Table 1 The Redox Potential ($E°$ in V) of Plastocyanin, Calculated with Either QM/MM or QTCP (the Latter with Four Different Methods to Treat Long-Range Electrostatics), with Different QM Systems (a–e, Shown in Fig. 9), and with the Protein Surface Charges Neutralized or not

QM System	Method	Neutral Protein	$E°$	$\Delta E°$
a	QM/MM	No	−3.18	1.50
a	QTCP/Born	No	−2.10	0.12
	QTCP/Ewald		0.70	0.07
	QTCP/GB		0.46	0.02
	QTCP/PB		−1.18	0.08
a	QTCP/Born	Yes	0.23	0.32
	QTCP/Ewald		0.76	0.17
	QTCP/GB		0.57	0.31
	QTCP/PB		0.46	0.29
a	QM/MM	Yes	1.28	2.00
b	QM/MM	Yes	−0.95	1.48
c	QM/MM	Yes	1.24	1.00
d	QM/MM	Yes	0.49	0.95
e	QM/MM	Yes	0.02	1.81

$\Delta E°$ is the difference in reduction potentials of structures with the MM system optimized for the reduced or the oxidized state. All results are averages over 10 QM/MM structures (Van Severen et al., unpublished data).

several putative intermediates in the O_2 reduction cycle with both QTCP and QM/MM-PBSA (Li, Farrokhnia, et al., 2015). The results still showed some variation with the continuum solvation model, but the QTCP results obtained with a Born/Onsager term or with Ewald summation gave consistent results. Since the redox potential of the blue copper site in this protein is known from experiments, the absolute potentials of the trinuclear site can be estimated. Moreover, the results showed a significant communication between the two redox sites. For example, the redox potential of the blue copper site varied by at least 0.18 V depending on the state of the trinuclear Cu site.

Similar methods can be used also to estimate acidity constants in proteins. Again, calculations of absolute potentials are extremely hard to converge. However, more stable estimates can be obtained by studying the proton transfer to a solvent-exposed reference site, especially if solvent-exposed charges are neutralized (Li, Farrokhnia, et al., 2015). The results showed some dependence on the choice of the reference site (± 4 pK_a units), but this dependence disappears if the relative acidity is considered. QM/MM-PBSA and -GBSA results differed by over 20 pK_a units, but the QTCP results were quite stable and intermediate between the former two results. By considering the acidity constants of the various putative intermediates, the most stable structure at pH 7 could typically be determined. Combining these results with redox potentials, isomerization energies as well as O_2-binding and water-dissociation energies allowed for the suggestion of a full reaction cycle of the multicopper oxidases. In most cases, the various states were well separated in energies, allowing for a reliable prediction of the most stable states, even if the accuracy of the QTCP predictions was not better than $\sim$30 kJ/mol.

These applications have illustrated the problems encountered when the net charge of the simulated system changes and how they can be solved or reduced by studying only internal electron or proton transfer.

4. SUGGESTED APPROACH FOR QM/MM INVESTIGATIONS

Based on the methods and examples described in the previous sections, as well as our other investigations of protein structure, function, and mechanisms (Delcey et al., 2014; Dong & Ryde, 2016; Hedegård et al., 2015; Heimdal et al., 2011; Hu, Farrokhnia, et al., 2011; Hu et al., 2013; Kaukonen et al., 2008a, 2008b; Li, Farrokhnia, et al., 2015; Rod & Ryde, 2005a, 2005b; Söderhjelm & Ryde, 2006; Sumner et al., 2013), I suggest the following approach for a QM/MM investigation of a protein:

1. Optimize the geometries of the various intermediates and transition states with EE-QM/MM, using a quite small QM system, a pure functional with DFT-D3 dispersion corrections, and an SVP basis set, eg, TPSS-D3/def2-SV(P). This should be done with both the MM system fixed at the crystal structure and the MM system relaxed.
2. Calculate single-point QM energies with the same QM system and a point-charge model using a QZP basis set and both pure and hybrid functionals (eg, B3LYP/def2-QZVP). If the reaction involves anions, diffuse basis sets should be used.
3. Run big-QM calculations with a pure functional and an SVP or (if it can be afforded) a TZP basis set (we typically use TPSS/def2-TZVP), using a point-charge model of the surroundings. To this energy, the

MM correction in Eq. (2) and a DFT-D3 correction should be added, the latter using Becke–Johnson damping (Grimme, Ehrlich, & Goerigk, 2011) and including three-body terms (Sure & Grimme, 2015). The big-QM system should include all residues within 4.5–6 Å of the minimal QM system, all buried charged groups in the protein, as well as two capped residues around each residue of the minimal QM system.

4. Calculate QTCP free energies at the same level and with the same QM system as the geometry optimization. These calculations employ structures with a fixed MM system.

5. Calculate thermal and entropy corrections for the isolated QM system, freely optimized in vacuum (the same QM system and method as in the geometry optimization). In parallel, the reaction should be fully characterized by a QM-cluster study, to get a feeling of the reaction and to estimate the catalytic effect of the protein.

6. If pure and hybrid DFT functionals give significant different results, high-level QM calculations, eg, LCCSD(T0)/CBS, should be used to decide which is most reliable.

7. If relaxation of the MM system significantly changes the structures and energies, the approach needs to be partly modified. The first step is to optimize the MM system for one state and employ this MM structure for all the other states. If the results still depend on whether the MM system is relaxed or not, or if the results depend on the state for which the MM system is relaxed, the QM system needs to be extended until the energies are stable.

8. Currently, our final energies are obtained from the big-QM energies (step 3), extrapolated with the single-point energies in step 2 or corrected with the LCCSD(T0) in step 6, and including zero-point energies, as well as thermal and entropy corrections from step 5, and a free-energy correction from the QTCP calculations in step 5 (Dong & Ryde, 2016). However, it would be more satisfactorily if the QM system in all calculations could be extended to the point that it reproduces the big-QM results. Then, the big-QM calculations would become superfluous (besides guiding the size of the QM system) and the energies could be based on the QTCP free energies (including the DFT-D3 correction in step 2), extrapolated to the large basis set, and including the thermal corrections from step 5.

9. The results should be checked and interpreted by considering various energy components: First, it is divided into MM, electrostatic, and

geometric components. Next, each of these components can be understood by further calculations, as was described in Section 3.1. EE and ME results should be compared, as well as the effect of relaxation of the protein.

10. Preferably, all these calculations should be repeated on a number of snapshots (3–10) from an MD simulation. The QTCP calculations are expected to reduce the starting-point dependence (cf. Table 1), but probably not completely.
11. If QTCP energies are hard to calculate (eg, for the binding of a ligand from the solution) or more approximate results can be accepted, QM/MM-PBSA energies can be employed (they are more stable and give a more accurate account of solvation effects than the raw QM/MM energies).
12. If you want to interpret a special state for which experimental X-ray, neutron, NMR, or EXAFS structural data are available, the combination of these methods and QM/MM geometry optimization is extremely powerful, as described in Section 2.6. Optimize all tentative structures and select the one that fits the experimental raw data best and also gives a small deviation in energy and structure from the vacuum-optimized structure.
13. Avoid studying reactions that change the net charge of the studied system. It is typically better to study internal electron or proton transfer, as described in Section 3.3. Further improvement can be obtained by neutralizing solvent-exposed residues. Still, such reactions give a lower accuracy than charge-conserving reactions.

By this approach, you solve or minimize the three most important problems with QM/MM calculations, the junctions (move them as far as possible from the reactive center), the size of the QM system (employ the big-QM approach), and the local-minima problem (calculate free energies with QTCP). Throughout this chapter, I have emphasized the importance of ensuring that the calculations are converged. The fact that some experimental data are reproduced does not guarantee that the calculations are reliable. Instead, you need to run several test calculations to ensure that the results can be trusted, as described in Section 3. Undoubtedly, QM/MM calculations are more complicated and error-prone than QM-cluster calculations. However, when carefully performed and tested, they can give accurate and reliable results for reactions in protein. In particular, they can be employed to understand the catalytic effect of the protein.

ACKNOWLEDGMENTS
This investigation has been supported by grants from the Swedish research council (Project 2014-5540) and from Knut and Alice Wallenberg Foundation (KAW 2013.0022). The computations were performed on computer resources provided by the Swedish National Infrastructure for Computing (SNIC) at Lunarc at Lund University and HPC2N at Umeå University.

REFERENCES
André, I., Kesvatera, T., Jönsson, B., Åkerfeldt, K. S., & Linse, S. (2004). The role of electrostatic interactions in calmodulin-peptide complex formation. *Biophysical Journal, 87*, 1929–1938.
Bashford, D., & Case, D. A. (2000). Generalized Born models of macromolecular solvation effects. *Annual Review of Physical Chemistry, 51*, 129–152.
Becke, A. D. (1988). Density-functional exchange-energy approximation with correct asymptotic behavior. *Physical Review A, 38*, 3098–3100.
Becke, A. D. (1993). Density-functional thermochemistry. III. The role of exact exchange. *The Journal of Chemical Physics, 98*, 5648–5652.
Bennett, C. H. (1976). Efficient estimation of free energy differences from Monte Carlo data. *Journal of Computational Physics, 22*, 245–268.
Blomberg, M. R. A., Borowski, T., Himo, F., Liao, R.-Z., & Siegbahn, P. E. M. (2014). Quantum chemical studies of mechanisms for metalloenzymes. *Chemical Reviews, 114*, 3601–3658.
Blumberger, J. (2010). Free energies for biological electron transfer from QM/MM calculation: Method, application and critical assessment. *Physical Chemistry Chemical Physics, 10*, 5651–5667.
Caldararu, O., Andrejic, M., Cioloboc, D., Mata, R. A., & Ryde, U. (2016). QM/MM study of the reaction mechanism of sulfite oxidase. *Journal of Biological Inorganic Chemistry*. submitted.
Cavanagh, J., Fairbrother, W. J., Palmer, A. G., & Skelton, N. J. (1996). *Protein NMR spectroscopy. Principles and practice*. London: Academic Press.
Chen, J., Liu, X., VandeVondele, J., Sulpizi, M., & Sprik, M. (2014). Redox potentials and acidity constants from density functional theory based molecular dynamics. *Accounts of Chemical Research, 47*, 3522–3529.
Cobb, N., Conrads, T., & Hille, R. (2005). Mechanistic studies of Rhodobacter sphaeroides Me$_2$SO reductase. *The Journal of Biological Chemistry, 280*, 11007.
Cooper, A. M., & Kästner, J. (2014). Averaging techniques for reaction barriers in QM/MM simulations. *Chemphyschem, 15*, 3264–3269.
Cossi, M., Tomasi, J., & Cammi, R. (1995). Analytical expressions of the free energy derivatives for molecules in solution. Application to the geometry optimization. *International Journal of Quantum Chemistry, 56*, 695–702.
Cramer, C. J. (2006). *Essentials of computational chemistry: Theories and models*. Chichester: J. Wiley & Sons.
Darden, T., York, D., & Pedersen, L. (1993). Particle mesh Ewald: An N-log(N) method for Ewald sums in large systems. *The Journal of Chemical Physics, 98*, 10089–10092.
Delcey, M. G., Pierloot, K., Phung, Q. M., Vancoillie, S., Lindh, R., & Ryde, U. (2014). Accurate calculations of geometries and singlet–triplet energy differences for active-site models of [NiFe] hydrogenase. *Physical Chemistry Chemical Physics, 16*, 7927–7938.
Dewar, M. J. S., Zoebisch, E. G., Healy, E. F., & Stewart, J. J. P. (1985). AM1: A new general purpose quantum mechanical molecular model. *Journal of the American Chemical Society, 107*, 3902–3909.

Dölker, N., Maseras, F., & Siegbahn, P. E. M. (2004). Stabilization of the adenosyl radical in coenzyme B 12—A theoretical study. *Chemical Physics Letters, 386*, 174–178.
Dong, G., & Ryde, U. (2016). Protonation states of intermediates in the reaction mechanism of [NiFe] hydrogenase studied by computational methods. *Journal of Biological Inorganic Chemistry, 21*, 2016, 383–394.
Ferré, N., Assfeld, X., & Rivail, J. L. (2002). Specific force field determination for the hybrid ab initio QM/MM LSCF method. *Journal of Computational Chemistry, 23*, 610–624.
Flaig, D., Beer, M., & Ochsenfeld, C. J. (2012). Convergence of electronic structure with the size of the QM region: Example of QM/MM NMR Shieldings. *Journal of Chemical Theory and Computation, 8*, 2260–2271.
Fock, V. A. (1930). An approximate method for quantum mechanical calculations. *Zeitsch. Phys., 62*, 795.
Freindorf, M., Shao, Y., Furlani, T. R., & Kong, J. (2005). Lennard–Jones parameters for the combined QM/MM method using the B3LYP/6-31G*/AMBER potential. *Journal of Computational Chemistry, 26*, 1270–1278.
Gao, J., Amara, P., Alhambra, C., & Field, M. J. (1998). A generalized hybrid orbital (GHO) method for the treatment of boundary atoms in combined QM/MM calculations. *The Journal of Physical Chemistry. A, 102*, 4714–4721.
Genheden, S., & Ryde, U. (2015). The MM/PBSA and MM/GBSA methods to estimate ligand-binding affinities. *Expert Opinion on Drug Discovery, 10*, 449–461.
Götz, A. W., Clark, M. A., & Walker, R. C. (2014). An extensible interface for QM/MM molecular dynamics simulations with AMBER. *Journal of Computational Chemistry, 35*, 95–108.
Gräter, F., Schwarzl, S. M., Dejaegere, A., Fischer, S., & Smith, J. C. (2005). Protein/ligand binding free energies calculated with quantum mechanics/molecular mechanics. *The Journal of Physical Chemistry. B, 109*, 10474–10483.
Grimme, S. (2006). Semiempirical GGA-type density-functional constructed with a long-range dispersion correction. *Journal of Computational Chemistry, 27*, 1787–1799.
Grimme, S. (2011). Density functional theory with London dispersion corrections. *WIREs Computational Molecular Science, 1*, 211–228.
Grimme, S., Antony, J., Ehrlich, S., & Krieg, H. (2010). A consistent and accurate ab initio parametrization of density functional dispersion correction (DFT-D) for the 94 elements H–Pu. *The Journal of Chemical Physics, 132*, 154104. 19 pages.
Grimme, S., Ehrlich, S., & Goerigk, L. (2011). Effect of the damping function in dispersion corrected density functional theory. *Journal of Computational Chemistry, 32*, 1456–1465.
Hadt, R. G., Sun, N., Marshall, N. M., Hodgson, K. O., Hedman, B., Lu, Y., et al. (2012). Spectroscopic and DFT studies of second-sphere variants of the Type 1 copper site in Azurin: Covalent and nonlocal electrostatic contributions to reduction potentials. *Journal of the American Chemical Society, 134*, 16701–16716.
Hansen, N., & van Gunsteren, W. F. (2014). Practical aspects of free-energy calculations: A review. *Journal of Chemical Theory and Computation, 10*, 2632–2647.
Hartree, D. R. (1928). The wavemechanics of an atom with a non-Coulomb central field. I. Theory and methods. *Proceedings of the Cambridge Philological Society, 24*, 89.
Hay, B. P., & Finke, R. G. J. (1986). Thermolysis of the cobalt–carbon bond of adenosylcobalamin. 2. Products, kinetics, and cobalt–carbon bond dissociation energy in aqueous solution. *American Chemical Society, 108*, 4820–4829.
Hedegård, E. D., Kongsted, J., & Ryde, U. (2015). Multiscale modelling of the active site of [Fe]-hydrogenase: The H_2 binding site in open and closed protein conformations. *Angewandte Chemie, 54*, 6246–6250.
Heimdal, J., Kaukonen, M., Srnec, M., Rulíšek, L., & Ryde, U. (2011). Reduction potentials and acidity constants of Mn superoxide dismutase calculated by QM/MM free-energy methods. *ChemPhysChem, 12*, 3337–3347.

Heimdal, J., Rydberg, P., & Ryde, U. (2008). Protonation of the proximal histidine ligand in haem peroxidases. *The Journal of Physical Chemistry. B, 112,* 2501–2510.

Heimdal, J., & Ryde, U. (2012). Convergence of QM/MM free-energy perturbations based on molecular-mechanics or semiempirical simulations. *Physical Chemistry Chemical Physics, 14,* 12592–12604.

Hersleth, H.-P., Ryde, U., Rydberg, P., Görbitz, C. H., & Andersson, K. K. (2006). Structures of the high-valent metal-ion haem–oxygen intermediates in peroxidases, oxygenases and catalases. *Journal of Inorganic Biochemistry, 100,* 460–476.

Hille, R., Hall, J., & Basu, P. (2014). The mononuclear molybdenum enzymes. *Chemical Reviews, 114,* 3963–4038.

Himo, F., & Siegbahn, P. E. (2009). Recent developments of the quantum chemical cluster approach for modeling enzyme reactions. *Journal of Biological Inorganic Chemistry, 14,* 643–651.

Hohenberg, P., & Kohn, W. (1964). Inhomogeneous electron gas. *Physical Review, 136*(3B), B864–B871.

Honig, B., Sharp, K., & Yang, A.-S. (1997). Macroscopic models of aqueous solutions: Biological and chemical applications. *The Journal of Physical Chemistry, 97,* 1101–1109.

Hopmann, K. H., & Himo, F. J. (2008). Quantum chemical modeling of the dehalogenation reaction of haloalcohol dehalogenase. *Journal of Chemical Theory and Computation, 4,* 1129–1137.

Hsiao, Y.-W., Drakenberg, T., & Ryde, U. (2005). NMR structure determination of proteins supplemented by quantum chemical calculations: Detailed structure of the Ca^{2+} sites in the EGF34 fragment of protein S. *Journal of Biomolecular NMR, 31,* 97–114.

Hsiao, Y.-W., & Ryde, U. (2006). Interpretation of EXAFS spectra for sitting-atop complexes with the help of computational methods. *Inorganica Chimica Acta, 359,* 1081–1092.

Hsiao, Y.-W., Tao, Y., Shokes, J. E., Scott, R. A., & Ryde, U. (2006). EXAFS structure refinement supplemented by computational chemistry. *Physical Review B, 74,* 214101.

Hu, L., Eliasson, J., Heimdal, J., & Ryde, U. (2009). Do quantum mechanical energies calculated for small models of protein active sites converge? *The Journal of Physical Chemistry. A, 113,* 11793–11800.

Hu, L., Farrokhnia, M., Heimdal, J., Shleev, S., Rulíšek, L., & Ryde, U. (2011). Reorganisation energy for internal electron transfer in multicopper oxidases. *The Journal of Physical Chemistry. B, 115,* 13111–13126.

Hu, L., Söderhjelm, P., & Ryde, U. (2011). On the convergence of QM/MM energies. *Journal of Chemical Theory and Computation, 7,* 761–777.

Hu, L., Söderhjelm, P., & Ryde, U. (2013). Accurate reaction energies in proteins obtained by combining QM/MM and large QM calculations. *Journal of Chemical Theory and Computation, 9,* 640–649.

Hu, H., & Yang, W. (2008). Free energies of chemical reactions in solution and in enzymes with ab initio quantum mechanics/molecular mechanics methods. *Annual Review of Physical Chemistry, 59,* 573–601.

Iftimie, R., Salahub, D., Wei, D., & Schofield, J. (2000). Using a classical potential as an efficient importance function for sampling from an *ab initio* potential. *Journal of Chemical Physics, 113,* 4852–4862.

Jensen, F. (2007). *Introduction to computational chemistry.* Chichester: J. Wiley & Sons.

Jensen, K. P., & Ryde, U. (2003). Theoretical prediction of the Co-C bond strength in cobalamins. *The Journal of Physical Chemistry. A, 107,* 7539–7545.

Jensen, K. P., & Ryde, U. (2005). How the Co-C bond is cleaved in coenzyme B12 enzymes, a theoretical study. *Journal of the American Chemical Society, 127,* 9117–9128.

Jensen, K. P., & Ryde, U. (2009). Cobalamins uncovered by modern electronic structure calculations. *Coordination Chemistry Reviews, 253,* 769–778.

Kaukonen, M., Söderhjelm, P., Heimdal, J., & Ryde, U. (2008a). Proton transfer at metal sites in proteins studied by quantum mechanical free-energy perturbations. *Journal of Chemical Theory and Computation, 4*, 985–1001.

Kaukonen, M., Söderhjelm, P., Heimdal, J., & Ryde, U. (2008b). A QM/MM-PBSA method to estimate free energies for reactions in proteins. *The Journal of Physical Chemistry. B, 112*, 12537–12548.

Kirkwood, J. G. (1935). Statistical mechanics of fluid mixtures. *The Journal of Chemical Physics, 3*, 300–313.

Klamt, A., & Schüürmann, G. (1994). COSMO: A new approach to dielectric screening in solvents with explicit expressions for the screening energy and its gradient. *Journal of the Chemical Society, Perkin Transactions, 2*, 799–805.

Kleywegt, G. J., & Jones, T. A. (1997). Model building and refinement practice. *Methods in Enzymology, 227*, 208.

Kollman, P. A., Massova, I., Reyes, C., Kuhn, B., Huo, S., Chong, L., et al. (2000). Calculating structures and free energies of complex molecules: Combining Molecular mechanics and continuum models. *Accounts of Chemical Research, 33*, 889–897.

König, G., & Boresch, S. (2011). Non-Boltzmann sampling and Bennett's acceptance ratio method: How to profit from bending the rules. *Journal of Computational Chemistry, 32*, 1082–1090.

König, P. H., Hoffman, M., Trauenheim, T., & Cui, Q. J. (2005). A critical evaluation of different QM/MM frontier treatments with SCC-DFTB as the QM method. *Physical Chemistry B, 109*, 9082–9095.

Kozlowski, P. M., Kumar, M., Piecuch, P., Li, W., Bauman, N. P., Hansen, J. A., et al. (2012). The cobalt–methyl bond dissociation in methylcobalamin: New benchmark analysis based on density functional theory and completely renormalized coupled-cluster calculations. *Journal of Chemical Theory and Computation, 8*, 1870–1894.

Kwiecien, R. A., Khavrutskii, I. V., Musaev, D. G., Morokuma, K., Banerjee, R., & Paneth, P. (2006). Computational insights into the mechanism of radical generation in B_{12}-dependent methylmalonyl-CoA mutase. *Journal of the American Chemical Society, 128*, 1287.

Lee, C. T., Yang, W. T., & Parr, R. G. (1988). Development of the Colic-Salvetti correlation-energy formula into a functional of the electron density. *Physical Review B, 37*, 785–789.

Li, J., Andrejic, M., Mata, R. A., & Ryde, U. (2015). A computational comparison of oxygen atom transfer catalyzed by DMSO reductase with Mo and W. *European Journal of Inorganic Chemistry, 2015*, 3580–3589.

Li, J., Farrokhnia, M., Rulíšek, L., & Ryde, U. (2015). Catalytic cycle of multicopper oxidases studied by combined quantum- and molecular-mechanical free-energy perturbation methods. *The Journal of Physical Chemistry. B, 119*, 8268–8284.

Li, J.-L., Mata, R. A., & Ryde, U. (2013). Large density-functional and basis-set effects for the DMSO reductase catalyzed oxo-transfer reaction. *Journal of Chemical Theory and Computation, 9*, 1799–1807.

Li, X., Siegbahn, P. E. M., & Ryde, U. (2015). A simulation of the isotropic EXAFS spectra for the S2 and S3 structures of the oxygen evolving complex in photosystem II. *Proceedings of the National Academy of Sciences of the United States of America, 112*, 3979–3984.

Li, X., Sproviero, E. M., Ryde, U., Batista, V. S., & Chen, G. (2013). Theoretical EXAFS studies of a model of the oxygen-evolving complex of photosystem II obtained with the quantum cluster approach. *International Journal of Quantum Chemistry, 113*, 474–478.

Liao, R.-Z., & Thiel, W. (2013). Convergence in the QM-Only and QM/MM modeling of enzymatic reactions: A case study for acetylene hydratase. *Journal of Combinatorial Chemistry, 34*, 2389–2397.

Liao, R.-Z., Yu, G., & Himo, F. J. (2011). Quantum chemical modeling of enzymatic reactions: The case of decarboxylation. *Journal of Chemical Theory and Computation, 7*, 1494–1501.

Lin, H., & Truhlar, D. G. J. (2005). Redistributed charge and dipole schemes for combined quantum mechanical and molecular mechanical calculations. *Physical Chemistry A, 109*, 3991–4004.

Lin, H., & Truhlar, D. G. (2007). QM/MM: What have we learned, where are we, and where do we go from here? *Theoretical Chemistry Accounts, 117*, 185–199.

Liu, J., Chakraborty, S., Hosseinzadeh, P., Yu, Y., Tian, S., Petrik, I., et al. (2014). Metalloproteins containing cytochrome, iron–sulfur, or copper redox centers. *Chemical Reviews, 114*, 4366–4469.

Lonsdale, R., Harvey, J. N., & Mulholland, A. J. (2010). Compound I reactivity defines alkene oxidation selectivity in cytochrome P450cam. *The Journal of Physical Chemistry. B, 114*, 1156–1162.

Luzhkov, V., & Warshel, A. (1992). Microscopic models for quantum mechanical calculations of chemical processes in solutions: LDIAMPAC and SCAAS/AMPAC calculations of solvation energies. *Journal of Computational Chemistry, 13*, 199–213.

Mackerell, A. D. (2004). Empirical force fields for biological macromolecules: Overview and issues. *Journal of Computational Chemistry, 25*, 1584–1604.

Manzoni, F., Oksanen, E., Logan, D., & Ryde, U. (2016). Structural of galectin 3 by a combination of neutron and X-ray crystallography and quantum mechanics. *Journal of Chemical Theory and Computation*. submitted.

Marsh, E. N. G. (2000). Coenzyme-B_{12}-dependent glutamate mutase. *Bioorganic Chemistry, 28*, 176–189.

Martin, B. D., & Finke, R. G. (1992). Methylcobalamin's full- vs. half-strength cobalt-carbon sigma bonds and bond dissociation enthalpies: A>10^{15} Co-CH$_3$ homolysis rate enhancement following one-antibonding-electron reduction of methylcobalamin. *Journal of the American Chemical Society, 114*, 585.

Metz, S., & Thiel, W. (2009). A combined QM/MM study on the reductive half-reaction of xanthine oxidase: Substrate orientation and mechanism. *Journal of the American Chemical Society, 2009*(131), 14885–14902.

Mikulskis, P., Genheden, S., & Ryde, U. (2014). Large-scale test of free-energy simulation estimates of protein-ligand binding affinities. *Journal of Chemical Information and Modeling, 54*, 2794–2806.

Møller, C., & Plesset, M. S. (1934). Note on an approximation treatment for many-electron systems. *Physics Review, 46*, 618.

Murphy, R. B., Philipp, D. M., & Friesner, R. A. (2000). A mixed quantum mechanics/molecular mechanics (QM/MM) method for large-scale modeling of chemistry in protein environments. *Journal of Computational Chemistry, 21*, 1442–1457.

Nicoll, R. M., Hindle, S. A., MacKenzie, G., Hillier, I. H., & Burton, N. A. (2001). Quantum mechanical/molecular mechanical methods and the study of kinetic isotope effects: Modelling the covalent junction region and application to the enzyme xylose isomerase. *Theoretical Chemistry Accounts, 106*, 105–112.

Nilsson, K., & Ryde, U. (2004). Protonation status of protein ligands can be determined by quantum refinement. *Journal of Inorganic Biochemistry, 98*, 1539–1546.

Olsson, M. A., Söderhjelm, P., & Ryde, U. (2016). Converging ligand-binding free energies obtained with free-energy perturbations at the quantum mechanical level. *Journal of Computational Chemistry, 37*, 1589–1600.

Olsson, M. H. A., Hong, G., & Warshel, A. (2003). Frozen density functional free energy simulations of redox proteins: Computational studies of the reduction potential of plastocyanin and rusticyanin. *Journal of the American Chemical Society, 125*, 5025–5039.

Padmakumar, R., Padmakumar, R., & Banerjee, R. (1997). Evidence that cobalt−carbon bond homolysis is coupled to hydrogen atom abstraction from substrate in methylmalonyl-CoA mutase. *Biochemistry, 36*, 3713–3718.

Perdew, J. P. (1986). Density-functional approximation for the correlation energy of the inhomogeneous electron gas. *Physical Review B, 33*, 8822–8824.

Perdew, J. P., Burke, K., & Ernzerhof, M. (1996). Generalized gradient approximation made simple. *Physical Review Letters, 77*, 3865–3868.

Philipp, D. M., & Friesner, R. A. (1999). *Journal of Computational Chemistry, 20*, 1468–1494.

Plotnikov, N. V., Kamerlin, S. C. L., & Warshel, A. (2011). Paradynamics: An effective and reliable model for ab initio QM/MM free-energy calculations and related tasks. *The Journal of Physical Chemistry. B, 115*, 7950–7962.

Poulsen, T. D., Kongsted, J., Osted, A., Ogilby, P. R., & Mikkelsen, K. V. (2001). The combined multiconfigurational self-consistent-field molecular mechanics wave function approach. *The Journal of Chemical Physics, 115*, 2393–2400.

Pu, J., Gao, J., & Truhlar, D. G. (2005). Generalized hybrid-orbital method for combining density functional theory with molecular mechanicals. *Chemphyschem, 6*, 1853–1865.

Raghavachari, K., Tucks, G. W., Pople, J. A., & Head-Gordon, M. (1989). A fifth-order perturbation comparison of electron correlation theories. *Chemical Physics Letters, 157*, 479–483.

Rappoport, D., & Furche, F. (2010). Property-optimized Gaussian basis sets for molecular response calculations. *The Journal of Chemical Physics, 133*, 134105.

Reddy, M. R., & Erion, M. D. (2007). Relative binding affinities of fructose-1,6-bisphosphatase inhibitors calculated using a quantum mechanics-based free energy perturbation method. *Journal of the American Chemical Society, 129*, 9296–9297.

Retegan, M., Milet, A., & Jamet, H. (2009). Exploring the binding of inhibitors derived from tetrabromobenzimidazole to the CK2 protein using a QM/MM-PB/SA approach. *Journal of Chemical Information and Modeling, 49*, 963–971.

Riahi, S., & Rowley, C. N. (2014). The CHARMM–TURBOMOLE interface for efficient and accurate QM/MM molecular dynamics, free energies, and excited state properties. *Journal of Computational Chemistry, 35*, 2076–2086.

Riccardi, D., Li, G., & Cui, Q. (2004). Importance of van der Waals interactions in QM/MM simulations. *The Journal of Physical Chemistry. B, 108*, 6467–6478.

Rod, T. H., & Ryde, U. (2005a). Quantum mechanical free energy barrier for an enzymatic reaction. *Physical Review Letters, 94*, 138302. 4 pages.

Rod, T. H., & Ryde, U. (2005b). Free energy barriers at the density functional theory level: Methyl transfer catalyzed by catechol O-methyltransferase. *Journal of Chemical Theory and Computation, 1*, 1240–1251.

Román-Meléndez, G. D., von Glehn, P., Harvey, J. N., Mulholland, A. J., & Marsh, E. N. G. (2014). Role of active site residues in promoting cobalt–carbon bond homolysis in adenosylcobalamin-dependent mutases revealed through experiment and computation. *Biochemistry, 53*, 169–177.

Rulíšek, L., & Ryde, U. (2006). Structure of reduced and oxidised manganese superoxide dismutase—A combined computational and experimental approach. *The Journal of Physical Chemistry. B, 110*, 11511–11518.

Ryde, U. (1995). On the role of Glu68 in alcohol dehydrogenase. *Protein Science, 4*, 1124–1132.

Ryde, U. (1996). The coordination of the catalytic zinc ion in alcohol dehydrogenase studied by combined quantum chemical and molecular mechanical calculations. *Journal of Computer-Aided Molecular Design, 10*, 153–164.

Ryde, U. (2003). Combined quantum and molecular mechanics calculations on metalloproteins. *Current Opinion in Chemical Biology, 7*, 136–142.

Ryde, U. (2007). Accurate metal-site structures in proteins obtained by combining experimental data and quantum chemistry. *Dalton Transactions, 2007,* 607–625.

Ryde, U., Dong, G., Li, J., Andrejic, M., & Mata, R. A. (2016). Computational studies of molybdenum and tungsten enzymes. In R. Hille, M. Kirk, & C. Schulzke (Eds.), *Molybdenum and tungsten enzymes.* in press.

Ryde, U., Hsiao, Y.-W., Rulíšek, L., & Solomon, E. I. (2007). Identification of the peroxy adduct in multicopper oxidases by a combination of computational chemistry and extended X-ray absorption fine-structure measurements. *Journal of the American Chemical Society, 129,* 726–727.

Ryde, U., Mata, R. A., & Grimme, S. (2011). Does DFT-D estimate accurate energies for the binding of ligands to metal complexes? *Dalton Transactions, 40,* 11176–11183.

Ryde, U., & Nilsson, K. (2003). Quantum chemistry can improve protein crystal structures locally. *Journal of the American Chemical Society, 125,* 14232–14233.

Ryde, U., Olsen, L., & Nilsson, K. (2002). Quantum chemical geometry optimisations in proteins using crystallographic raw data. *Journal of Combinatorial Chemistry, 23,* 1058–1070.

Ryde, U., & Olsson, M. H. M. (2001). Structure, strain, and reorganization energy of blue copper models in the protein. *International Journal of Quantum Chemistry, 81,* 335–347.

Ryde, U., Olsson, M. H. M., & Pierloot, K. (2001). The structure and function of blue copper proteins. In L. A. Eriksson (Ed.), *Theoretical and computational chemistry: Vol. 9. Theoretical biochemistry. Processes and properties of biological systems* (pp. 1–56). Amsterdam: Elsevier.

Ryde, U., & Söderhjelm, P. (2016). Ligand-binding affinity estimates supported by quantum-mechanical methods. *Chemical Reviews, 116,* 5520–5566.

Senn, H. M., & Thiel, W. (2009). QM/MM methods for biomolecular systems. *Angewandte Chemie, International Edition, 48,* 1198–1229.

Senn, H. M., Thiel, S., & Thiel, W. (2005). Enzymatic hydroxylation in p-hydroxybenzoate hydroxylase: A case study for QM/MM molecular dynamics. *Journal of Chemical Theory and Computation, 1,* 494–505.

Sevastik, R., & Himo, F. (2007). Quantum chemical modeling of enzymatic reactions: The case of 4-oxalocrotonate tautomerase. *Bioorganic Chemistry, 35,* 444–457.

Sharma, P. K., Chu, Z. T., Olsson, M. H., & Warshel, A. (2007). A new paradigm for electrostatic catalysis of radical reactions in vitamin B_{12} enzymes. *Proceedings of the National academy of Sciences of the United States of America, 104,* 9661.

Sherwood, P., de Vries, A. H., Collins, S. J., Greatbanks, S. P., Burton, N. A., Vincent, M. A., et al. (1997). Computer simulation of zeolite structure and reactivity using embedded cluster methods. *Faraday Discussions, 106,* 79–92.

Sherwood, P., de Vries, A. H., Guest, M. F., Schreckenbach, G., Catlow, C. R. A., French, S. A., et al. (2003). QUASI: A general purpose implementation of the QM/MM approach and its application to problems in catalysis. *Journal of Molecular Structure, 632,* 1–28.

Siegbahn, P. E. M., Blomberg, M. R. A., & Chen, S.-L. (2010). Significant van der Waals effects in transition metal complexes. *Journal of Chemical Theory and Computation, 6,* 2040–2044.

Singh, U. C., & Kollman, P. A. (1986). A combined ab initio quantum mechanical and molecular mechanical method for carrying out simulations on complex molecular systems: Applications to the CH_3Cl + Cl exchange reaction and gas phase protonation of polyethers. *Journal of Combinatorial Chemistry, 7,* 718–730.

Söderhjelm, P., Husberg, C., Strambi, A., Olivucci, M., & Ryde, U. (2009). Protein influence on electronic spectra modelled by multipoles and polarisabilities. *Journal of Chemical Theory and Computation, 5,* 649–658.

Söderhjelm, P., & Ryde, U. (2006). Combined computational and crystallographic study of the oxidised states of [NiFe] hydrogenase. *Journal of Molecular Structure (THEOCHEM)*, *770*, 199–219.
Sousa, S. F., Fernandes, P. A., & Ramos, M. J. (2012). Computational enzymatic catalysis—Clarifying enzymatic mechanisms with the help of computers. *Physical Chemistry Chemical Physics*, *14*, 12431–12441.
Stewart, J. J. P. (1989). Optimization of parameters for semiempirical methods. I: Method. *Journal of Combinatorial Chemistry*, *10*, 209–220.
Stewart, J. J. P. (2007). Optimization of parameters for semiempirical methods. V: Modification of NDDO approximations and application to 70 elements. *Journal of Molecular Modeling*, *13*, 1173–1213.
Sumner, S., Söderhjelm, P., & Ryde, U. (2013). Effect of geometry optimisations on QM-cluster and QM/MM studies of reaction energies in proteins. *Journal of Chemical Theory and Computation*, *9*, 4205–4214.
Sumowski, C. V., & Ochsenfeld, C. J. (2009). A convergence study of QM/MM isomerization energies with the selected size of the QM region for peptidic systems. *Physical Chemistry A*, *113*, 11734–11741.
Sure, R., & Grimme, S. (2015). Comprehensive benchmark of association (free) energies of realistic host−guest complexes. *Journal of Chemical Theory and Computation*, *11*, 3785–3801.
Svensson, M., Humbel, S., Froese, R. D. J., Matsubara, T., Sieber, S., & Morokuma, K. (1996). ONIOM: A multilayered integrated MO + MM method for geometry optimizations and single point energy predictions. A test for Diels-Alder reactions and Pt(P(t-Bu)$_3$)$_2$ + H2 oxidative addition. *The Journal of Physical Chemistry*, *100*, 19357–19363.
Tao, J., Perdew, J. P., Staroverov, V. N., & Scuseria, G. E. (2003). Climbing the density functional ladder: Nonempirical meta-generalized gradient approximation designed for molecules and solids. *Physical Review Letters*, *91*, 146401. 4 pages.
Théry, V., Rinaldi, D., Rivail, J. L., Maigret, B., & Ferenczy, G. J. (1994). Quantum mechanical computations on very large molecular systems: The local self-consistent field method. *Computers & Chemistry*, *15*, 269–282.
Thiel, W. (2014). Semiempirical quantum−chemical methods. *WIREs Computational Molecular Science*, *4*, 145–157.
Tian, B., Strid, Å., & Eriksson, L. A. (2011). Catalytic roles of active-site residues in 2-methyl-3-hydroxypyridine-5-carboxylic acid oxygenase: An ONIOM/DFT study. *The Journal of Physical Chemistry. B*, *115*, 1918–1926.
Tomasi, J., Mennucci, B., & Cammi, R. (2005). Quantum mechanical continuum solvation models. *Chemical Reviews*, *105*, 2999–3094.
Van den Bosch, M., Swart, M., Snijders, J. G., Berendsen, H. J. C., Mark, A. E., Oostenbrink, C., et al. (2005). Calculation of the redox potential of the protein azurin and some mutants. *Chembiochem*, *6*, 738–746.
VandeVondele, J., & Rothlisberger, U. (2002). Canonical adiabatic free energy sampling (CAFES): A novel method for the exploration of free energy surfaces. *The Journal of Physical Chemistry. B*, *106*, 203–208.
Van Severen, M.-C., Andrejic, M., Li, J.-L., Starke, K., Mata, R. A., Nordlander, E., et al. (2014). A quantum-mechanical study of the reaction mechanism of sulfite oxidase. *Journal of Biological Inorganic Chemistry*, *19*, 1165–1179.
Von Lilienfeld, O. A., Tavernelli, I., & Rothlisberger, U. (2005). Variational optimization of effective atom centered potentials for molecular properties. *The Journal of Chemical Physics*, *122*, 014113. 6 pages.
Vreven, T., Byun, K. S., Komáromi, I., Dapprich, S., Montgomery, J. A., Morokuma, K., et al. (2006). Combining quantum mechanics methods with molecular mechanics methods in ONIOM. *Journal of Chemical Theory and Computation*, *2*, 815–826.

Warshel, A., & Levitt, M. (1976). Theoretical studies of enzymic reactions: Electrostatic and steric stabilization of the carbonium ion in the reaction of lysozyme. *Journal of Molecular Biology, 103*, 227–249.

Weigend, F., & Ahlrichs, R. (2005). Balanced basis sets of split valence, triple zeta valence and quadruple zeta valence quality for H to Rn: Design and assessment of accuracy. *Physical Chemistry Chemical Physics, 7*, 3297–3305.

Werner, H.-J., & Schütz, M. (2011). An efficient local coupled cluster method for accurate thermochemistry of large systems. *The Journal of Chemical Physics, 135*, 144116. 15 pages.

Wood, R. H., Yezdimer, E. M., Sakane, S., Barriocanal, J. A., & Doren, D. J. (1999). Free energies of solvation with quantum mechanical interaction energies from classical mechanical simulations. *Journal of Chemical Physics, 110*, 1329–1337.

Woods, C. J., Shaw, K. E., & Mulholland, A. J. (2015). Combined quantum mechanics/molecular mechanics (QM/MM) simulations for protein–ligand complexes: Free energies of binding of water molecules in influenza neuraminidase. *The Journal of Physical Chemistry. B, 119*, 997–1001.

Zhang, Y. (2006). Pseudobond ab initio QM/MM approach and its applications to enzyme reactions. *Theoretical Chemistry Accounts, 116*, 43–50.

Zhang, Y., Liu, H., & Yang, W. J. (2000). Free energy calculation on enzyme reactions with an efficient iterative procedure to determine minimum energy paths on a combined ab initio QM/MM potential energy surface. *Chemical Physics, 112*, 3483–3492.

Zwanzig, R. W. (1954). High-temperature equation of state by a perturbation method. I. Nonpolar gases. *The Journal of Chemical Physics, 22*, 1420–1427.

CHAPTER SEVEN

Enzymatic Cleavage of Glycosidic Bonds: Strategies on How to Set Up and Control a QM/MM Metadynamics Simulation

L. Raich*, A. Nin-Hill*, A. Ardèvol[†], C. Rovira*,[‡],[1]

*Institut de Química Teòrica i Computacional (IQTCUB), Universitat de Barcelona, Barcelona, Spain
[†]Max-Planck Institut für Biophysik, Frankfurt am Main, Germany
[‡]Institució Catalana de Recerca i Estudis Avançats (ICREA), Barcelona, Spain
[1]Corresponding author: e-mail address: c.rovira@ub.edu

Contents

1. Introduction 160
2. Carbohydrate Structures and Glycoside Hydrolases 160
   2.1 Carbohydrate Structures 160
   2.2 Glycoside Hydrolase Catalytic Mechanisms 162
   2.3 The Conformational Catalytic Itinerary 163
3. QM/MM Molecular Dynamics Simulation of the Catalytic Mechanism 167
   3.1 Before You Start: Search for a Reliable X-Ray Structure 167
   3.2 Classical MD Equilibration 169
   3.3 QM/MM Setup and Equilibration 171
   3.4 Choosing of Collective Variables: Metadynamics Simulation 174
   3.5 Analysis of the Reaction Mechanism and Conformational Itinerary 176
4. Conclusions 179
Acknowledgments 180
References 180

Abstract

Carbohydrates play crucial roles in many biological processes, from cell–cell adhesion to chemical signaling. Their complexity and diversity, related to α/β anomeric configuration, ring substituents, and conformational variations, require a diverse set of enzymes for their processing. Among them, glycoside hydrolases (GHs) are responsible for the hydrolysis of one of the strongest bonds in nature: the glycosidic bond. These highly specialized biological catalysts select particular conformations their carbohydrate substrates to enhance catalysis. The evolution of this conformation during the reaction of glycosidic bond cleavage, known as the *conformational catalytic itinerary*, is of fundamental interest in glycobiology, with impact on inhibitor and drug design. Here we review some of the aspects and the main strategies one needs to take into account when simulating a reaction in a GH enzyme using QM/MM metadynamics. Several specific aspects are highlighted, from the importance of the distortion of the substrate at the Michaelis complex to the variable control during the metadynamics simulation or the analysis of the reaction mechanism and conformational itinerary. The increasing speed of computer power and methodological advances have added a vital tool to the study of GH mechanisms, as shown here and recent reviews. It is hoped that this chapter will serve as a first guide for those attempting to perform a metadynamics simulation of these relevant and fascinating enzymes.

1. INTRODUCTION

Enzymatic catalysis has attracted the attention of chemists and biologists since long ago due to their molecular complexity and tremendous efficiency. Enzymes are able to enhance the rate of chemical reactions by several to many orders of magnitude (Garcia-Viloca, Gao, Karplus, & Truhlar, 2004). For instance, glycoside hydrolases (GHs, also named glycosidases) can break the glycosidic bond (with an estimated half-life of about 5 million years (Wolfenden & Snider, 2001)), in a few milliseconds. Whereas it is difficult to settle on how enzymes work in general, with hot debates over many years (Garcia-Viloca et al., 2004; Hammes-Schiffer, 2013; Moliner, 2011; Olsson, Parson, & Warshel, 2006), significant progress has been made in elucidating the catalytic mechanisms of specific enzymes and enzyme families by means of experimental and computational approaches. Among the latter, first-principles QM/MM simulations are of remarkable relevance because they allow to simulate a reaction in the active site considering explicitly the enzymatic environment with a reduced computational cost. Here we focus on GHs and provide a practical guide to analyze their catalytic mechanisms by means of the QM/MM metadynamics approach, which has received considerable interest in the last few years (Barducci, Bonomi, & Parrinello, 2011; Ensing, De Vivo, Liu, Moore, & Klein, 2006).

2. CARBOHYDRATE STRUCTURES AND GLYCOSIDE HYDROLASES

2.1 Carbohydrate Structures

Carbohydrates are present in nature forming several types of glycosidic linkages (eg, β-1,4 in cellulose or α-1,4/α-1,6 in starch and glycogen). In addition, their constituent sugar units may adopt different ring sizes (eg, furanoses, pyranoses, or heptanoses) and a variety of ring substituents (eg, hydroxyl, hydroxymethyl, or N-acetyl), which may be present in two different orientations (axial or equatorial, Fig. 1). This structural diversity makes them highly complex, as simple switch in the orientation of one substituent changes completely their properties. In addition, sugar rings can adopt a myriad of conformations, which makes carbohydrate structures even more complex and challenging for both experiment and theory. Ring flexibility in carbohydrates is very important because different conformers display the exocyclic groups in distinct orientations and this confers them different reactivity. Enzymes make use of carbohydrate flexibility to control the substrate specificity and accelerate the reaction by inducing a certain ring conformation (see later; Biarnés, Nieto, Planas, & Rovira, 2006; Davies, Planas, & Rovira, 2012). Because they are the most abundant in nature, we focus here on carbohydrates based on six-membered ring sugars (pyranoses) and provide hints on how to model the reactivity of their enzyme complexes, but similar considerations could be applied to other sugar ring sizes.

Following the IUPAC (1980) nomenclature, pyranoses exhibit 38 canonical conformations that can be classified into chair, half-chair, boat, skew-boat, and envelope conformations (represented by the letters C, H, B, S, and E, respectively). Each conformation type displays four atoms on the same plane, and two out-of-plane atoms (except E conformers, which have only one out-of-plane atom). The out-of-plane atoms are indicated by superscript and subscript indexes that refer to their position with respect to the reference ring plane (above or below, respectively; Fig. 2A).

In 1971, Stoddart introduced two diagrams (Stoddart, 1971; Fig. 2B) integrating all possible conformations of a pyranose ring, together with their connectivity, that has become very popular for representing conformational itineraries of glycosides during catalysis (Davies, Ducros, Varrot, & Zechel, 2003; Vocadlo & Davies, 2008). Stoddart diagrams, and their rectangular or Mercator counterpart (Ardèvol & Rovira, 2015), are very useful to discuss

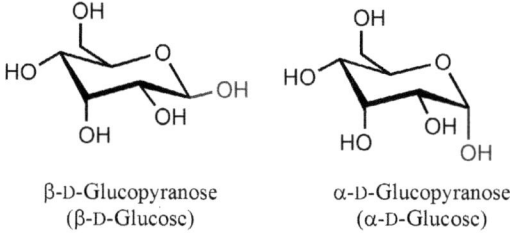

β-D-Glucopyranose   α-D-Glucopyranose
(β-D-Glucose)   (α-D-Glucose)

Fig. 1 α and β isomers of the biologically active D-glucose.

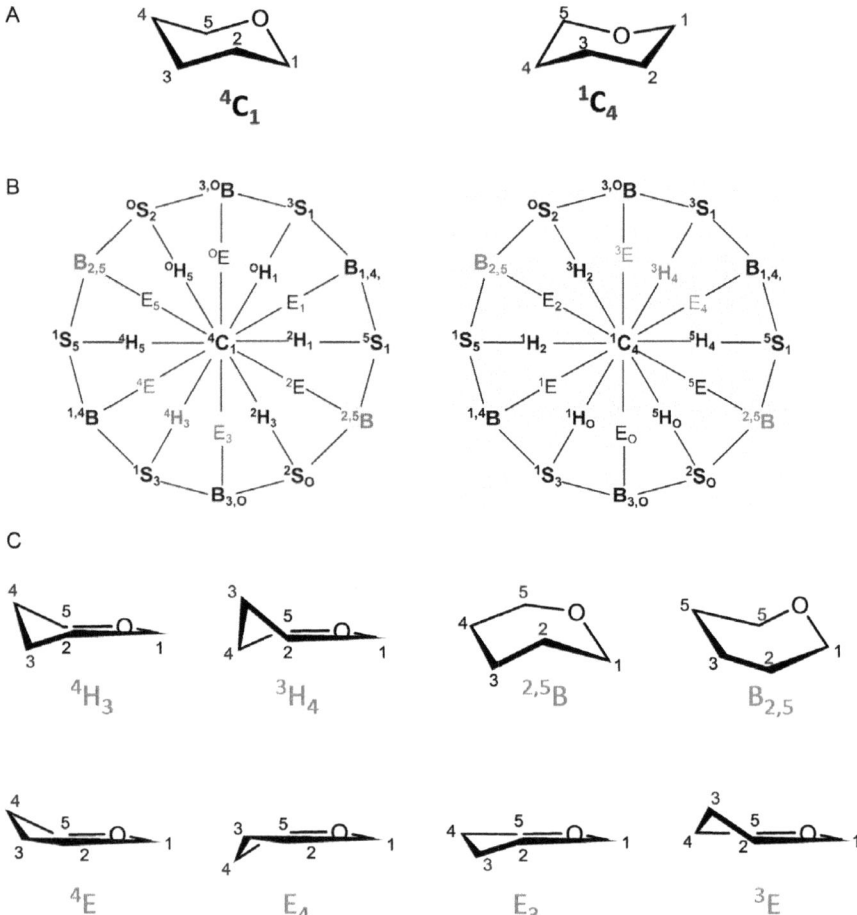

Fig. 2 (A) The two chair conformations of a pyranose. (B) Stoddart diagrams, centered at $^4C_1$ and $^1C_4$, respectively. (C) Favored conformations for a pyranose oxocarbenium ion (potential TS conformations of the GH reaction).

substrate distortions in GHs (see Davies et al., 2003, 2012; Davies & Williams, 2016; Satoh & Manabe, 2013; Speciale, Thompson, Davies, & Williams, 2014; Vocadlo & Davies, 2008 for research monographs on the topic).

2.2 Glycoside Hydrolase Catalytic Mechanisms

GHs are classified into 135 families (by Feb. 2016) according to their sequence similarity (http://www.cazy.org; Cantarel et al., 2009; Lombard, Golaconda Ramulu, Drula, Coutinho, & Henrissat, 2014). Enzymes from the same family usually act on a similar substrate and share the same catalytic mechanism. Despite the large number of GH families known, most GHs follow a common acid/base catalytic mechanism with retention or inversion of the anomeric configuration. The reaction mechanism of GHs is either one or two classical S_N2 displacement reactions, assisted by two essential residues: a proton donor (or acid/base residue) and a general base (or nucleophile residue; White & Rose, 1997). The former is usually a glutamic or aspartic acid, whereas the latter is a glutamate or aspartate conjugate base (remarkable exceptions are sialidases, in which the nucleophile is an activated tyrosine (Amaya et al., 2004; Pierdominici-Sottile, Horenstein, & Roitberg, 2011), or chitinases, in which the nucleophile is the 2-N-acetyl substituent of the −1 saccharide).

Inverting enzymes operate by a single nucleophilic substitution (Fig. 3A), while retaining glycosidases follow a double-displacement mechanism via formation and hydrolysis of a glycosyl-enzyme intermediate (or covalent intermediate; Fig. 3B). Regardless of the type of mechanism, each reaction step involves an oxocarbenium ion-like transition state (TS; Rye & Withers, 2000; Zechel & Withers, 2000), as evidenced by kinetic measurements of the isotope effect and by quantum mechanical calculations (Zechel & Withers, 2000). As the C1–O1 bond breaks, the oxocarbenium ion-like TS displays an sp^2 hybridization and development of positive charge at the anomeric carbon. The excess of positive charge is partially stabilized by electron donation from the ring oxygen, shown by a shrinkage in the C1–O5 distance (Biarnés et al., 2006). In this structure, the sugar ring is distorted from a relaxed chair conformation—the most stable in the gas phase (Biarnés et al., 2007) or in aqueous solution (Spiwok, Kralova, & Tvaroska, 2010)—into a conformation in which the atoms C2, C1, O5, and C5 are as coplanar as possible, as required for a stable oxocarbenium ion (Ardèvol & Rovira, 2015). Interestingly, only eight pyranose conformations fulfill this criteria (Fig. 2C): two boat ($B_{2,5}$, $^{2,5}B$), two half-chair ($^4H_3$, $^3H_4$), and four envelope conformations (^3E, E_3, ^4E, and E_4). Therefore, the TS of GH-catalyzed enzymatic reaction is characterized by the sugar ring adopting one of these eight conformations (Ardèvol & Rovira, 2015). This, in turn, restricts the possible conformations of the sugar in the reactants and the products states to those that are directly connected to the TS in the Stoddart's diagram.

2.3 The Conformational Catalytic Itinerary

The pyranose ring bearing the anomeric carbon of the glycosidic bond to be cleaved changes conformation during catalysis. These conformational changes (from the reactants to the products in inverting GHs and from reactants to the glycosyl-enzyme intermediate in retaining GHs) are known as the conformational catalytic itinerary of the reaction (Vocadlo & Davies, 2008).

A **Inverting GHs** — Single displacement

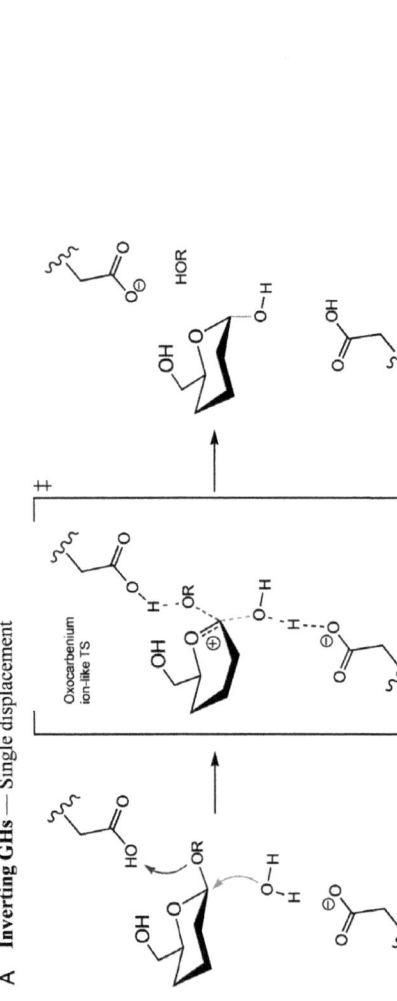

B **Retaining GHs** — Double displacement

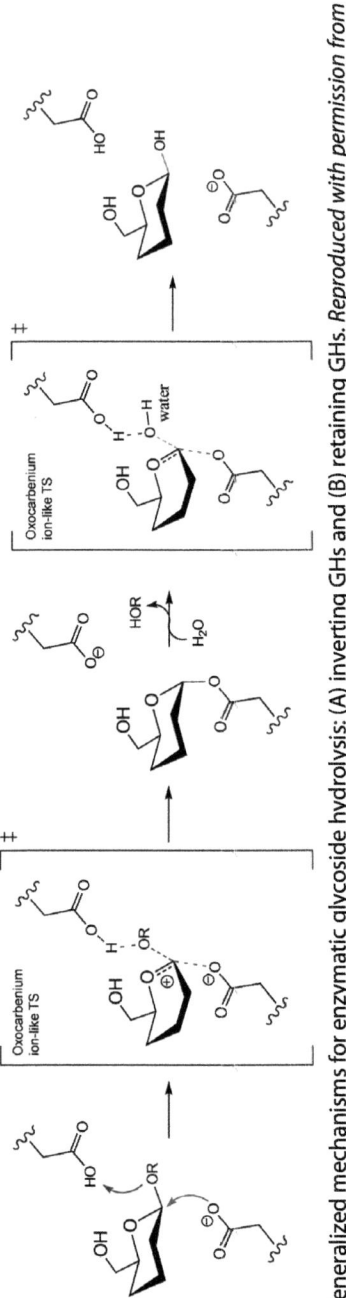

Fig. 3 Generalized mechanisms for enzymatic glycoside hydrolysis: (A) inverting GHs and (B) retaining GHs. *Reproduced with permission from Ardèvol, A., & Rovira, C. (2015). Reaction mechanisms in carbohydrate-active enzymes: Glycoside hydrolases and glycosyltransferases. Insights from ab initio quantum mechanics/molecular mechanics dynamic simulations. Journal of the American Chemical Society, 137, 7528–7547. Copyright 2015 American Chemical Society.*

Understanding it for a given GH is of paramount importance when designing potent and selective inhibitors (Caines et al., 2007). This is because the strongest binding inhibitors are those that mimic the structure and properties of the TS, and an inhibitor designed for a GH with a specific conformational catalytic itinerary will not be active (or to a less extent) with other GHs that follow a different itinerary. Catalytic itineraries of GHs are usually drawn as radial or azimuthal lines in Stoddart diagrams. For instance, retaining β-D-glucosidases are known to operate via a $^4H_3$-type TS, preferentially following a $^1S_3 \rightarrow {}^4H_3 \rightarrow {}^4C_1$ itinerary, whereas retaining α-D-mannosidases follow a $^1S_5 \rightarrow B_{2,5} \rightarrow {}^OS_2$ itinerary (Fig. 4; Davies et al., 2012).

Experimentally, the conformational catalytic itinerary is typically inferred from the X-ray structure of an analog of the Michaelis complex (Offen, Zechel, Withers, Gilbert, & Davies, 2009), the TS (Thompson et al., 2012), or the glycosyl–enzyme adduct (White, Tull, Johns, Withers, & Rose, 1996). In addition, computational simulations have proven to be very valuable to predict it, since (i) they can reveal the complete conformational itinerary (Fig. 5), (ii) they can be done in the catalytically competent enzyme–substrate complex, and (iii) they can provide further information on electronic rearrangements and energy changes during the reaction (Ardèvol & Rovira, 2015).

The prediction of conformational catalytic itineraries has received a considerable interest in the last few years, as new enzyme structures are

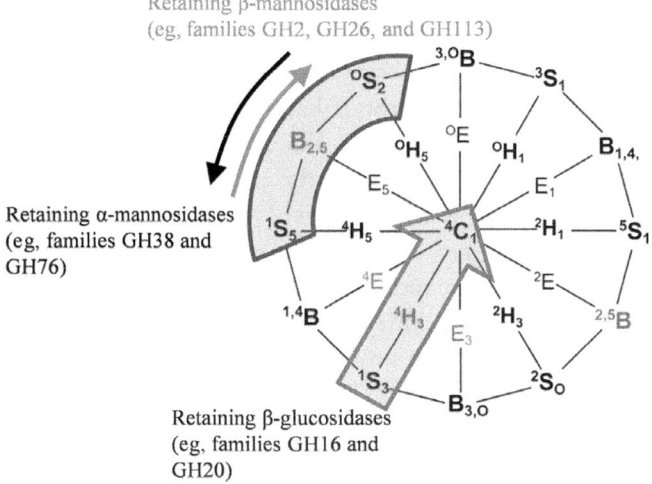

Fig. 4 Stoddart's diagram (centered at $^4C_1$) illustrating the proposed itineraries followed by selected GH families. (See the color plate.)

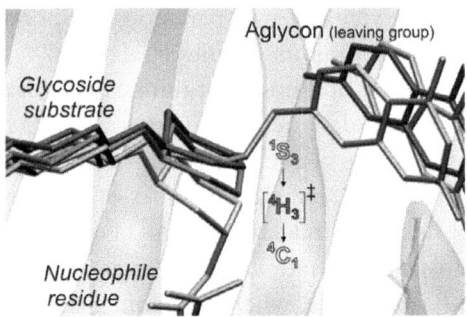

Fig. 5 Gradual change in ring conformation upon forming the glycosyl-enzyme intermediate for 1,3-1,4-β-glucanase, a family 16 retaining GH, computed by QM/MM metadynamics. Adapted with permission from Biarnés, X., Ardèvol, A., Iglesias-Fernández, J., Planas, A., & Rovira, C. (2011). Catalytic itinerary in 1,3-1,4-beta-glucanase unraveled by QM/MM metadynamics. Charge is not yet fully developed at the oxocarbenium ion-like transition state. Journal of the American Chemical Society, 133, 20301–20309. Copyright 2011 American Chemical Society.

available, showing a more profuse scenario of molecular mechanisms and conformational itineraries than it was initially thought. It is nowadays assumed that the catalytic conformational itinerary is generally unique for a given type of enzyme acting on a given substrate (eg, retaining α-mannosidases and inverting β-glucosidases), but there are several exceptions (eg, β-xylanases can adopt $B_{2,5}$ or $^1S_3$ TS depending on the GH family; Iglesias-Fernández, Raich, Ardèvol, & Rovira, 2015).

Modeling of GH catalytic itineraries complexes is challenging for computational approaches as both the enzyme and carbohydrate substrate are highly flexible. Therefore, methods based on molecular dynamics (MD) such as metadynamics (Ensing et al., 2006; Laio & Gervasio, 2008; Laio & Parrinello, 2002) are particularly useful to sketch these itineraries, as one can dynamically follow the evolution of the chemical reaction through the minimum free energy path, even when the structure of the products state is unknown experimentally. We would like to point out that conformational catalytic itineraries can be also determined with several other free energy methods, such as umbrella sampling (US; Torrie & Valleau, 1977) or transition path sampling (TPS; Dellago & Bolhuis, 2009), in which both reactants and product states need to be defined beforehand. Due to their high computational cost, US and TPS simulations usually rely on semiempirical methods to describe the QM atoms (Knott et al., 2014; Pierdominici-Sottile et al., 2011). In this work, we focus on

QM(DFT)/MM metadynamics simulations and describe some strategies and tactics that are essential to study glycosidase reaction mechanisms.

3. QM/MM MOLECULAR DYNAMICS SIMULATION OF THE CATALYTIC MECHANISM

3.1 Before You Start: Search for a Reliable X-Ray Structure

Because of the scarcity of structures of GH complexes with natural substrates, the choice of the initial structure is crucial for a successful simulation. Ideally, one should find a starting structure that corresponds to the Michaelis complex of the wild-type GH with its natural glycoside substrate. However, this ideal structure is rarely available, as experiments are usually carried out under conditions in which the reaction does not advance, such as working with an enzyme variant (eg, mutating one of the two catalytically essential residues) or with a modified substrate (eg, TS-like inhibitors). As a consequence, the sugar at the -1 enzyme subsite (named as the -1 sugar hereafter) might not exhibit any distortion or, even worse, a wrong one. For instance, in the X-ray structure of the nucleophile mutant (D204A) of Golgi α-mannosidase II in complex with a GlcMan5GlcNAc, the substrate features a $^{4}C_{1}$ conformation of the Man5 substrate mannosyl ring (the -1 sugar; Fig. 6A). In contrast, a mixture of E_{5} and $^{O}H_{5}$ conformations is observed for the acid/base mutant (D341A) of the same enzyme. E_{5} and $^{O}H_{5}$ are unprecedented conformations in GHs, and they are most likely a consequence of the mutations used to trap the Michaelis complex in the crystal.

Starting from an incorrect conformation can influence the conformational itinerary and the energy barrier. An example of that can be seen in Liu et al., in which the initial structure had the -1 glucose in a chair conformation and an unlikely pathway for cellulose hydrolysis was found (Liu, Wang, & Xu, 2010). Because sugar conformations have been often overlooked in the literature, it is not surprising to find saccharide rings with erroneous sugar conformations (Agirre, Davies, Wilson, & Cowtan, 2015). For instance, a crystallographic analysis of a GH11 retaining xylanase reported a $^{2}S_{O}$ conformation for the -1 xylose, whereas the structure deposited in the PDB displays a mixture of $^{4}C_{1}/^{O}H_{5}$ conformations (even though the electronic density at the -1 subsite was very poor).

As a rule of thumb, one should verify that the initial structure fulfills: (i) that it conforms to the electron density and (ii) that it is the expected conformation for the particular GH family. To aid to check this, Agirre,

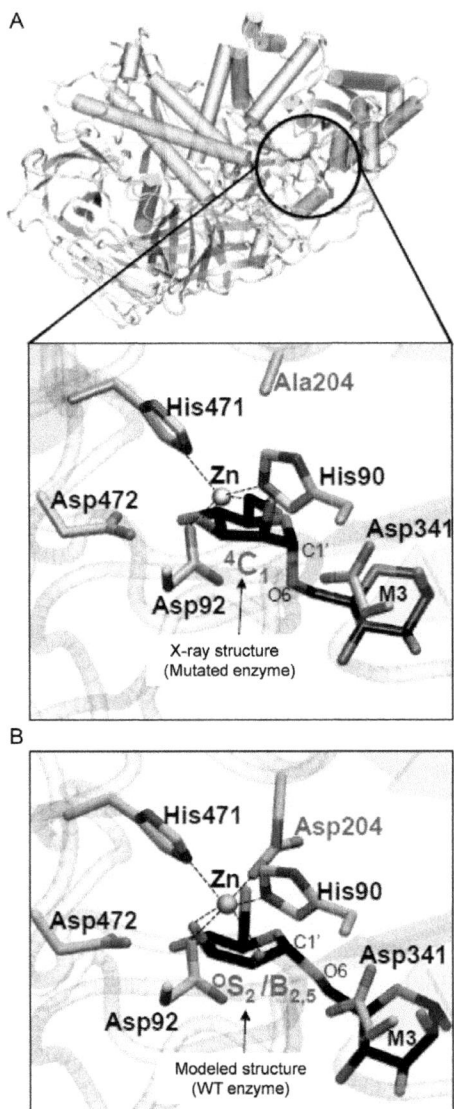

Fig. 6 Active site of α-mannosidase II complexed with GlcNAcMan5GlcNAc. (A) X-ray structure of the D204A mutant (PDB 3CZN), featuring a $^4C_1$ conformation of the Man5 substrate mannosyl ring. (B) Optimized structure of the native enzyme obtained from the QM/MM simulation after reverting the D204A mutation. The Man5 substrate mannosyl ring adopts a $^OS_2/B_{2,5}$ conformation. *Reproduced with permission from Petersen, L., Ardèvol, A., Rovira, C., & Reilly, P. J. (2010). Molecular mechanism of the glycosylation step catalyzed by Golgi alpha-mannosidase II: A QM/MM metadynamics investigation.* Journal of the American Chemical Society, 132, *8291–8300. Copyright 2010 American Chemical Society.*

Iglesias-Fernandez, et al. (2015) recently developed a software that validates the conformation of carbohydrate structures inside enzymes.

3.2 Classical MD Equilibration

The standard protocol of a QM/MM calculation on an enzyme complex is illustrated in Fig. 7. An initial MM equilibration step is necessary at least for two reasons. First, because the crystal structure does not correspond to a catalytically competent form of the enzyme (eg, due to the use of inactive mutants or the lack of a bound substrate). Second, because it represents a static view of the protein, as it is a time average of the structure at a given temperature (usually around 100 K). Thus, one needs to equilibrate the structure to adapt it to the changes made by the model (eg, reversion of mutations or docking of the substrate) and generate an ensemble of structures that capture the different active-site configurations accessible at a given temperature. A direct QM/MM calculation from the crystal structure will be biased by this static view of the structure, losing possible side-chain interactions that might appear after a proper equilibration at the enzyme working temperature (Raich et al., 2016). Nowadays, these steps need to be performed with a force field-based MD, as the timescales needed to reach thermal equilibrium (nanoseconds) are far from the timescales that are affordable with ab initio and QM/MM MD.

Different ring conformations and exocyclic group stereoisomers usually have remarkably different chemical and thermodynamic properties, and this represents a big challenge for the classical force-fields. For instance, the charge of the anomeric carbon in a distorted $^1S_3/^{1,4}B$ conformation is higher than in the $^4C_1$, the C1–O5 bond is shorter and the C1–O1 bond is longer (Biarnés et al., 2006). It is hard for standard force fields, which rely on fixed atom charges and bond parameters, to capture the properties of the sugar in the enzymatic environment. There are several recent carbohydrate-specific force fields (GROMOS.56A6$_{CARBO-R}$ (Plazinski, Lonardi, & Hünenberger, 2016), GLYCAM06 (Kirschner et al., 2008), CHARMM (Guvench et al., 2011)), but all of them suffer to reproduce some properties of the saccharides. Importantly, they often fail to stabilize the correct distortion of the −1 sugar ring in the Michaelis complex of GHs.

Thompson et al. characterized the $^{3,O}B/^3S_1$-distorted conformation of the −1 sugar in a GH47 α-mannosidase using a thioderivative substrate with X-ray crystallography, but classical MD was not able to reproduce it unless conformational restraints were used (Thompson et al., 2012).

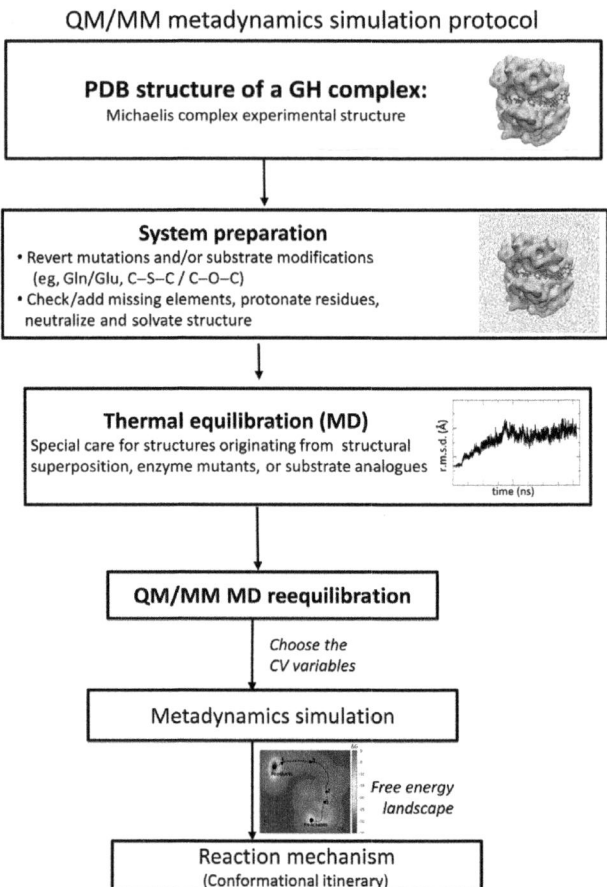

Fig. 7 Standard protocol of a QM/MM metadynamics simulation of a GH starting from the pristine crystallographic coordinates. *Adapted with permission from Ardèvol, A., & Rovira, C. (2015). Reaction mechanisms in carbohydrate-active enzymes: Glycoside hydrolases and glycosyltransferases. Insights from ab initio quantum mechanics/molecular mechanics dynamic simulations. Journal of the American Chemical Society, 137, 7528–7547. Copyright 2015 American Chemical Society.*

Likewise, classical MD simulations on 1,3-1,4-β-glucanase (Biarnés et al., 2006) did not reproduce the distortion of the substrate, as only the relaxed $^4C_1$ conformation was stable in the classical MD simulations. In our experience, sugar conformations obtained in classical MD simulations should be taken with care. In case the -1 sugar ring evolves spontaneously to an unexpected conformation during energy minimization or MD equilibration (eg, a $^4C_1$ for the -1 sugar in a Michaelis complex), one should wonder whether this is a "real" transition or an artifact. In case of having no previous information on the sugar conformation in the enzyme, a QM/MM calculation of the free energy landscape with respect to ring distortion is a useful (albeit computationally expensive) strategy to find the most stable sugar conformation and possible conformational transitions (Biarnés, Ardèvol, Iglesias-Fernández, Planas, & Rovira, 2011).

3.3 QM/MM Setup and Equilibration

Once the system is MM equilibrated, at least one MD snapshot should be selected and a set of QM-described atoms need to be defined for the QM/MM simulation. It is important to start with a suitable snapshot since ring conformational changes might involve sizable energy barriers (>5 kcal/mol) and thus are unlikely in the picosecond timescale at room temperature. The quantum region needs to include at least all the atoms and molecules that are Molecular Dynamics involved in the reaction, which in the case of GHs are the −1 and +1 sugars (enclosing the glycosidic bond to be broken) and the side chains of the two catalytic residues. This is the partition used in most studies of GH mechanisms (Ardèvol & Rovira, 2015). However, residues forming relevant interactions with the sugar might also need to be included if there is any doubt on the force-field parameters for these residues. For instance, in metal-dependent enzymes such as Golgi α-mannosidase II, the QM region included the coordination sphere of the metal atom (Zn^{2+}), as the electrostatic environment of the ion is likely to influence catalysis (Petersen, Ardèvol, Rovira, & Reilly, 2010).

In most cases, the QM–MM boundary necessarily divides a covalent bond. This is very common for long carbohydrate substrates or when residue side chains need to be included in the QM region. In the CPMD program (http://www.cpmd.org), two options are available to treat the QM–MM boundary: monovalent pseudopotentials and capping hydrogens. The first is easier to set up and does not imply the inclusion of extra atoms in the QM subsystem, but requires a specific pseudopotential. The latter has a trickier setup (inputs and topologies of the MM part need to be modified and an exclusion list of electrostatic interactions involving the dummy atom needs to be specified) but does not need any special pseudopotential. To minimize problems with these cutting schemes, the frontier bonds should be nonpolar (ideally C–C bonds). Cutting side-chain residues is easily done through the Cα–Cβ, but it is not obvious how to divide the carbohydrate substrate (either cutting the glycosidic bond or a sugar ring). Tests performed in our group on several GHs showed that there is no best choice and the solution is system dependent. Once the QM region is determined, the electrostatic QM–MM interaction must be carefully set before QM/MM MD equilibration is initiated, as the final shape of the carbohydrate, in particular the distortion of the -1 sugar ring, might be influenced by nonbonded interactions with the environment.

Sugar distortion in GHs is induced by the shape of the enzyme binding pocket (Biarnés et al., 2006) and the interactions of the -1 sugar hydroxyl groups with the nucleophile (specially the 2-OH; Raich et al., 2016). The electrostatic environment in the enzyme cavity also has an important effect in catalysis. As it is not possible to include a large number of residues in the QM region, it is important to have a rigorous treatment of the QM–MM electrostatic interactions. In CPMD, these interactions between the QM and MM regions are handled via a Hamiltonian coupling scheme based on a multilayer approach (Laio, VandeVondele, & Rothlisberger, 2002). In this approach, the Coulombic interaction between the QM and the closest MM atoms (named as nearest neighbor or NN atoms) is treated exactly (Eq. 1 and Fig. 8A), whereas the MM atoms at an intermediate distances (named as ESP atoms) interact with the QM atoms treated as electrostatic potential (ESP) derived point charges (Eq. 2 and Fig. 8A). The more distant QM–MM electrostatic interactions are treated via a multipole expansion (Laio et al., 2002).

$$H_{QM-MM_{NN}}^{electrostatic} = \sum_{i=1}^{NN_{atoms}} q_i \int_0^\infty d\vec{r}\, \frac{\rho(\vec{r})}{|\vec{r}-\vec{r_i}|} \quad (1)$$

$$H_{QM-MM_{ESP}}^{electrostatic} = \sum_{i=1}^{ESP_{atoms}} \sum_{j=1}^{QM_{atoms}} \frac{q_i Q_j^{RESP}}{|\vec{r_i}-\vec{r_j}|} \quad (2)$$

The NN and ESP regions, defined as the MM atoms within a radius from each QM region, can be determined by the following procedure:

(i) A series of single-point calculations are performed using equal and increasing values for R_{NN} and R_{ESP} until convergence of the total energy of the system is reached. An optimal R_{ESP} is chosen.

(ii) The above R_{ESP} is kept fixed and R_{NN} is gradually diminished. An optimal R_{NN} value is selected.

A balance between accuracy and computational cost needs to be achieved. As shown in Fig. 8B, the maximum accuracy is obtained when large radii values are chosen. For instance, using $R_{NN}=R_{ESP}=60$ a.u. the

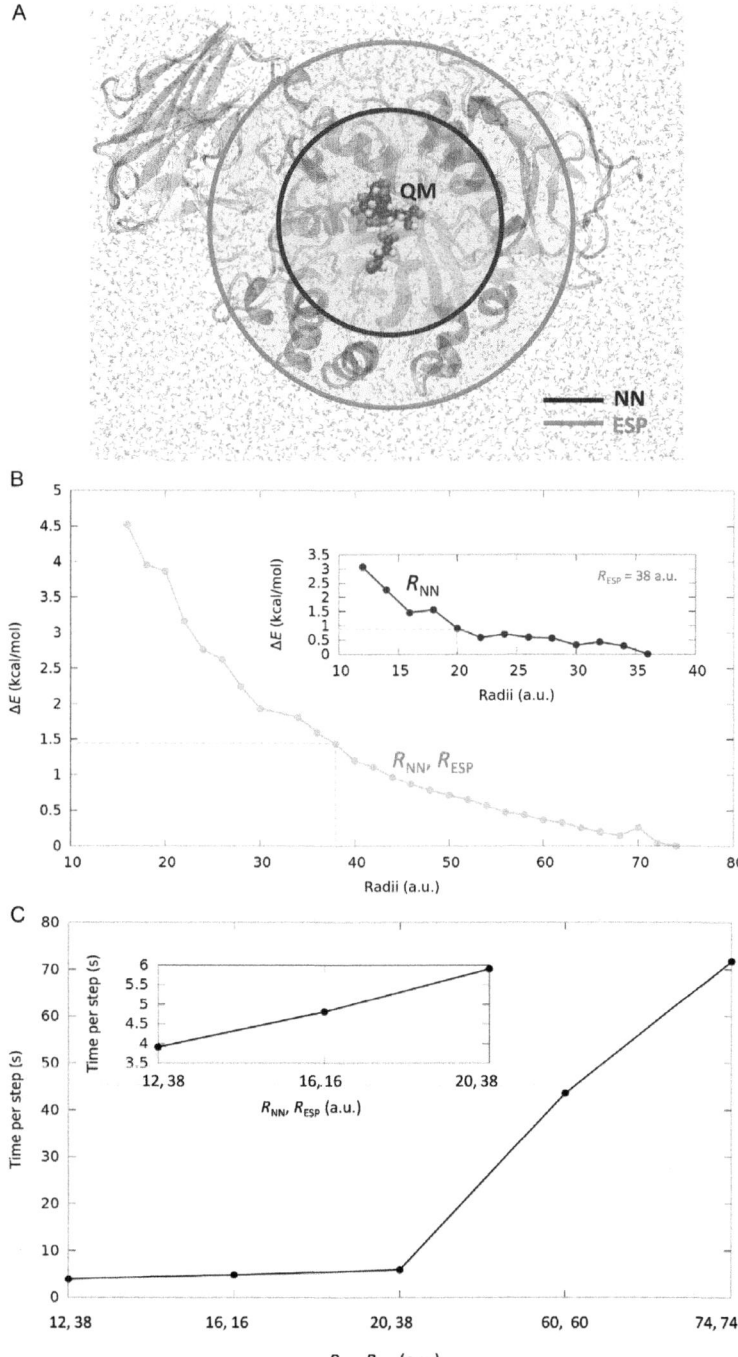

Fig. 8 (A) Electrostatic regions illustrated for a retaining GH. Values correspond to $R_{NN}=20$ a.u. and $R_{ESP}=38$ a.u. The water box has been reduced for visualization purposes. (B) Energy differences computed to tune the optimal radii of ESP and NN regions. The plot corresponds to equal values of NN and ESP radii whereas the inset illustrates how to obtain the optimal R_{NN} for a fixed R_{ESP} of 38 a.u. (C) Computational cost for various values of R_{NN} and R_{ESP} (calculations performed using 80 Intel SandyBridge processors).

total energy of the system is converged (<1.5 kcal/mol from the reference value). In contrast, choosing $R_{NN}=R_{ESP}=16$ a.u. results in a significant deviation (>4.5 kcal/mol). Obviously, one could choose the first option ($R_{NN}=R_{ESP}=60$ a.u.), but, as shown in Fig. 8C, the CPU time per integration step increases by a factor of 8 (the number of NN atoms increases from 1161 to 22,596). Values $R_{NN}=20$ a.u., $R_{ESP}=38$ a.u. (Fig. 8B and C) represent a good balance between accuracy and computational cost.

Once the radii of the electrostatic regions are defined, the QM/MM system needs to be optimized and equilibrated at room temperature before activating the chemical reaction by an enhanced sampling technique.

3.4 Choosing of Collective Variables: Metadynamics Simulation

Metadynamics (Laio & Parrinello, 2002; and its variants) is an MD method that aims to accelerate rare events. The MD trajectories are projected onto a set of well-chosen collective variables (CVs) or reaction coordinates which define an effective energy landscape of the process of interest. As the simulation runs, a historical biasing potential is built as a sum of Gaussian-shaped functions centered on the previously visited configurations in the CV space. As such, the system is forced to escape the local minima along the minimum free energy paths. It has been demonstrated (Dama, Parrinello, & Voth, 2014) that for long simulation times, the bias potential will be proportional to the negative of the free energy. Noteworthy, as with all similar methods, the choice of the CVs is crucial for obtaining a good physical description of the process.

The most obvious choice of CVs to simulate the hydrolysis of a glycosidic bond is to consider the covalent bonds that need to be broken or formed, as they correspond to the slowest motions along the reaction coordinate. In the case of retaining GHs (glycosylation step, Fig. 3), this includes the glycosidic bond between the C1 and O1 atoms (bond to be cleaved, d_1), the distance between the nucleophile side chain and the anomeric carbon (bond to be formed, d_2), the O–H distance of the acid/base (bond to be broken, d_3), and the distance between the glycosidic oxygen and the acid/base proton (bond to be formed, d_4). The pairs of distances d_1/d_2 (S_N2 displacement) and d_3/d_4 (proton transfer) are related; when one increases the other decreases. Therefore, it is reasonable to consider a distance difference or, alternatively, a difference of coordination numbers. The coordination number (CN; Iannuzzi, Laio, & Parrinello, 2003) between two atoms is given by:

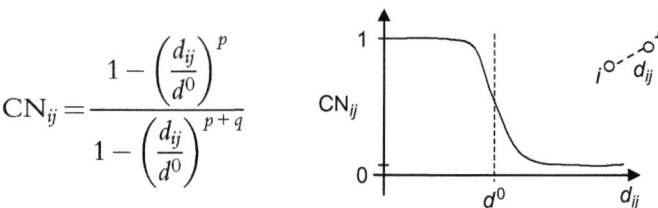

$$CN_{ij} = \frac{1 - \left(\frac{d_{ij}}{d^0}\right)^p}{1 - \left(\frac{d_{ij}}{d^0}\right)^{p+q}}$$

where d_{ij} is the internuclear distance of the atoms involved, d^0 is the threshold distance for bonding, and p and q are exponents that determine the steepness of CN_{ij} decay with respect to d_{ij}. CN values range from 0 (no bond) to 1 (a bond). The way to choose these parameters is to plot the CN vs the distance and vary p and q until the sigmoidal curve covers the distance range in which the bond can be considered broken and formed. Obviously, these values are system dependent and a given set of parameters cannot be used for all GHs. As an example, two CVs were used in a metadynamics simulation of Golgi α-mannosidase II (Fig. 9A): CV1 is the difference in CN between the scissile glycosidic bond and the bond between the nucleophile and the anomeric carbon atom C1 ($CN_{C1-O6}-CN_{C1-OAsp204}$), while CV2 is the difference in CN between the O–H of the acid/base residue and the distance of the glycosidic oxygen and the acid/base hydrogen ($CN_{HAsp341'-OAsp341}-CN_{HAsp341'-O6}$). CV1 and CV2 describe the nucleophilic attack and the proton transfer between the catalytic acid/base and the glycosidic oxygen atom, respectively.

One way to save computational time is to omit the proton transfer variable. This is a good approximation when the aglycon is a good leaving group such as 4-nitrophenol, as in this case the cleavage of the glycosidic bond and formation of the glycosyl-enzyme intermediate do not need acid catalysis. For instance, the simulation of the reaction mechanism for the complex GH1 β-mannosidase with a 4-nitrophenyl β-D-glycoside (Tankrathok et al., 2015) could be performed using just one CV, corresponding to the difference of coordination numbers of the glycosidic bond and the nucleophile–sugar bond. However, for Golgi α-mannosidase II with a GlcMan5GlcNAc substrate (ie, a sugar-leaving group), omitting the proton transfer variable led to a failed metadynamics in which the system could not exit the reactants well.

In general, one starts by selecting the smallest set of possible variables and using chemical intuition, but it should also be taken into account that there is also much to be learned from failed attempts. If the mechanism is unphysical or the barriers are larger than experimental values, other sets of CVs or MD snapshots might need to be considered. For instance, in the case of *Saccharomyces cerevisiae* Gas2 (*Sc*Gas2), a GH of family 72, a set of two CVs involving the proton transfer and the nucleophilic attack (each one taken as a difference of distances) were initially chosen (Raich et al., 2016). However, the system did not escape the reactants well. Analysis of the CV evolution, and in particular the single distances involving each CVs, showed that the distance corresponding to the bond to be cleaved was not being activated. To solve this problem, this CV was split into two separate distances, which led to a well-behaved CV oscillation and reaction mechanism.

3.5 Analysis of the Reaction Mechanism and Conformational Itinerary

The main output from the metadynamics simulation is the trajectory and the history-dependent bias. Analysis of the trajectory, the evolution of the CVs, and the different energy terms give information about the quality of the simulation. To obtain a good estimate of the reaction energy barrier, one must ensure that recrossing over the TS is observed at least once (Ensing, Laio, Parrinello, & Klein, 2005). Convergence test calculations on GHs have shown that in this case the error on the free energy barriers is of the order of 1 kcal/mol (Raich et al., 2016).

As an example, Fig. 9 shows the results obtained in the simulation of glycosidic bond cleavage for Golgi α-mannosidase II, considering two CVs (Fig. 9A). Fig. 9B and C displays the evolution of the CVs and the reconstructed free energy surface (FES) obtained from the metadynamics simulation, respectively. Starting from the Michaelis complex, the scissile glycosidic bond was cleaved and the mannosyl–Asp204 covalent intermediate was formed after ~4.8 ps. Then, the system sampled the products and the states with oxocarbenium ion-like character (the left and the middle region of the FES in Fig. 9C, respectively) for ~31 ps before the mannosyl glycosidic bond was reformed and the system reached the reactant state again. Notice, however, that due to the biasing potential, these times do not correspond to the reaction rate. The relevant stationary points corresponding to reactant, TS (determined by an isocommittor analysis; Biarnés et al., 2011), and product (the covalent intermediate) states can be easily identified in the FES. The calculated reaction free energy of activation is 23 kcal/mol, which

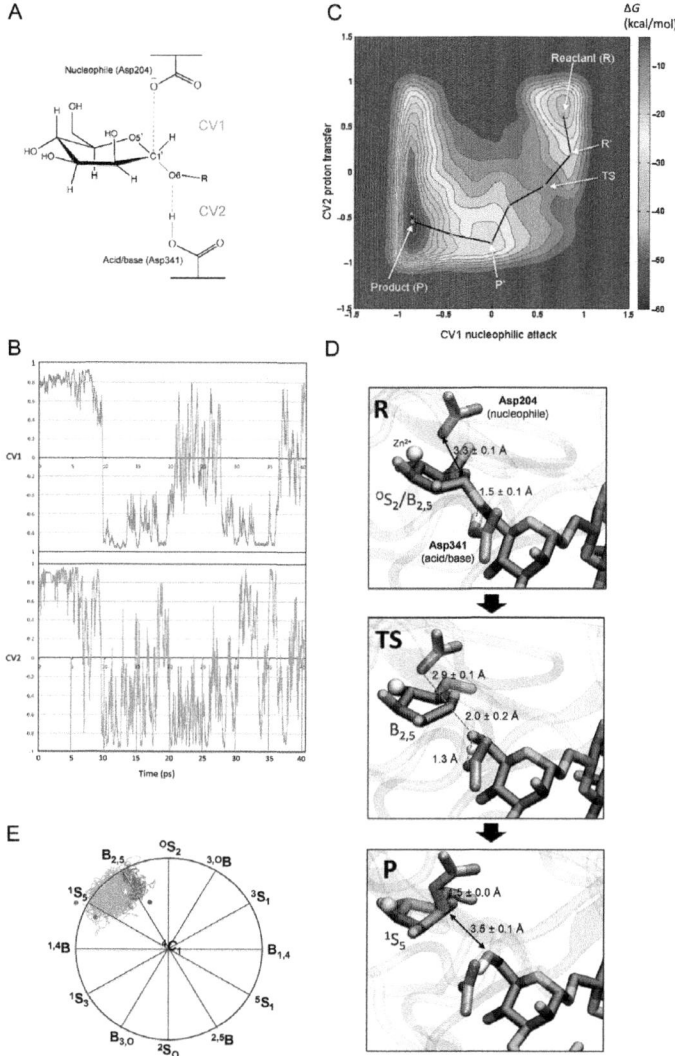

Fig. 9 Results obtained for the QM/MM metadynamics simulation of glycosidic bond cleavage in α-mannosidase II. (A) Collective variables definition. (B) Evolution of the CVs during the metadynamics simulation. (C) Reconstructed free energy landscape. The minimal free energy pathway is shown by *gray dots* connected with *black lines*. Contour lines are separated by 4 kcal/mol intervals. (D) Structures of the reactants (R), transition state (TS), and glycosyl-enzyme intermediate (P). The Zn ligands have been omitted for clarity. (E) Metadynamics simulation, mapped onto a Cremer–Pople sphere. The conformations visited before the TS are shown in *blue*, while those visited after the TS are shown in *red*. The average TS conformation is shown with a *yellow star*. The experimentally observed conformation for the Michaelis complex in a mutated (D341N) enzyme is shown with a *purple dot* (PDB 3BUP). Three conformations corresponding to glycosyl-enzyme intermediate structures are shown as *brown* (PDB 1QX1), *green* (PDB 1QWU), and *light blue* (PDB 1QWN) dots. Adapted with permission from Petersen, L., Ardèvol, A., Rovira, C., & Reilly, P. J. (2010). Molecular mechanism of the glycosylation step catalyzed by Golgi alpha-mannosidase II: A QM/MM metadynamics investigation. *Journal of the American Chemical Society, 132*, 8291–8300. Copyright 2010 American Chemical Society. (See the color plate.)

agrees well with the experimentally measured free energy of activation of 20 kcal/mol (obtained from the hydrolysis rate constant) for GH38 *Aspergillus fischeri* α-mannosidase hydrolyzing the α-Man-(1→6)-Man bond (Shashidhara & Gaikwad, 2009).

A detailed description of the glycosylation reaction in Golgi α-mannosidase II can be obtained by following the minimum free energy pathway, as given by the intrinsic reaction coordinate (Fukui, 1981). Snapshots of average structures for the reactant, TS, and product states are shown in Fig. 9D. At the reactant state, the catalytic acid/base (Asp341) is in its protonated state, forming a hydrogen bond (1.74 Å) with the glycosidic oxygen atom (O6). When the reaction starts, the nucleophile (Asp204) is in its charged state and is significantly separated (3.25 Å) from the C1 atom (Table 1). The reaction begins with the Asp341 O–H bond lengthening along the minimal free energy pathway from the reactant state (R) to R′ (Fig. 6C), indicating the partial transfer of the Asp341 proton. At R′, the Asp341 proton is 1.35 Å from the O6 glycosidic oxygen atom (Table 1). The system then reaches the TS when the scissile bond elongates to 2 Å

Table 1 Lengths (Å) of Important Bonds Along the Reaction Pathway of Golgi α-Mannosidase II

Bond	Reactant (R)	R′	TS	P′	Product (P)
$C1'-O6$	1.53 ± 0.05[a]	1.56 ± 0.06	2.00 ± 0.23	2.57 ± 0.10	3.47 ± 0.11
$C1'-O_{Asp204}$	3.25 ± 0.06	3.25 ± 0.12	2.93 ± 0.11	2.56 ± 0.14	1.50 ± 0.04
$O_{Asp341}-H_{Asp341}$	1.03 ± 0.03	1.12 ± 0.06	1.27 ± 0.02	1.73 ± 0.05	1.56 ± 0.03
$H_{Asp341}-O6$	1.74 ± 0.05	1.35 ± 0.04	1.19 ± 0.03	1.02 ± 0.03	1.03 ± 0.02
$C1'-O5'$	1.41 ± 0.03	1.40 ± 0.03	1.32 ± 0.03	1.29 ± 0.02	1.41 ± 0.04
$O2'-H2'$	1.04 ± 0.02	1.04 ± 0.03	1.03 ± 0.04	1.06 ± 0.05	1.03 ± 0.03
$H2'-O_{Asp92}$	1.55 ± 0.11	1.56 ± 0.10	1.56 ± 0.09	1.51 ± 0.13	1.60 ± 0.14
$Zn-O_{Asp204}$	2.12 ± 0.06	2.10 ± 0.08	2.11 ± 0.09	2.19 ± 0.14	3.00 ± 0.08
$Zn-O_{Asp92}$	2.37 ± 0.13	2.28 ± 0.12	2.23 ± 0.10	2.26 ± 0.12	2.08 ± 0.07
$Zn-N_{His90}$	2.17 ± 0.05	2.15 ± 0.07	2.18 ± 0.07	2.10 ± 0.07	2.12 ± 0.06
$Zn-N_{His471}$	2.13 ± 0.08	2.15 ± 0.07	2.18 ± 0.08	2.13 ± 0.07	2.11 ± 0.06
$Zn-O2'$	2.27 ± 0.06	2.35 ± 0.13	2.46 ± 0.16	2.29 ± 0.11	2.24 ± 0.10
$Zn-O3'$	2.32 ± 0.07	2.34 ± 0.16	2.37 ± 0.11	2.39 ± 0.11	2.26 ± 0.11

[a]Standard deviation.

and the distance between the nucleophile, Asp204, and the C1′ atom shortens from 3.25 Å in the reactant state to 2.93 Å in the TS. The Asp341 proton transfers further to within 1.19 Å of O6. In addition, the ring C1′–O5 bond shrinks significantly from 1.41 Å in the reactant state to 1.32 Å in the TS, indicating the formation of a partial double bond and the increasing oxocarbenium ion character of the mannosyl substrate. The substrate is expected to reach its maximal oxocarbenium ion character at CV1 = 0, where both the C1′–OAsp204 and C1′–O6 bonds are broken (ie, their coordination numbers vanish). This corresponds to P′ in the FES (Fig. 9C). At this point, both the nucleophile and leaving group are at ∼2.5 Å from the C1′ atom. The C1′–O5 distance reaches a minimum (1.29 Å) and the C1 atom becomes sp^2 hybridized. Interestingly, the TS does not coincide with the point of maximal oxocarbenium ion character (P′) but is closer to the reactants (Fig. 9C). Therefore, the simulation shows that the glycosylation reaction in Golgi α-manosidase II follows a $D_N A_N$ dissociative mechanism, where the nucleophilic attack occurs before the scissile glycosidic bond is fully broken (Schramm & Shi, 2001).

The conformations of the −1 glycosyl unit of the substrate during the reaction can be analyzed using the puckering coordinates of all configurations along the metadynamics trajectory and project them over Stoddart's diagram (Fig. 9E). The mannosyl ring conformations visited before the TS (shown in blue) lie around the $B_{2,5}$ conformation, slightly protruding into the $^OS_2$ region. The conformations visited after the TS are shown in red, populating the region between the $B_{2,5}$ and $^1S_5$ conformations. Experimental points for the Michaelis complex and enzyme–mannosyl covalent intermediate are also plotted onto the Stoddart diagram for comparison. The average conformation found for the TS structures is almost a perfect $B_{2,5}$ (the yellow star in Fig. 9E), which conforms to one of the six conformations that stabilize an oxocarbenium ion that were mentioned previously. Therefore, the enzyme follows a $^OS_2/B_{2,5} \rightarrow [B_{2,5}]^{\ddagger} \rightarrow {}^1S_5$ itinerary, supporting an earlier proposal based on comparing α- and β-mannanases (Numao, Kuntz, Withers, & Rose, 2003).

4. CONCLUSIONS

GHs are tremendously efficient biological catalysts that induce particular conformations on the sugar substrates to enhance reaction rates. In this chapter, we have shown how important is to equilibrate the crystallographic structure by classical MD to relax the modified models and obtain several structures of the catalytic system. Special care has to be taken with the sugar conformations during this step, as current force fields often fail to capture sugar distortion in GHs. Once equilibrium is reached, the selection of a reliable initial structure and a large enough QM region is crucial for obtaining meaningful results. A good accuracy/cost balance for the QM/MM electrostatic embedding needs to be considered before the QM/MM equilibration step. We also showed how to control and select CVs of the metadynamics simulation and solving possible technical problems. The simulation trajectory and the evolution of the CVs can give information on how to improve the variables, as well as detecting slow modes that were not previously taken into account. Considering all these aspects is necessary to fully capture the mechanistic complexity of this important class of enzymes.

ACKNOWLEDGMENTS
This work was supported by the Generalitat de Catalunya (Grant 2014SGR-987) and by the Ministerio de Economía y Competitividad (MINECO; Grant CTQ2014-55174). We acknowledge the computer support, technical expertise, and assistance provided by the Barcelona Supercomputing Center-Centro Nacional de Supercomputación (BSC-CNS). L.R. acknowledges the University of Barcelona for a Ph.D. studentship (APIF 2013-2014).

REFERENCES
Agirre, J., Davies, G., Wilson, K., & Cowtan, K. (2015). Carbohydrate anomalies in the PDB. *Nature Chemical Biology, 11*, 303.
Agirre, J., Iglesias-Fernandez, J., Rovira, C., Davies, G. J., Wilson, K. S., & Cowtan, K. D. (2015). Privateer: Software for the conformational validation of carbohydrate structures. *Nature Structural & Molecular Biology, 22*, 833–834.
Amaya, M. F., Watts, A. G., Damager, I., Wehenkel, A., Nguyen, T., Buschiazzo, A., et al. (2004). Structural insights into the catalytic mechanism of *Trypanosoma cruzi* trans-sialidase. *Structure, 12*, 775–784.
Ardèvol, A., & Rovira, C. (2015). Reaction mechanisms in carbohydrate-active enzymes: Glycoside hydrolases and glycosyltransferases. Insights from ab initio quantum mechanics/molecular mechanics dynamic simulations. *Journal of the American Chemical Society, 137*, 7528–7547.
Barducci, A., Bonomi, M., & Parrinello, M. (2011). Metadynamics. *WIREs Computational Molecular Science, 1*, 826–843.
Biarnés, X., Ardèvol, A., Iglesias-Fernández, J., Planas, A., & Rovira, C. (2011). Catalytic itinerary in 1,3-1,4-beta-glucanase unraveled by QM/MM metadynamics. Charge is not yet fully developed at the oxocarbenium ion-like transition state. *Journal of the American Chemical Society, 133*, 20301–20309.
Biarnés, X., Ardèvol, A., Planas, A., Rovira, C., Laio, A., & Parrinello, M. (2007). The conformational free energy landscape of beta-D-glucopyranose. Implications for substrate

preactivation in beta-glucoside hydrolases. *Journal of the American Chemical Society, 129*, 10686–10693.

Biarnés, X., Nieto, J., Planas, A., & Rovira, C. (2006). Substrate distortion in the Michaelis complex of *Bacillus* 1,3-1,4-beta-glucanase. Insight from first principles molecular dynamics simulations. *The Journal of Biological Chemistry, 281*, 1432–1441.

Caines, M. E., Hancock, S. M., Tarling, C. A., Wrodnigg, T. M., Stick, R. V., Stutz, A. E., et al. (2007). The structural basis of glycosidase inhibition by five-membered iminocyclitols: The clan a glycoside hydrolase endoglycoceramidase as a model system. *Angewandte Chemie International Edition in English, 46*, 4474–4476.

Cantarel, B. L., Coutinho, P. M., Rancurel, C., Bernard, T., Lombard, V., & Henrissat, B. (2009). The Carbohydrate-Active EnZymes database (CAZy): An expert resource for glycogenomics. *Nucleic Acids Research, 37*, D233–D238.

CPMD Program, Copyright MPI für Festkörperforschung, Stuttgart 1997–2001. http://www.cpmd.org.

Dama, J. F., Parrinello, M., & Voth, G. A. (2014). Well-tempered metadynamics converges asymptotically. *Physical Review Letters, 112*, 240602.

Davies, G. J., Ducros, V. M., Varrot, A., & Zechel, D. L. (2003). Mapping the conformational itinerary of beta-glycosidases by X-ray crystallography. *Biochemical Society Transactions, 31*, 523–527.

Davies, G. J., Planas, A., & Rovira, C. (2012). Conformational analyses of the reaction coordinate of glycosidases. *Accounts of Chemical Research, 45*, 308–316.

Davies, G. J., & Williams, S. J. (2016). Carbohydrate-active enzymes: Sequences, shapes, contortions and cells. *Biochemical Society Transactions, 44*, 79–87.

Dellago, C., & Bolhuis, P. (2009). Transition path sampling and other advanced simulation techniques for rare events. In C. Holm & K. Kremer (Eds.), *Advanced computer simulation approaches for soft matter sciences III: Vol. 221.* (pp. 167–233). Berlin, Heidelberg: Springer.

Ensing, B., De Vivo, M., Liu, Z. W., Moore, P., & Klein, M. L. (2006). Metadynamics as a tool for exploring free energy landscapes of chemical reactions. *Accounts of Chemical Research, 39*, 73–81.

Ensing, B., Laio, A., Parrinello, M., & Klein, M. L. (2005). A recipe for the computation of the free energy barrier and the lowest free energy path of concerted reactions. *The Journal of Physical Chemistry B, 109*, 6676–6687.

Fukui, K. (1981). The path of chemical reactions—The IRC approach. *Accounts of Chemical Research, 14*, 363–368.

Garcia-Viloca, M., Gao, J., Karplus, M., & Truhlar, D. G. (2004). How enzymes work: Analysis by modern rate theory and computer simulations. *Science, 303*, 186–195.

Guvench, O., Mallajosyula, S. S., Raman, E. P., Hatcher, E., Vanommeslaeghe, K., Foster, T. J., et al. (2011). CHARMM additive all-atom force field for carbohydrate derivatives and its utility in polysaccharide and carbohydrate–protein modeling. *Journal of Chemical Theory and Computation, 7*, 3162–3180.

Hammes-Schiffer, S. (2013). Catalytic efficiency of enzymes: A theoretical analysis. *Biochemistry, 52*, 2012–2020.

Iannuzzi, M., Laio, A., & Parrinello, M. (2003). Efficient exploration of reactive potential energy surfaces using Car-Parrinello molecular dynamics. *Physical Review Letters, 90*, 238302.

Iglesias-Fernandez, J., Raich, L., Ardèvol, A., & Rovira, C. (2015). The complete conformational free-energy landscape of β-xylose reveals a two-fold catalytic itinerary for β-xylanases. *Chemical Science, 6*, 1167–1177.

IUPAC. (1980). IUPAC-IUB Joint Commission on Biochemical Nomenclature (JCBN). Conformational nomenclature for five and six-membered ring forms of monosaccharides and their derivatives: Recommendations 1980. *European Journal of Biochemistry, 111,* 295–298.

Kirschner, K. N., Yongye, A. B., Tschampel, S. M., Gonzalez-Outeirino, J., Daniels, C. R., Foley, B. L., et al. (2008). GLYCAM06: A generalizable biomolecular force field. Carbohydrates. *Journal of Computational Chemistry, 29,* 622–655.

Knott, B. C., Haddad Momeni, M., Crowley, M. F., Mackenzie, L. F., Gotz, A. W., Sandgren, M., et al. (2014). The mechanism of cellulose hydrolysis by a two-step, retaining cellobiohydrolase elucidated by structural and transition path sampling studies. *Journal of the American Chemical Society, 136,* 321–329.

Laio, A., & Gervasio, F. L. (2008). Metadynamics: A method to simulate rare events and reconstruct the free energy in biophysics, chemistry and material science. *Reports on Progress in Physics, 71,* 126601.

Laio, A., & Parrinello, M. (2002). Escaping free-energy minima. *Proceedings of the National Academy of Sciences of the United States of America, 99,* 12562–12566.

Laio, A., VandeVondele, J., & Rothlisberger, U. (2002). A Hamiltonian electrostatic coupling scheme for hybrid Car-Parrinello molecular dynamics simulations. *The Journal of Chemical Physics, 116,* 6941–6947.

Liu, J., Wang, X., & Xu, D. (2010). QM/MM study on the catalytic mechanism of cellulose hydrolysis catalyzed by cellulase Cel5A from *Acidothermus cellulolyticus*. *The Journal of Physical Chemistry B, 114,* 1462–1470.

Lombard, V., Golaconda Ramulu, H., Drula, E., Coutinho, P. M., & Henrissat, B. (2014). The carbohydrate-active enzymes database (CAZy) in 2013. *Nucleic Acids Research, 42,* D490–D495.

Moliner, V. (2011). "Eppur si muove" (Yet it moves). *Proceedings of the National Academy of Sciences of the United States of America, 108*(37), 15013–15014.

Numao, S., Kuntz, D. A., Withers, S. G., & Rose, D. R. (2003). Insights into the mechanism of *Drosophila melanogaster* Golgi alpha-mannosidase II through the structural analysis of covalent reaction intermediates. *The Journal of Biological Chemistry, 278,* 48074–48083.

Offen, W. A., Zechel, D. L., Withers, S. G., Gilbert, H. J., & Davies, G. J. (2009). Structure of the Michaelis complex of beta-mannosidase, Man2A, provides insight into the conformational itinerary of mannoside hydrolysis. *Chemical Communications,* 2484–2486.

Olsson, M. H., Parson, W. W., & Warshel, A. (2006). Dynamical contributions to enzyme catalysis: Critical tests of a popular hypothesis. *Chemical Reviews, 106,* 1737–1756.

Petersen, L., Ardèvol, A., Rovira, C., & Reilly, P. J. (2010). Molecular mechanism of the glycosylation step catalyzed by Golgi alpha-mannosidase II: A QM/MM metadynamics investigation. *Journal of the American Chemical Society, 132,* 8291–8300.

Pierdominici-Sottile, G., Horenstein, N. A., & Roitberg, A. E. (2011). Free energy study of the catalytic mechanism of *Trypanosoma cruzi* trans-sialidase. From the Michaelis complex to the covalent intermediate. *Biochemistry, 50,* 10150–10158.

Plazinski, W., Lonardi, A., & Hünenberger, P. H. (2016). Revision of the GROMOS 56A6CARBO force field: Improving the description of ring-conformational equilibria in hexopyranose-based carbohydrates chains. *Journal of Computational Chemistry, 37,* 354–365.

Raich, L., Borodkin, V., Fang, W., Castro-Lopez, J., van Aalten, D. M., Hurtado-Guerrero, R., et al. (2016). A trapped covalent intermediate of a glycoside hydrolase on the pathway to transglycosylation. Insights from experiments and QM/MM simulations. *Journal of the American Chemical Society, 138*(10), 3325–3332.

Rye, C. S., & Withers, S. G. (2000). Glycosidase mechanisms. *Current Opinion in Chemical Biology, 4,* 573–580.

Satoh, H., & Manabe, S. (2013). Design of chemical glycosyl donors: Does changing ring conformation influence selectivity/reactivity? *Chemical Society Reviews, 42,* 4297–4309.

Schramm, V. L., & Shi, W. (2001). Atomic motion in enzymatic reaction coordinates. *Current Opinion in Structural Biology, 11,* 657–665.
Shashidhara, K. S., & Gaikwad, S. M. (2009). Class II alpha-mannosidase from Aspergillus fischeri: Energetics of catalysis and inhibition. *International Journal of Biological Macromolecules, 44,* 112–115.
Speciale, G., Thompson, A. J., Davies, G. J., & Williams, S. J. (2014). Dissecting conformational contributions to glycosidase catalysis and inhibition. *Current Opinion in Structural Biology, 28C,* 1–13.
Spiwok, V., Kralova, B., & Tvaroska, I. (2010). Modelling of beta-D-glucopyranose ring distortion in different force fields: A metadynamics study. *Carbohydrate Research, 345*(4), 530–537.
Stoddart, J. F. (1971). *Stereochemistry of carbohydrates*. New York: Wiley-Interscience.
Tankrathok, A., Iglesias-Fernández, J., Williams, R. J., Pengthaisong, S., Baiya, S., Hakki, Z., et al. (2015). A single glycosidase harnesses different pyranoside ring transition state conformations for hydrolysis of mannosides and glucosides. *ACS Catalysis, 5,* 6041–6051.
Thompson, A. J., Dabin, J., Iglesias-Fernández, J., Ardèvol, A., Dinev, Z., Williams, S. J., et al. (2012). The reaction coordinate of a bacterial GH47 alpha-mannosidase: A combined quantum mechanical and structural approach. *Angewandte Chemie International Edition in English, 51*(44), 10997–11001.
Torrie, G. M., & Valleau, J. P. (1977). Nonphysical sampling distributions in Monte Carlo free-energy estimation: Umbrella sampling. *Journal of Computational Physics, 23,* 187–199.
Vocadlo, D. J., & Davies, G. J. (2008). Mechanistic insights into glycosidase chemistry. *Current Opinion in Chemical Biology, 12,* 539–555.
White, A., & Rose, D. R. (1997). Mechanism of catalysis by retaining beta-glycosyl hydrolases. *Current Opinion in Structural Biology, 7,* 645–651.
White, A., Tull, D., Johns, K., Withers, S. G., & Rose, D. R. (1996). Crystallographic observation of a covalent catalytic intermediate in a beta-glycosidase. *Nature Structural Biology, 3,* 149–154.
Wolfenden, R., & Snider, M. J. (2001). The depth of chemical time and the power of enzymes as catalysts. *Accounts of Chemical Research, 34,* 938–945.
Zechel, D. L., & Withers, S. G. (2000). Glycosidase mechanisms: Anatomy of a finely tuned catalyst. *Accounts of Chemical Research, 33,* 11–18.

CHAPTER EIGHT

Toward Determining ATPase Mechanism in ABC Transporters: Development of the Reaction Path–Force Matching QM/MM Method

Y. Zhou, P. Ojeda-May, M. Nagaraju, J. Pu[1]

Indiana University-Purdue University Indianapolis, Indianapolis, IN, United States
[1]Corresponding author: e-mail address: jpu@iupui.edu

Contents

1. Introduction — 186
2. Biological Questions — 187
   2.1 Overview — 187
   2.2 How Many ATPs Are Hydrolyzed? — 189
   2.3 What Is the Precise ATP Hydrolysis Mechanism? — 190
   2.4 Is H-Loop His a "Chemical Linchpin"? — 191
3. QM/MM Simulations of the ABC-Transporter HlyB — 193
   3.1 QM/MM Potential Energy Study of ATP Hydrolysis in HlyB — 193
   3.2 GAC Mechanism Identified as the Operative Mechanism — 194
   3.3 QM/MM Minimum Free Energy Path Study of HlyB Using the String Method — 195
4. Computational Challenges — 196
   4.1 Quantum Mechanics and Statistical Sampling: Can One Avoid the Other? — 196
   4.2 Existing SE(-SRP) Methods Are Not Reliable for Simulating ATP Hydrolysis — 199
5. A Multiscale QM/MM Method: RP–FM — 200
   5.1 Overall Strategy — 200
   5.2 Generic Procedure — 201
   5.3 Case Study: Proton Transfer — 203
   5.4 Relation to Other Methods — 205
6. Concluding Remarks — 207
Acknowledgments — 208
References — 208

Abstract

Adenosine triphosphate (ATP)-binding cassette (ABC) transporters are ubiquitous ATP-dependent membrane proteins involved in translocations of a wide variety of substrates across cellular membranes. To understand the chemomechanical coupling mechanism as well as functional asymmetry in these systems, a quantitative description of how ABC transporters hydrolyze ATP is needed. Complementary to experimental approaches, computer simulations based on combined quantum mechanical and molecular mechanical (QM/MM) potentials have provided new insights into the catalytic mechanism in ABC transporters. Quantitatively reliable determination of the free energy requirement for enzymatic ATP hydrolysis, however, requires substantial statistical sampling on QM/MM potential. A case study shows that brute force sampling of ab initio QM/MM (AI/MM) potential energy surfaces is computationally impractical for enzyme simulations of ABC transporters. On the other hand, existing semiempirical QM/MM (SE/MM) methods, although affordable for free energy sampling, are unreliable for studying ATP hydrolysis. To close this gap, a multiscale QM/MM approach named reaction path–force matching (RP–FM) has been developed. In RP–FM, specific reaction parameters for a selected SE method are optimized against AI reference data along reaction paths by employing the force matching technique. The feasibility of the method is demonstrated for a proton transfer reaction in the gas phase and in solution. The RP–FM method may offer a general tool for simulating complex enzyme systems such as ABC transporters.

1. INTRODUCTION

Found in organisms of all three kingdoms of life, adenosine triphosphate (ATP)-binding cassette (ABC) transporters represent a family of molecular motor proteins that enable translocations of various substrates, including small organic/inorganic molecules, ions, lipids, peptides, and toxins, across the cell membranes, at the expense of ATP hydrolysis (Davidson, Dassa, Orelle, & Chen, 2008; Hollenstein, Dawson, & Locher, 2007; Schneider & Hunke, 1998). ABC transporters are biomedically important because mutations in these systems can cause an array of human diseases and clinical problems (Borst & Oude Elferink, 2002). All ABC transporters share a common architecture of two basic building blocks: a dimeric transmembrane domain (TMD) composed of transmembrane helices and a pair of intracellular nucleotide-binding domains (NBDs), which is a highly conserved motor ATPase that energizes the transport (Oswald, Holland, & Schmitt, 2006). Forming a channel for substrate passage, the TMD exists at least two distinct states, termed the inward- and outward-facing conformations (Rees, Johnson, & Lewinson, 2009). Despite the general knowledge that the TMD cycles between the inward- and outward-facing conformations as a result of ATP binding and hydrolysis in the NBDs (Newstead et al., 2009), the precise mechanism of ATP hydrolysis in ABC transporters as well as the extent to which enzyme catalysis is coupled to conformational dynamics during substrate translocation have largely remained unknown (Davidson et al., 2008).

One intriguing question staying controversial is how many ATP molecules are hydrolyzed per translocation cycle. Although the dimeric arrangement of NBDs appears to imply that two ATP molecules are consumed, a great body of experimental evidence suggests that the two NBDs have intrinsic functional asymmetry and may be catalytically nonequivalent. If both sites are able to hydrolyze ATP, are their actions of catalysis concerted or sequential? Answers to these questions can provide important insights into understanding the sequence of events that controls conformational coupling between TMD and NBD. A quantitative model to describe the hydrolysis process also offers an important step toward understanding how enzyme catalysis in ABC transporters is coupled to protein dynamics especially when NBDs undergo large conformational changes.

Given the mechanistic complexity of ABC transporters, which are large-sized enzymes that involve multiple active sites, multiscale computer simulations based on the combined quantum mechanical and molecular mechanical (QM/MM) approach are especially useful to complement experiments by providing a unique molecular-level understanding of enzyme mechanism that synthesizes structural, dynamic, and energetic information. In this chapter, we discuss progress and challenges in QM/MM simulations of ABC transporters, which have also led to our recent development of a new multiscale approach aimed at overcoming the limitations found in the existing methods.

To highlight the problem-driven nature of this development, we organize our discussion by first describing the biological questions we seek to answer (Section 2), followed by a brief summary of our QM/MM simulations of ATP hydrolysis in a representative ABC transporter (Section 3). In Section 4, we discuss challenges in those simulations and limitations in the existing computational methods. The newly developed Reaction Path–Force Matching (RP–FM) method is presented in Section 5. Concluding remarks are provided in Section 6.

2. BIOLOGICAL QUESTIONS

2.1 Overview

Based on the direction of translocation, the ABC-transporter family can be divided into two functional categories: importers and exporters

(Davidson et al., 2008). To illustrate a typical transport cycle, below we use the maltose transporter (an importer) as an example.

The maltose transporter is responsible for uptake of maltose in bacteria. Its TMD (MalFG) and NBDs (MalK2) are organized as separate polypeptide chains and assembled into a membrane-bound complex. The maltose transport cycle is regulated under a peripheral maltose-binding protein (MBP) (Davidson & Chen, 2004). Crystal structures of isolated NBDs showed that MalK2 can adopt at least three distinct NBD conformations, referred to as the open, semiopen, and closed states (Chen, Lu, Lin, Davidson, & Quiocho, 2003). These NBD conformations were also observed in crystal structures of intact maltose transporter captured at various stages of its working cycle (Fig. 1), including an inward-facing "resting state" (Khare, Oldham, Orelle, Davidson, & Chen, 2009), an outward-facing "catalytic intermediate" (Oldham, Khare, Quiocho, Davidson, & Chen, 2007), and a pretranslocation state (denoted "pre-T") (Oldham & Chen, 2011). In the resting state structure, NBD is found in the open conformation and interacts with MalFG that faces the cytosolic side of the membrane. The inward- to outward-facing conformational change of TMD during the "pre-T" to "catalytic intermediate" transition is primarily driven by MalK2 dimer closure upon ATP binding. After hydrolysis and product release, the system returns to the resting state, and the motor cycle is reset.

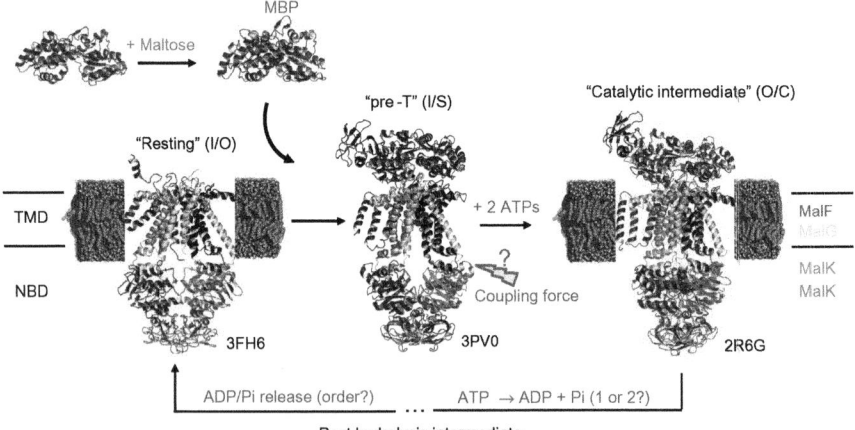

Fig. 1 Crystal structures captured for key intermediate states in the working cycle of maltose transporter. The TMD/NBD conformations are given in parentheses (TMD: I or O, for inward- or outward-facing conformations; NBD: O, S, or C, for open, semiopen, or closed conformations). (See the color plate.)

Key questions to be answered for ABC transporters include how TMD and NBD are mechanically coupled and to what extent enzyme catalysis is coupled to conformation transitions of NBD. For a thorough understanding of the chemomechanical coupling mechanism in ABC transporters, catalysis needs to be examined as a function of conformational change. As the first step, we focus on the ATP hydrolysis mechanism in a single-state conformation, ie, the closed conformer of NBD, which is believed to be catalytically more active than the open and semiopen conformers. Specifically, we seek to answer the following questions.

2.2 How Many ATPs Are Hydrolyzed?

One fundamental question regarding the mechanisms of action in ABC transporters is how many ATP molecules are hydrolyzed per translocation cycle. At first glance, the fact that most functional NBDs in bacterial ABC transporters are homodimers seems to suggest that two copies of ATP need to be hydrolyzed to provide the chemical free energy for substrate translocation. Contrary to this intuition, direct measurements of ATP/substrate ratio in ABC transporters, however, have given greatly diverging results ranging from 1 to 25 (Davidson et al., 2008), indicating that a definitive answer to this question is perhaps difficult to obtain experimentally.

A great body of evidence has implied that the two NBDs in ABC transporters are functionally nonequivalent for ATP hydrolysis. In many eukaryotic exporters, such as the transporter associated with antigen processing (Procko, Ferrin-O'Connell, Ng, & Gaudet, 2006), the multidrug resistance protein P-glycoprotein (Urbatsch, Sankaran, Weber, & Senior, 1995), and the cystic fibrosis transmembrane conductance regulator (Atwell et al., 2010), NBDs are heterodimeric in that only one of the two NBD active sites contains all essential residues required for catalysis, suggesting that only one ATP is hydrolyzed. Interestingly, evidence for functional asymmetry has also been reported for ABC transporters that contain homodimeric NBDs. For example, vanadate-trapping experiments showed that the hydrolysis product is found in only one subunit of the NBD homodimer in maltose transporter (Sharma & Davidson, 2000). Intriguingly, mutation to a single NBD active site has been shown to abolish ATPase and transport activities in histidine permease (Nikaido & Ames, 1999; Shyamala, Baichwal, Beall, & Ames, 1991), maltose transporter (Davidson & Sharma, 1997), and vitamin B12 transporter (Tal, Ovcharenko, & Lewinson, 2013). Recent crystal structures of NBDs of HlyB, a bacterial exporter, suggested that the two NBDs, although identical in sequence, may release their hydrolysis products differently (Zaitseva et al., 2006). For the exporter MJ0796, it was found that disabling ATP hydrolysis in a single active site leaves dimer dissociation essentially unchanged (Zoghbi & Altenberg, 2013). Finally, computer simulations of dimeric NBDs of maltose transporter suggested that hydrolysis in a single site can trigger dimer interface opening (Wen & Tajkhorshid, 2008).

Quantitative characterization of such functional asymmetry, especially in terms of catalytic activity in the dimeric NBD sites, may also help resolve mechanistic controversy existing for ABC transporters. Catalytically asymmetric dimer would enhance the chance that only one ATP is hydrolyzed at a time, an essential element in the alternating catalysis mechanism (Senior, Al-Shawi, & Urbatsch, 1995). Without significant catalytic nonequivalence, the transporter would more likely operate under the processive clamp model (Janas et al., 2003), which requires both ATP molecules to be hydrolyzed before NBD dimer dissociation.

2.3 What Is the Precise ATP Hydrolysis Mechanism?

The first step to determine whether the two NBD sites are equivalent in catalysis is to obtain a quantitative description of their ATP hydrolysis mechanism. The observation that NBDs across the ABC-transporter family share a highly conserved active site architecture suggests that they hydrolyze ATP by a common mechanism. Below we use the bacterial exporter HlyB as an example to describe the enzyme mechanistic questions we seek to answer by computation.

Haemolysin B (HlyB) is the inner membrane component of the type I protein secretion apparatus that transports the 107 kDa pore-forming toxin haemolysin A out of Gram-negative bacterial cells (Holland, Schmitt, & Young, 2005). It is well established that HlyB functions as an ABC transporter, consisting of two cytosolic NBDs and two TMDs. As the molecular motor component of the transporter, HlyB-NBD works as an ATPase that catalyzes ATP hydrolysis at a rate constant of $k_{cat} = 0.2 \text{ s}^{-1}$ (Zaitseva, Jenewein, Jumpertz, Holland, & Schmitt, 2005), corresponding to a rate acceleration of more than six orders of magnitude compared to the solution phase reaction ($k = 8 \times 10^{-8} \text{ s}^{-1}$) (Khan & Mohan, 1973).

Despite a wealth of information accumulated from sequence analyses (Geourjon et al., 2001) as well as from biochemical and structural characterizations (Zaitseva, Holland, & Schmitt, 2004; Zaitseva, Jenewein, Jumpertz, et al., 2005; Zaitseva, Jenewein, Wiedenmann, et al., 2005; Zaitseva et al., 2006), the precise mechanism of ATP hydrolysis in HlyB remains unavailable. Like many members of ABC transporters, HlyB has a pair of NBD active sites that contain a collection of highly conserved sequence motifs (Fig. 2A), including the Walker A motif (or P-loop: GXXXXGKST, for phosphate tail binding) and the Walker B motif ($\phi\phi\phi\phi$D, for Mg^{2+} binding), the signature-loop (or C-loop: LSGGQ, from the opposite subunit), a conserved Glu (E631, immediately following the Walker B motif), and the H-loop that contains a His (H662) (Zaitseva, Jenewein, Jumpertz, et al., 2005). The functional HylB-NBD dimer adopts a "head-to-tail" arrangement such that each of the two ATP molecules bound at the dimer interface is sandwiched between the P-loop of one subunit and the C-loop of the other subunit.

2.4 Is H-Loop His a "Chemical Linchpin"?

Unlike the relatively well-defined role of the Walker motifs, the role of the H-loop His (H662 in HlyB) in catalysis has remained elusive. The H-loop His is located in the switch II region of HlyB-NBD (Schmitt, Benabdelhak, Blight, Holland, & Stubbs, 2003). Sequence and structural comparisons of ABC transporters and helicases have suggested that the H-loop His may act as a sensor for the γ-phosphate in ATP (Geourjon et al., 2001), therefore reminiscent of the conserved Gln in RecA (Story & Steitz, 1992). Mutagenesis studies showed that the mutation of this H-loop His to Ala (H662A) reduces the ATPase activity of the NBD in HlyB to background levels (<0.1% residual ATPase activity) (Zaitseva et al., 2004; Zaitseva, Jenewein, Jumpertz, et al., 2005). The hydrogen bonding interactions between the H-loop and the P-loop within the same NBD as well as the D-loop (SALD) in the opposite NBD suggest roles of H662 in ATP binding and in conformational signaling across the NBD dimer interface (Zaitseva, Jenewein, Jumpertz, et al., 2005). Based on a crystal structure of ATP/Mg^{2+} bound dimeric NBDs of HlyB that contains Ala mutation at the position of H662, Schmitt and coworkers proposed that H662 and E631 form a catalytic dyad, in which the H662 acts as a "linchpin" (Fig. 2A) that holds other active site residues at their catalytically competent configurations (Zaitseva, Jenewein, Jumpertz, et al., 2005).

One key question we asked is whether the H-loop His can explicitly participate in the enzyme mechanism, by acting as a "chemical linchpin" (Zhou, Ojeda-May, & Pu, 2013). Specifically, two different hydrolysis

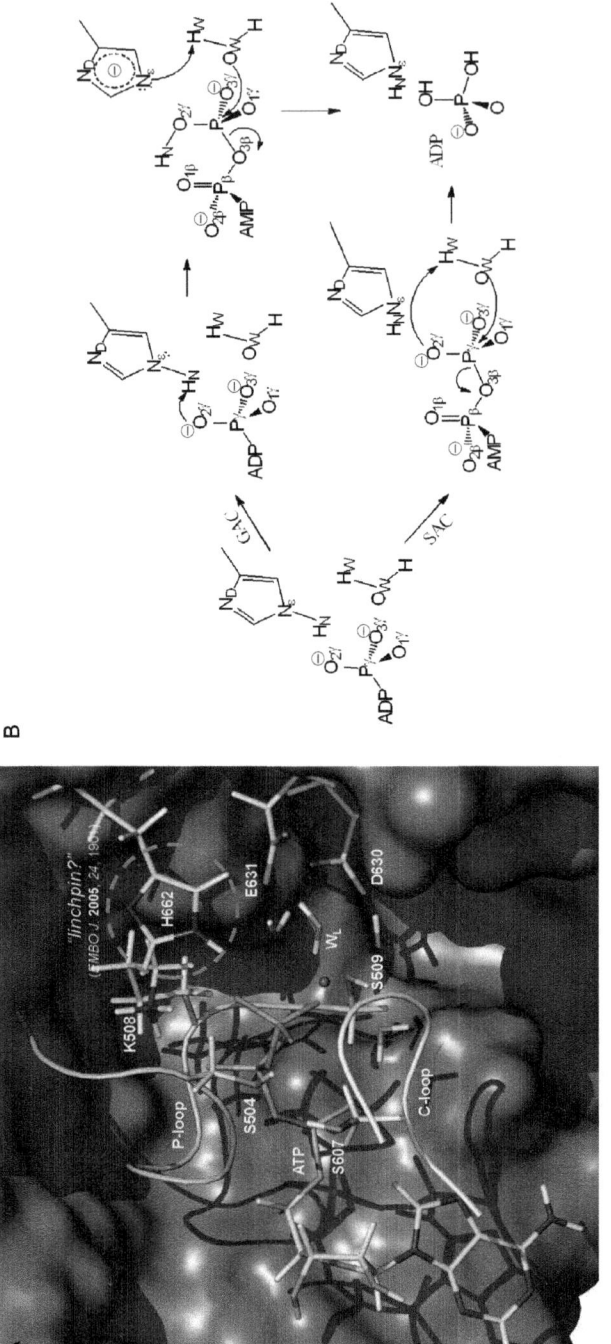

Fig. 2 (A) Active site of HlyB-NBD. In the "mechanical linchpin" proposal, H662 holds active site residues at their catalytically competent configurations. (B) Proposed enzyme mechanisms for ATP hydrolysis in ABC transporters. In the GAC mechanism, H662 serves as a "chemical linchpin" that explicitly participates in catalysis by providing proton relay. By contrast, in the SAC mechanism, a proton is directly transferred from the catalytic water to the γ-phosphate of ATP.

mechanisms may exist (Fig. 2B). In the first mechanism, which is referred to as the general acid catalysis (GAC) mechanism, the H662 initially serves as a general acid by donating its proton at the N_ε position to the γ-phosphate of ATP and subsequently accepts a proton from the lytic water. The second mechanism utilizes substrate-assisted catalysis (SAC) (Zaitseva, Jenewein, Jumpertz, et al., 2005). The two mechanisms differ in the role of the H-loop H662 residue: in the GAC mechanism H662 explicitly participates in catalysis through proton relay, whereas in the SAC mechanism a direct proton transfer takes place in the presence of a spectator H662.

To provide quantitative answers to the questions concerning ATP hydrolysis in ABC transporters, free energy requirements need to be examined for various catalytic mechanisms in different active sites. To date, the majority of work done in simulating ABC transporters is based on classical mechanical force fields (Jones, O'Mara, & George, 2009; Li, Wen, Moradi, & Tajkhorshid, 2015; Moradi & Tajkhorshid, 2013; Oloo, Kandt, O'Mara, & Tieleman, 2006; Weng, Fan, & Wang, 2010). Classical mechanical simulations, however, are not suitable for studying enzyme mechanisms in ABC transporters due to involvement of bond rearrangements during catalysis. Combined QM/MM (Gao & Thompson, 1998; Warshel & Levitt, 1976) methods provide an especially appealing approach in closing this gap. In QM/MM treatment of enzymes, a small-sized reactive subsystem containing active site molecules is described by QM, and the rest of the system, including nonreactive protein fragments and bulk solvent, is modeled by efficient MM force fields. Despite the popularity of the method, QM/MM had not been applied to simulating any ABC-transporter system only until recently by our group.

3. QM/MM SIMULATIONS OF THE ABC-TRANSPORTER HlyB

3.1 QM/MM Potential Energy Study of ATP Hydrolysis in HlyB

In our recent study (Zhou et al., 2013), we have examined both the GAC and SAC mechanisms (see Fig. 2B for definition) for ATP hydrolysis in HlyB-NBD by QM/MM potential energy calculations. The computer model was constructed based on two crystal structures of *E. coli* HlyB (PDB: 1XEF and 2FGK). The HlyB-NBD dimer was solvated in a water sphere with a radius of 30.4 Å (Fig. 3A). The selected active site is partitioned into two regions (see Fig. 3B). The QM region, described by

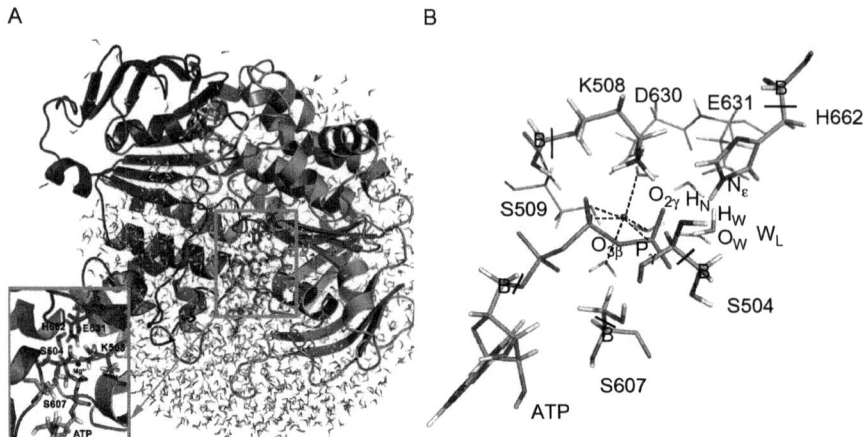

Fig. 3 (A) Simulation setup for studying ATP hydrolysis mechanism in HlyB. (B) QM/MM partition of the active site of HlyB ("B" refers to QM boundary atoms treated by the GHO method). *Reproduced from Zhou, Y., Ojeda-May, P., & Pu, J. (2013). H-loop histidine catalyzes ATP hydrolysis in the E. coli ABC-transporter HlyB. Physical Chemistry Chemical Physics, 15, 15811, doi: 10.1039/C3CP50965F with permission from the PCCP Owner Societies.*

the semiempirical AM1 (Dewar, Zoebisch, Healy, & Stewart, 1985) Hamiltonian, consists of the three phosphate groups of ATP, side chains of S504, K508, and H662, the side chain of S607 in the opposite monomer, the lytic water, and five boundary carbon atoms that are treated by the generalized hybrid orbital (GHO) method (Gao, Amara, Alhambra, & Field, 1998; Pu, Gao, & Truhlar, 2004) (Fig. 3B). The MM region, described by the CHARMM22 (MacKerell et al., 1998) force field with CMAP corrections (MacKerell, Feig, & Brooks, 2004), contains the rest of the system. The CHARMM program (Brooks et al., 2009) was employed for both the system setup and QM/MM energy calculations.

3.2 GAC Mechanism Identified as the Operative Mechanism

In the same study (Zhou et al., 2013), we obtained two-dimensional (2D) QM/MM potential energy surfaces (PESs) for both the GAC and SAC mechanisms (Fig. 4). For the GAC mechanism (Fig. 4A), RC1 represents a proton relay reaction coordinate (RC) involving the H-loop residue H662. For the SAC mechanism (Fig. 4B), RC1 describes a direct proton transfer (RC1 = O_W–H_W − H_W–$O_{2\gamma}$). The phosphoryl transfer reaction coordinate RC2 in both mechanisms is the difference between the $O_{3\beta}$–P_γ and P_γ–O_W bond distances. Based on the reaction pathways

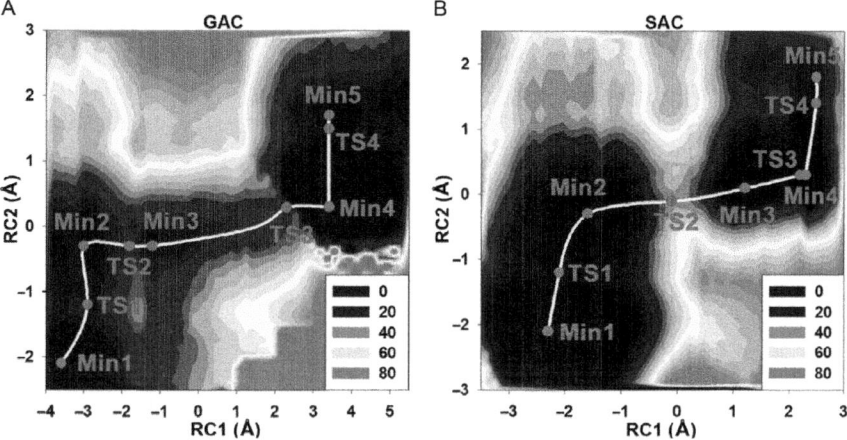

Fig. 4 2D potential energy surfaces (AM1/CHARMM) of ATP hydrolysis catalyzed by HlyB-NBD: (A) the GAC and (B) the SAC mechanisms. The highest barrier found in the GAC mechanism (22.1 kcal/mol; GAC:TS3) is substantially lower than that in the SAC mechanism (32.1 kcal/mol; SAC:TS2). *Reproduced from Zhou, Y., Ojeda-May, P., & Pu, J. (2013). H-loop histidine catalyzes ATP hydrolysis in the E. coli ABC-transporter HlyB.* Physical Chemistry Chemical Physics, 15, *15811, doi: 10.1039/C3CP50965F with permission from the PCCP Owner Societies.* (See the color plate.)

identified on these 2D QM/MM PESs, we found that the highest barrier in the GAC mechanism is only 22.1 kcal/mol (GAC:TS3). By contrast, the SAC mechanism gives a substantially higher barrier of 32.1 kcal/mol (SAC:TS2). These results suggest that the GAC mechanism, which involves H-loop assisted proton relay, is energetically more favorable than the direct proton transfer pathway in the SAC mechanism.

3.3 QM/MM Minimum Free Energy Path Study of HlyB Using the String Method

To incorporate entropic effects, we have recently obtained QM/MM free energy profiles for both the GAC and SAC mechanisms (Zhou, Ojeda-May, Nagaraju, & Pu, 2016). By employing the finite temperature string method (Maragliano, Fischer, Vanden-Eijnden, & Ciccotti, 2006), we represented minimum free energy paths (MFEPs) by discrete images that connect the reactant and product states based on a set of collective variables (CVs) (Zhou et al., 2016). After mean forces of the free energy with respect to the CVs were determined from restrained QM/MM molecular dynamics (MD) simulations, steepest descent dynamics was carried out to minimize the free energy path. To prevent images along the path from falling into the nearest minima, the reparametrization step was performed to evenly distribute the images along the path (Maragliano et al., 2006). The minimization/reparametrization cycles were repeated until the string paths converge. This QM/MM MFEP study confirmed our earlier conclusion based on QM/MM PESs. We found that the overall free energy of activation for the SAC mechanism is about 10 kcal/mol higher than that for the GAC mechanism (Zhou et al., 2016), supporting our "chemical linchpin" proposal (Zhou et al., 2013).

4. COMPUTATIONAL CHALLENGES

While our exploratory QM/MM simulations provided valuable new insights into the enzyme mechanism in ABC transporters, the computational accuracy needed for characterizing functional asymmetry in dimeric NBDs remains to be achieved. As we explain below, quantitatively reliable QM/MM free energy simulations of enzymatic ATP hydrolysis have posed a grand challenge to existing methods in computational enzymology. Consequently, development of new cost-effective QM/MM approach is highly desirable.

4.1 Quantum Mechanics and Statistical Sampling: Can One Avoid the Other?

Hydrolysis of ATP (or guanosine triphosphate, GTP) is an important subject that has been investigated extensively by computation in the past for many non-ABC transporter systems [see a recent review by Warshel and coworkers (Kamerlin, Sharma, Prasad, & Warshel, 2013) and references therein]. Reliable free energy simulations of enzymatic ATP hydrolysis, however, require sufficient statistical sampling on an accurate QM PES. Such a requirement makes combined QM/MM (Gao & Thompson, 1998; Warshel & Levitt, 1976) an attractive approach, with which the enzyme active site can be treated by QM, and the rest of the protein plus bulk solvents by efficient MM.

In terms of sampling the configuration space, although PES scans can be used to quickly explore the major features of the reaction mechanism, such potential energy information is often considered unreliable due to the lack of entropic contributions and has been associated with various pitfalls in enzyme simulations (Klahn, Braun-Sand, Rosta, & Warshel, 2005). Reliable results can be obtained by sampling the system, often along a few selected RCs, using MD simulations, based on which proper ensemble averages can be made. For reaction mechanism studies, this can be done by using umbrella sampling (Torrier & Valleau, 1974) or MFEP techniques, such as the finite temperature string method (E, Ren, & Vanden-Eijnden, 2005; Maragliano et al., 2006).

For the choice of the QM method in QM/MM, although ab initio molecular orbital (Hehre, Radom, Schleyer, & Pople, 1986) methods and density function theory (DFT) (Parr & Yang, 1994), here collectively referred to as ab initio QM (AI) methods, are considered more accurate and reliable, the computational costs associated with AI/MM calculations preclude these simulations to be performed with sufficient statistical sampling. Even aided with extended Lagrangian (EL) techniques, such as Car-Parrinello (Car & Parrinello, 1985) and density matrix propagation (Schlegel et al., 2002) MD, which eliminate self-consistent iterations of solving the Schrödinger equation, directly sampling AI/MM PESs for extended period of time (eg, ns and beyond), as conventionally done in pure molecular mechanical simulations, remains practically infeasible.

By contrast, semiempirical QM (SE) methods (Thiel, 2014) are several orders of magnitude more efficient than AI methods. Their efficiency stems from the use of simplified electronic integrals under the neglect of diatomic differential overlap approximation (Pople, Santry, & Segal, 1965) as well as the use of minimal basis sets. In spite of being considered less reliable, SE methods have been widely used in combined QM/MM simulations, with which moderate sampling of the reactive systems becomes feasible. For example, one- or two-dimensional potential of mean force (PMF) free energy profiles can be routinely obtained through umbrella sampling simulations at SE/MM levels (Garcia-Viloca, Truhlar, & Gao, 2003; Poulsen, Garcia-Viloca, Gao, & Truhlar, 2003).

When both accuracy and efficiency are demanded, one is facing with a dilemma of choosing between them. Ideally, one would like to conduct umbrella sampling or MFEP simulations on an AI/MM surface for its reliability. However, such a combination is often computationally prohibitive. To illustrate the magnitude of this challenge, we tabulate in Table 1 the computational costs for QM/MM simulations of ATP hydrolysis in HlyB by using various methods. For simulating HlyB-mediated ATP hydrolysis, both proton transfer and phosphoryl transfer RCs are needed; explicit inclusion of more than two RCs becomes impractical in umbrella sampling simulations, but can be done by using the string method, which optimizes MFEPs on the basis of multidimensional CVs (Maragliano et al., 2006; Rosta, Nowotny, Yang, & Hummer, 2011).

Table 1 Computational Costs for Various Combinations of QM/MM and Sampling Methods Based on Stochastic Boundary Simulations of ATP Hydrolysis in the HlyB System, Which Consists of ~60 QM Atoms and ~13,600 MM Atoms

QM Method in QM/MM	N_{cpu}	SPE[a]	MD (200 ps)[b]	2D-PES scan[c]	2D-US[d]	String MFEP[e]
AM1[f]	1	0.9 s	2 days	1 days	10 days	10 days
AM1/d-PhoT[f]	1	1.5 s	3.5 days	2.5 days	17.5 days	18 days
HF/6-31G[g]	16	1.5 m	**202 days**	**125 days**	**2.7 years**	**2.7 years**
B3LYP/6-31+G(d,p)[g]	16	7.0 m	**2.7 years**	**1.6 years**	**13.5 years**	**13.5 years**

[a]SPE denotes single-point energy.
[b]200,000 time steps.
[c]Potential energy scan for ~5,000 configurations.
[d]US denotes umbrella sampling; 400 windows in parallel, each with 500 ps sampling.
[e]Minimum free energy paths (MFEP) using the string method; 25 images in parallel, each with 20 ps sampling for mean force evaluation; 50 cycles of string path optimizations.
[f]SE(-SRP)/MM methods.
[g]AI/MM methods, for which b–e (numbers in bold) are rough estimates, based on their SPE costs and ratios of the costs of these calculations in AM1 with respect to the corresponding AM1 SPE cost.

From Table 1, we can see that attempts of using AI/MM methods to simulate ATP hydrolysis in a typical ABC-transporter system are infeasible with commonly available computing resources. Although employing the EL techniques (Car & Parrinello, 1985; Schlegel et al., 2002) can alleviate the problem, these techniques unlikely resolve the computational dilemma outlined earlier; the speedup of using ELMD would become less significant when smaller integration steps are used to maintain adiabatic separation between the electronic and nuclear degrees of freedom. To break even, parallel AI/MM MD simulations need to be accelerated by about three orders of magnitude.

Ideally, however, one would like to perform sampling at the cost of SE/MM methods (which take ~10 days to complete the free energy profile for the HlyB system; see Table 1) to achieve the accuracy of AI/MM. In practice, this goal can be realized, at least in part, by improving SE with specific reaction parameters (SRPs) (Gonzalez-Lafont, Truong, & Truhlar, 1991). Optimization of SRPs for an SE method typically involves one or more of the following steps: (a) construct a training set of gas-phase molecular models that mimic the actual reactive systems; (b) obtain benchmark data, such as reaction energies, barrier heights, proton affinities, vibrational frequencies, atomic charges, dipole moments, geometries, etc., from AI calculations (or from experiments if AI is inaccurate), based on the geometries of stationary species, such as separate reactants/products, reactant/product complexes, and transition states, optimized at the corresponding AI level; (c) adjust selected parameters in the SE method such that the resulting SE-SRP method satisfactorily reproduces the AI results; and finally (d) apply the SE-SRP method to gas-phase or condensed-phase dynamics simulations.

The case study presented here led us to conclude that the practical solution to meet the competing demands for accuracy and efficiency in enzyme simulations of ABC transporters is to use SE(-SRP)/MM methods in conjunction with MFEP samplings. However, as we will show below, the existing SE(-SRP) methods have their own limitations.

4.2 Existing SE(-SRP) Methods Are Not Reliable for Simulating ATP Hydrolysis

The standard parametrizations of popular SE methods, such as AM1, are generally considered unreliable for studying phosphoryl transfer reactions, owing to the empirical nature of its Gaussian treatment of core–core repulsion and the lack of d-orbital description (Lopez & York, 2003). Artificial stabilization of pentacovalent phosphorane intermediates has been reported when AM1 was used to study model reactions for RNA catalysis (Lopez & York, 2003; Nam, Cui, Gao, & York, 2007). There are several SE-SRP methods aimed at improved performance for biological phosphoryl transfers, by including d-orbitals and/or specific parametrizations. Examples of such SE-SRP methods include MNDO + G-SRP (Arantes & Loos, 2006) and AM1/d-PhoT (Nam et al., 2007). Most of these methods were optimized with a training set of data primarily composed of monophosphate esters, which are considerably different from biological phosphate anhydrides, such as ATP. Data in Table 2 provide a glimpse of the challenge of using these existing SE(-SRP) methods to study ATP hydrolysis, based

Table 2 Reaction Energy of MTP Hydrolysis from SE(-SRP) and AI Methods[a]

Method	ΔE (H_2O Attack)	ΔE (OH^- Attack)
AM1	−168.1	−103.2
MNDO + G-SRP	−144.1	−89.7
AM1/d-PhoT	−152.4	−90.7
DFT[b]	−168.0	−103.5
DFT + PCM solvation	−8.6	−25.1

[a]SE method: AM1; SE-SRP methods: MNDO + G-SRP and AM1/d-PhoT; AI method: DFT (see footnote b); reaction energies (ΔE) are in kcal/mol and calculated in the gas phase (except data in the last row) without Mg^{2+}.
[b]B3LYP/6-31 + G(d,p).

on the reaction energy of hydrolyzing methyl triphosphate (MTP), which serves as a reasonable model for ATP hydrolysis:

Table 2 shows that when SE or SE-SRP methods optimized for other chemical situations are used to study ATP hydrolysis, extra care needs to be taken. Both AM1/d-PhoT and MNDO+G-SRP display large errors (15–25 kcal/mol) in reaction energies compared to the DFT benchmark results. Although AM1 seems to agree with the DFT method for the reaction energy of MTP hydrolysis, the agreement relies more on fortuitous error cancelation in AM1, rather than indicating that AM1 is reliable.

5. A MULTISCALE QM/MM METHOD: RP–FM

The analyses in Section 4 suggest that although SE/MM should be chosen over AI/MM for sampling efficiency, currently available SE methods, including those special-purpose SE-SRP methods obtained using the traditional gas-phase fitting procedures, are unlikely to provide the accuracy and reliability needed for determining the ATPase mechanism in ABC transporters. To address this challenge, we recently introduced a new multiscale strategy for reparametrizing SE-SRP within the framework of QM/MM free energy simulations. The new method enables iterative refinements of a selected SE method by fitting accurate forces from AI/MM calculations along reaction paths. Because reaction paths are always sampled at efficient SE-SRP/MM levels, direct sampling of the expensive AI/MM surface is avoided.

5.1 Overall Strategy

The stationary-point-based gas-phase SE-SRP fitting procedure outlined in Section 4.1 may suffer from several limitations. First, only a very limited amount AI data are used for SRP fitting. Second, transferability of SRPs relies on the hope that the gas-phase model systems closely resemble the condensed-phase systems. Third, gas-phase properties may be overrepresented in acquiring SRPs. For MTP hydrolysis, the gas-phase reaction energy is -168.0 kcal/mol (Table 2), in comparison with a reaction free energy of -8.6 kcal/mol in solution [based on PCM solvation (Miertus, Scrocco, & Tomasi, 1981)]. This difference highlights the importance of taking solvation free energy changes into account when fitting SRPs for this reaction. In addition, conformations of molecules can change significantly from the gas phase to condensed phases. For example, the β- and γ-phosphates in GTP tend to adopt a staggered configuration in the gas phase and in water but are often found in the eclipsed configuration when bound in enzyme active sites especially when Mg^{2+} is present (Rudack, Xia, Schlitter, Kotting, & Gerwert, 2012).

Due to these reasons, more robust SE-SRP fitting is expected when condensed-phase samples are taken into account. Given that directly sampling AI/MM PESs in condensed phases is such a formidable task, an alternative strategy is to parametrize SE-SRPs based on condensed-phase reaction-path configurations sampled at efficient SE/MM levels (Zhou & Pu, 2014). In addition, we employed the force matching technique (Arkin-Ojo, Song, & Wang, 2008; Csanyi, Albaret, Payne, & De Vita, 2004; Doemer, Maurer, Campomanes, Tavernelli, & Rothlisberger, 2014; Ercolessi & Adams, 1994; Izvekov, Parrinello, Burnham, & Voth, 2004; Izvekov & Voth, 2005; Knight, Maupin, Izvekov, & Voth, 2010; Laio, Bernard, Chiarotti, Scandolo, & Tosatti, 2000; Maurer, Laio, Hugosson, Colombo, & Rothlisberger, 2007) pioneered by Voth and coworkers (Izvekov et al., 2004; Izvekov & Voth, 2005; Knight et al., 2010) for SRP optimization, so that the resulting SE-SRP method reproduces the atomic forces computed at a selected target AI/MM level. Because of the combined use of reaction path (RP) and force matching (FM) techniques in our approach, we named this method as Reaction Path–Force Matching (RP–FM) (Zhou & Pu, 2014).

The RP–FM method can be viewed as repetition of a two-stage process. In the RP stage, an ensemble of configurations along a specific reaction path are collected using SE/MM simulations; examples of such reaction paths include minimum energy paths (MEPs) [or paths along the intrinsic reaction coordinate (IRC) (Fukui, 1981)] and MFEPs obtained by the string method (E et al., 2005; Maragliano et al., 2006). In the subsequent FM stage, the SE-SRP method is calibrated against the target AI/MM method through force matching. Once the AI/MM forces on the selected atoms are reproduced, the resulting SE-SRP method is used to resample the configuration space to obtain a new pool of reaction-path configurations for the next iteration of force matching. The cycle of "reaction-path configuration sampling" (predictor) and "force matching" (corrector) is repeated iteratively until convergence is established.

5.2 Generic Procedure

A generic procedure for implementing RP–FM within the QM/MM framework is illustrated in Fig. 5. In this flowchart, the steps associated with

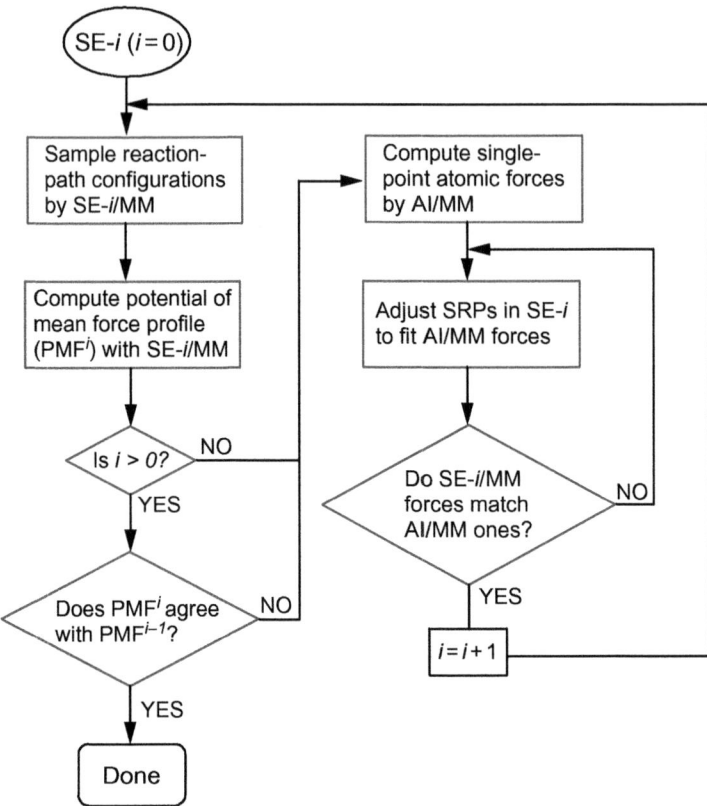

Fig. 5 Flowchart of a generic procedure for implementing RP–FM. *Reprinted with permission from Zhou, Y., & Pu, J. (2014). Reaction path-force matching: A new strategy of fitting specific reaction parameters for semiempirical methods in combined QM/MM simulations.* Journal of Chemical Theory and Computation, 10, 3038–3054. Copyright (2014) American Chemical Society.

the RP and FM stages are grouped in the red and blue boxes, respectively. For convenience, we define the union of a RP stage and the FM stage that immediately follows as a single RP–FM cycle, based on which iterations are carried out. For the initial SE method, referred to as SE-0, we assign a token iteration, ie, "iteration 0." For each time a single RP–FM cycle is completed, the iteration number is increased by 1. As a convention, the SE-SRP method obtained at the end of the ith RP–FM iteration is labeled as SE-i, which will be used to update the reaction-path configurations in the next iteration. In our notation, we define the reaction path sampled consistently on the SE-i/MM PES as the SE-i path. Following the same convention, we label the PMF profile determined from the SE-i/MM simulations as PMFi, which can be used to monitor convergence.

5.3 Case Study: Proton Transfer

In a proof-of-concept study (Zhou & Pu, 2014), we have demonstrated the feasibility of the RP–FM method for a proton transfer reaction between ammonium (NH_4^+) and ammonia (NH_3). The Hartree–Fock (HF) (Roothaan, 1951) and PM3 (Stewart, 1989) methods were selected as the target AI and SE levels, respectively. The PM3 method has been shown to overestimate the barrier height for this reaction by ~9 kcal/mol compared to HF/3-21G (Mo & Gao, 2000).

5.3.1 QM/MM Setup and Computer Programs

For a QM/MM treatment of the system, the solute molecules are treated by QM (nine QM atoms) at either the PM3 or HF/3-21G level and a box (40 Å in size) of solvent molecules are represented by the modified TIP3P model (Brooks et al., 1983). The MNDO97 (Thiel, 1998) package incorporated in the CHARMM (Brooks et al., 2009) program was used for PM3 (-SRP)/MM calculations. The Q-Chem (Shao et al., 2006) program combined with CHARMM was used for AI/MM single point force calculations. For force matching, a local code of microgenetic algorithm (μGA) (Carroll, 1998) was employed to minimize the mean force deviation between the SE-SRP/MM and target AI/MM levels; a total of 21 SE parameters in PM3 were adjusted for N and H (Zhou & Pu, 2014).

5.3.2 RP-FM Based On Gas-Phase Reaction Path

For testing purpose, the RP–FM method was first employed to parametrize PM3 along a gas-phase reaction path (Zhou & Pu, 2014). In this case, we chose to fit AI (HF/3-21G) forces along MEP; this represents the "zero-temperature" version of the algorithm. The MEP was determined approximately based on a RC defined as $RC = r(N^+-H) - r(N-H)$. Restrained geometry optimizations were carried out to scan the PES along RC in the range of 0.0–0.8 Å, resulting in an MEP of nine configurations. Fig. 6A shows the potential energies along MEPs optimized during five RP–FM iterations. The HF single point energy (SPE) profile calculated based on the PM3-n path (ie, HF/3-21G // PM3-n) are also shown for comparison; these "double-slash" profiles are labeled "HF-n."

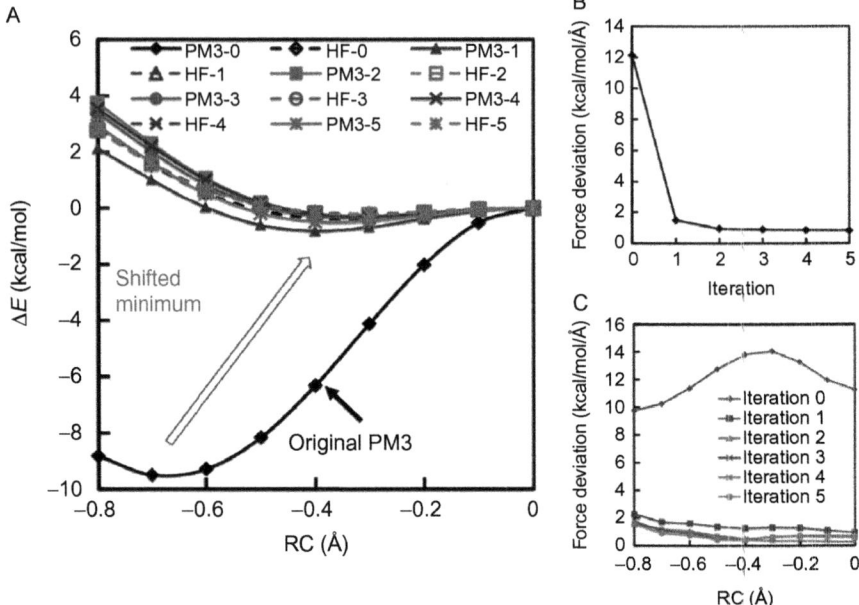

Fig. 6 Gas-phase RP–FM applied to proton transfer between ammonium and ammonia: (A) Energy profiles along MEPs of PM3-SRPs optimized using the RP–FM procedure against atomic forces obtained at the HF/3-21 level; (B) Average force deviations between PM3-SRPs and HF/3-21G over RP–FM iterations; and (C) Force deviations decomposed to individual configurations along MEP. *Reprinted with permission from Zhou, Y., & Pu, J. (2014). Reaction path-force matching: A new strategy of fitting specific reaction parameters for semiempirical methods in combined QM/MM simulations.* Journal of Chemical Theory and Computation, 10, 3038–3054. Copyright (2014) American Chemical Society.

It can be seen from Fig. 6A that the reaction barrier height given by the original PM3 (PM3-0) is 9.5 kcal/mol, which is substantially higher than the barrier of 0.2 kcal/mol given by the corresponding HF/3-21 energy profile (HF-0). After the first RP–FM iteration, the new PM3 energy profile (PM3-1) is in a closer agreement with the associated HF SPE profile (HF-1). Such agreements are steadily improved over RP–FM iterations. After five iterations, the optimized PM3-SRP energy files (PM3-5) and the energy profile at the target level (HF-5) are almost identical. The RP–FM procedure also successfully shifts the product complex minimum from RC = 0.7 Å, located on the original PM3 path, to RC = 0.3 Å as determined on the HF/3-21 path. As a result, both height and shape of the HF/3-21G barrier are reproduced at the optimized PM3-SRP level.

The force deviations between the SE-SRP and AI levels are displayed in Fig. 6B. The original PM3 yields an average force deviation of 12 kcal/mol/Å per Cartesian coordinate; the deviation is successfully reduced to ~2 kcal/mol/Å after the first round of force matching and stabilizes at <1 kcal/mol/Å during the rest of RP–FM iterations.

Decompositions of the force deviations to individual configurations are shown in Fig. 6C. Our first observation is that the deviations between the PM3 and HF/3-21G forces vary along the MEP. For the PM3-0 method, the transition state (TS) configuration (RC=0.0 Å) gives a force deviation of 11.3 kcal/mol/Å per Cartesian coordinate, which is slightly higher than that of 10.3 kcal/mol/Å at the reactant configuration (RC=−0.7 Å); a maximal force deviation of 14.0 kcal/mol/Å is found at a nonstationary structure (RC=−0.3 Å). After RP–FM optimization, excellent agreements were obtained between the PM3-SRP and HF/3-21G forces on all configurations; force deviations were reduced to <1.5 kcal/mol/Å throughout the MEP.

5.3.3 RP–FM for Solution-Phase Proton Transfer

We have also applied the RP–FM method for the same proton transfer in solution (Zhou & Pu, 2014). Umbrella sampling (Torrier & Valleau, 1974) simulations were used to sample along the reaction path on the PM3-SRP/MM PES; seven umbrella windows were placed between the reactant and the transition state. PMF profiles were determined by the weighted histogram analysis method (Kumar, Bouzida, Swendsen, Kollman, & Rosenberg, 1992). For each window, 10 configurations were randomly selected over 200 ps of PM3-SRP/MM MD simulations, resulting in 70 configurations that correspond to 1,890 Cartesian force components for force matching. The PMFs obtained from solution-phase RP–FM, referred to as PM3-iS, are shown in Fig. 7, where the PMF determined separately at the HF/3-21G level (with shorter sampling of 10 ps/window) is also shown for comparison. One can see from Fig. 7 that although initially the original PM3 PMF differs substantially from the HF/3-21G PMF, the RP–FM optimization dramatically improves the agreement between the two levels. At the end of the fifth RP–FM iteration, PMFs obtained for PM3-5S and HF become essentially identical.

5.4 Relation to Other Methods

Before we conclude, we discuss the differences between RP–FM and other related methods. The FM technique has been employed in the QM/MM

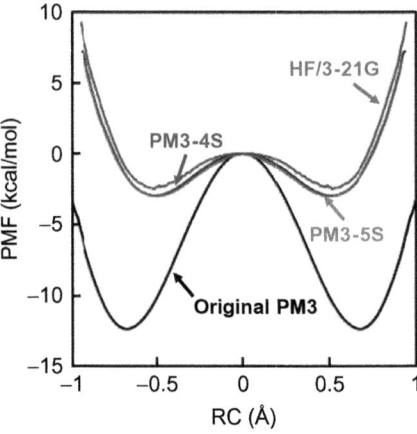

Fig. 7 PMFs of solution-phase proton transfer between ammonium and ammonia obtained from PM3-SRPs/MM and benchmark HF/3-21G/MM simulations; PM3-SRP methods were optimized by RP–FM in solution to reproduce atomic forces at the HF/3-21 level.

context by others (Arkin-Ojo et al., 2008; Csanyi et al., 2004; Maurer et al., 2007). These investigations, however, focused on using QM/MM as a target level to fit MM force fields or empirical potentials. By contrast, RP–FM is designed to simulate chemical and enzymatic reactions involving bond rearrangements, for which classical potentials cannot be used. The RP–FM method, in spirit similar to the pioneering work of Voth and coworkers (Knight et al., 2010), matches forces between a tailored reactive potential and a reference QM potential. In the force-matched multistate empirical valence bond (FM-MS-EVB) method developed by Voth and coworkers, the reactivity of the tailored potential is introduced through a reactive force field (Knight et al., 2010). In our RP–FM approach, the SE potential being tailored is instead an electronic structure-based QM potential. A "reactive" scheme is employed in RP–FM so that forces at the two QM levels are matched along reaction paths. The combined use of QM/MM methods in both the tailored and reference potentials and force fitting along reaction paths makes RP–FM a distinctive approach that enables multiscale "reactive" force matching.

The RP–FM method also differs from the "learn-on-the-fly" (LOTF) approach (Csanyi et al., 2004) in that the tailored potentials in the LOTF approach are empirical functions made "reactive" by employing time-dependent parameters, whereas in RP–FM we parametrize a true QM potential using semiglobal force fitting along reaction path. The LOTF method has been demonstrated for simulating diffusion of point defects on solid surfaces, whereas RP–FM has a focus on simulating solution-phase and enzymatic reactions.

The iterative nature of fitting the force and resampling the PES in our work resembles the spirit of the optimal potential (OP) method (Laio et al., 2000) in solid-phase simulations and that of the adaptive force matching method developed by Wang and coworkers (Arkin-Ojo et al., 2008). A similar adaptive potential refining procedure has also been employed in the "paradynamics" method pioneered by Warshel and coworkers (Plotnikov, Kamerlin, & Warshel, 2011; Plotnikov & Warshel, 2012), in which EVB potentials in conjunction with Gaussian-type corrections are parametrized against QM/MM data in the free energy perturbation framework.

6. CONCLUDING REMARKS

In this chapter, we have discussed recent progress in QM/MM simulations of ATPase mechanism in ABC transporters and how the challenges in these simulations fertilize the development of the multiscale RP–FM method. As a paradigm molecular motor system, ABC transporter may have evolved to couple ATP hydrolysis with conformational dynamics through functional asymmetry encoded in its dimeric active sites. Although our exploratory QM/MM simulations have provided exciting new insights into the related enzyme mechanism, conducting such simulations in a quantitatively reliable manner to achieve the ab initio QM/MM accuracy has remained a challenging task. The RP–FM method has been developed to offer a multiscale solution to the challenge. We have successfully demonstrated the effectiveness of this approach in modeling a proton transfer reaction in the gas phase and in solution. To achieve the ultimate goal of determining ATPase mechanism in ABC transporters by RP–FM, the next step is to validate the method on more complex reactions such as solution-phase ATP hydrolysis and a number of well-characterized enzymatic reactions; these investigations are underway.

As we have discussed in this chapter, the RP–FM strategy aims at resolving the dilemma of choosing between accuracy and efficiency in QM/MM simulations, which is a long-standing challenge in computational enzymology. Therefore, development of the RP–FM method is expected to provide not only a feasible way to characterize ATP hydrolysis in ABC transporters but also a general tool for chemical and enzyme simulations in condensed phases.

ACKNOWLEDGMENTS
This work is supported by a startup and RSFG grants from Indiana Univ.–Purdue Univ. Indianapolis (IUPUI), and the National Institutes of Health (R15GM116057). Computing time was provided through a High-Performance Computing Cluster funded by the School of Science at IUPUI and the BigRed2 supercomputing facilities at Indiana University.

REFERENCES
Arantes, G. M., & Loos, M. (2006). Specific parametrisation of a hybrid potential to simulate reactions in phosphatases. *Physical Chemistry Chemical Physics, 8,* 347–353.
Arkin-Ojo, O., Song, Y., & Wang, F. (2008). Developing ab initio quality force field from condensed phase quantum-mechanics/molecular-mechanics calculations through the adaptive force matching method. *The Journal of Chemical Physics, 129,* 064108.
Atwell, S., Brouillette, C. G., Conners, K., Emtage, S., Gheyi, T., Guggino, W. B., et al. (2010). Structures of a minimal human CFTR first nucleotide-binding domain as a monomer, head-to-tail homodimer, and pathogenic mutant. *Protein Engineering, Desing & Selection, 23,* 375–384.
Borst, P., & Oude Elferink, R. (2002). Mammalian ABC transporters in health and disease. *Annual Review of Biochemistry, 71,* 537–592.
Brooks, B. R., Brooks, C. L., III, MacKerell, A. D., Jr., Nilsson, L., Petrella, R. J., Roux, B., et al. (2009). CHARMM: The biomolecular simulation program. *Journal of Computational Chemistry, 30,* 1545–1614.
Brooks, B. R., Bruccoleri, R. E., Olafson, B. D., States, D. J., Swaminathan, S., & Karplus, M. (1983). CHARMM: A program for macromolecular energy, minimization, and dynamics calculations. *Journal of Computational Chemistry, 4,* 187–217.
Car, R., & Parrinello, M. (1985). Unified approach for molecular dynamics and density-functional theory. *Physical Review Letters, 55,* 2471–2474.
Carroll, D. L. (1998). *Genetic algorithm driver 1.70.* Urbana–Champaign: University of Illinois.
Chen, J., Lu, G., Lin, J., Davidson, A. L., & Quiocho, F. A. (2003). A tweezers-like motion of the ATP-binding cassette dimer in an ABC transport cycle. *Molecular Cell, 12,* 651–661.
Csanyi, G., Albaret, T., Payne, M. C., & De Vita, A. (2004). "Learn on the fly": A hybrid classical and quantum-mechanical molecular dynamics simulation. *Physical Review Letters, 93,* 175503.
Davidson, A. L., & Chen, J. (2004). ATP-binding cassette transporters in bacteria. *Annual Review of Biochemistry, 73,* 241–268.
Davidson, A. L., Dassa, E., Orelle, C., & Chen, J. (2008). Structure, function, and evolution of bacterial ATP-binding cassette systems. *Microbiology and Molecular Biology Reviews, 72,* 317–364.
Davidson, A. L., & Sharma, S. (1997). Mutation of a single MalK subunit severely impairs maltose transport activity in Escherichia coli. *Journal of Bacteriology, 179,* 5458–5464.
Dewar, M. J. S., Zoebisch, E. G., Healy, E. F., & Stewart, J. J. P. (1985). AM1: A new general purpose quantum mechanical molecular model. *Journal of the American Chemical Society, 107,* 3902–3909.
Doemer, M., Maurer, P., Campomanes, P., Tavernelli, I., & Rothlisberger, U. (2014). A generalized QM/MM force matching approach applied to the 11-cis protonated Schiff base chromophore of rhodopsin. *Journal of Chemical Theory and Computation, 10,* 412–422.
E, W., Ren, W., & Vanden-Eijnden, E. (2005). Finite temperature string method for the study of rare events. *The Journal of Physical Chemistry B, 109,* 6688–6693.

Ercolessi, F., & Adams, J. B. (1994). Interatomic potentials from first-principles calculations: The force-matching method. *Europhysics Letters, 26*, 583–588.

Fukui, K. (1981). The path of chemical-reactions—The IRC approach. *Accounts of Chemical Research, 14*, 363–368.

Gao, J., Amara, P., Alhambra, C., & Field, M. J. (1998). A generalized hybrid orbital (GHO) method for the treatment of boundary atoms in combined QM/MM calculations. *The Journal of Physical Chemistry. A, 102*, 4714–4721.

Gao, J., & Thompson, M. A. (Eds.), (1998). *Combined quantum mechanical and molecular mechanical methods (ACS Symposium Series)*. Washington, DC: American Chemical Society.

Garcia-Viloca, M., Truhlar, D. G., & Gao, J. (2003). Reaction-path energetics and kinetics of the hydride transfer reaction catalyzed by dihydrofolate reductase. *Biochemistry, 42*, 13558–13575.

Geourjon, C., Orelle, C., Steinfels, E., Blanchet, C., Deleage, G., Di Pietro, A., et al. (2001). A common mechanism for ATP hydrolysis in ABC transporter and helicase superfamilies. *Trends in Biochemical Sciences, 26*, 539–544.

Gonzalez-Lafont, A., Truong, T. N., & Truhlar, D. G. (1991). Direct dynamics calculations with neglect of diatomic differential overlap molecular orbital theory with specific reaction parameters. *The Journal of Physical Chemistry, 95*, 4618–4627.

Hehre, W. J., Radom, L., Schleyer, P. v. R., & Pople, J. A. (1986). *Ab initio molecular orbital theory*. New York: John Wiley.

Holland, I. B., Schmitt, L., & Young, J. (2005). Type 1 protein secretion in bacteria, the ABC-transporter dependent pathway. *Molecular Membrane Biology, 22*, 29–39.

Hollenstein, K., Dawson, R. J. P., & Locher, K. P. (2007). Structure and mechanism of ABC transporter proteins. *Current Opinion in Structural Biology, 17*, 412–418.

Izvekov, S., Parrinello, M., Burnham, C. J., & Voth, G. A. (2004). Effective force fields for condensed phase systems from ab initio molecular dynamics simulation: A new method for force-matching. *The Journal of Chemical Physics, 120*, 10896–10913.

Izvekov, S., & Voth, G. A. (2005). A multiscale coarse-graining method for biomolecular systems. *The Journal of Physical Chemistry. B, 109*, 2469–2473.

Janas, E., Hofacker, M., Chen, M., Gompf, S., van der Does, C., & Tampe, R. (2003). The ATP hydrolysis cycle of the nucleotide-binding domain of the mitochondrial ATP-binding cassette transporter Mdl1p. *The Journal of Biological Chemistry, 278*, 26862–26869.

Jones, P. M., O'Mara, M. L., & George, A. M. (2009). ABC transporters: A riddle wrapped in a mystery inside an enigma. *Trends in Biochemical Sciences, 34*, 520–531.

Kamerlin, S. C. L., Sharma, P. K., Prasad, R. B., & Warshel, A. (2013). Why nature really chose phosphate. *Quarterly Reviews of Biophysics, 46*, 1–132.

Khan, M. M. T., & Mohan, M. S. (1973). The metal chelates of riboflavin and riboflavin monophosphate. *Journal of Inorganic and Nuclear Chemistry, 35*, 1749–1755.

Khare, D., Oldham, M. L., Orelle, C., Davidson, A. L., & Chen, J. (2009). Alternating access in maltose transporter mediated by rigid-body rotations. *Molecular Cell, 33*, 528–536.

Klahn, M., Braun-Sand, S., Rosta, E., & Warshel, A. (2005). On possible pitfalls in ab initio quantum mechanics/molecular mechanics minimization approaches for studies of enzymatic reactions. *The Journal of Physical Chemistry. B, 109*, 15645–15650.

Knight, C., Maupin, C. M., Izvekov, S., & Voth, G. A. (2010). Defining condensed phase reactive force fields from ab initio molecular dynamics simulations: The case of the hydrated excess proton. *Journal of Chemical Theory and Computation, 6*, 3223–3232.

Kumar, S., Bouzida, D., Swendsen, R. H., Kollman, P. A., & Rosenberg, J. M. (1992). The weighted histogram analysis method for free-energy calculations on biomolecules. I. The method. *Journal of Computational Chemistry, 13*, 1011–1021.

Laio, A., Bernard, S., Chiarotti, G. L., Scandolo, S., & Tosatti, E. (2000). Physics of iron at Earth's core conditions. *Science, 287*, 1027–1030.

Li, J., Wen, P.-C., Moradi, M., & Tajkhorshid, E. (2015). Computational characterization of structural dynamics underlying function in active membrane transporters. *Current Opinion in Structural Biology, 31*, 96–105.

Lopez, X., & York, D. M. (2003). Parameterization of semiempirical methods to treat nucleophilic attacks to biological phosphates: AM1/d parameters for phosphorus. *Theoretical Chemistry Accounts, 109*, 149–159.

MacKerell, A. D., Jr., Bashford, D., Bellott, M., Dunbrack, R. L., Jr., Evanseck, J. D., Field, M. J., et al. (1998). All-atom empirical potential for molecular modeling and dynamics studies of proteins. *The Journal of Physical Chemistry. B, 102*, 3586–3616.

MacKerell, A. D., Jr., Feig, M., & Brooks, C. L., III. (2004). Extending the treatment of backbone energetics in protein force fields: Limitations of gas-phase quantum mechanics in reproducing protein conformational distributions in molecular dynamics simulations. *Journal of Computational Chemistry, 25*, 1400–1415.

Maragliano, L., Fischer, A., Vanden-Eijnden, E., & Ciccotti, G. (2006). String method in collective variables: Minimum free energy paths and isocommittor surfaces. *The Journal of Chemical Physics, 125*, 024106.

Maurer, P., Laio, A., Hugosson, H. W., Colombo, M. C., & Rothlisberger, U. (2007). Automated parametrization of biomolecular force fields from quantum mechanics/molecular mechanics (QM/MM) simulations through force matching. *Journal of Chemical Theory and Computation, 3*, 628–639.

Miertus, S., Scrocco, E., & Tomasi, J. (1981). Electrostatic interaction of a solute with a continuum. A direct utilization of ab initio molecular potentials for the prevision of solvent effects. *Chemical Physics, 55*, 117–129.

Mo, Y., & Gao, J. (2000). An ab initio molecular orbital-valence bond (MOVB) method for simulating chemical reactions in solution. *The Journal of Physical Chemistry. A, 104*, 3012–3020.

Moradi, M., & Tajkhorshid, E. (2013). Mechanistic picture for conformational transition of a membrane transporter at atomic resolution. *Proceedings of the National Academy of Sciences of the United States of America, 110*, 18916–18921.

Nam, K., Cui, Q., Gao, J., & York, D. M. (2007). Specific reaction parameterization of the AM1/d hamiltonian for phosphoryl transfer reactions: H, O, and P atoms. *Journal of Chemical Theory and Computation, 3*, 486–504.

Newstead, S., Fowler, P. W., Bilton, P., Carpenter, E. P., Sadler, P. J., Campopiano, D. J., et al. (2009). Insights into how nucleotide-binding domains power ABC transport. *Structure, 17*, 1213–1222.

Nikaido, K., & Ames, G. F.-L. (1999). One intact ATP-binding subunit is sufficient to support ATP hydrolysis and translocation in an ABC transporter, the histidine permease. *The Journal of Biological Chemistry, 274*, 26727–26735.

Oldham, M. L., & Chen, J. (2011). Crystal structure of the maltose transporter in a pretranslocation intermediate state. *Science, 332*, 1202–1205.

Oldham, M. L., Khare, D., Quiocho, F. A., Davidson, A. L., & Chen, J. (2007). Crystal structure of a catalytic intermediate of the maltose transporter. *Nature, 450*, 515–522.

Oloo, E. O., Kandt, C., O'Mara, M. L., & Tieleman, D. P. (2006). Computer simulations of ABC transporter components. *Biochemistry and Cell Biology, 84*, 900–911.

Oswald, C., Holland, I. B., & Schmitt, L. (2006). The motor domains of ABC-transporters: What can structures tell us? *Naunyn-Schmiedeberg's Archives of Pharmacology, 372*, 385–399.

Parr, R. G., & Yang, W. (1994). *Density-functional theory of atoms and molecules*. New York: Oxford University Press.

Plotnikov, N. V., Kamerlin, S. C. L., & Warshel, A. (2011). Paradynamics: An effective and reliable model for ab initio QM/MM free-energy calculations and related tasks. *The Journal of Physical Chemistry B, 115*, 7950–7962.

Plotnikov, N. V., & Warshel, A. (2012). Exploring, refining, and validating the paradynamics QM/MM sampling. *The Journal of Physical Chemistry. B, 116*, 10342–10356.

Pople, J. A., Santry, D. P., & Segal, G. A. (1965). Approximate self-consistent molecular orbital theory. I. Invariant procedures. *The Journal of Chemical Physics, 43*, S129–S135.

Poulsen, T. D., Garcia-Viloca, M., Gao, J., & Truhlar, D. G. (2003). Free energy surface, reaction paths, and kinetic isotope effect of short-chain Acyl-CoA dehydrogenase. *The Journal of Physical Chemistry. B, 107*, 9567–9578.

Procko, E., Ferrin-O'Connell, I., Ng, S.-L., & Gaudet, R. (2006). Distinct structural and functional properties of the ATPase sites in an asymmetric ABC transporter. *Molecular Cell, 24*, 51–62.

Pu, J., Gao, J., & Truhlar, D. G. (2004). Generalized hybrid orbital (GHO) method for combining ab initio Hartree-Fock wave functions with molecular mechanics. *The Journal of Physical Chemistry. A, 108*, 632–650.

Rees, D. C., Johnson, E., & Lewinson, O. (2009). ABC transporters: The power to change. *Nature Reviews. Molecular Cell Biology, 10*, 218–227.

Roothaan, C. C. J. (1951). New developments in molecular orbital theory. *Reviews of Modern Physics, 23*, 69–89.

Rosta, E., Nowotny, M., Yang, W., & Hummer, G. (2011). Catalytic mechanism of RNA backbone cleavage by ribonuclease H from quantum mechanics/molecular mechanics simulations. *Journal of the American Chemical Society, 133*, 8934–8941.

Rudack, T., Xia, F., Schlitter, J., Kotting, C., & Gerwert, K. (2012). Ras and GTPase-activating protein (GAP) drive GTP into a precatalytic state as revealed by combining FTIR and biomolecular simulations. *Proceedings of the National Academy of Sciences of the United States of America, 109*, 15295–15300.

Schlegel, H. B., Iyengar, S. S., Li, X., Millam, J. M., Voth, G. A., Scuseria, G. E., et al. (2002). Ab initio molecular dynamics: Propagating the density matrix with Gaussian orbitals. III. Comparison with Born–Oppenheimer dynamics. *The Journal of Chemical Physics, 117*, 8694–8704.

Schmitt, L., Benabdelhak, H., Blight, M. A., Holland, I. B., & Stubbs, M. T. (2003). Crystal structure of the nucleotide-binding domain of the ABC-transporter haemolysin B: Identification of a variable region within ABC helical domains. *Journal of Molecular Biology, 330*, 333–342.

Schneider, E., & Hunke, S. (1998). ATP-binding-cassette (ABC) transport systems: Functional and structural aspects of the ATP-hydrolyzing subunits/domains. *FEMS Microbiology Reviews, 22*, 1–20.

Senior, A. E., Al-Shawi, M. K., & Urbatsch, I. L. (1995). The catalytic cycle of P-glycoprotein. *FEBS Letters, 377*, 285–289.

Shao, Y., Molnar, L. F., Jung, Y., Kussmann, J., Ochsenfeld, C., Brown, S. T., et al. (2006). Advances in methods and algorithms in a modern quantum chemistry program package. *Physical Chemistry Chemical Physics, 8*, 3172–3191.

Sharma, S., & Davidson, A. L. (2000). Vanadate-induced trapping of nucleotides by purified maltose transport complex requires ATP hydrolysis. *Journal of Bacteriology, 182*, 6570–6576.

Shyamala, V., Baichwal, V., Beall, E., & Ames, G. F.-L. (1991). Structure-function analysis of the histidine permease and comparison with cystic fibrosis mutations. *The Journal of Biological Chemistry, 266*, 18714–18719.

Stewart, J. J. P. (1989). Optimization of parameters for semiempirical methods I. Method. *Journal of Computational Chemistry, 10*, 209–220.

Story, R. M., & Steitz, T. A. (1992). Structure of the recA protein-ADP complex. *Nature, 355*, 374–376.

Tal, N., Ovcharenko, E., & Lewinson, O. (2013). A single intact ATPase site of the ABC transporter BtuCD drives 5% transport activity yet supports full in vivo vitamin B12 utilization. *Proceedings of the National Academy of Sciences of the United States of America, 110*, 5434–5439.

Thiel, W. (1998). *MNDO97*. Zurich, Switzerland: University of Zurich.

Thiel, W. (2014). Semiempirical quantum-chemical methods. *WIREs Computational Molecular Science, 4*, 145–157.

Torrier, G. M., & Valleau, J. P. (1974). Monte Carlo free energy estimates using non-Boltzmann sampling: Application to the sub-critical Lennard-Jones fluid. *Chemical Physics Letters, 28*, 578–581.

Urbatsch, I. L., Sankaran, B., Weber, J., & Senior, A. E. (1995). P-glycoprotein is stably inhibited by vanadate-induced trapping of nucleotide at a single catalytic site. *The Journal of Biological Chemistry, 270*, 19383–19390.

Warshel, A., & Levitt, M. (1976). Theoretical studies of enzymic reactions. *Journal of Molecular Biology, 103*, 227–249.

Wen, P.-C., & Tajkhorshid, E. (2008). Dimer opening of the nucleotide binding domains of ABC transporters after ATP hydrolysis. *Biophysical Journal, 95*, 5100–5110.

Weng, J.-W., Fan, K.-N., & Wang, W.-N. (2010). The conformational transition pathway of ATP binding cassette transporter MsbA revealed by atomistic simulations. *The Journal of Biological Chemistry, 285*, 3053–3063.

Zaitseva, J., Holland, I. B., & Schmitt, L. (2004). The role of CAPS buffer in expanding the crystallisation space of the nucleotide binding domain of the ABC-transporter from E. coli. *Acta Crystallographica. Section D, Biological Crystallography, 60*, 1076–1084.

Zaitseva, J., Jenewein, S., Jumpertz, T., Holland, I. B., & Schmitt, L. (2005a). H662 is the linchpin of ATP hydrolysis in the nucleotide-binding domain of the ABC transporter HlyB. *The EMBO Journal, 24*, 1901–1910.

Zaitseva, J., Jenewein, S., Wiedenmann, A., Benabdelhak, H., Holland, I. B., & Schmitt, L. (2005b). Functional characterization and ATP-induced dimerization of the isolated ABC-domain of the haemolysin B transporter. *Biochemistry, 44*, 9680–9690.

Zaitseva, J., Oswald, C., Jumpertz, T., Jenewein, S., Wiedenmann, A., Holland, I. B., et al. (2006). A structural analysis of asymmetry required for catalytic activity of an ABC-ATPase domain dimer. *The EMBO Journal, 25*, 3432–3443.

Zhou, Y., Ojeda-May, P., Nagaraju, M., & Pu, J. (2016). Mapping free energy paths of ATP hydrolysis in the E. coli ABC-transporter HlyB by the finite temperature string method. to be submitted.

Zhou, Y., Ojeda-May, P., & Pu, J. (2013). H-loop histidine catalyzes ATP hydrolysis in the E. coli ABC-transporter HlyB. *Physical Chemistry Chemical Physics, 15*, 15811–15815.

Zhou, Y., & Pu, J. (2014). Reaction path-force matching: A new strategy of fitting specific reaction parameters for semiempirical methods in combined QM/MM simulations. *Journal of Chemical Theory and Computation, 10*, 3038–3054.

Zoghbi, M. E., & Altenberg, G. A. (2013). Hydrolysis at one of the two nucleotide-binding sites drives the dissociation of ATP-binding cassette nucleotide-binding domain dimers. *The Journal of Biological Chemistry, 288*, 34259–34265.

CHAPTER NINE

QM/MM Analysis of Transition States and Transition State Analogues in Metalloenzymes

D. Roston[1], Q. Cui[1]

Theoretical Chemistry Institute, University of Wisconsin–Madison, Madison, WI, United States
[1]Corresponding authors: e-mail address: droston@chem.wisc.edu; cui@chem.wisc.edu

Contents

1. Introduction	214
2. Background on Computational Methods	215
2.1 Levels of Theory	216
2.2 Types of Calculations	220
2.3 Summary	222
3. Case Study: The Transition State of AP	223
3.1 TSA Binding	223
3.2 Computational Benchmarks	226
3.3 Active-Site Optimizations	226
3.4 QM/MM Optimizations	229
3.5 QM/MM PES Scan	230
3.6 Tests of Protonation States	233
3.7 Comparison with Experiment	236
3.8 TSAs vs TSs	240
4. Summary/Conclusions	241
Acknowledgments	242
References	243

Abstract

Enzymology is approaching an era where many problems can benefit from computational studies. While ample challenges remain in quantitatively predicting behavior for many enzyme systems, the insights that often come from computations are an important asset for the enzymology community. Here we provide a primer for enzymologists on the types of calculations that are most useful for mechanistic problems in enzymology. In particular, we emphasize the integration of models that range from small active-site motifs to fully solvated enzyme systems for cross-validation and dissection of specific contributions from the enzyme environment. We then use a case study of the enzyme alkaline phosphatase to illustrate specific application of the methods. The case study involves examination of the binding modes of putative transition state analogues (tungstate and vanadate) to the enzyme. The computations predict covalent binding of these ions to the enzymatic nucleophile and that they adopt the trigonal bipyramidal geometry of the expected transition state. By comparing these structures with transition states found through free energy simulations, we assess the degree to which the transition state analogues mimic the true transition states. Technical issues worth treating with care as well as several remaining challenges to quantitative analysis of metalloenzymes are also highlighted during the discussion.

1. INTRODUCTION

Enzymologists can now use computations to understand a wide variety of problems. Phenomena such as ligand binding (Halperin, Ma, Wolfson, & Nussinov, 2002), chemical mechanisms (Gao et al., 2006), transition states (Pu, Gao, & Truhlar, 2006; Warshel et al., 2006), and enzyme dynamics (Benkovic & Hammes-Schiffer, 2003; Kerns et al., 2015; Schwartz & Schramm, 2009; Shukla, Meng, Roux, & Pande, 2014) are all accessible through computations of one form or another. In recent years, computational design of novel enzymes (Blomberg et al., 2013; Rothlisberger et al., 2008; Siegel et al., 2010) has emerged as one of the most promising avenues of biotechnology research. The ability to readily design novel enzymes from basic principles would herald in an era where enzymologists could truly claim to understand enzyme catalysis. Not long ago, the field of computational enzymology was still in its infancy, with relatively limited claims of successful predictive power (Kraut, Carroll, & Herschlag, 2003). The computational design of novel activity, therefore, is quite promising, but as of yet, computationally designed enzymes are sluggish relative to naturally evolved ones (Kries, Blomberg, & Hilvert, 2013; Lassila, Baker, & Herschlag, 2010). Thus, despite much progress in recent years, it remains challenging to predict behavior for an arbitrary enzyme, suggesting that there are missing pieces in general theories of enzyme catalysis (Benkovic & Hammes-Schiffer, 2003; Garcia-Viloca, Gao, Karplus, & Truhlar, 2004) or in specific computational models.

From a technical point of view, metalloenzymes are particularly challenging because the metal cofactors are generally more difficult to treat reliably with computational approaches. Therefore, balancing computational efficiency and accuracy, which is the key to computational studies of complex systems, is most essential to the analysis of metalloenzymes. To this end, it is important to choose the computational model and methods carefully based on the problem in hand. Moreover, we find that it is instructive to integrate models of different complexity and methods at different levels of theory; comparing the results from different calculations helps to cross-validate robustness of mechanistic models and to evaluate the importance of specific factors (eg, protein motion) to the properties of interest.

Another challenge to the analysis of complex enzymes is a gap between practitioners of experimental techniques and computational methods for studying enzymes. Experimentalists may assert that there is a dearth of computational models making "blind" predictions, while the opposite side of that coin is that rather rarely are experimentalists willing to test predictions that emerge from computational studies. Much of the problem likely lies in the seeming lack of communication and understanding between experimentalists and theoreticians. Computational chemists often have little sense of what experimental tests are feasible or how to interpret experimental measurements properly. Similarly, experimental chemists may not grasp what or how something has been calculated in a computational study, or indeed how calculated properties ought to correspond to "reality" as measured in a lab. Enhancing the communication between experimental and computational enzymologists is essential to pushing computational enzymology out of its adolescence.

Bearing those thoughts in mind, in the present work, we aim to provide an understanding of the types of calculations that can illuminate various problems in enzymology that should be accessible to those with limited background on computational methods, but still useful for more experienced computational chemists. While some types of calculations are fairly specialized, others are readily accomplished with user-friendly software and can therefore be adopted and used effectively by experimental chemists themselves to obtain the desired information. We begin by providing a brief survey of the types of calculations commonly used for studying (metallo)enzymes; more details can be found in several recent review articles by us (Gaus et al., 2014; Lu et al., 2016) and others (Brunk & Rothlisberger, 2015; Senn & Thiel, 2009), as well as in the other chapters in this volume. We will then use a case study of the metalloenzyme alkaline phosphatase (AP) to illustrate details of how one may use calculations for understanding enzyme chemistry and make clear connections to experimental measurements.

2. BACKGROUND ON COMPUTATIONAL METHODS

Many considerations go into choosing an appropriate approach for studying a biomolecular system. First, there is the question of what property one wants to calculate. The goal of a calculation should be to obtain an experimentally testable property that arises out of a model for biological behavior. In principle, calculations can predict kinetic and thermodynamic properties, geometrical properties, spectral signatures, etc. But even given a particular experimentally observable property, there are many considerations that go into how to calculate it. To begin, one must choose a level of theory that is appropriate for the specific system of interest. This usually entails balancing reliability of calculations with computational cost in order to achieve the highest accuracy possible for the model size and timescale one needs. Biological systems of interest can range in size from small peptides in solution to something as large as a ribosome translating a gene that is kilobases long. Furthermore, enzymological processes occur on timescales ranging from femtoseconds to many seconds (Schramm, 2015), and for many processes, one cannot easily separate these timescales. For example, a chemical reaction's passage through a transition state (TS) could occur in a matter of tens of femtoseconds, but the vibrations that are necessary to align active-site dipoles in order to allow for that barrier crossing may occur in picoseconds. In turn, this alignment might be dependent upon other conformational transitions that may range from a side-chain isomerization to domain reorientations, which may occur in nanoseconds or microseconds and beyond. Thus the fact that a TS only lasts a matter of tens of femtoseconds has little bearing on how long one must simulate a system in order to understand properties of a TS because motions at various timescales facilitate TS crossing. The size of a system and the timescale of the processes, therefore, are important considerations in choosing a level of theory to use for a computational study.

2.1 Levels of Theory

Due to practical limits on computational power, the most reliable computational methods (ie, ab initio quantum mechanics) cannot be used for large systems or for long simulations. Thus, one generally has to consider how best to balance computational cost and accuracy. When chemistry (ie, bond breaking/formation) is not involved, oftentimes one can use a molecular mechanics (MM) force field to model an enzymological system (MacKerell et al., 1998; Ponder et al., 2010; Salomon-Ferrer, Case, & Walker, 2013). For many purposes, a well-parameterized force field can provide very reliable results for a question of interest, such as binding affinity of ligands to an enzyme active site (Jiao, Golubkov, Darden, & Ren, 2009).

Force fields have their limits, though. For one, a force field must be parameterized for a specific system, which is no simple task despite some recent advances (Huang & Roux, 2013; Vanommeslaeghe, Raman, & MacKerell, 2012). If one wishes to know how a promising new drug binds to a particular protein, for example, one needs to find a set of force field parameters (vibrational force constants, electrostatic charges, etc.) specific to that drug. Second, most force fields for biomolecules involve rather simple functional forms and physical terms (eg, even polarizable force fields (Cui, Meuwly, & Ren, 2016) often assume atom-centered isotropic polarizabilities), which limit the transferability and accuracy, especially when metal ions are involved (Hu & Ryde, 2011).

For a process that involves bond reorganization, one generally needs to include electrons in the calculation using some QM method. In addition to being capable of modeling bond reorganization events, QM calculations have the advantage that they are more general and need not be parameterized for a specific purpose (although some QM methods can be specifically parameterized to achieve greater accuracy (Cui & Karplus, 2002; Doron, Major, Kohen, Thiel, & Wu, 2011; Rossi & Truhlar, 1995)). Different QM methods have different domains where they are either more or less reliable; for example, density functional theory methods (Becke, 2014; Parr & Yang, 1989) are useful in most metalloenzyme studies (Blomberg, Borowski, Himo, Liao, & Siegbahn, 2014), although their limitations are also well documented (Cohen, Mori-Sanchez, & Yang, 2012) and highly correlated ab initio QM methods (Chan & Sharma, 2011; Claeyssens et al., 2006) are needed even for qualitative insights in some cases (Kurashige, Chan, & Yanai, 2013). A drawback is that QM calculations can be very costly in terms of computational time, and even the cheapest QM methods are 100- to 1000-fold slower than MM calculations, depending on the size of the system. Thus, even using a semiempirical QM method (eg, AM1: Dewar, Zoebisch, Healy, & Stewart, 1985; PM3: Stewart, 1989; DFTB: Cui & Elstner, 2014; Elstner et al., 1998; etc.) for an entire enzyme system is not feasible, although recent work (Kulik, Leuhr, Ufimtsev, & Martinez, 2012; Liao & Thiel, 2013) suggests that an era may be approaching when computers are capable of treating a wide range of biological problems strictly quantum mechanically. Nonetheless, it is not obvious that approximate QM methods will necessarily achieve more accurate results than those of well-parameterized MM methods in describing, for example, protein conformations. At any rate, one solution to the cost of QM calculations is to use a small model that includes just the active-site atoms

(Blomberg et al., 2014). Oftentimes one can extract very useful information from small models, especially if there are good experimental constraints to guide the model. This requires restraints on certain atoms, though, and otherwise ignores the role of the enzyme environment outside of the active site.

A more sophisticated approach is to use hybrid QM/MM calculations. QM/MM calculations have been developing for a few decades now (Brunk & Rothlisberger, 2015; Field, Bash, & Karplus, 1990; Friesner & Guallar, 2005; Gao, 1995; Hu & Yang, 2008; Kamerlin, Haranczyk, & Warshel, 2009; Monard & Merz, 1999; Riccardi et al., 2006; Senn & Thiel, 2009; van der Kamp & Mulholland, 2013; Warshel & Levitt, 1976) and the value of these calculations was recently recognized by the 2013 Nobel Prize in chemistry. QM/MM calculations take advantage of the best aspects of the two methods within the same calculation. In these calculations, the parts of the system that must be treated quantum mechanically (eg, the active site of a chemical reaction or a ligand for which no force field exists) are treated with QM, but the remainder of the system is treated with MM. This allows one to use a reliable and versatile method for the active-site atoms, while still accounting for the effects of the enzyme environment. There have been many recent and excellent review articles on QM/MM methods for biological applications (Brunk & Rothlisberger, 2015; Gao et al., 2006; Hu & Yang, 2008; Senn & Thiel, 2009; van der Kamp & Mulholland, 2013); thus we will not repeat them here. We only hope to emphasize several technical points often not emphasized in the literature.

First, it is important to balance QM–MM and MM–MM interactions. Since QM–MM interaction terms (Field et al., 1990; Gao, 1995) are an inexpensive component of QM/MM calculations, they are often computed without any cutoff distance; MM–MM interactions, by contrast, in many nonperiodic setups are usually treated with a cutoff, meaning that interactions beyond a certain distance are ignored to limit computational cost. Such a combination can lead to overpolarization of the MM region and thus structural artifacts in QM/MM simulations (Schaefer, Riccardi, & Cui, 2005). One solution is to employ a periodic boundary condition with particle-mesh-Ewald type of treatment (Darden, York, & Pedersen, 1993) for MM–MM interactions, which divides the infinite summation of interactions into a short-range component and a long-range component and the latter is computed in Fourier space to achieve fast convergence. Another solution is to use a finite-spherical setup with extended electrostatics (Stote, States, &

Karplus, 1991) for MM–MM interactions and implicit solvation at the edge of the sphere (Benighaus & Thiel, 2011; Im, Berneche, & Roux, 2001; Schaefer et al., 2005; Zienau & Cui, 2012); the size of the spherical region needs to be carefully chosen to avoid too much structural constraint in the reactive site (Lu & Cui, 2013). Second, QM–MM interactions need to be carefully calibrated (Freindorf & Gao, 1996; Giese & York, 2007; Hou, Zhu, Elstner, & Cui, 2012), even when a rather large QM region is used. As illustrated in Fig. 1, relatively small errors in QM–MM interactions favor an altered hydrogen-bonding pattern at the QM–MM boundary; although being >10 Å away from the center of interest, the altered hydrogen-bonding pattern ultimately propagates to the active site. Therefore, QM–MM interactions need to be calibrated such that artifactual structural changes can be distinguished from realistic structural transitions. Finally, as emphasized by several recent studies (Ghosh, Prat-Resina, & Cui, 2009; Goyal, Lu, Yang, Gunner, & Cui, 2013; Hu, Boone, & Yang, 2008), it is essential to carefully determine the water occupancy of active sites and the titration states of residues; in some cases, cumulative effects of titration state of residues that are distant from the active site may still have a notable impact on the structural and energetic properties of the reactive center.

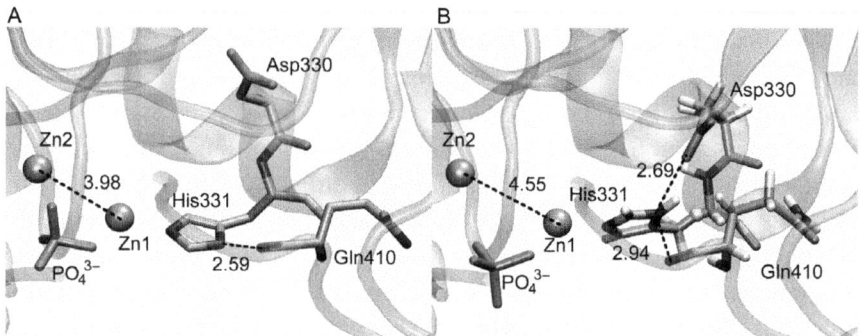

Fig. 1 Artifacts may arise from uncalibrated QM–MM interactions in enzyme simulations. (A) The crystal structure of alkaline phosphatase (AP) with bound PO_4^{3-} (PDB: 1ED8) shows a H-bond between the side chain of His331 and the backbone of Gln410. (B) A representative snapshot from a DFTB3/MM simulation of the system shows that the His331–Gln410 H-bond is broken and a new one has formed between the side chains of His331 and Asp330. This lengthens the Zn–Zn distance and perturbs the binding mode of the PO_4^{3-}. In (B), the *yellow* (*light gray* in the print version) *ball* in the His331 side chain indicates the location of the QM link atom separating the QM and MM subsystems.

2.2 Types of Calculations

For mechanistic studies, the simplest type of calculation is a local geometry optimization that leads the system to a local minimum on the potential energy surface (PES). Starting from a local energy minimum, one can further explore nearby saddle points and minimum energy pathways (MEPs) that lead to, for example, intermediate and product for the catalyzed reaction. This type of analysis is most effective for active-site models of enzymes (Blomberg et al., 2014), for which the number of important stationary points on the PES is quite limited. For full enzyme models, minimization-based calculations are generally less useful due to the enormously large number of stationary points and therefore MEPs on the high-dimensional PES, and it is challenging (if not impossible) to estimate the proper statistical weights from these MEPs (Riccardi et al., 2006; Wales, 2003). Another important limitation of minimization type of calculations is the lack of thermal fluctuations and therefore entropic contribution to the reaction energetics; along this line, it is important to recall that protein and solvent fluctuations (not just the reactive motifs) cannot be ignored when considering the entropic contribution (Kazemi, Himo, & Aqvist, 2016; Warshel, 1991). Therefore, it is generally difficult to compare the results from a limited number of MEP calculations to experimental free energy data, although qualitative insights may still be obtained from the analysis of MEP calculations.

To obtain free energies, it is important to conduct adequate sampling of the PES, which may involve simulations lasting nanoseconds to microseconds and beyond, depending on the characteristics of the system. There is a vast literature on free energy calculations for biomolecules, some of which is also discussed in chapters in this volume. A difficulty with free energy simulations is often the problem of sampling rare configurations. For example, in a typical protein system, even the most advanced computers can only run simulations lasting on the order of microseconds (using classical potential functions) (Shaw et al., 2010), yet enzymes typically have turnover numbers in the ms to seconds timescales (Fersht, 1999). These slow events correspond to high barriers on the free energy surface and to observe them requires enhanced sampling techniques, such as umbrella sampling (Torrie & Valleau, 1977) or metadynamics (Barducci, Bonomi, & Parrinello, 2011). Simulations using enhanced sampling add biasing potentials of one form or another to the PES with the goal of flattening the surface so that all relevant regions are sampled evenly and efficiently. In these cases, one has to carefully choose an "order parameter," often corresponding to the reactive degrees of freedom of interest, along which to bias the PES and sample the system. Some problems benefit from using collective order parameters such as energy gap (Warshel, 1991) or local level of hydration (Lu et al., 2016); identifying the proper degrees of freedom to thoroughly sample is indeed the major challenge to free energy simulations, especially when the bottleneck is entropic in nature. Other types of simulations use completely nonphysical order parameters; for example, an "alchemical" mutation (Simonson, Archontis, & Karplus, 2002; Straatsma & McCammon, 1992) that changes one atom to another (or even one residue or ligand to another) can assess properties like proton affinities (Li & Cui, 2003) or relative binding affinities of ligands (Deng & Roux, 2009; Simonson et al., 2002; Straatsma & McCammon, 1992).

Alchemical mutations highlight the fact that often times simulations are limited only by a chemist's creativity, but difficulty can arise if the question of interest does not contain an obvious choice for an order parameter. In such cases, a viable alternative is transition path sampling (TPS) (Bolhuis, Chandler, Dellago, & Geissler, 2002), which finds dynamic trajectories connecting two states without the biasing of umbrella sampling or metadynamics; closely related techniques are the finite-temperature string methods (Weinan & Vanden-Eijnden, 2010). If one is interested in dynamic aspects of an enzyme, and not just properties of reactants and transition states, TPS may be vital, and some have suggested that understanding enzyme reactivity requires understanding dynamic trajectories (Schwartz & Schramm, 2009). Likewise, others have found that the ensemble of transition states differs in important ways when examined by umbrella sampling and TPS (Doron, Kohen, Nam, & Major, 2014). TPS is ideal for transition events (such as barrier crossing) that are intrinsically fast; for processes that involve slower or diffusive events, other techniques such as string (Weinan & Vanden-Eijnden, 2010) and Markov state modeling (Bowman, Beauchamp, Boxer, & Pande, 2009) are likely more effective. The controversial question of the role of dynamics in enzyme catalysis has been argued extensively (Benkovic & Hammes-Schiffer, 2003; Boehr, Dyson, & Wright, 2006; Hay & Scrutton, 2012; Henzler-Wildman et al., 2007; Kamerlin & Warshel, 2010; Nagel & Klinman, 2009), and the practical concern of which computational methods work best for specific step(s) in the catalytic cycle may play a role in guiding that debate.

In addition to structures and energetics for the catalytic cycle, computational studies are also able to calculate various experimental observables such as NMR chemical shifts, vibrational spectra, and isotope effects. These are important quantities that allow one to make quantitative connection to experiments and provide validation/test of computational results; in many cases, comparison between computation and experiments is essential to the assignment and molecular interpretation of the experimental data (Gao & Truhlar, 2002; Molina & Jensen, 2003; Phatak, Ghosh, Yu, Cui, & Elstner, 2008; Wolf, Freier, Cui, & Gerwert, 2014; Wolf, Freier, & Gerwert, 2014). Depending on the system and observable(s) of interest, the degree of required sampling varies; generally speaking, however, averaging over a set of structures is likely important even to observables (eg, nuclear quadrupole coupling constant) that are often regarded as reporting "static" structural properties (Riccardi, Yang, & Cui, 2010).

The type of calculation dictates what level of theory should be used. For minimization type of calculations, it is feasible to use ab initio QM or QM/MM potentials. For simulations that require much more extensive sampling (eg, free energy and TPS), it is usually unrealistic to use ab initio QM/MM potentials in the framework of standard molecular dynamics. The solution is to either decouple the sampling of QM and MM regions (Hu & Yang, 2008; Rod & Ryde, 2005), so that QM is effectively treated with minimization while the MM region is sampled using molecular dynamics with frozen QM structure, or effectively integrate results from low-level QM/MM and ab initio QM/MM calculations (Claeyssens et al., 2006; Lu et al., 2016; Marti, Moliner, & Tuñón, 2005; Plotnikov, Kamerlin, & Warshel, 2011; Polyak, Benighaus, Boulanger, & Thiel, 2013). For spectral observables or isotope effects, results are usually most robust with ab initio QM(/MM) potentials, although carefully calibrated semiempirical QM(/MM) methods can be useful for specific properties, such as vibrational spectra (Phatak et al., 2008; Wolf, Freier, Cui, & Gerwert, 2014; Wolf, Freier, & Gerwert, 2014) and UV adsorption spectra (Hoffman et al., 2006; Wanko et al., 2005).

2.3 Summary

Just as there are a wide variety of experimental methods one can use to study an enzyme, there is a wide variety of computational methods. Here we have provided an outline of some of the most useful strategies for calculating properties and behavior of enzymes. Some techniques, such as active-site modeling, can be used successfully by (primarily) experimental enzymologists to add a layer of quantitative understanding to the interpretation of experimental results (Kipp, Hirschi, Wakata, Goldstein, & Schramm, 2012; Roston & Kohen, 2010). Other methods, such as free energy simulations, require more specialized training and until programs like CHARMM and AMBER become even more transparent to general users, computational chemists will have to communicate the results and implications of simulations to the experimental community as clearly as possible. We now turn our attention to a case study that demonstrates the usefulness of some of the earlier techniques.

3. CASE STUDY: THE TRANSITION STATE OF AP

Here we describe some recent studies on TS structure in the enzyme AP to illustrate the use of a few of the methods earlier. The studies involve QM calculations using active-site models, as well as QM/MM calculations and simulations of the entire enzyme. AP catalyzes the hydrolysis of phosphate monoesters, its native substrates, in addition to exhibiting promiscuous activity toward phosphate diesters (O'Brien & Herschlag, 2001; Zalatan & Herschlag, 2006), phosphorylated pyridines (Labow, Herschlag, & Jencks, 1993), phosphorothioates (Hollfelder & Herschlag, 1995), and even sulfonic esters (O'Brien & Herschlag, 1998). The chemical mechanism of hydrolysis by AP (Fig. 2) is similar to phosphoryl transfer in many enzymes (Cleland & Hengge, 2006): the substrate phosphorylates an enzymatic nucleophile (S102), and this phosphorylated enzyme intermediate is subsequently hydrolyzed to release phosphate. Important questions remain about how AP and other phosphoryl transferases catalyze the first phosphorylation step.

We have examined the binding mode(s) of putative TS analogue (TSA) inhibitors in AP using geometry optimizations and PES scans, and compared those geometries with the geometries of TS structures for phosphate monoester hydrolysis recently found by free energy simulations (Roston, Demapan, & Cui, in press). In addition to illustrating some of the computational methods available to enzymologists, we provide insights into the ability of TSAs to mimic the actual TS of the enzyme and the scope and limits of what can be gleaned from experimental observations of TSAs in this enzyme.

3.1 TSA Binding

TSAs serve as important inhibitors for enzymes and are currently used clinically as treatments for a variety of diseases (Schramm, 2011, 2015). In

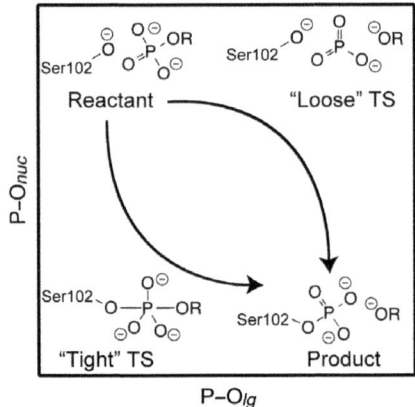

Fig. 2 The hydrolysis of phosphate esters in AP occurs by phosphorylation of the S102 nucleophile followed by hydrolysis of that phosphorylated enzyme intermediate. Phosphoryl transfer pathways vary as a function of the lengths of the breaking and forming P–O bonds (P–O_{lg} and P–O_{nuc}, respectively) and can occur through tight or loose pathways, or somewhere in between (synchronous). Here we explore how different TSs and different TS analogue inhibitors vary in their geometries and interactions with active-site motifs.

addition to their clinical relevance, TSAs are useful in many experimental probes of enzyme structure and dynamics, including X-ray crystallography (Holtz, Stec, & Kantrowitz, 1999), NMR (Boehr et al., 2006), IR (Bandaria et al., 2010), and others. Furthermore, TSAs have been useful in the development of catalytic antibodies (Benkovic, 1992; Mader & Bartlett, 1997; Schultz, 1989) and other biomimetic catalysts. One can imagine a variety of experimental criteria that would define a "good" TSA ligand: the ligand binds tightly to the enzyme and competitively inhibits catalysis (Schramm, 2011); mutation affects the inhibition constant of the TSA in the same way that it affects the rate of the native enzymatic reaction (Peck, Sunden, Andrews, Pande, & Herschlag, 2016); and even the absence of certain kinds of dynamics has been proposed to indicate that an enzyme–ligand complex resembles the TS (Bandaria et al., 2010). Perhaps the most obvious criterion is that the TSA and TS have similar structural properties and charge distributions. While crystallographic studies of TSA complexes illuminate the structures of TSAs bound to active sites, crystallographic studies cannot probe actual TSs for proper comparison. Despite sensationalistic claims of "trapping" TSs (Chen, Sharma, Quiocho, & Davidson, 2001; Pearson et al., 2015), actual TSs are unstable points on a free energy surface and thus cannot be trapped. Their structures can, however, be obtained through computational methods and our recent study of TS structure in AP (Roston et al., in press) allows us to compare TSAs with actual TSs.

Here we use a variety of computational methods to understand the binding modes of the tungstate (WO_4^{2-}) and vanadate (VO_4^{3-}) ions in the active site of AP, with an eye to answering the question of how well TSAs truly mimic TSs. In the previous work, crystallographers used vanadate as a TSA for AP (Holtz et al., 1999) as well as other phosphatases (Lindqvist, Schneider, & Vihko, 1994; Zalatan, Fenn, Brunger, & Herschlag, 2006; Zhang, Zhou, VanEtten, & Stauffacher, 1997). VO_4^{3-} binds covalently to the enzymatic nucleophile (S102, O_{nuc}), but it has no leaving group (LG) (or at least a very poor one), so the ligand adopts the trigonal bipyramidal geometry one expects (Lassila, Zalatan, & Herschlag, 2011) in the TS of a concerted transphosphorylation reaction. Vanadate therefore fulfills many roles of a good TSA, but it has limitations. The magnitude of the charge of VO_4^{3-} is larger than native substrates (phosphate monoesters, charge $= -2$) as well as promiscuous substrates (diesters and sulfate esters, charge $= -1$) and this difference in charge could result in misleading experimental results. The magnitude of this charge depends on the protonation state of the ion (Cruywagen, 2000), and uncertainty as to the protonation state when bound the enzyme clouds the interpretation of experimental results. WO_4^{2-}, on the other hand, has the same charge as the native substrate and may have a low enough pK_a to avoid ambiguity of charge (Cruywagen, 2000; Griffith & Lesniak, 1969). Until recently (Peck et al., 2016), however, WO_4^{2-} had not been used as a TSA for AP or any other enzyme with a serine nucleophile. We note that since initiating this project, Herschlag and coworkers have solved a crystal structure of AP with bound tungstate (Peck et al., 2016). The results of their crystallographic study were not made available to us, though, until we deposited our results with a third-party. Thus, the results we present here on tungstate should be viewed as a "blind" prediction except where noted.

In this work, we use a variety of computational methods to test whether WO_4^{2-} could serve as a TSA in wt *Escherichia coli* AP. We will move back and forth between methods and results/discussion in order to highlight the use of specific computational methods, the results that they yield, and how that information might inspire additional calculations (or experiments). We describe ab initio QM calculations on a small model of the enzyme active site, as well as ab initio QM/MM calculations on the full enzyme. We also test the computational methods using vanadate, which we were able to compare to a published crystal structure. Overall, our calculations support the use of WO_4^{2-} as a TSA that binds covalently to the serine nucleophile. After comparing our prediction with the recently solved crystal structure, we find that the geometry of the predicted covalent complex aligns reasonably well with the crystallographic model, but we consider whether protonation state is, in fact, an issue with tungstate as well as vanadate. Our original assumption of a deprotonated tungstate has similar bonding to the nucleophile and LG oxygen (O_{nuc} and O_{lg}, respectively) as the TS for phosphate monoesters with good LGs. The modeled ligands also interact with nearby enzymatic residues (eg, R166) in a similar manner to the TS of those monoesters, though there are still questions about the protonation states of the ligands—especially vanadate—which may be manifested in the effects of mutations on binding constants.

3.2 Computational Benchmarks

Prior to employing any computational method to predict unknown behavior it is necessary to examine the accuracy of the method for the property of interest. In this case, we wished to find a QM method that could accurately model geometric properties of tetra- and pentacoordinate tungstate and vanadate species; the method also needs to be efficient for QM/MM calculations of full enzyme models. To this end, we tested the ability of density functional theory (Parr & Yang, 1989) at the B3LYP/6-31+G* level (Becke, 1993; Lee, Yang, & Parr, 1988) with an effective core potential (Hay & Wadt, 1985) on the central metal atom to reproduce the geometries of small-molecule crystals. The species chosen were WO_4^{2-}, VO_4^{3-}, and two entries in the Cambridge Structural Database (CSD), CINMES and YIQLEP, which are WO_5 and VO_5 derivatives, respectively. We performed geometry optimizations of these species using Gaussian (Frisch et al., 2004) and the optimized geometries are listed in Table 1. We find that the computations reproduce the structural properties quite well and that the QM method is well suited for modeling the ligands in the AP active site.

3.3 Active-Site Optimizations

In an initial assessment of using WO_4^{2-} as a TSA in wt AP we sought to compute the binding mode of the ligand in a small model of the enzyme active site. The starting structure of the active-site model was the crystal structure of wt AP with bound PO_4^{3-} at 1.75 Å resolution (PDB: 1ED8) (Stec, Holtz, & Kantrowitz, 2000) and consisted of the groups in Fig. 3.

Table 1 Optimized Geometries of Small-Molecule Crystals[a]

	M–O1		M–O2		M–O3		M–O4		M–O5	
	Calc.	Exp.	Calc.	Exp.	Calc.	Exp.	Calc.	Exp.	Calc.	Exp.
WO_4^{2-}	1.79	1.80	1.77	1.80	1.78	1.80	1.81	1.80	NA	NA
VO_4^{3-}	1.76	1.72	1.76	1.72	1.72	1.72	1.72	1.72	NA	NA
CINMES[b]	1.91	1.87	1.91	1.87	1.91	1.87	1.91	1.87	1.71	1.70
YIOLEP[c]	1.83	1.87	1.88	1.86	1.91	1.89	2.15	2.09	2.11	2.10

[a]Each entry is the distance from the central metal to the oxygen (in Å), calculated at the B3LYP/6-31+G* level using an effective core potential for the metal ion.
[b]A pentacoordinate square pyramidal WO_5 derivative.
[c]A pentacoordinate trigonal bipyramidal VO_5 derivative. Experimental geometries are from the Inorganic Crystal Structure Database (ICSD) and the Cambridge Structural Database (CSD).

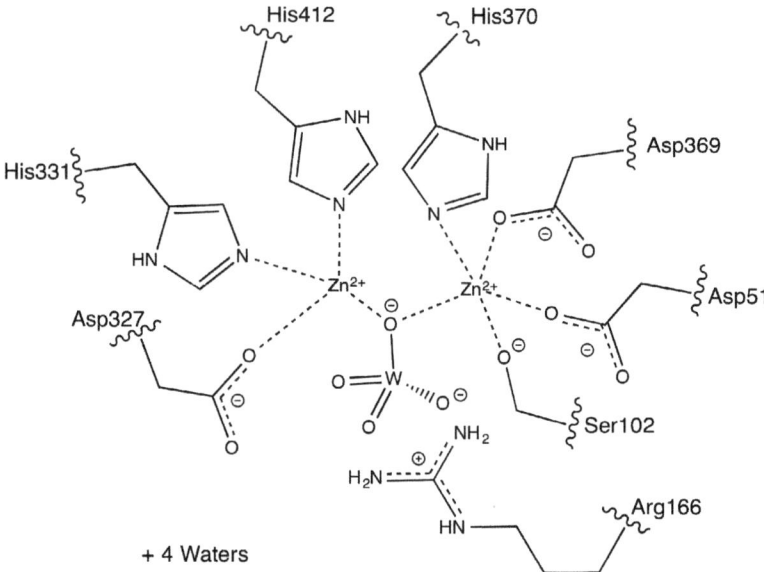

Fig. 3 The atoms in the active-site model of AP as well as the QM region used for the QM/MM calculations of the full enzyme. Note that in the B3LYP/MM optimizations of the full enzyme model, all waters were treated as MM.

The enzymatic residues were truncated at their α carbons, which were converted to methyl groups and frozen during optimizations, so the overall model contained 124 atoms and had a net charge of -1. The structure was optimized using the same ab initio QM method as earlier. In addition to the vacuum model, the optimization was performed using implicit solvent models (Tomasi, Mennucci, & Cammi, 2005) corresponding to a range of solvents in order to understand the magnitude and nature of any environmental effects.

Our optimizations of tungstate bound to the active-site model (Fig. 4) suggest that the ligand could, indeed, serve as a TSA. Optimized W–O_{nuc} distances range from 2.13 to 2.16 Å, depending on the solvent model, indicating a clear covalent nature to the bonding between the two atoms. For reference, the sum of the two atoms' van der Waals radii is ca. 4.1 Å (Alvarez, 2013). Additionally, the geometry around the W is trigonal bipyramidal, which is expected (Lassila et al., 2011) for the TS of transphosphorylation reactions. For reference, the other W–O bonds (ie, bonds to oxygens besides O_{nuc}) range from 1.75 to 1.90 Å, so the bond to the nucleophile is somewhat longer than a typical W–O bond. This is a desirable characteristic if tungstate is to serve as a TSA, because the bond to the central atom is presumably not fully formed at the TS of the reaction (Lassila et al., 2011).

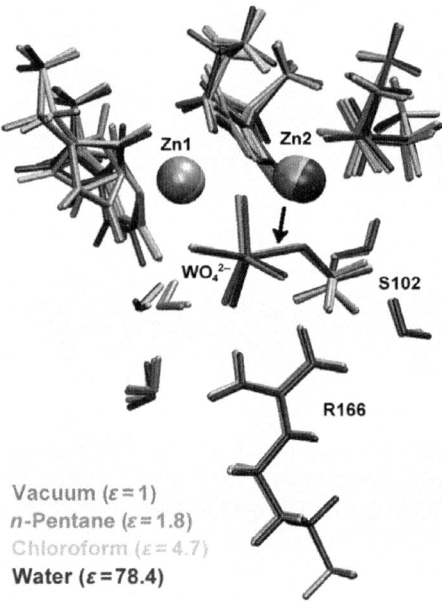

Fig. 4 Structures of the active-site model of AP with tungstate using different solvent models (indicated by *color*). Structures were optimized at the B3LYP/6-31+G* level, using an effective core potential for the metal atoms. The bond from the W to the nucleophilic serine alkoxide (O_{nuc}) is indicated by the *arrow*. Alpha carbons were frozen to their crystal structure (Stec et al., 2000) positions during the optimizations. (See the color plate.)

3.4 QM/MM Optimizations

To test how the ligand would behave in a more realistic model, we conducted optimizations of WO_4^{2-} in the full enzymatic environment in stages using CHARMM (Brooks et al., 2009). Since initial configuration could affect the optimized structure, we obtained a variety of starting structures for the optimizations, both from crystal structures with PO_4^{3-} (PDB: 1ED8; Stec et al., 2000) and VO_4^{3-} (PDB: 1B8J; Holtz et al., 1999), and from snapshots taken from QM/MM molecular dynamics simulations of the enzyme with PO_4^{3-} or with p-nitrophenyl phosphate (pNPP, a monoester substrate) bound to the active site. The simulations were conducted in accordance with the general methods described in Hou and Cui (2012), Hou and Cui (2013), and Roston et al. (in press). The starting structure was the crystal structure with PO_4^{3-} used earlier, and the ligand was changed to pNPP in silico for the simulations with that ligand. The active site was overlaid with a spherical droplet of water with 25 Å radius centered at one of the Zn^{2+} ions and water molecules within 2.8 Å of any crystallographic atoms were removed. The 25 Å radius sphere ("inner region") was treated using fully flexible molecular dynamics, and outside of that sphere ("outer region"), the protein was frozen and the outer region was treated with the generalized solvent boundary potential (GSBP) (Im et al., 2001; Schaefer et al., 2005). A 4-Å shell that was treated by Langevin dynamics served as a buffer region connecting the outer region with the inner region (Brooks & Karplus, 1989). The simulations used DFTB3/3OB as the QM method (Gaus, Cui, & Elstner, 2011; Gaus, Goez, & Elstner, 2012; Gaus, Lu, Elstner, & Cui, 2014; Lu, Gaus, Elstner, & Cui, 2015) with the CHARMM force field (MacKerell et al., 1998; MacKerell, Feig, & Brooks, 2004) for the MM region. The QM region consisted of the atoms shown in Fig. 3 with link atoms between the α and β carbons of QM side chains. The system underwent a short optimization and then the system was heated from 48 to 298 K over the course of 100 ps using 1 fs timesteps, after which it was equilibrated at 298 K for at least 50 ps before snapshots were taken to use as starting structures for optimizations with WO_4^{2-}. During the simulations, mild restraints helped to tune the QM–MM boundary, which must be managed carefully as discussed earlier (Goyal et al., 2014). This included restraints on the H-bond between the side chain of Asp330 and the backbone of Ser347 that is apparent in the crystal structure and a restraint on the C–O bond of Asp51 (QM region) that interacts with the nearby Mg^{2+} (MM region). The snapshots used for subsequent optimizations were all taken at intervals of at least 100 ps.

The snapshots obtained from the simulations as well as the original crystal structures (1ED8 and 1B8J) were first optimized at the DFTB3/MM level since that method is 100× faster than B3LYP/MM. Unfortunately, a complete parameterization of DFTB3 for tungsten is not yet available (Wahiduzzaman et al., 2013), so these initial optimizations used PO_4^{3-} as the ligand with the goal of optimizing the enzyme environment. Snapshots with pNPP as the ligand were converted to PO_4^{3-} and S102 was protonated for the optimizations with PO_4^{3-}. Thus, the overall charge remained constant. These structures underwent 10,000 steps of adapted-basis Newton–Raphson (ABNR) optimization (Brooks et al., 1983). At that point, the PO_4^{3-} was converted to WO_4^{2-}, S102 was deprotonated, and optimization using B3LYP/MM calculations began using the Gaussian09 interface with CHARMM (Zienau & Cui, 2012). All waters were treated as MM during this stage of the optimization. The QM region was treated with the same QM method as in the active-site model. ABNR optimization continued for 200–500 steps using the solvent macromolecular boundary potential (Benighaus & Thiel, 2011; Zienau & Cui, 2012) instead of GSBP. Gradients for the QM atoms were used as the primary criterion for determining convergence of the optimizations.

The optimized structures (Fig. 5) are consistent with the active-site models, suggesting that within the enzyme environment, tungstate can form a (long) covalent bond with the serine nucleophile and adopt the trigonal bipyramidal geometry of the TS. Given the range of starting structures for these calculations, the resulting structures are significantly more heterogeneous than the active-site models (Table 2). The $W-O_{nuc}$ distances, for example, range from 2.12 to 4.42 Å. Some of these structures, therefore, cannot be thought of as exhibiting covalent bonds to the ligand; they are clearly in the van der Waals region (Alvarez, 2013). The structures that do not form covalent complexes add ambiguity to the ability of tungstate to serve as a TSA, but we were able to clarify this ambiguity with a PES scan.

3.5 QM/MM PES Scan

To clarify the question of why some structures converged to geometries without a covalent bond, we conducted a PES scan along the $W-O_{nuc}$ coordinate using a structure that optimized to one of the longest $W-O_{nuc}$ distances (4.34 Å). At each point in the scan, the system was optimized with a harmonic restraining potential (force constant = 1000 kcal/mol Å^2) maintaining the $W-O_{nuc}$ distance. Each point along this coordinate was

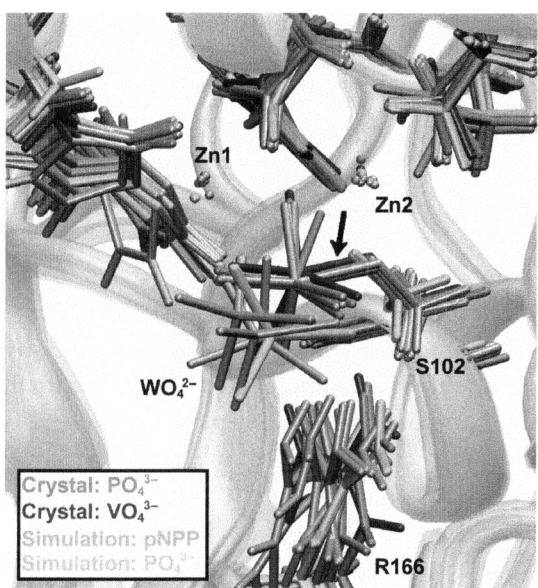

Fig. 5 Optimized structures from B3LYP/MM calculations of tungstate in the active site of AP. Starting structures were obtained from crystal structures and simulations with other ligands as indicated by *color*. In cases where the W–O_{nuc} distance is less than 2.5 Å, a bond is drawn between the two atoms, indicated by the *arrow*. (See the color plate.)

Table 2 Initial and Final W–O_{nuc} Distance (in Å) for Each Optimized Structure

Source[a]	Initial Value[b]	Final Value
VO$_4$-xtal	1.72[c]	2.12
PO$_4$-xtal	3.11	2.16
pNPP MD	4.08	4.26
pNPP MD	4.06	4.34
pNPP MD	2.86	2.24
pNPP MD	2.94	2.25
pNPP MD	3.10	2.22
pNPP MD	3.42	3.26
pNPP MD	4.05	4.42
pNPP MD	3.22	3.08

[a]Source of initial coordinates for optimization.
[b]Following the initial optimization with PO$_4^{3-}$ in the active site.
[c]Since a covalent bond is visible in the crystal structure but is unlikely to remain when the V is replaced with P, the initial optimization with PO$_4^{3-}$ had a restraint on this distance to maintain the interactions in the crystal structure.

generated following 10 steps of ABNR optimization at the neighboring point, and all points were subsequently optimized for an additional 100 steps; a modest number of optimization steps was used because the enzyme structure was thoroughly optimized prior to the PES scan. In the initial scan, points were calculated every 0.25 Å, but additional points were added in the region of the barrier separating the van der Waals complex and the covalently bonded complex and those points were treated the same way as the other points.

The PES scan (Fig. 6) demonstrates that—at least in the case of the structure chosen to scan—the noncovalently bound complex was simply trapped in a local minimum, and that in an experimental setting with thermal energies, it would be able to form the covalent complex. The true minimum along the W–O_{nuc} distance is at around 2.25 Å, which is in accordance with the structures that originally optimized to covalent complexes. This covalently bound state is more stable than the van der Waals complex by around 15 kcal/mol. Furthermore, the barrier to forming the covalently bound state from the van der Waals complex is only around 8 kcal/mol, and it is therefore accessible under crystallographic or other experimental conditions. As mentioned previously, a limitation of PES scans like this is that they calculate only the relative energies of the covalent and van der Waals complexes, and the results can be sensitive to subtle structural changes (eg, a water flip) during geometry optimization (Tao, Hodoscek, Larkin, Shao, & Brooks, 2012). Therefore, free energies are desirable for making more quantitative predictions of the relative stability of the covalent and van der Waals complexes. Unfortunately,

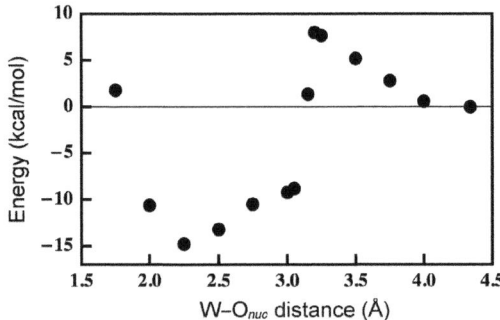

Fig. 6 PES scan at the B3LYP/MM level of the W–O_{nuc} distance, starting from a structure that converged to one of the longest W–O_{nuc} distances (4.34 Å). The covalently bound complex is approximately 15 kcal/mol more stable than the van der Waals complex, and the barrier between the two is around 8 kcal/mol. Note that these should be regarded as semiquantitative estimates since only potential energy is included in the computations.

conducting free energy simulations of the tungstate complex is not straightforward because the B3LYP method with about 120 QM atoms is very costly and there is no semiempirical alternative that is reliable for tungsten; we are currently pursuing B3LYP/MM calculations using the QM/MM minimum free energy path approach (Hu, Lu, Parks, Burger, & Yang, 2008) (D. Fang & Q. Cui, work in progress). We note, however, that testing the results of such free energy calculations experimentally would be difficult. Measurement of binding constants, for example, does not distinguish between the covalent and van der Waals complex, while crystallographic measurements do not contain meaningful information on relative free energies (for example, there are crystal structures of AP available with PO_4 either covalently (Stec et al., 2000) or noncovalently (Bobyr et al., 2012) bound, but information on the relative stabilities of the two binding modes is not available from the crystals. Given that both complexes can be measured, all that can be said is that the relative stabilities are dependent on crystallographic conditions). Further spectroscopic measurements might provide additional insights into the potential diversity in the binding mode of tungstate.

3.6 Tests of Protonation States

An important question to ask at this point is how reliable are B3LYP/MM calculations of the interactions between tungstate and AP? We note that previous calculations of the interaction between vanadate and AP (Hou & Cui, 2012) found that B3LYP/MM calculations predicted a significantly longer $V-O_{nuc}$ bond than that observed in the crystal structure (2.42 Å calculated vs 1.72 Å observed). A $V-O_{nuc}$ length of 2.42 Å would indicate an interaction with very little covalent character (Alvarez, 2013), and further concern comes from the fact that the geometry of the optimized complex appeared to be more tetrahedral than trigonal bipyramidal. Those calculations, therefore, did not precisely reproduce the experimentally resolved crystal structure (Holtz et al., 1999), which demonstrated a much shorter $V-O_{nuc}$ bond and trigonal bipyramidal geometry. One could question, then, whether B3LYP/MM calculations underestimate the strength of $M-O_{nuc}$ bonds. Considering the overall satisfying performance of B3LYP for several model compounds shown in Table 1, another possibility is that the protonation state of the vanadate played a role in that discrepancy. The previous calculations were with VO_4^{3-}, but it is very possible that this species is protonated under crystallographic conditions since HVO_4^{2-} has a pK_a of 13.3 in solution (Cruywagen, 2000). The positively charged groups (two Zn^{2+}, an arginine, and a Mg^{2+}-bound water) in the immediate vicinity of this ligand in the active site almost certainly lower its pK_a substantially, but it is difficult to know by precisely how much. The serine nucleophile, for example, is believed to be deprotonated in the apo enzyme and have a pK_a <5 (Andrews, Deng, & Herschlag, 2011), suggesting that the active site lowers its pK_a relative to that in solution by at least 10 pH units. Given that vanadate binds covalently, however, the additional negative charge from O_{nuc} may raise the pK_a. Thus, it is not obvious which species will predominate under experimental conditions. To test this question, we calculated structures of multiple protonation states for vanadate in the active-site model and in the full enzyme model.

In both the active-site model and the full enzyme model, optimizations with vanadate (VO_4^{3-}, HVO_4^{2-}, or $H_2VO_4^-$) were initiated from an optimized structure with tungstate covalently bound to O_{nuc}. The W was replaced with V and where applicable, the O opposite the S102 nucleophile (O_{lg}) was protonated. In the doubly protonated species, an additional phosphoryl oxygen was protonated. The active-site model was optimized as earlier. In the full enzyme model, water molecules within 15 Å of the vanadate were first optimized for 1000 steps of ABNR with all other atoms frozen and treated at the MM level, in order to allow the hydrogen-bonding network to adapt to the different protonation state. The system then underwent 100–200 steps of ABNR optimization using the B3LYP/MM method used for tungstate. Again, the gradients on the ligand and the nucleophile were the primary criteria for convergence. Both the active-site model and the full enzyme model (Fig. 7) find that the singly protonated species converges to a structure most consistent with that found experimentally, suggesting that vanadate is singly protonated in the active site of AP. In contrast to the deprotonated form calculated previously, the protonated form converges to a structure where the V–O_{nuc} bond is shorter than the V–O_{lg} bond (1.97 Å vs 2.14 Å), which is the case in the crystal structure (1.72 Å vs 1.92 Å). Additionally, as is the case in the crystal structure, the protonated structure exhibits trigonal bipyramidal geometry. We note that the doubly protonated species in the active-site model had large distortions of the active-site geometry (eg, Zn–Zn distance of 5.3 Å vs 4.1 Å in the crystal). Together, these support the hypothesis that HVO_4^{2-} is the form of vanadate that predominates in the crystal structure and that if one chooses the correct protonation state, B3LYP/MM calculations can adequately reproduce the binding modes of MO_4^{n-} ligands. Thus, knowing nothing about the recent X-ray studies of tungstate in AP, we were able to predict that tungstate is able

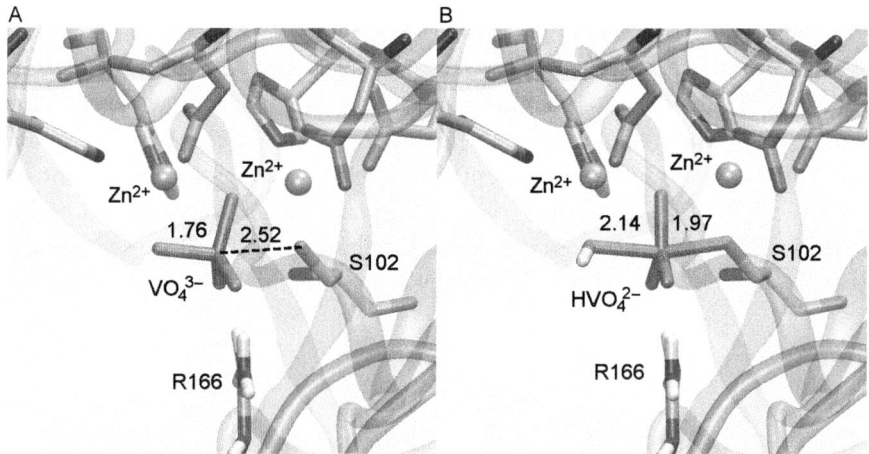

Fig. 7 Optimized structures with VO_4^{3-} (A) and HVO_4^{2-} (B) in AP. Bond lengths between V–O_{nuc} and V–O_{lg} are indicated in the figure. For reference, those bond lengths are 1.72 and 1.92 Å, respectively, in the crystal structure (see Fig. 9) (Holtz et al., 1999). The bond lengths in (A) differ slightly from those calculated previously for that form 79 because the starting structures for the optimizations differed. Also visible in these structures is the notable difference in geometry around the V, which is tetrahedral in (A), but trigonal bipyramidal in (B). These optimizations were both initiated from the same optimized structure with tungstate covalently bound to the serine nucleophile.

to bind covalently in a trigonal bipyramidal geometry reminiscent of the TS for phosphoryl transfer.

A potential criticism of the vanadate analysis could be that it is, in a sense, circular: we have sampled different protonation states in order to find one that fits the experimental results, and then claimed that the computational method is accurate if we choose the correct protonation state. Another possibility is that the method is inaccurate, but if we choose an incorrect protonation state, it gives results in agreement with the crystal structure. Experimental measurements of the pH dependence of the crystal structure, therefore, are necessary to test the model. Another issue with this analysis is that we are fitting our computational model to an experimental model (ie, atomic coordinates from refinement of the X-ray diffraction data); a better comparison would have been with the raw experimental data (ie, the electron density). When judging the veracity of a computational model, one should ask whether it correctly predicts/reproduces experimental results, not whether it correctly predicts/reproduces an experimentalist's interpretation of those results. Unfortunately, electron density is not available in the PDB for the vanadate structure.

3.7 Comparison with Experiment

At this point, we can examine whether the recently solved structure with tungstate (Peck et al., 2016) supports or refutes our findings. We have overlaid our calculated structures and the experimental structures in Fig. 8 and we find relatively good agreement: the crystallographic structure contains a covalent W–O_{nuc} bond and trigonal bipyramidal geometry. We do note, however, that the crystallographic structure has a substantially longer W–O_{lg} bond than we obtain (Table 3). Initially we assumed that tungstate would be

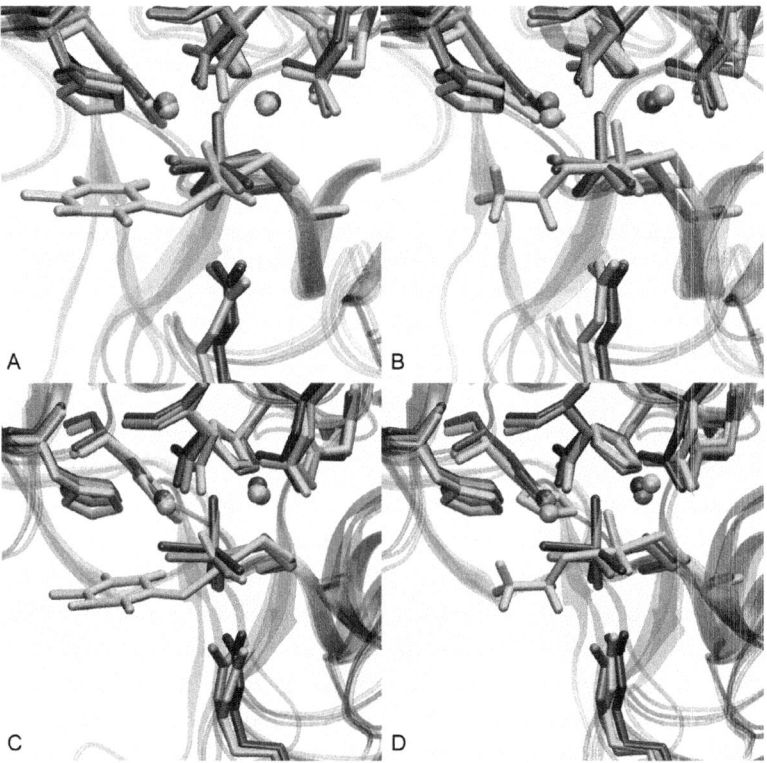

Fig. 8 Overlays of representative structures of TSAs (calculated, *red* (*gray* in the print version); crystallographic, *blue* (*dark gray* in the print version)) and TS structures (computed with DFTB3/MM free energy simulations (Roston et al., in press), in *green* (*light gray* in the print version)) for two different phosphate monoesters. (A) Phenyl phosphate and vanadate. (B) Ethyl phosphate and vanadate. (C) Phenyl phosphate and tungstate. (D) Ethyl phosphate and tungstate. These structures suggest that in terms of overall geometry especially ligation by the Zn^{2+} ions the TSAs are more representative of the TS for ethyl phosphate, but in terms of M–O_{nuc} and M–O_{lg} bond lengths, tungstate more resembles phenyl phosphate (cf. Fig. 9). The structures were aligned based on the alpha carbons of the residues composing the QM region.

Table 3 Geometries[a] of Calculated and Experimental TSAs and Calculated TSs

	M(P)–O$_{lg}$	M(P)–O$_{nuc}$	Zn–Zn	Zn1–O$_{lg}$	Zn2–O$_{nuc}$	Zn1–O$_{nb}$	Zn2–O$_{nb}$
WO$_4$-xtal (5C66)	2.2	2.1	4.0	2.0	2.1	2.1	2.0
WO$_4^{2-}$-active site	1.9	2.2	4.1	2.2	2.1	2.1	2.2
HWO$_4^-$-active site	2.1	2.0	4.2	2.2	2.2	2.2	2.2
WO$_4^{2-}$-QM/MM[b]	1.8	2.2	4.1	2.4	2.0	2.1	2.1
VO$_4$-xtal (1B8J)	1.9	1.7	4.1	2.4	1.9	2.1	2.2
VO$_4^{3-}$-active site	1.8	3.6	4.3	2.1	2.0	2.4	2.2
HVO$_4^{2-}$-active site	2.2	2.0	4.1	2.1	2.1	2.1	2.2
H$_2$VO$_4^-$-active site	2.2	1.9	5.2	2.1	2.1	2.2	3.3
VO$_4^{3-}$-QM/MM	1.8	2.5	4.0	2.2	2.0	2.1	2.1
HVO$_4^{2-}$-QM/MM	2.1	2.0	4.1	2.1	2.0	2.1	2.2
PhOP-QM/MM[c]	1.9	2.1	4.2	3.5	2.0	2.0	3.1
EtOP-QM/MM[c]	2.1	1.8	4.2	2.1	2.3	2.3	2.1

[a] Distances are in Å. "Active site" indicates active site model optimized at the pure QM level; "QM/MM" indicates QM/MM optimized result.
[b] Average of the structures that converged to covalent W–O$_{nuc}$ bonds.
[c] From Roston et al. (in press).
EtOP, ethyl phosphate; *PhOP*, phenyl phosphate.

deprotonated in the active site because the pK_a of HWO_4^- is 3.6 in solution (Cruywagen, 2000; Griffith & Lesniak, 1969) and the positively charged moieties in the active site would lower that even further. The large deviation with the crystal structure made us question this assumption and so we calculated the structure of the protonated species in the active-site model. As Table 3 and Fig. 9 indicate, the results for active-site models do not differ in substantial ways from the full enzyme models of the same species. The structure of HWO_4^- in the active-site model optimized to a structure whose $W-O_{lg}$ bond was more similar to that in the crystal structure, but that model has a somewhat shorter $W-O_{nuc}$ bond than the crystal structure. Thus, the question of the protonation state of WO_4-AP is not as clear as in VO_4-AP.

Further adding to this uncertainty, the electron density and the anomalous diffraction in the crystal structure with tungstate (Peck et al., 2016) have peculiarities and ambiguities that warrant discussion. In contrast to most crystal structures of wt E. coli AP, which have two Zn^{2+} and one Mg^{2+} in the active site, the Mg^{2+} site had too much electron density to model a Mg^{2+} there, so the model contains an octahedrally coordinated Zn^{2+} at

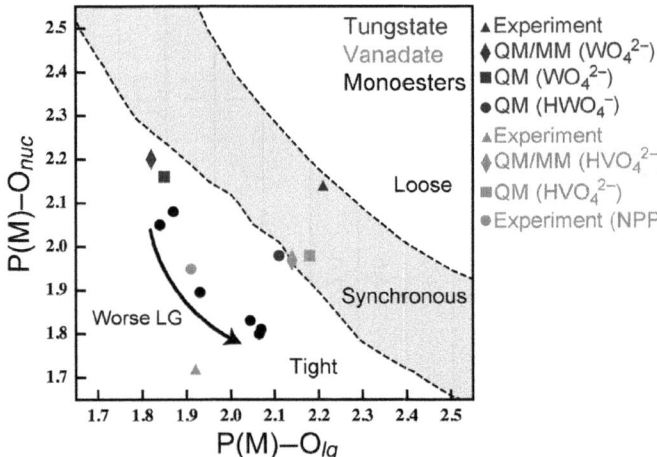

Fig. 9 The TS region of the phosphate ester hydrolysis reaction, showing the various TS structures for phosphate monoesters found in free energy simulations (Roston et al., in press), as well as the crystallographic TSA structures (including one vanadate structure bound to a closely related enzyme, NPP (Zalatan et al., 2006)) and the most relevant calculated TSA structures. The guides as to the tight, synchronous, and loose regions are determined from the sum of bond order of the P–O_{lg} and P–O_{nuc} bonds as described in Roston et al. (in press). The calculated structure of tungstate here is the average of the structures that converged to a covalent bond in the initial optimizations. The points labeled QM/MM refer to the full enzyme model and those labeled QM refer to the active-site model.

that site. While octahedral Zn^{2+} is not unheard of, it is relatively rare in proteins (Alberts, Nadassy, & Wodak, 1998) and is not typically observed in AP, although one structure from the same lab also reported a Zn in place of the Mg (Andrews, Fenn, & Herschlag, 2013). Furthermore, the tungsten atom in the crystal had weaker electron density than expected, despite normal density for three oxygens around the tungsten (of the five oxygens expected around tungsten). The authors interpreted this as indicating a half-occupied tungstate covalently bound to S102. The unoccupied active site contains waters in similar positions as two of the tungstate oxygens and the position of S102 is independent of tungstate binding. This configuration of waters differs from a structure of the *apo* enzyme (Stec et al., 2000) and results in a linear $Zn-OH_2-Zn$, which is a surprising configuration based on structures with similar di-Zn^{2+} moieties (eg, metallo β-lactamase; PDB: 4HL2). In the end, the way the authors chose to fit the electron density may be the most reasonable given the ambiguities. Nonetheless, we discuss these ambiguities, as well as those with the protonation state of the computational structure, to urge caution in the way computationalists interpret experimental results and the way experimentalists judge the validity of computational results. Often times the situation is not as clear-cut as crystallographic or computational structures may imply and human choices like protonation states or the best way to fit electron density may lower one's confidence in the particular model's applicability.

Despite the overall similarity between the calculated and experimental TSA structures, Fig. 9 shows that there are potentially important differences in bond lengths in the calculated TSA structures and their crystallographic counterparts. For example, the $W-O_{lg}$ bond is nearly 0.5 Å longer in the crystallographic model than in the calculated WO_4^{2-} model. The protonated structure, HWO_4^-, achieves closer agreement, but the ambiguities mentioned earlier with the crystal structure make it difficult to use the crystal structure to rule out the deprotonated structure as a possibility. The long bond length in the crystal structure, for example, may result from an artifact of deconvoluting electron density contributions from both O_{lg} and a water in similar positions. We could find no other structures in the PDB containing a Ser-bound tungstate to determine what typical bond lengths might be. A few structures are available with Asp-bound tungstate (eg, 3ET5, 2I34, 1J9K) and these have $W-O_{nuc}$ bonds in the range of 2.4–2.8 Å indicating that $W-O_{nuc}$ bonds in enzymes might generally be longer than our model predicts. Clearly, additional work will be necessary to determine where the discrepancy in bond lengths between the calculated and crystallographic models arises.

The differences in bond lengths with vanadate are also intriguing. In this case, the proper protonation state appears to be HVO_4^{2-}, given that VO_4^{3-} does not bind covalently and $H_2VO_4^-$ results in large deformations of the di-Zn^{2+} moiety (Table 3). The difference between the calculated structures and the experimental structure is ca. 0.2–0.3 Å in both the V–O_{nuc} and V–O_{lg} lengths. As mentioned, the electron density for the vanadate structure is not available to offer any hints from where this discrepancy may originate. A high resolution structure of a closely related enzyme, nucleotide pyrophosphatase (PDB: 2GSO) (Zalatan et al., 2006), has V–O_{nuc} and V–O_{lg} lengths of 1.9 and 2.0 Å, respectively, which is closer to what we obtain computationally, but at this time we cannot suggest that there are clear problems with the vanadate-AP structure.

3.8 TSAs vs TSs

Now we wish to ask whether either tungstate or vanadate is a good mimic of the actual TS. To answer this computationally, we must examine models for the actual TS, which we have recently done using DFTB3/MM free energy simulations (Roston et al., in press); the models of the TS are qualitatively consistent with diverse experimental data that include free energy relations, mutation effects, and kinetic isotope effects. We note that DFTB3 was required because the active site of AP is very solvent accessible and therefore adequate sampling is crucial to the proper modeling of the TS structure (cumulatively more than 10 ns is used for each substrate in the DFTB3/MM umbrella sampling simulations); the DFTB3 method was also calibrated by a set of relevant phosphoryl transfer reactions and pK_a calculations against higher level calculations as well as relevant experimental data (see SI of Roston et al., in press). What we found in that study is that the structure of the TS depends strongly on the LG for both phosphate mono- and diesters. The reaction pathways can all be characterized as tight, where nucleophilic attack precedes cleavage of the bond to the LG, but worse LGs have TSs that are later in the reaction coordinate and involve significantly more P–O_{lg} cleavage. Fig. 8 shows the TSAs overlaid with representative snapshots from the TS region for phosphate monoesters with a good LG and a poor LG. The overall active-site architectures are quite similar for the TSs and TSAs, but in terms of ligation to the Zn^{2+} ions, the TSAs seem to better resemble the TS for poor LGs. In Fig. 9, which contains a summary of the relationships between enzymatic TSs and the structures of the TSAs in terms of bonding to the central atom (P, V, or W), one can see that in terms of the primary components of the reaction coordinate, the calculated structure of WO_4^{2-} appears to be most similar to the TS of phosphate monoesters with good LGs, despite differences in Zn^{2+} ligation apparent in Fig. 8 (and Table 3). If, however, HWO_4^- is the relevant species, it may be more similar to TSs with poor LGs, similar to the calculated structures for HVO_4^{2-}.

While the ligation pattern to the Zn^{2+} ions is somewhat different at the TS for phenyl phosphate than for tungstate (or vanadate), the interactions with the H-bonding network in the active site (Sunden, Peck, Salzman, Ressl, & Herschlag, 2015), and most notably R166, are quite similar. Thus, the model predicts that mutations of this network will affect tungstate and vanadate binding in a similar manner as they will affect catalysis of phosphate ester hydrolysis. Measurements (Peck et al., 2016) showed that mutations (including double- and triple-mutants, etc.) cause similar changes in tungstate binding and catalytic rate for hydrolysis of *p*-nitrophenyl phosphate. While many mutations also show such a relationship with vanadate binding, most of the mutants that involve the E322Y mutation, which disrupts Mg^{2+} binding, do not fit the trend. The E322Y mutation appears to have little effect on vanadate binding, despite the fact that it hinders both tungstate binding and catalytic rate. It is not obvious why this should be the case from our calculated geometries or the geometries in the crystal structures. The catalytic role of Mg^{2+} appears to be to position a water to donate a hydrogen bond to the phosphoryl group at the TS of the reaction (Zalatan, Fenn, & Herschlag, 2008). The fact that ablation of this Mg^{2+} does not disrupt vanadate binding may indicate that the H-bond donated by the Mg^{2+}-bound water is itself disrupted by protonation of the vanadate at that position.

4. SUMMARY/CONCLUSIONS

Despite the many challenges in using computational methods to understand enzymes, when they are applied carefully they can make reliable predictions of behavior in a variety of contexts. We have shown that active-site modeling using ab initio QM calculations can be used to understand ligand binding, and this can be supplemented by QM/MM calculations using the full enzyme. Our results indicate the importance of sampling many configurations, as the starting configuration can bias the outcome of optimizations. In many cases the most reliable and predictive method for understanding behavior will be free energy simulations, but for systems that contain heavy metals, computational cost generally prohibits straightforward ab initio QM/MM free energy simulations and force fields are generally unavailable. Thus combining insights from ab initio QM/MM-based optimization and PES scans with semiempirical QM/MM (eg, DFTB3/MM)-based molecular dynamics can provide valuable insights on questions such as binding mode, geometry and similarity between TSA and TS. Several important directions of development for metalloenzyme studies include improving the efficiency and robustness of ab initio QM/MM-based free energy simulations, developing semiquantitative semiempirical QM methods for metal ions, and better integrating multilevel QM/MM methods for quantitative computations. These developments will make it possible to make more quantitative and testable predictions from computational studies of metalloenzymes. Along this line, we echo the suggestion from experimental community that computational studies ought to aim at not only analyzing published experimental results but also making clearly spelled out predictions that can be tested by future experiments. One format might be to summarize such predictions explicitly in a publication (Roston et al., in press).

Our comparison of the binding geometry of TSAs with the structures of actual TS models provides some confidence in using those TSAs for experimental studies of TS structure. Still, the variation of TS structure as a function of LG ability (Roston et al., in press) urges caution in which TSA should be used to model which TS. Tungstate is apparently a good mimic of the TS of monoesters with good LGs, but vanadate may better model those with poor LGs. Measurements of inhibition by the TSAs vs rate for different LGs can test this prediction (Peck et al., 2016). Additionally, the fact that AP binds to tungstate with the geometry of the TS speaks to the great catalytic power of AP. Many structures in the PDB, for example, contain tungstate bound to the active site of a phosphatase but not covalently. The fact that AP stabilizes this ligand in the conformation of the TS is likely related to APs extraordinary ability to stabilize the TS. Continued study of how AP is so successful at stabilizing its TS will provide important insights into the general mechanisms of biological catalysis.

ACKNOWLEDGMENTS

We thank Dr. Dan Herschlag and coworkers (A. Peck and F. Sunden) for providing the coordinates to the crystal structure with tungstate prior to publication. We thank James Fraser for serving as an intermediary to verify that the computational work was accomplished prior to sharing of X-ray crystallography results. D.R. is supported by an NIH NRSA Postdoctoral Fellowship (1F32GM112371-01A1). This research was also supported by an NIH grant (R01GM106443) and an XSEDE allocation (TG-MCB110014) to Q.C. Computational resources from the Extreme Science and

Engineering Discovery Environment (XSEDE), which is supported by NSF Grant No. OCI-1053575, are greatly appreciated; computations are also partly supported by the National Science Foundation through a major instrument grant (CHE-0840494).

REFERENCES

Alberts, I. L., Nadassy, K., & Wodak, S. J. (1998). Analysis of zinc binding sites in protein crystal structures. *Protein Science, 7*, 1700–1716.

Alvarez, S. (2013). A cartography of the van der Waals territories. *Dalton Transactions, 42*, 8617–8636.

Andrews, L. D., Deng, H., & Herschlag, D. (2011). Isotope-edited FTIR of alkaline phosphatase resolves paradoxical ligand binding properties and suggests a role for ground-state destabilization. *Journal of the American Chemical Society, 133*, 11621–11631.

Andrews, L. D., Fenn, T. D., & Herschlag, D. (2013). Ground state destabilization by anionic nucleophiles contributes to the activity of phosphoryl transfer enzymes. *PLoS Biology, 11*, e1001599.

Bandaria, J. N., Dutta, S., Nydegger, M. W., Rock, W., Kohen, A., & Cheatum, C. M. (2010). Characterizing the dynamics of functionally relevant complexes of formate dehydrogenase. *Proceedings of the National Academy of Sciences of the United States of America, 107*, 17974–17979.

Barducci, A., Bonomi, M., & Parrinello, M. (2011). Metadynamics. *WIREs Computational Molecular Science, 1*, 826–843.

Becke, A. D. (1993). Density-functional thermochemistry. III. The role of exact exchange. *Journal of Chemical Physics, 98*, 5648–5652.

Becke, A. D. (2014). Perspective: Fifty years of density-functional theory in chemical physics. *Journal of Chemical Physics, 140*, 18A301.

Benighaus, T., & Thiel, W. (2011). Long-range electrostatic effects in QM/MM studies of enzymatic reactions: Application of the solvated macromolecule boundary potential. *Journal of Chemical Theory and Computation, 7*, 238–249.

Benkovic, S. J. (1992). Catalytic antibodies. *Annual Review of Biochemistry, 61*, 29–54.

Benkovic, S. J., & Hammes-Schiffer, S. (2003). A perspective on enzyme catalysis. *Science, 301*, 1196–1202.

Blomberg, M. R. A., Borowski, T., Himo, F., Liao, R. Z., & Siegbahn, P. E. M. (2014). Quantum chemical studies of mechanisms for metalloenzymes. *Chemical Reviews, 114*, 3601–3658.

Blomberg, R., Kries, H., Pinkas, D. M., Mittl, P. R. E., Grutter, M. G., Privett, H. K., ... Hilvert, D. (2013). Precision is essential for efficient catalysis in an evolved Kemp eliminase. *Nature, 503*, 418.

Bobyr, E., Lassila, J. K., Wiersma-Koch, H. I., Fenn, T. D., Lee, J. J., Nikolic-Hughes, I., ... Herschlag, D. (2012). High-resolution analysis of Zn2+ coordination in the alkaline phosphatase superfamily by EXAFS and X-ray crystallography. *Journal of Molecular Biology, 415*, 102–117.

Boehr, D. D., Dyson, H. J., & Wright, P. E. (2006). An NMR perspective on enzyme dynamics. *Chemical Reviews, 106*, 3055–3079.

Bolhuis, P. G., Chandler, D., Dellago, C., & Geissler, P. L. (2002). Transition path sampling: Throwing ropes over rough mountain passes, in the dark. *Annual Review of Physical Chemistry, 53*, 291–318.

Bowman, G. R., Beauchamp, K. A., Boxer, G., & Pande, V. S. (2009). Progress and challenges in the automated construction of Markov state models for full protein systems. *Journal of Chemical Physics, 131*, 124101.

Brooks, B. R., Brooks, C. L., III, Mackerell, A. D., Nilsson, L., Petrella, R. J., Roux, B., ... Karplus, M. (2009). CHARMM: The biomolecular simulation program. *Journal of Computational Chemistry, 30*, 1545–1614.

Brooks, B. R., Bruccoleri, R. E., Olafson, B. D., States, D. J., Swaminathan, S., & Karplus, M. (1983). CHARMM: A program for macromolecular energy, minimization and dynamics calculations. *Journal of Computational Chemistry, 4*(2), 187–217.

Brooks, C. L., III, & Karplus, M. (1989). Solvent effects on protein motion and protein effects on solvent motion: Dynamics of the active site region of lysozyme. *Journal of Molecular Biology, 208*(1), 159–181.

Brunk, E., & Rothlisberger, U. (2015). Mixed quantum mechanical/molecular mechanical molecular dynamics simulations of biological systems in ground and electronically excited states. *Chemical Reviews, 115*, 6217–6263.

Chan, G. K. L., & Sharma, S. (2011). The density matrix renormalization group in quantum chemistry. *Annual Review of Physical Chemistry, 62*, 465–481.

Chen, J., Sharma, S., Quiocho, F. A., & Davidson, A. L. (2001). Trapping the transition state of an ATP-binding cassette transporter: Evidence for a concerted mechanism of maltose transport. *Proceedings of the National Academy of Sciences of the United States of America, 98*, 1525–1530.

Claeyssens, F., Harvey, J. N., Manby, F. R., Mata, R. A., Mulholland, A. J., Ranaghan, K. E., … Werner, H. J. (2006). High-accuracy computation of reaction barriers in enzymes. *Angewandte Chemie International Edition in English, 45*, 6856–6859.

Cleland, W. W., & Hengge, A. C. (2006). Enzymatic mechanisms of phosphate and sulfate transfer. *Chemical Reviews, 106*, 3252–3278.

Cohen, A. J., Mori-Sanchez, P., & Yang, W. T. (2012). Challenges for density functional theory. *Chemical Reviews, 112*, 289–320.

Cruywagen, J. J. (2000). Protonation, oligomerization, and condensation reactions of vanadate(V), molybdate(VI), and tungstate(VI). *Advances in Inorganic Chemistry, 49*, 127–182.

Cui, Q., & Elstner, M. (2014). Density functional tight binding: Values of semi-empirical methods in an ab initio era. *Physical Chemistry Chemical Physics, 16*, 14368–14377.

Cui, Q., & Karplus, M. (2002). QM/MM studies of the triosephosphate isomerase (TIM) catalyzed reactions: The effect of geometry and tunneling on proton transfer rate constants. *Journal of the American Chemical Society, 124*, 3093–3124.

Cui, Q., Meuwly, M., & Ren, P. Y. (Eds.), (2016). *Many-body effects and electrostatics in biomolecules*. Singapore: Pan Stanford Publishing.

Darden, T., York, D., & Pedersen, L. (1993). Particle mesh Ewald—An N.log(N) method for Ewald sums in large systems. *Journal of Chemical Physics, 98*, 10089–10092.

Deng, Y. Q., & Roux, B. (2009). Computations of standard binding free energies with molecular dynamics simulations. *Journal of Physical Chemistry B, 113*, 2234–2246.

Dewar, M. J. S., Zoebisch, E. G., Healy, E. F., & Stewart, J. J. P. (1985). Development and use of quantum mechanical molecular models. 76. AM1: A new general purpose quantum mechanical molecular model. *Journal of the American Chemical Society, 107*(13), 3902–3909. http://dx.doi.org/10.1021/ja00299a024.

Doron, D., Kohen, A., Nam, K., & Major, D. T. (2014). How accurate are transition states from simulations of enzymatic reactions? *Journal of Chemical Theory and Computation, 10*, 1863–1871.

Doron, D., Major, D. T., Kohen, A., Thiel, W., & Wu, X. (2011). Hybrid quantum and classical simulations of the dihydrofolate reductase catalyzed hydride transfer reaction on an accurate semi-empirical potential energy surface. *Journal of Chemical Theory and Computation, 7*, 3420–3437.

Elstner, M., Porezag, D., Jungnickel, G., Elsner, J., Haugk, M., Frauenheim, T., … Seifert, G. (1998). Self-consistent-charge density-functional tight-binding method for simulations of complex materials properties. *Physical Review B, 58*(11), 7260–7268.

Fersht, A. (1999). *Structure and mechanism in protein science: A guide to enzyme catalysis and protein folding*. New York: W.H. Freeman and Company.

Field, M. J., Bash, P. A., & Karplus, M. (1990). A combined quantum-mechanical and molecular mechanical potential for molecular-dynamics simulations. *Journal of Computational Chemistry, 11*(6), 700–733.
Freindorf, M., & Gao, J. L. (1996). Optimization of the Lennard-Jones parameters for a combined ab initio quantum mechanical and molecular mechanical potential using the 3–21G basis set. *Journal of Computational Chemistry, 17*, 386–395.
Friesner, R. A., & Guallar, V. (2005). Ab intio QM and QM/MM methods for studying enzyme catalysis. *Annual Review of Physical Chemistry, 56*, 389–427.
Frisch, M. J., Trucks, G. W., Schlegel, H. B., Scuseria, G. E., Robb, M. A., Cheeseman, J. R., ... Pople, J. A. (2004). *Gaussian 03, Revision C.01*. Wallingford, CT: Gaussian, Inc.
Gao, J. (1995). Methods and applications of combined quantum mechanical and molecular mechanical potentials. In K. B. Lipkowitz & D. B. Boyd (Eds.), *Reviews in computational chemistry: Vol. 7* (pp. 119–185). New York: VCH.
Gao, J. L., Ma, S. H., Major, D. T., Nam, K., Pu, J. Z., & Truhlar, D. G. (2006). Mechanisms and free energies of enzymatic reactions. *Chemical Reviews, 106*, 3188–3209.
Gao, J. L., & Truhlar, D. G. (2002). Quantum mechanical methods for enzyme kinetics. *Annual Review of Physical Chemistry, 53*, 467–505.
Garcia-Viloca, M., Gao, J., Karplus, M., & Truhlar, D. G. (2004). How enzymes work: Analysis by modern rate theory and computer simulations. *Science, 303*, 186–195.
Gaus, M., Cui, Q., & Elstner, M. (2011). Dftb-3rd: Extension of the self-consistent-charge density-functional tight-binding method scc-dftb. *Journal of Chemical Theory and Computation, 7*, 931–948.
Gaus, M., Goez, A., & Elstner, M. (2012). Parametrization and benchmark of DFTB3 for organic molecules. *Journal of Chemical Theory and Computation, 9*, 338–354.
Gaus, M., Goyal, P., Hou, G., Lu, X., Pang, X., Zienau, J., ... Cui, Q. (2014). Toward quantitative analysis of metalloenzyme function using MM and hybrid QM/MM methods: Challenges, methods and recent applications. In R. Zhou (Ed.), *Molecular modeling at the atomic scale: Methods and applications in quantitative biology* (pp. 33–82): New York: CRC Press.
Gaus, M., Lu, X., Elstner, M., & Cui, Q. (2014). Parameterization of DFTB3/3OB for sulfur and phosphorus for chemical and biological applications. *Journal of Chemical Theory and Computation, 10*, 1518–1537.
Ghosh, N., Prat-Resina, X., & Cui, Q. (2009). Towards a reliable molecular model of cytochrome c oxidase: Insights from microscopic pKa calculations. *Biochemistry, 48*, 2468–2485.
Giese, T. J., & York, D. M. (2007). Charge-dependent model for many-body polarization, exchange, and dispersion interactions in hybrid quantum mechanical/molecular mechanical calculations. *Journal of Chemical Physics, 127*, 194101.
Goyal, P., Lu, J., Yang, S., Gunner, M. R., & Cui, Q. (2013). Changing hydration level in an internal cavity modulates the proton affinity of a key glutamate in cytochrome c oxidase. *Proceedings of the National Academy of Sciences of the United States of America, 110*, 18886–18891.
Goyal, P., Qian, H. J., Irle, S., Lu, X., Roston, D., Mori, T., ... Cui, Q. (2014). Feature article: Molecular simulation of water and hydration effects in different environments: Challenges and developments for DFTB based models. *Journal of Physical Chemistry B, 118*, 11007–11027.
Griffith, W. P., & Lesniak, P. J. B. (1969). Raman studies on species in aqueous solutions. 3. Vanadates molybdates and tungstates. *Journal of the Chemical Society A: Inorganic, Physical, Theoretical*, 1066–1071.
Halperin, I., Ma, B. Y., Wolfson, H., & Nussinov, R. (2002). Principles of docking: An overview of search algorithms and a guide to scoring functions. *Proteins: Structure, Function, and Genetics, 47*, 409–443.

Hay, S., & Scrutton, N. S. (2012). Good vibrations in enzyme-catalysed reactions. *Nature Chemistry*, *4*, 161–168.

Hay, P. J., & Wadt, W. R. (1985). Ab initio effective core potentials for molecular calculations—Potentials for the transition-metal atoms Sc to Hg. *Journal of Chemical Physics*, *82*, 270–283.

Henzler-Wildman, K. A., Lei, M., Thai, V., Kerns, S. J., Karplus, M., & Kern, D. (2007). A hierarchy of timescales in protein dynamics is linked to enzyme catalysis. *Nature*, *450*, 913–916.

Hoffman, M., Wanko, M., Strodel, P., Konig, P. H., Frauenheim, T., Schulten, K., ... Elstner, M. (2006). Color tuning in rhodopsins: The mechanism for the spectral shift between bacteriorhodopsin and sensory rhodopsin II. *Journal of the American Chemical Society*, *128*, 10808–10818.

Hollfelder, F., & Herschlag, D. (1995). The nature of the transition-state for enzyme-catalyzed phosphoryl transfer—Hydrolysis of O-aryl phosphorothioates by alkaline-phosphatase. *Biochemistry*, *38*, 12255–12264.

Holtz, K. M., Stec, B., & Kantrowitz, E. R. (1999). A model of the transition state in the alkaline phosphatase reaction. *Journal of Biological Chemistry*, *274*, 8351–8354.

Hou, G. H., & Cui, Q. (2012). QM/MM analysis suggests that alkaline phosphatase (AP) and Nucleotide pyrophosphatase/phosphodiesterase (NPP) slightly tighten the transition state for phosphate diester hydrolysis relative to solution: Implication for catalytic promiscuity in the AP superfamily. *Journal of the American Chemical Society*, *134*, 229–246.

Hou, G. H., & Cui, Q. (2013). Stabilization of different types of transition states in a single enzyme active site: QM/MM analysis of enzymes in the alkaline phosphatase superfamily. *Journal of the American Chemical Society*, *135*, 10457–10469.

Hou, G., Zhu, X., Elstner, M., & Cui, Q. (2012). A modified QM/MM Hamiltonian with the self-consistent-charge density-functional-tight-binding theory for highly charged QM regions. *Journal of Chemical Theory and Computation*, *8*, 4293–4304.

Hu, H., Boone, A., & Yang, W. T. (2008). Mechanism of OMP decarboxylation in orotidine 5′-monophosphate decarboxylase. *Journal of the American Chemical Society*, *130*, 14493–14503.

Hu, H., Lu, Z. Y., Parks, J. M., Burger, S. K., & Yang, W. T. (2008). Quantum mechanics/molecular mechanics minimum free-energy path for accurate reaction energetics in solution and enzymes: Sequential sampling and optimization on the potential of mean force surface. *Journal of Chemical Physics*, *128*, 034105.

Hu, L. H., & Ryde, U. (2011). Comparison of methods to obtain force-field parameters for metal sites. *Journal of Chemical Theory and Computation*, *7*, 2425–2463.

Hu, H., & Yang, W. T. (2008). Free energies of chemical reactions in solution and in enzymes with ab initio quantum mechanics/molecular mechanics methods. *Annual Review of Physical Chemistry*, *59*, 573–601.

Huang, L., & Roux, B. (2013). Automated force field parameterization for nonpolarizable and polarizable atomic models based on ab initio target data. *Journal of Chemical Theory and Computation*, *9*, 3543–3556.

Im, W., Berneche, S., & Roux, B. (2001). Generalized solvent boundary potential for computer simulations. *Journal of Chemical Physics*, *114*, 2924–2937.

Jiao, D., Golubkov, P. A., Darden, T. A., & Ren, P. (2009). Calculation of protein-ligand binding free energy by using a polarizable potential. *Proceedings of the National Academy of Sciences of the United States of America*, *105*, 6290–6295.

Kamerlin, S. C. L., Haranczyk, M., & Warshel, A. (2009). Progress in ab initio QM/MM free-energy simulations of electrostatic energies in proteins: Accelerated QM/MM studies of pK(a), redox reactions and solvation free energies. *Journal of Physical Chemistry B*, *113*, 1253–1272.

Kamerlin, S. C. L., & Warshel, A. (2010). At the dawn of the 21st century: Is dynamics the missing link for understanding enzyme catalysis? *Proteins: Structure, Function, and Bioinformatics, 78*, 1339–1375.

Kazemi, M., Himo, F., & Aqvist, J. (2016). Enzyme catalysis by entropy without Circe effect. *Proceedings of the National Academy of Sciences of the United States of America, 113*, 2406–2411.

Kerns, S. J., Agafonov, R. V., Cho, Y. J., Pontiggia, F., Otten, R., Pachov, D. V., ... Kern, D. (2015). The energy landscape of adenylate kinase during catalysis. *Nature Structural & Molecular Biology, 22*, 124–131.

Kipp, D. R., Hirschi, J. S., Wakata, A., Goldstein, H., & Schramm, V. L. (2012). Transition states of native and drug-resistant HIV-1 protease are the same. *Proceedings of the National Academy of Sciences of the United States of America, 109*, 6543–6548.

Kraut, D. A., Carroll, K. S., & Herschlag, D. (2003). Challenges in enzyme mechanism and energetics. *Annual Review of Biochemistry, 72*, 517–571.

Kries, H., Blomberg, R., & Hilvert, D. (2013). De novo enzymes by computational design. *Current Opinion in Chemical Biology, 17*, 221–228.

Kulik, H. J., Luehr, N., Ufimtsev, I. S., & Martinez, T. J. (2012). Ab initio quantum chemistry for protein structures. *Journal of Physical Chemistry B, 116*, 12501–12509.

Kurashige, Y., Chan, G. K. L., & Yanai, T. (2013). Entangled quantum electronic wavefunctions of the Mn4CaO5 cluster in photosystem II. *Nature Chemistry, 5*, 660–666.

Labow, B. I., Herschlag, D., & Jencks, W. P. (1993). Catalysis of the hydrolysis of phosphorylated pyridines by alkaline-phosphatase has little or no dependence on the pK(a) of the leaving group. *Biochemistry, 32*, 8737–8741.

Lassila, J. K., Baker, D., & Herschlag, D. (2010). Origins of catalysis by computationally designed retroaldolase enzymes. *Proceedings of the National Academy of Sciences of the United States of America, 107*, 4937–4942.

Lassila, J. K., Zalatan, J. G., & Herschlag, D. (2011). Biological phosphoryl transfer reactions: Understanding mechanism and catalysis. *Annual Review of Biochemistry, 80*, 669–702.

Lee, C., Yang, W., & Parr, R. G. (1988). Development of the Colle-Salvetti correlation-energy formula into a functional of the electron density. *Physical Review B, 37*(2), 785–789. http://dx.doi.org/10.1103/PhysRevB.37.785.

Li, G., & Cui, Q. (2003). pKa calculations with QM/MM free energy perturbations. *Journal of Physical Chemistry B, 107*, 14521–14528.

Liao, R. Z., & Thiel, W. (2013). Convergence in the QM-only and QM/MM modeling of enzymatic reactions: A case study for acetylene hydratase. *Journal of Computational Chemistry, 34*, 2389–2397.

Lindqvist, Y., Schneider, G., & Vihko, P. (1994). Crystal-structures of rat acid-phosphatase complexed with the transition-state analogs vanadate and molybdate—Implications for the reaction-mechanism. *European Journal of Biochemistry, 221*, 139–142.

Lu, X., & Cui, Q. (2013). Charging free energy calculations using the generalized solvent boundary potential (gsbp) and periodic boundary condition: A comparative analysis using ion solvation and reduction potential in proteins. *Journal of Physical Chemistry B, 117*, 2005–2018.

Lu, X., Fang, D., Ito, S., Okamoto, Y., Ovchinnikov, V., & Cui, Q. (2016). QM/MM free energy simulations: Recent progress and challenges. *Molecular Simulation*, (in press).

Lu, X., Gaus, M., Elstner, M., & Cui, Q. (2015). Parameterization of DFTB3/3OB for magnesium and zinc for chemical and biological applications. *Journal of Physical Chemistry B, 119*, 1062–1082.

MacKerell, A. D., Jr., Bashford, D., Bellott, M., Dunbrack, R. L., Jr., Evenseck, J. D., Field, M. J., ... Karplus, M. (1998). All-atom empirical potential for molecular modeling and dynamics studies of proteins. *Journal of Physical Chemistry B, 102*, 3586–3616.

MacKerell, A. D., Jr., Feig, M., & Brooks, C. L., III. (2004). Extending the treatment of backbone energetics in protein force fields: Limitations of gas-phase quantum mechanics in reproducing protein conformational distributions in molecular dynamics simulations. *Journal of Computational Chemistry, 25*, 1400–1415.

Mader, M. M., & Bartlett, P. A. (1997). Binding energy and catalysis: The implications for transition-state analogs and catalytic antibodies. *Chemical Reviews, 97*, 1281–1301.

Marti, S., Moliner, V., & Tuñón, I. (2005). Improving the QM/MM description of chemical processes: A dual level strategy to explore the potential energy surface in very large systems. *Journal of Chemical Theory and Computation, 1*, 1008–1016.

Molina, P. A., & Jensen, J. H. (2003). A predictive model of strong hydrogen bonding in proteins: The Nδ-H-Oδ hydrogen bond in low-pH alpha-chymotrypsin and alpha-lytic protease. *Journal of Physical Chemistry B, 107*, 6226–6233.

Monard, G., & Merz, K. M., JR. (1999). Combined quantum mechanical/molecular mechanical methodologies applied to biomolecular systems. *Accounts of Chemical Research, 32*, 904–911.

Nagel, Z. D., & Klinman, J. P. (2009). A 21(st) century revisionist's view at a turning point in enzymology. *Nature Chemical Biology, 5*, 543–550.

O'Brien, P. J., & Herschlag, D. (1998). Sulfatase activity of E. coli alkaline phosphatase demonstrates a functional link to arylsulfatases, an evolutionarily related enzyme family. *Journal of the American Chemical Society, 120*, 12369–12370.

O'Brien, P. J., & Herschlag, D. (2001). Functional interrelationships in the alkaline phosphatase superfamily: Phosphodiesterase activity of Escherichia coli alkaline phosphatase. *Biochemistry, 40*, 5691–5699.

Parr, R. G., & Yang, W. T. (1989). *Density-functional theory of atoms and molecules*. New York: Oxford University Press.

Pearson, A. D., Mills, J. H., Song, Y., Nasertorabi, F., Han, G. W., Baker, D., ... Schultz, P. G. (2015). Trapping a transition state in a computationally designed protein bottle. *Science, 347*, 863–867.

Peck, A., Sunden, F., Andrews, L. D., Pande, V. S., & Herschlag, D. (2016). Tungstate as a transition state analog for catalysis by alkaline phosphatase. *Journal of Molecular Biology, 428*(13), 2758–2768. http://dx.doi.org/10.1016/j.jmb.2016.05.007.

Phatak, P., Ghosh, N., Yu, H., Cui, Q., & Elstner, M. (2008). Amino acids with an intermolecular proton bond as the proton storage site in bacteriorhodopsin. *Proceedings of the National Academy of Sciences of the United States of America, 105*, 19672–19677.

Plotnikov, N. V., Kamerlin, S. C. L., & Warshel, A. (2011). Paradynamics: An effective and reliable model for ab Initio QM/MM free-energy calculations and related tasks. *Journal of Physical Chemistry B, 115*, 7950–7962.

Polyak, I., Benighaus, T., Boulanger, E., & Thiel, W. (2013). Quantum mechanics/molecular mechanics dual Hamiltonian free energy perturbation. *Journal of Chemical Physics, 139*, 064105.

Ponder, J. W., Wu, C. J., Ren, P. Y., Pande, V. S., Chodera, J. D., Schnieders, M. J., ... Head-Gordon, T. (2010). Current status of the AMOEBA polarizable force field. *Journal of Physical Chemistry B, 114*, 2549–2564.

Pu, J. Z., Gao, J. L., & Truhlar, D. G. (2006). Multidimensional tunneling, recrossing, and the transmission coefficient for enzymatic reactions. *Chemical Reviews, 106*, 3140–3169.

Riccardi, D., Schaefer, P., Yang, Y., Yu, H., Ghosh, N., Prat-Resina, X., ... Cui, Q. (2006). Feature article: Development of effective quantum mechanical/molecular mechanical (QM/MM) methods for complex biological processes. *Journal of Physical Chemistry B, 110*, 6458–6469.

Riccardi, D., Yang, S., & Cui, Q. (2010). Proton transfer function of carbonic anhydrase: Insights from QM/MM simulations. *Biochimica et Biophysica Acta, 1804*, 342–351.

Rod, T. H., & Ryde, U. (2005). Accurate QM/MM free energy calculations of enzyme reactions: Methylation by catechol O-methyltransferase. *Journal of Chemical Theory and Computation, 1*, 1240–1251.

Rossi, I., & Truhlar, D. G. (1995). Parameterization of NDDO wave-functions using genetic algorithms—An evolutionary approach to parameterizing potential-energy surfaces and direct dynamics calculations for organic-reactions. *Chemical Physics Letters, 233*, 231–236.

Roston, D., Demapan, D., & Cui, Q. (in press). Leaving group ability affects transition state structure for phosphoryl transfer in a single enzyme active site, *Journal of the American Chemical Society*. http://dx.doi.org/10.1021/jacs.6b03156.

Roston, D., & Kohen, A. (2010). Elusive transition state of alcohol dehydrogenase unveiled. *Proceedings of the National Academy of Sciences of the United States of America, 107*, 9572–9577.

Rothlisberger, D., Khersonsky, O., Wollacott, A. M., Jiang, L., DeChancie, J., Betker, J., ... Baker, D. (2008). Kemp elimination catalysts by computational enzyme design. *Nature, 453*, 190–194.

Salomon-Ferrer, R., Case, D., & Walker, R. (2013). An overview of the Amber biomolecular simulation package. *WIREs Computational Molecular Science, 3*, 198–210. http://dx.doi.org/10.1002/wcms.1121.

Schaefer, P., Riccardi, D., & Cui, Q. (2005). Reliable treatment of electrostatics in combined QM/MM simulation of macromolecules. *Journal of Chemical Physics, 123*. Article No. 014905.

Schramm, V. L. (2011). Enzymatic transition states, transition-state analogs, dynamics, thermodynamics, and lifetimes. *Annual Review of Biochemistry, 80*, 703–732.

Schramm, V. L. (2015). Transition states and transition state analogue interactions with enzymes. *Accounts of Chemical Research, 48*, 1032–1039.

Schultz, P. G. (1989). Catalytic antibodies. *Angewandte Chemie International Edition in English, 28*, 1283–1295.

Schwartz, S. D., & Schramm, V. L. (2009). Enzymatic transition states and dynamic motion in barrier crossing. *Nature Chemical Biology, 5*, 552–559.

Senn, H. M., & Thiel, W. (2009). QM/MM methods for biomolecular systems. *Angewandte Chemie International Edition in English, 48*, 1198–1229.

Shaw, D. E., Maragakis, P., Lindorff-Larsen, K., Piana, S., Dror, R. O., Eastwood, M. P., ... Wriggers, W. (2010). Atomic-level characterization of the structural dynamics of proteins. *Science, 330*, 341–346.

Shukla, D., Meng, Y. L., Roux, B., & Pande, V. S. (2014). Activation pathway of Src kinase reveals intermediate states as targets for drug design. *Nature Communications, 5*, 3397.

Siegel, J. B., Zanghellini, A., Lovick, H. M., Kiss, G., Lambert, A. R., Clair, J. L. S., ... Baker, D. (2010). Computational design of an enzyme catalyst for a stereoselective bimolecular Diels-Alder reaction. *Science, 329*, 309–313.

Simonson, T., Archontis, G., & Karplus, M. (2002). Free energy simulations come of age: Protein-ligand recognition. *Accounts of Chemical Research, 35*, 430–437.

Stec, B., Holtz, K. M., & Kantrowitz, E. R. (2000). A revised mechanism for the alkaline phosphatase reaction involving three metal ions. *Journal of Molecular Biology, 299*, 1303–1311.

Stewart, J. J. P. (1989). Optimization of parameters for semiempirical methods. ii. applications. *Journal of Computational Chemistry, 10*(2), 221–264. Retrieved from, http://dx.doi.org/10.1002/jcc.540100209.

Stote, R. H., States, D. J., & Karplus, M. (1991). On the treatment of electrostatic interactions in biomolecular simulation. *Journal of Chimie Physique, 88*, 2419–2433.

Straatsma, T. P., & McCammon, J. A. (1992). Computational alchemy. *Annual Review of Physical Chemistry, 43*, 407–435.

Sunden, F., Peck, A., Salzman, J., Ressl, S., & Herschlag, D. (2015). Extensive site-directed mutagenesis reveals interconnected functional units in the alkaline phosphatase active site. *eLife, 4*, e06181.
Tao, P., Hodoscek, M., Larkin, J. D., Shao, Y. H., & Brooks, B. R. (2012). Comparison of three chain-of-states methods: Nudged elastic band and replica path with restraints or constraints. *Journal of Chemical Theory and Computation, 8*, 5035–5051.
Tomasi, J., Mennucci, B., & Cammi, R. (2005). Quantum mechanical continuum solvation models. *Chemical Reviews, 105*, 2999–3093.
Torrie, G. M., & Valleau, J. P. (1977). Non-physical sampling distributions in Monte-Carlo free-energy estimation: Umbrella sampling. *Journal of Computational Physics, 23*, 187–199.
van der Kamp, M. W., & Mulholland, A. J. (2013). Combined quantum mechanics/molecular mechanics (QM/MM) methods in computational enzymology. *Biochemistry, 52*, 2708–2728.
Vanommeslaeghe, K., Raman, E. P., & MacKerell, A. D., Jr. (2012). Automation of the CHARMM general force field (CGenFF). II: Assignment of bonded parameters and partial atomic charges. *Journal of Chemical Information and Modeling, 52*, 3155–3168.
Wahiduzzaman, M., Oliveira, A. F., Philipsen, P., Zhechkov, L., van Lenthe, E., Witek, H. A., & Heine, T. (2013). DFTB parameters for the periodic table: Part 1. Electronic structure. *Journal of Chemical Theory and Computation, 9*, 4006–4017.
Wales, D. J. (2003). *Energy landscapes, with applications to clusters, biomolecules and glasses*. Cambridge, UK: Cambridge University Press.
Wanko, M., Hoffman, M., Strodel, P., Koslowski, A., Thiel, W., Neese, F., ... Elstner, M. (2005). Calculating absorption shifts for retinal proteins: Computational challenges. *Journal of Physical Chemistry B, 109*, 3606–3615.
Warshel, A. (1991). *Computer modeling of chemical reactions in enzymes and solution*. New York: Wiley.
Warshel, A., & Levitt, M. (1976). Theoretical studies of enzymic reactions—Dielectric, electrostatic and steric stabilization of carbonium-ion in reaction of lysozyme. *Journal of Molecular Biology, 103*, 227–249.
Warshel, A., Sharma, P. K., Kato, M., Xiang, Y., Liu, H. B., & Olsson, M. H. M. (2006). Electrostatic basis for enzyme catalysis. *Chemical Reviews, 106*, 3210–3235.
Weinan, E., & Vanden-Eijnden, E. (2010). Transition-path theory and path-finding algorithms for the study of rare events. *Annual Review of Physical Chemistry, 61*, 391–420.
Wolf, S., Freier, E., Cui, Q., & Gerwert, K. (2014). Infrared spectral marker bands characterizing a transient water wire inside a hydrophobic membrane protein. *Journal of Chemical Physics, 141*, 22D524.
Wolf, S., Freier, E., & Gerwert, K. (2014). A delocalized proton-binding site within a membrane protein. *Biophysical Journal, 107*, 174–184.
Zalatan, J. G., Fenn, T. D., Brunger, A. T., & Herschlag, D. (2006). Structural and functional comparisons of nucleotide pyrophosphatase/phosphodiesterase and alkaline phosphatase: Implications for mechanism and evolution. *Biochemistry, 45*, 9788–9803.
Zalatan, J. G., Fenn, T. D., & Herschlag, D. (2008). Comparative enzymology in the alkaline phosphatase superfamily to determine the catalytic role of an active-site metal ion. *Journal of Molecular Biology, 384*, 1174–1189.
Zalatan, J. G., & Herschlag, D. (2006). Alkaline phosphatase mono- and diesterase reactions: Comparative transition state analysis. *Journal of the American Chemical Society, 128*, 1293–1303.
Zhang, M., Zhou, M., VanEtten, R. L., & Stauffacher, C. V. (1997). Crystal structure of bovine low molecular weight phosphotyrosyl phosphatase complexed with the transition state analog vanadate. *Biochemistry, 36*, 15–23.
Zienau, J., & Cui, Q. (2012). Implementation of the solvent macromolecule boundary potential and application to model and realistic enzyme systems. *Journal of Physical Chemistry B, 116*, 12522–12534.

CHAPTER TEN

Practical Aspects of Multiscale Classical and Quantum Simulations of Enzyme Reactions

M. Dixit, S. Das, A.R. Mhashal, R. Eitan, D.T. Major[1]

Lise Meitner-Minerva Center of Computational Quantum Chemistry, Bar-Ilan University, Ramat Gan, Israel
[1]Corresponding author: e-mail address: majort@biu.ac.il

Contents

1. Introduction — 252
2. Enzyme System Modeling — 253
    2.1 Embarking an In Silico Enzyme-Modeling Project — 253
    2.2 Choice of Protonation State and Tautomers — 256
    2.3 Correction of Initial Protein Structure and Generation of Mutant Forms — 256
    2.4 Choice of System Boundary Conditions — 257
3. The Potential Energy Surface — 260
    3.1 Hybrid Quantum Mechanical–Molecular Mechanical Approaches — 260
    3.2 Choosing a Method for the QM Region — 262
    3.3 Choosing a Method for the MM Region — 264
    3.4 Nonbonded Interactions Between the QM and MM Regions — 265
    3.5 Covalent Connections Between QM and MM Regions — 266
    3.6 Size of QM Region — 266
4. Reaction Coordinates and Classical Free Energy Simulations — 267
    4.1 Reaction Rates and Chemical Progress Coordinates — 267
    4.2 Classical Free Energy Simulations — 269
    4.3 Quantum Free Energy Simulations — 271
5. Concluding Words — 274
Acknowledgments — 277
References — 277

Abstract

This chapter aims to present some basic multiscale approaches available for enzyme simulations, and to point out practical details and pitfalls that are not often discussed in the literature, but can greatly influence the outcome of any in silico enzyme study. We cover principle methodological steps of multiscale studies of general enzyme reactions. This includes choice of starting structures, boundary conditions, potential energy surfaces, reaction coordinates, simulation methods, as well as the choice of method for the treatment of nuclear quantum effects. Together, these and additional steps are crucial for the success of enzyme-modeling projects and should be considered prior to embarking on multiscale modeling.

1. INTRODUCTION

Enzymes are the finest chemical catalysts known to mankind and have captured the imagination of scientists for centuries. These molecular machines constitute the molecular factories in Nature and are present in all organisms. Enzymes perform an astonishing range of different chemistries, ranging from transfer of single atoms to multistep synthesis of highly complex molecules with numerous stereochemical centers (Frey & Hegeman, 2007). These biocatalysts accelerate reaction rates by tens of orders of magnitude, when compared to analogous reactions in aqueous solution, with rates approaching the bimolecular diffusion limit (Hammes, 2002; Wolfenden & Snider, 2001). Enzymes have in many cases reached chemical perfection and have enthused generations of scientists to explore the underpinnings of enzyme catalysis.

One of the most promising avenues in the study of enzyme catalysis is computational enzymology. This field was conceived once innovative minds realized that it is possible to tackle very complex systems, such as enzymes, in a divide-and-conquer manner. Specifically, in order to treat the chemistry of enzymes, one must use quantum mechanics (QM), whereas the general enzyme environment can be represented by classical molecular mechanics (MM). The beginning of this field dates back to the pioneering work of the 2013 Nobel laureates Karplus, Levitt, and Warshel in the early 1970s. In 1970, Warshel and Bromberg published a QM+MM Note the difference between QM+MM and QM/MM: The latter includes coupling between the QM and MM regions (ie, polarization), whereas the former does not. The study of the oxidation of 4a,4b-dihydrophenanthrene (Warshel & Bromberg, 1970), is the first published QM+MM simulation of a chemical process. In 1971, Karplus and Honig studied the retinal chromophore and employed a hybrid Hamiltonian relying on a Hückel one-electron term for the π-electrons, and a pairwise nonbonded energy function for the sigma bond framework (Honig & Karplus, 1971). Subsequently, Warshel and Karplus continued work on retinal, and in 1972, published a hybrid method for planar systems that combined more expensive QM methods with the cheaper MM methods (Warshel & Karplus, 1972). In this work the π-electrons were described by QM, whereas the remaining system was treated classically, and the σ-electrons were incorporated into the nuclei. Significant improvements to the multiscale approach were presented in the classic Warshel and Levitt (1976) paper, where these authors described what may be considered to be the first realization of a QM/MM approach.

In spite of the many successes of multiscale modeling, there are many practical details that are not often discussed in the literature, but can greatly influence the outcome of an enzyme study. In the present review, we cover principle methodological steps of multiscale studies of general enzyme reactions and point out potential pitfalls that we believe require special attention. This includes choice of starting structures, boundary conditions, potential energy surfaces, reaction coordinates, simulation methods, as well as the choice of method for the treatment of nuclear quantum effects (NQE).

2. ENZYME SYSTEM MODELING
2.1 Embarking an In Silico Enzyme-Modeling Project

At the outset of an enzyme-modeling project, one needs to decide on the structural starting point. In most cases, a natural initial protein model is comprised of the Cartesian coordinates of an experimentally determined structure, such as from X-ray crystallography (Sherwood & Cooper, 2015) or nuclear magnetic resonance (NMR) (Cavanagh, Fairbrother, Palmer, Rance, & Skelton, 2006). However, these methods have inherent strengths and limitations that can potentially affect the modeling outcome (van den Bedem & Fraser, 2015).

Crystal structures provide precise positions of the atoms, which are essential for modeling, but these are average structures typically determined in the solid state. In reality proteins are dynamic entities that mostly thrive in aqueous or lipid environments, continuously shifting between conformational substates (Boehr, McElheny, & Dyson, 2006; Henzler-Wildman et al., 2007). For instance, the crystallized protein, although still flexible (Wall et al., 2014) might have crystal contacts (ie, packing) that are not significantly populated in solution, and hence might not be representative of the active state of the enzyme (van den Bedem & Fraser, 2015). Moreover, crystal structures are often determined at low, cryogenic temperatures, which might freeze out certain conformations. For instance, it has been observed that side chain positions in crystal structures (Keedy et al., 2014), as well as ligand binding (Fischer, Shoichet, & Fraser, 2015), are temperature dependent (Keedy et al., 2014).

NMR experiments, on the other hand, do not produce a single, unique structure, but rather an ensemble of models that agree with the measured restraints (eg, nuclear Overhauser effect, NOE, restraints) (Cavanagh et al., 2006). These ensembles are underdetermined, as there is not sufficient information from the experimental restraints and the energy function employed (Nilsson, Clore, Gronenborn, Brunger, & Karplus, 1986), to determine the relative populations of the ensembles. Hence, it remains challenging to extract detailed phase-space information from the structural NMR ensemble (van den Bedem & Fraser, 2015).

Recent integrative experimental techniques that combine information from multiple techniques, such as NMR, X-ray, and theory (ie, integrative structural biology), show great promise (Ward, Sali, & Wilson, 2013) and are likely to be of use also in the field of enzyme modeling. This suggests a tight cooperation between experimentalists and computational enzymologists from the outset of the modeling endeavor.

If no experimental structural data exist, one could in principle also start from a theoretical protein model, such as de novo models that have been used in computational enzyme design (Ashworth et al., 2006; Jiang et al., 2008). Another possibility is generating homology models (Bjelic & Aqvist, 2004), although it is questionably whether such structures would be accurate enough in general for enzyme modeling, unless the homology between the target and template is very high.

If the initial model chosen is from X-ray, hydrogen atoms must be added to this crystal structure unless the resolution is very high. This can typically be done using energy-based search algorithms to identify the most probable rotamer (Brunger & Karplus, 1988), and such routines are standard in modern modeling packages. Often parts of residues or complete residues are missing in the crystal structure, due to their flexibility or because they were not expressed in the enzyme production system, and these need to be added to complete the enzyme model. In NMR structures, it is necessary to add water molecules, including tightly bound waters. Specialized methods exist that can locate tightly bound water molecules, as well as networks of bound waters (Ross, Bodnarchuk, & Essex, 2015).

Ideally, the initial structure is similar to the state one wants to model. This suggests that the enzyme active site is occupied by a ligand similar to the reactant state (RS), product state (PS), or ideally a transition state

(TS) analogue (Schramm, 2011). The latter case increases the likelihood that the enzyme was caught in a catalytically competent state. However, in many cases the experimentally determined enzyme is not in its fully closed, preorganized catalytic form. Additionally, the active site might be empty, or only include some of the ligands required for catalysis. Indeed, catalytic function often depends on the enzyme reaching a transient, veiled excited state, which might not be accessible experimentally (van den Bedem & Fraser, 2015). In such cases, great care must be taken in generating a complete and an active form of the enzyme. This may entail homology modeling or some form of de novo design, including loop modeling, and might have to be done in an iterative fashion with enzyme mechanistic simulations, until a properly closed and active form of the enzyme is obtained (Schrepfer et al., 2016). One, or more, missing catalytic components might have to be docked into the active site, and such docking might benefit from including restraints based on the chemical requirements of the reaction at hand (Major et al., 2009). For instance, one might want to assure that the docked substrate pose constitutes a reactive conformation and orientation, ready for reaction. In order to dock substrate(s) and/or cofactor(s) into the active site one may use any of the existing docking approaches (Morris & Lim-Wilby, 2008). However, chemical intuition is essential at this stage, and one needs to verify that the predicted complex makes chemical sense. In Fig. 1, we present an example where the substrate NO_2-ethane was docked into the active site of nitroalkane oxidase, using a combination of chemical intuition in the form of computational NOE restraints, an accurate QM/MM potential, and an efficient simulation protocol (Major et al., 2009). In particular, the NOE restraints limited the docking search space to configurations that placed the donor and acceptor atoms in close proximity. We stress that the docking

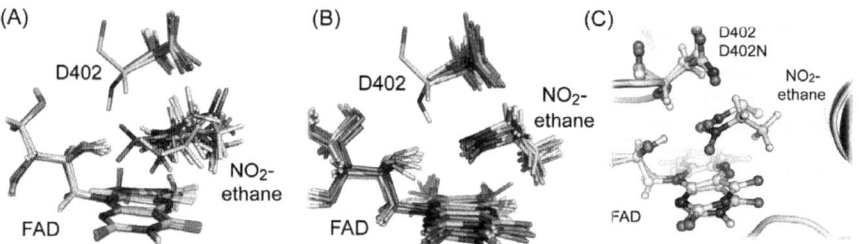

Fig. 1 QM/MM docking of NO_2-ethane in the active site of nitroalkane oxidase. (A) Randomly generated configurations of NO_2-ethane. (B) Final docked poses of NO_2-ethane. (C) Comparison of a docked pose (*green* carbons) and the crystal structure of the D402N mutant enzyme (*yellow* carbons). (See the color plate.)

studies were performed two years prior to publication of the relevant crystal structure (Fig. 1C). We conclude this section by stating the obvious, yet painful reality: It is difficult to overemphasize the importance of the initial docking step; an incorrect initial Michaelis complex will necessarily lead to inaccuracies in the modeling of the enzymatic process.

2.2 Choice of Protonation State and Tautomers

An additional crucial point related to the previous section is the choice of protonation states for protein amino acid residues, potential cofactors, and substrates. The titratable residues that require routine attention are Asp, Glu, His, Lys, and Tyr. Additionally, Cys, Ser, and Arg residues involved in catalysis might also adopt various protonation states. Numerous methods exist for the rapid prediction of pK_a values of amino acids in proteins, and these usually yield reasonable protonation states (Alexov et al., 2011). However, these methods are not ordinarily parameterized for actives sites that might include a delicate equilibrium between different protonation states, and extra care must be taken for residues in the vicinity of the active site. Indeed, internal amino acids might experience significant shifts in pK_a values (Isom, Castaneda, Cannon, & Garcia-Moreno, 2011). Additionally, the protonation state of cofactors and substrates in solution might not correspond to the bioactive form (Isom et al., 2011; Liu et al., 2014), and the enzyme-bound species might very well not be the most abundant tautomer in solution (ten Brink & Exner, 2009). Therefore, the pK_a and possible tautomers of any substrate and cofactor should be determined for the bound state and not in solution. Adding to the complexity, standard pK_a programs often do not include force field parameters for novel substrates and cofactors. In conclusion, due to the many possible pitfalls surrounding the titratable residues, it is usually sensible to manually inspect the hydrogen-bond pattern of all acid/base residues to verify that the chosen protonation states are reasonable. Nonetheless, it may often be necessary to carry out the same enzyme mechanistic study using different protonation states of the enzyme (Major & Gao, 2006).

2.3 Correction of Initial Protein Structure and Generation of Mutant Forms

It is important to realize that experimentally determined structures are only models, and may contain ambiguities, and possibly errors (Davis, Teague, & Kleywegt, 2003). This is due to uncertainties introduced in the process of deriving an atomic model from experimental electron densities or restraints.

Both crystal and NMR structure determination software rely on some form of force field-based energy minimization or molecular dynamics (MD) simulations in conjunction with the experimental data as restraints (Cavanagh et al., 2006; Sherwood & Cooper, 2015). Moreover, in crystal structures, hydrogen atoms may be added employing riding models; unless the structure was resolved at a very high resolution, the titratable hydrogens should be carefully inspected. Hence, in addition to determining the most plausible protonation state and tautomers for amino acid side chains, substrates, and cofactors, one might also need to correct for occasional errors in the crystal structures. Examples of common problems is the orientations of side chains in His, and the identity of the heteroatoms in the side chain of Asn and Gln residues (Word, Lovell, Richardson, & Richardson, 1999). Additionally, one should be aware of subtle differences in seemingly identical structures. To underscore this latter point of sensitivity of enzyme simulations on the initial structure, we note that a reported study showed how identical setup protocols, but different crystal structures for the same enzyme, can result in rather different results (Garcia-Viloca, Poulsen, Truhlar, & Gao, 2004).

If one wants to study a mutant form of an enzyme, and assuming an experimental model for the mutant structure does not exist, one must generate the point mutation based on the wild-type (wt) enzyme. In the case of a mutation that does not change the overall protein structure significantly, this is a practical strategy (Ishida, 2010), although it might be necessary to allow the mutant enzyme to relax using relatively long MD simulations. However, if the enzyme undergoes large-scale structural changes, such as movement of secondary motifs, it is questionable whether in silico mutations of the wt enzyme can generate the correct mutant enzyme (Doron, Stojkovic, Gakhar, Kohen, & Major, 2015).

As a final note in this section, it is helpful for the modelist to perform thorough visual inspection of the enzyme to get familiar with the details of the structure he or she is about to model (eg, active site interactions and hydrogen bonding networks through the enzyme). This is likely to be very helpful in later stages of the work when one attempts to draw conclusions of biological significance based on the numerical data obtained.

2.4 Choice of System Boundary Conditions

In order to complete the enzyme model, one needs to immerse the protein in aqueous solution. This is often taken to be an explicit solvent–ion mix, although one can also employ an implicit solvent model for the QM/MM simulations.

An important question that arises is what kind of boundary to choose for the system. The most common and reliable options are periodic boundary conditions (PBC) (Allen & Tildesley, 1989) or stochastic boundary conditions (SBC) (Fig. 2) (Brooks, Brünger, & Karplus, 1985; Brünger, Brooks, & Karplus, 1984). In a PBC setup, the enzyme–ligand complex is embedded in a preequilibrated periodic water system, while within SBC it is embedded in a sphere of preequilibrated water molecules. The arguably preferred option is PBC, which allows full flexibility of the enzyme and in conjunction with the Ewald summation method (Darden, York, & Pedersen, 1993; Nam, Gao, & York, 2004) accounts for long-range electrostatics. SBC can in principle also include full enzyme flexibility, if a sufficiently large water sphere is used. However, there will still be some boundary effects (Brooks et al., 1985; Brünger et al., 1984). Other methods include the spherical solvent boundary potential model (Beglov & Roux, 1994), the generalized solvent boundary potential (GSBP), where not only solvent molecules but also part of the macromolecule atoms can be treated implicitly (Halgren & Damm, 2001), and the QM/MM general boundary potential of Benighaus and Thiel (Benighaus & Thiel, 2009, 2010; Boulanger & Thiel, 2012; Zienau & Cui, 2012). The effect

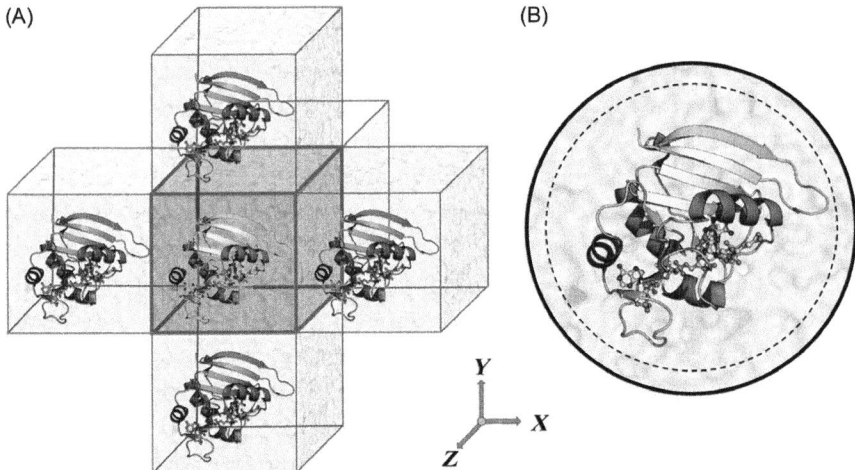

Fig. 2 Possible setups for enzyme systems (ie, dihydrofolate reductase): (A) Periodic boundary conditions where the enzyme is immersed in an orthorhombic cell, which is surrounded by an infinite number of replicas of the unit cell. (B) Stochastic boundary conditions where the enzyme is immersed in a spherical water droplet. Within the *dotted circle*, the atoms are propagated by unrestrained MD, whereas beyond it the system is restrained and propagated with Langevin dynamics. (See the color plate.)

of long-range electrostatics for SBC may be mimicked via the extended Ewald method of Kuwajima and Warshel (Kuwajima & Warshel, 1988) or the analytic continuum electrostatic method of Schaefer and Karplus (1996). The most appropriate choice of method depends on a number of factors: the size of the system, the type of potential energy surface employed, the software used for the study, and more.

A subtle concern that arises in constructing the complete enzyme solvent system is the number of water molecules to include in the system, and hence the density. Presumably, the density of the initial pure water sphere or water box is correct within the restrictions of the water model employed. However, the ambiguity arises because of the need to delete water molecules overlapping with the enzyme–substrate complex, and it is not always clear which waters one should retain. If one includes too many waters, the system becomes too dense, which in turn will inhibit protein motion. If one includes too few water molecules, the protein might be too flexible and charges may not be sufficiently well screened. In particular, the precise position of water molecules in the active site can be crucial, as was shown by Yang and coworkers (Hu, Boone, & Yang, 2008). The question of water density is particularly acute when using SBC, as the system volume is fixed from the outset, and hence the water density cannot equilibrate. An additional practical point is that too small a sphere within the SBC approach might not be appropriate for enzymes that are very flexible, as the boundary restraints in this approach inhibits motion. For instance, in our study of the enzyme alanine racemase, numerous different SBC systems were constructed and modeled to address these concerns (Major & Gao, 2006).

An additional point that requires some attention is the addition of ions to the simulation sphere or box. In general, when using the Ewald summation method, counter ions should be added to neutralize the system so that the summation over electrostatic terms converges (Hub, de Groot, Grubmüller, & Groenhof, 2013). If possible, the number of ions added to the system should match the experimental concentration (Kastenholz & Hünenberger, 2003). This is however not always trivial, as lab experiments usually contain some buffer solution, which often includes ionized species, hence contributing to the total ionic strength. It is currently not common practice to include buffer ingredients other than simple alkali and halide ions. It is our experience that increasing the ionic strength toward the experimental conditions leads to more stable simulations, with less fluctuations in the computed free energy profiles (unpublished results).

3. THE POTENTIAL ENERGY SURFACE
3.1 Hybrid Quantum Mechanical–Molecular Mechanical Approaches

Once the enzyme system has been constructed, one needs to choose a potential that models the interactions between the particles. Since one is interested in chemical reactions occurring within a complex heterogeneous environment, one typically needs to resort to multiscale approaches. According to this strategy, one divides the system into regions that will be treated at different levels of theory. Clearly, to treat the chemical event one must use quantum chemistry; however, the regions surrounding the reacting fragments may be treated by classical methods such as MM. The QM/MM approach was originally pioneered by Karplus, Levitt, and Warshel in the 1970s (Warshel & Bromberg, 1970; Warshel & Karplus, 1972; Warshel & Levitt, 1976), and has seen significant advances over the past four decades (Friesner & Guallar, 2005; Gao, 1995; Hu & Yang, 2008; Senn & Thiel, 2009; van der Kamp & Mulholland, 2013; Warshel, 1991).

Within the QM/MM formalism, an effective Hamiltonian of the complete system may be defined as (Fig. 3):

$$\hat{H} = \hat{H}_{QM} + \hat{H}_{MM} + \hat{H}_{QM/MM} \tag{1}$$

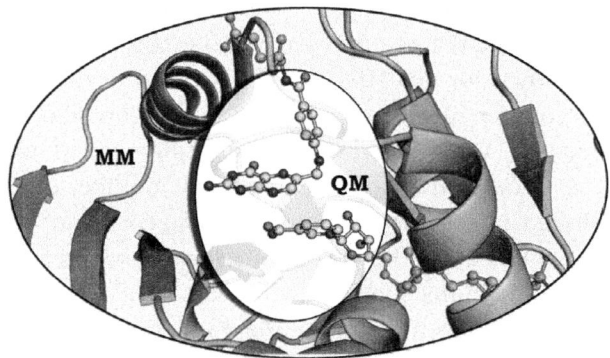

Fig. 3 Partitioning the enzyme system (ie, dihydrofolate reductase) into QM and MM regions. The QM region contains a reactive fragment composed of parts of the substrate and cofactor, whereas the MM region includes the remaining substrate/cofactor fragment, enzyme, water, and ions.

Here, the Hamiltonian of the full system is obtained by combining the QM operator describing the inner region with the analogous MM term representing the outer region. Furthermore, an explicit coupling term is added that describes the interaction between both layers. The coupling term is crucial (Warshel & Levitt, 1976), and includes the electrostatic, van der Waals, and bonded interactions between the atoms in both regions.

$$\hat{H}_{\text{QM/MM}} = -\sum_{i,n} \frac{q_n}{|R_i - r_n|} + \sum_{N,n} \frac{Z_N q_n}{|R_N - r_n|}$$
$$+ \sum_{N,n} 4\varepsilon_{Nn} \left(\left(\frac{\sigma_{Nn}}{R_N - r_n} \right)^{12} - \left(\frac{\sigma_{Nn}}{R_N - r_n} \right)^{6} \right) \quad (2)$$
$$+ \hat{H}_{\text{QM/MM}}^{\text{Bonded}}$$

Here, the first term accounts for the interaction between the partial charges of the MM atoms (n), q_n, and the electrons of the QM region, i. The second term accounts for the interaction between the MM partial charges and the nuclei, N, of the QM region, while the third term accounts for the van der Waals interactions between the QM and MM atoms. This latter term is needed to account for nonelectrostatic interactions between QM and MM atoms. The terms of Eq. (2) are added to the Fock (ab initio (AI) or semiempirical (SE) methods) or Kohn-Sham (density functional theory (DFT) methods) matrix, and the wavefunction (WF) and electron density are polarized via the first term. If the QM/MM boundary extends across chemical bonds, it is necessary to account for this in a way that minimizes the quantum-classical boundary effect. Various approaches have been suggested for this, including the link atom (Field, Bash, & Karplus, 1990) and boundary atom approaches (Antes & Thiel, 1999; Zhang, Lee, & Yang, 1999), as well as local self-consistent field (Monari, Rivail, & Assfeld, 2013) and generalized hybrid orbital (Gao, Amara, Alhambra, & Field, 1998) approaches (Fig. 4). These will be discussed later.

In the AI and SE formalisms, the total energy may be obtained by solving the Schrödinger equation in the field of the MM partial charges

$$\hat{H}\Psi = E_{\text{QM}}\Psi \quad (3)$$

and the total potential energy is then defined as

$$U_{\text{Tot}} = \langle \Psi | \hat{H}_{\text{QM}} | \Psi \rangle + \langle \Psi | \hat{H}_{\text{QM/MM}} | \Psi \rangle + E_{\text{MM}} = E_{\text{QM}} + E_{\text{MM}} \quad (4)$$

Similar expressions arise for Kohn-Sham versions of DFT.

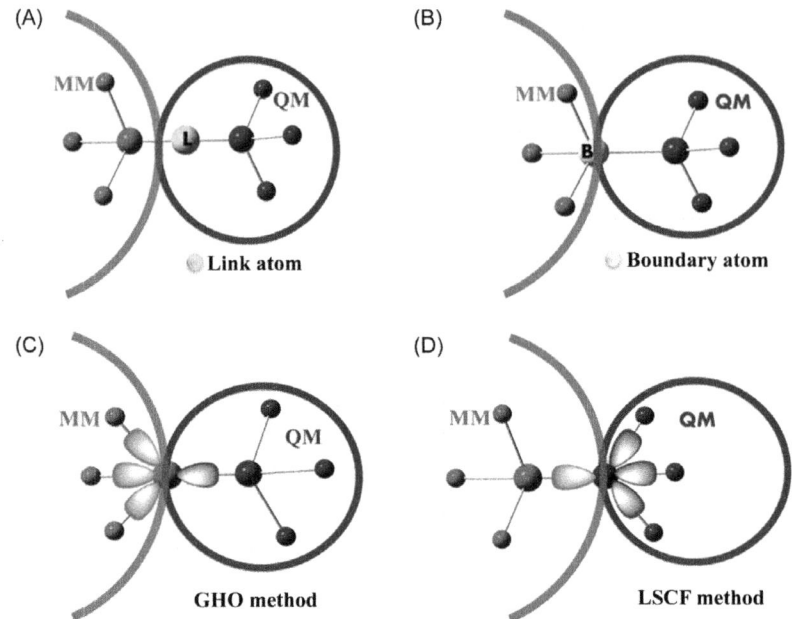

Fig. 4 QM/MM boundary treatments discussed in the text.

3.2 Choosing a Method for the QM Region

One of the principle challenges incumbent to *all* QM/MM approaches is the high computational cost associated with the QM part of the calculation. This problem is amplified if one is interested in QM/MM free energies, where extensive conformational sampling is required in order to obtain stable and converged results. Ideally, one would like to employ high-level AI or DFT methods for the QM region. However, this is often not practical due to the high computational cost. Certain implementations can enhance the efficiency of QM calculations, such as Car–Parrinello methods (Senn & Thiel, 2009) or linear scaling DFT approaches (Lever et al., 2014). Clearly, one should choose an AI or DFT method that describes the chemistry catalyzed by the enzyme at a satisfactory level of accuracy. For instance, specialized functionals have been proposed that perform well for thermochemistry and kinetics (Zhao & Truhlar, 2008). Even after making a reasonable quantum chemistry choice for the QM region, it is a good idea to perform benchmark calculations at an even higher level of theory if possible, to estimate possible errors in the underlying PES (Weitman & Major, 2010).

One may also employ a SE Hamiltonian to describe the QM region, and there have been an increasing number of recent studies utilizing such approaches in studies of enzymatic reactivity (Lin & Gao, 2011; Vardi-Kilshtain, Doron, & Major, 2013). However, standard SE methods are often not accurate enough to give quantitative insight and it is necessary to perform reparameterization of the canonical parameter set to fit a specific target reaction (Rossi & Truhlar, 1995). Such specific reaction parameter approaches have been applied with great success to numerous enzymes (Doron, Major, Kohen, Thiel, & Wu, 2011; Luk et al., 2013; Major et al., 2009; Major, Garcia-Viloca, & Gao, 2006a; Major, York, & Gao, 2005; Rubinstein & Major, 2009, 2010; van der Kamp, Chaudret, & Mulholland, 2013; Vardi-Kilshtain, Major, Kohen, Engel, & Doron, 2012). SE versions of DFT present another fairly accurate, yet fast, description of the QM region (Cui, Guo, & Karplus, 2002).

In spite of the usefulness of reparameterized SE methods or SE DFT, the QM calculation often remains the bottleneck for typical QM/MM calculations. One of the most successful approaches to QM/MM modeling of enzymes is the empirical valence bond (EVB) approach (Warshel, 1991; Warshel & Weiss, 1980). This method is an empirically based QM/MM method, which transforms a force field-based description of individual electronic ground states into a quantum chemical framework using valence bond theory.

We also note that recent work reporting minimization of small proteins at a pure QM level using code specially designed for graphical processing units is a promising direction (Kulik, Luehr, Ufimtsev, & Martinez, 2012). An additional avenue is linear scaling DFT, and a TS search using an all QM description of an enzyme was recently reported (Lever et al., 2014).

Realizing there is a plethora of possible QM/MM approaches one can choose between, the question is what is the optimal choice? The answer depends on the system being modeled and the software and hardware resources available. The primary concern is the accuracy of the QM method, as any simulation can only be as accurate as the underlying PES. Therefore, one must determine the level of accuracy required for the project. Once this has been determined, one must inquire the suitability of each quantum chemistry method for the chemistry at hand. In this regard, the requirements are the same as for gas-phase chemistry and the considerable experience accrued with the use of QM methods for the study of chemical reactions may be used. Hence, DFT methods that are suitable for thermodynamics and kinetics of chemical reactions may be used for QM/MM simulations (Zhao & Truhlar, 2008). One family of reactions were AI or DFT methods are essential is for terpenes, which include highly reactive intermediates. In this case, the quantum chemistry method must be able to treat a wide variety of unexpected bonding situations that might occur during the course of the reaction (Major & Weitman, 2012; Weitman & Major, 2010).

In cases where extensive sampling is required, it may be impractical to use DFT or AI methods, and one might have to resort to SE or empirical QM/MM methods. As mentioned earlier, in certain cases SE may work well without reparameterization of the Hamiltonian (eg, for decarboxylation reactions) (Lin & Gao, 2011; Vardi-Kilshtain et al., 2013). However, in general SE methods need to be reparameterized prior to use in enzymatic reactions. This may be done on a per-reaction basis (eg, SRP) (Doron et al., 2011; Major & Gao, 2006; Major et al., 2009; Vardi-Kilshtain et al., 2012), or per functional group. For instance, SE methods may be reparameterized to treat glycochemistry (Govender, Gao, & Naidoo, 2014). Nonetheless, specialized software is required to perform proper reparameterization of SE methods, and these programs are not generally available to the wider community. One may also generate an EVB potential for the reaction at hand, and this may yield very accurate free energy profiles. A caveat with this method is that you can only observe chemical changes that were predefined during the construction of the valence bond states.

3.3 Choosing a Method for the MM Region

In general, any of the many force fields commonly used for biosimulations may be adopted in a QM/MM study (eg, CHARMM (Best, Mittal, Feig, & MacKerell, 2012; Best, Zhu, et al., 2012; MacKerell, Feig, & Brooks, 2004), Amber (Case et al., 2005; Cornell et al., 1995), OPLS (Jorgensen, Maxwell, & Tirado-Rives, 1996; Jorgensen & Tirado-Rives, 1988), Gromos (Oostenbrink, Villa, Mark, & Gunsteren, 2005), Molaris (Lee, Chu, & Warshel, 1993; Warshel, Kato, & Pisliakov, 2007)). We note that the latter force field is polarizable. These force fields have been extensively tested (Beauchamp, Lin, Das, & Pande, 2012; Cerutti, Freddolino, Duke, & Case, 2010; Lindorff-Larsen et al., 2012; Piana, Klepeis, & Shaw, 2014; Warshel et al., 2007), and are continuously being improved. However, a built-in inconsistency appears in many QM/MM approaches, due to a lack of mutual polarization of the QM and MM regions. The charge distribution in the QM region may naturally be polarized by the surrounding environment, whereas the partial charges of the MM region are static in most implementations. This imbalance in the treatment of charge polarization may be overcome by employing a polarizable force field (Halgren & Damm, 2001; Warshel et al., 2007; Xie, Orozco, Truhlar, & Gao, 2009) and examples of polarizable QM/MM approaches exist (Boulanger & Thiel, 2012, 2014; Warshel & Levitt, 1976).

Numerous water models exist as well (Guillot, 2002; Jorgensen & Tirado-Rives, 2005), such as ST2 (Stillinger & Rahman, 1974), TIPnP ($n=3$ (Jorgensen, Chandrasekhar, Madura, Impey, & Klein, 1983), 4 (Jorgensen & Madura, 1985), 5 (Mahoney & Jorgensen, 2000; Vega, Abascal, Conde, & Aragones, 2009), SPC (Berendsen, Postma, van Gunsteren, & Hermans, 1981), AMOEBA (Ren & Ponder, 2003), and Langevin dipoles (Florian & Warshel, 1999). These models have undergone continuous improvements since their original implementations, hinting to the difficulty in reaching optimal models. The water model adopted should ideally be the same one employed as part of developing the force field parameters, to yield a well-balanced model. It is also necessary to validate the water model for QM/MM interactions (Gao, 1994; Gao & Xia, 1992).

3.4 Nonbonded Interactions Between the QM and MM Regions

The partial atomic charges of the MM region that polarize the QM region are usually part of a delicately balanced FF and should not be changed. The same is true for the vdW parameters. However, one can modulate the strength of the QM/MM interactions by adding a blurring function on the MM atoms (Das et al., 2002), or by modifying Gaussian type functions on the QM atoms (Field et al., 1990). Another way to fine-tune QM/MM interactions, is via the vdW parameters. For instance, by reducing the size of the QM atom by lowering the σ parameter (Eq. 2), one essentially increases the electrostatic interaction between the QM and nearby MM atoms by allowing them to approach closer. Increasing the σ value will result in a weakening of the QM/MM interactions. Typically, vdW parameters for the QM atoms must be parameterized for each enzyme system by careful comparison with benchmark data (Gao, 1994; Gao & Xia, 1992). As a rule of thumb, the QM atom vdW parameters should be derived using the same general philosophy as the force field it interacts with, and if a target QM method was used to develop the force field, the same method should be used to derive the vdW parameters.

3.5 Covalent Connections Between QM and MM Regions

In many cases, there are covalent boundaries between the QM and MM regions. This boundary may be treated by several alternative methods (Fig. 4). The simplest of these approaches is the link atom approach, where the QM open valence facing the MM region is saturated by a capping H-atom (Field et al., 1990). The advantage of this method is its simplicity, while a disadvantage of the approach is that it adds an artificial degree of freedom, which is not present in the real system. In the boundary atom approach (Antes & Thiel, 1999; Zhang et al., 1999), the boundary atom appears both as a QM and a MM atom, and is included in the calculation of each region. A somewhat different family of methods is the localized orbital (ie, LSCF) (Monari et al., 2013) and generalized hybrid orbital (GHO) (Gao et al., 1998; Pu, Gao, & Truhlar, 2005) methods where hybrid orbitals are placed at the boundary. More recent efforts have introduced a generalization of the GHO method to WF- and DFT-based QM/MM, which does not require system dependent parameters (Wang & Gao, 2015). Another recent advancement is the exact link orbital method (Sun & Chan, 2014). With any of the above boundary schemes artifacts may arise if the boundary between the QM and MM level regions is too close to the reacting atoms and care should taken when assigning partial MM charges in the vicinity of the boundary (Reuter, Dejaegere, Maigret, & Karplus, 2000).

3.6 Size of QM Region

Another important point that requires careful attention is the size of the QM region. This is a question of the utmost importance: if the QM region is too small, important physical interactions might be lost. However, a large QM region comes at a computational cost, which could slow down the entire project. As a minimum requirement the QM region must include the reactive fragment undergoing chemical change, including any part of the enzyme participating in the chemistry. There have been some enzyme studies addressing the issue of size of QM region (Kulik, Zhang, Klinman, & Martinez, 2015; Liao & Thiel, 2013; Sadeghian et al., 2014; Zhang, Kulik, Martinez, & Klinman, 2015), and usually this should be studied on a per case basis. However, in many cases there are practical considerations, such as the cost of the QM method and the need for configurational sampling that dictates the size of QM region. Considerable progress is being made in the area of GPU-accelerated quantum chemistry, facilitating increasingly larger QM regions (Kulik et al., 2012; Zhang et al., 2015).

Practically speaking, in many cases one might commence with a small QM region, and as the project progresses realize that one must enlarge it, for example due to physical interactions or a need to include additional acid/base residues or water molecules in the mechanism.

Dispersion interactions are important in proteins, and in ligand binding (Antony, Grimme, Liakos, & Neese, 2011; Lonsdale, Harvey, & Mulholland, 2012). Hence, the QM/MM method should be able to treat such interactions, and this is included in the form of an empirical expression (Eq. 2). Hence, if one assumes that dispersion interactions are important, it then makes little sense to use a large QM region with a DFT functional that does not account for dispersion, or include some form of dispersion correction.

4. REACTION COORDINATES AND CLASSICAL FREE ENERGY SIMULATIONS

4.1 Reaction Rates and Chemical Progress Coordinates

The rate of enzyme reactions is usually computed within the confines of transition state theory (TST). Specifically, the reaction rate may be written as (Eyring, 1935; Schenter, Garret, & Truhlar, 2003):

$$k_C^{TST} = \Omega e^{-\beta \Delta W_C^\ddagger(\zeta)} \tag{5}$$

Here, $\Delta W_C^\ddagger$ is the classical potential of mean force (PMF) barrier, $\beta = (k_B T)^{-1}$, k_B Boltzmann's constant, T is the temperature, and ζ is a chemical progress coordinate, termed the reaction coordinate. If the reaction coordinate is not rectilinear, a small correction termed needs to be included (Schenter et al., 2003). The prefactor, Ω, is defined as (Hinsen & Roux, 1997):

$$\Omega = \left\langle \left(\frac{Z_\zeta}{2\pi\beta}\right)^{1/2} \right\rangle_{\zeta^\ddagger} \tag{6}$$

where $\langle \cdots \rangle_{\zeta^\ddagger}$ is a configurational average performed at the TS, and Z_ζ is the Jacobian factor defined as:

$$Z_\zeta \equiv \sum_{i=1}^{3N} \frac{1}{m_i}\left(\frac{\partial \zeta}{\partial r_i}\right)^2 \tag{7}$$

Here, r_i is the coordinates of degree of freedom i and m_i is the mass associated with that degree of freedom.

In enzyme-modeling studies, it is typically necessary to choose the reaction coordinate, ζ, with some care. Several reaction coordinate options exist, and the results may vary depending on the choice made. Usually, the goal is to obtain the free energy profile for the enzymatic reaction as a function of some chemical progress coordinate (ie, the PMF). The difference between the free energy of the TS, $W(\zeta_{TS})$, and the RS, $W(\zeta_{RS})$, will then constitute the free energy barrier, $\Delta W_C^{\ddagger}$, which may be plugged into Eyring's equation (Eyring, 1935) to obtain the rate constant for the catalytic step. This rate constant may in turn, be compared with experimental rate constants. Here we note that the experimental rate constants usually constitute an upper limit to the rate constant for the chemical step, as the chemistry is often not the rate limiting step in enzymes (Hammes, 2002; Watt, Shimada, Kovrigin, & Loria, 2007).

The reaction coordinate should facilitate exploration of as much as possible of the configuration space relevant to the chemical change. In many cases, a simple geometric reaction coordinate, such as a bond distance, angle, dihedral angle, or difference between two bond distances (ie, antisymmetric stretch) may be employed (Rohrdanz, Zheng, & Clementi, 2013). Such reaction coordinates are usually sufficiently adequate to model gas-phase reactions and solution phase reactions where the solvent is treated as a dielectric continuum. Indeed, model gas-phase reactions can be very useful in giving insight into a useful reaction coordinate. However, in the study of enzyme reactions, where the protein and solvent are treated explicitly (ie, not as a continuum), one needs to assure that the reaction coordinate can correctly represent the response of the environment to changes in the charge distribution and conformation of the solute. In spite of these concerns, experience has shown that geometric coordinates are adequate for treating enzyme reactions (Gao et al., 2006). Using geometric metrics, one may also define the reaction progress in terms of rehybridization, which provides additional insight (Doron, Kohen, & Major, 2012).

A fundamentally different approach to the reaction coordinate definition is to use the energy gap between two diabatic states. Here, ζ is defined as the energy gap between the reactant and product diabatic states, $\zeta = \varepsilon_{RS} - \varepsilon_{PS}$. The energy gap reaction coordinate is a generalized reaction coordinate that does not require a predefined geometric path. Instead, it uses a difference in the reactant and product potentials to describe the reaction progress (Kamerlin & Warshel, 2011; Warshel, 1991). The main advantage of the energy gap reaction coordinate is that it captures much of the physics of the response of the environment to changes in the reactive fragment charge distribution and geometry.

The choice of reaction coordinate is often directly related to the quantum chemical method chosen. A geometric reaction coordinate is often a natural choice for AI, SE, or DFT methods, whereas in EVB approaches the energy gap coordinate is a good choice.

So one may ask oneself how important is the choice of reaction coordinate? Choosing a poor reaction coordinate will result in increased recrossing at the dividing surface. That is, trajectories that reach the TS from the RS will recross and return to the reactant side. This may be corrected for quantitatively by computing a recrossing coefficient (Pu, Gao, & Truhlar, 2006), which may be done by rotation of the dividing surface or by running activated dynamics simulations (McCammon & Karplus, 1979; Northrup, Pear, Lee, McCammon, & Karplus, 1982). The choice of reaction coordinate is of course also dependent on whether it is implemented in the software of choice or not.

4.2 Classical Free Energy Simulations

After having chosen a reaction coordinate, it is necessary to pick a method that will allow sampling of the most important degrees of freedom and to let the system climb the barrier for the chemical reaction in a reasonable simulation time. The reason for this is the difference in time scale of enzyme reactions in nature, and the time scales one can cover with QM/MM methods using modern computer facilities. Hence, current simulation methods almost universally employ some form of biasing potential, to facilitate barrier crossing. One of the most commonly used sampling methods is umbrella sampling (US) (Torrie & Valleau, 1977). The free energy difference along the reaction coordinate (ie, the PMF) is defined as:

$$\Delta W_C^{\ddagger}(\zeta) = W_{TS} - W_{RS} = -\beta^{-1} \ln \frac{P(\zeta^{\ddagger})}{\int_{-\infty}^{\zeta^{\ddagger}} P(\zeta) d\zeta} \quad (8)$$

In the above equation, P is the probability density along the reaction coordinate, ζ.

In the US approach, one applies an umbrella potential, which ideally equals $U_{Umb}(\zeta) = -\Delta W(\zeta)$. Since this function is not known a priori (ie, the goal is to find $\Delta W(\zeta)$), an iterative approach is usually adopted, wherein the umbrella potential is refined in subsequent free energy simulations (so-called "adaptive umbrella sampling") (Bartels & Karplus, 1997, 1998; Hooft, van Eijck, & Kroon, 1992; Mezei, 1987). Additionally, the reaction coordinate is usually divided into windows that allow focused sampling in designated regions along the progress coordinate. To keep the system in the simulation windows, a restraining potential is applied, usually in the form of a harmonic potential ($U_{\text{Res}}(\zeta) = k(\zeta - \zeta')^2$). The combined biasing potential is thus composed of the umbrella and restraining potentials. The effect of the bias is removed using statistical methods such as the weighted histogram analysis method (Kumar, Rosenberg, Bouzida, Swendsen, & Kollman, 1992).

Additional standard simulation techniques may be employed, such as free energy perturbation (FEP) and thermodynamic integration (TI). In the FEP approach, one calculates free energy differences, $\Delta G(\zeta)$, along the reaction coordinate using Zwanzig's perturbation formula (Zwanzig, 1954):

$$\Delta G^{\zeta' \to \zeta''} = -\beta^{-1} \ln \left\langle \exp\left\{ -\beta \Delta U_{\text{Tot}}^{\zeta' \to \zeta''} \right\} \right\rangle_{\zeta'} \qquad (9)$$

In Eq. (9), the brackets $\langle \ldots \rangle$ represent an ensemble average. Analogously, one may define the free energy using TI, where the mean force acting on the reaction coordinate is integrated:

$$\Delta G^{\zeta' \to \zeta''} = \int_{\zeta'}^{\zeta''} d\zeta \frac{dG(\zeta)}{d\zeta} = \int_{\zeta'}^{\zeta''} d\zeta \left\langle \frac{\partial U_{\text{Tot}}(\zeta)}{\partial \zeta} \right\rangle_{\zeta} \qquad (10)$$

Practically, in TI simulations, constrained dynamics may be run at discrete values of ζ and one samples the average force at each location, and the average forces are then integrated numerically over the reaction coordinate.

A simulation approach that does not require the definition of a reaction coordinate is transition path sampling (TPS) (Bolhuis, Dellago, Chandler, & Geissler, 2002). TPS requires the definition of reaction and product states (ie, basins), and the pathway between the basins is obtained by a stochastic hunt in position and momentum space. The TPS algorithm has been applied to several enzymatic reactions (Doron, Kohen, Nam, & Major, 2014; Schwartz & Schramm, 2009). The principle disadvantage of the TPS method is that it is not trivial to compute the rate of the enzymatic reactions (Dellago, Bolhuis, & Chandler, 1999; Dellago, Bolhuis, Csajka, & Chandler, 1998), including the waiting time needed to reach a reactive state from which the chemical step can take place. Additionally, the TPS method is rather expensive from a computational perspective.

The above-mentioned free energy simulation methods are usually combined with the MD technique (Alder & Wainwright, 1957, 1962), which is used to propagate the system coordinates and momenta based on the forces (where U is obtained from the QM/MM potential). This kind of simulation provides both ensemble-averaged information as well as time-dependent dynamic information. However, typically such simulations are propagated according to the classical Newtonian equations of motion, whereas in reality all particles are quantum in nature. In the next section we will discuss some approaches to quantum simulations.

4.3 Quantum Free Energy Simulations
4.3.1 Overview of Quantum Simulation Methods
In order to accurately predict reaction rates for proton, hydrogen, and hydride transfer reactions, it is necessary to including NQE in the simulations. Additionally, in order to predict isotope effects, such as kinetic, equilibrium, or binding isotope effects, inclusion of NQE is essential. The main source of NQE in enzymes is zero-point energy and tunneling (Layfield & Hammes-Schiffer, 2014). NQE may be included on the fly in direct quantum simulations (Layfield & Hammes-Schiffer, 2014), or as a postclassical simulation correction (Hwang & Warshel, 1996; Pu et al., 2006; Vardi-Kilshtain, Nitoker, & Major, 2015). Due to the great cost of quantum simulations, the latter approach has traditionally been the most commonly applied in the study of enzyme reactions. If the quantum effects are moderate, and the "quantum trajectory" does not deviate radically from the classical pathways, the quantum correction scheme is a reasonable approach. However, if the changes are significant then a direct approach might be most appropriate. In either case, one commonly chooses a region of the active site that is directly involved in the chemistry and treats these selected atoms quantum mechanically, while the remaining atoms are treated classically.

Numerous methods to include NQE are available: Feynman path-integral (PI) methods (Feynman & Hibbs, 1965; Hwang, Chu, Yadav, & Warshel, 1991; Hwang & Warshel, 1993; Major & Gao, 2005, 2007; Major et al., 2006a; Major, Nam, & Gao, 2006b), WF-based methods (Billeter, Webb, Agarwal, Iordanov, & Hammes-Schiffer, 2001; Billeter, Webb, Iordanov, Agarwal, & Hammes-Schiffer, 2001; Hammes-Schiffer, 2004), and semiclassical (SC) methods (Pu et al., 2006). PI methods are highly suitable for condensed phased simulations as they are readily applicable to multidimensional problems (ie, many atoms) and are naturally coupled with standard classical simulation strategies such as MD and Monte Carlo. WF-based methods, on the other hand, are usually limited to a few degrees of freedom (eg, a single atom treated in three dimensions). SC methods are also applicable to many particles, although usually entails an artificial separation of bound vibrations and tunneling effects.

4.3.2 PI Methods

In the PI approach (Feynman & Hibbs, 1965), classical particles are replaced by a necklace of beads that are connected via harmonic springs (ie, a ring polymer), which gives the particle a delocalized quantum behavior (Fig. 5). The theory provides the correct thermodynamic behavior of quantum particles. One of the first NQE methods to be implemented for enzyme simulations was the quantized classical path (QCP) due to Hwang and Warshel (Hwang & Warshel, 1993, 1996; Hwang et al., 1991). This method performs an initial classical simulation, which produces a classical path (ie, MD trajectory). Subsequently, in a separate simulation, the classical path is quantized. In practice, this is done by selecting several atoms that are either directly or indirectly involved in the chemical change and these are to be corrected for quantum behavior. The QCP methods and its extensions have been applied to a large number of enzyme reactions (Azuri, Engel, Doron, & Major, 2011; Doron et al., 2011; Hwang & Warshel, 1996; Liu & Warshel, 2009; Major et al., 2009; Major et al., 2006b; Major & Weitman, 2012; Mavri, Liu, Olsson, & Warshel, 2008; Olsson, Siegbahn, & Warshel, 2004a, 2004b; Rubinstein & Major, 2009; Vardi-Kilshtain et al., 2012; Wang & Hammes-Schiffer, 2006), including calculation of various kinds of isotope effects (Feierberg, Luzhkov, & Aqvist, 2000; Lin & Gao, 2011;

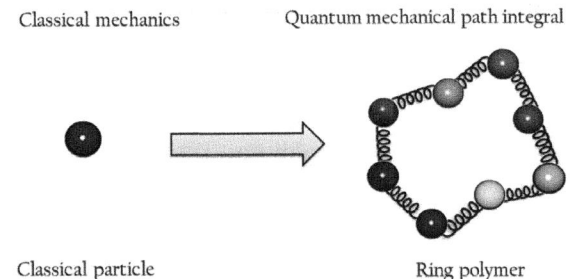

Fig. 5 Illustration of the path-integral (PI) concept. A classical particle is described as a single point, whereas in a PI depiction the particle is represented by a ring polymer of quasi-particles held together by harmonic springs, resulting in a delocalized nature.

Liu & Warshel, 2007; Major et al., 2009; Major et al., 2006b; Vardi-Kilshtain et al., 2013, 2015).

Care must be taken in the application of PI methods in general, and QCP in particular, as convergence of the free energy correction with respect to bead sampling can be challenging (Berne & Thirumalai, 1986). Hence, specialized sampling techniques should be applied, such as the bisection (Azuri et al., 2011; Major & Gao, 2005) or staging approaches (Azuri et al., 2011), and one must carefully check for convergence (Major & Gao, 2005; Major et al., 2006a). It is also essential to describe the quantum particles with a sufficient number of beads, and it seems like 16–32 beads are sufficient in order to describe quantized atoms (Major & Gao, 2005; Major et al., 2006a). Also the number of atoms to be quantized needs to be checked for convergence, as quantizing too few atoms can easily underestimate the extent of NQE (Major et al., 2006a). In conclusion, although PI may be the optimal way to perform quantum dynamics in enzymes, some care must be taken in applying various schemes.

4.3.3 WF-Based Methods
An elegant example of a WF-based approach is one developed by Hammes-Schiffer and coworkers, which describes the PES by an EVB potential, while the transferring hydrogen nucleus is treated by a three–dimensional QM WF on a grid (Billeter, Webb, Agarwal, et al., 2001; Billeter, Webb, Iordanov, et al., 2001; Hammes-Schiffer, 2004). Within this formalism, the reaction coordinate is defined as an energy gap coordinate. Hence the TST reaction rate is computed from the free energy barrier along this reaction coordinate, while recrossing of the diving surface may be estimated via a reactive flux method. The advantage of this method is that the nuclear WF is propagated on the fly; hence a true quantum trajectory is generated. A disadvantage is that only a single particle is quantized, and any quantum mechanical coupling between atoms is lost. Another limitation is that the method is designed to treat hydrogen nuclei only and is not currently a general approach for studying NQE.

4.3.4 Semiclassical Methods
A general approach to NQE was developed to Gao, Truhlar, and coworkers (Alhambra et al., 2001; Pu et al., 2006). This SC method is an extension of the methods developed by the Truhlar group for the study of gas-phase and homogenous-condensed-phase reactions (Fernandez-Ramos, Ellingson, Garret, & Truhlar, 2007). In this generalization of the method, which is

termed ensemble-averaged variational transition state theory with multidimensional tunneling (VTST/MT), one can readily study enzymatic reactions. VTST/MT is based on breaking the rate constant into several independent contributions: statistical barrier crossing, tunneling through the barrier, and recrossing of the barrier. These terms are computed in several separate stages. The former term includes the classical free energy barrier obtained from a classical free energy calculation (eg, US) and quantum vibrational contributions within the harmonic approximation. In its implementation within CHARMM the classical free energy is usually computed using US, while the quantization of the vibrations is computed via the Polyrate code (Corchado et al., 2005), which performs a normal mode analysis using the QM/MM potential energy supplied by CHARMM. The second term accounts for multidimensional tunneling via a transmission coefficient, and this calculation can account for tunneling along the minimum energy path, as well as along corner cutting paths. This part of the calculation is also computed using Polyrate. The recrossing term may be estimated by comparing the free energy difference of the TS determined by PMF and that along the minimum free energy path (Alhambra et al., 2001), or conversely via activated dynamics (McCammon & Karplus, 1979; Northrup et al., 1982).

5. CONCLUDING WORDS

A key question when planning a multiscale enzyme venture is the availability of various methods in the existing software platforms. The strategy of choice in an enzyme-modeling project will necessarily depend on the software available, and the nature of the methods implemented in the software. In Table 1, we present the current state of some common software packages as presented in recent literature and on the web, the force fields available therein, which quantum chemistry methods they interface with and more.

This review strived to cover principle methodological steps of multiscale studies of enzyme reactions. This includes choice of starting structures, boundary conditions, potential energy surfaces, reaction coordinates, simulation methods, as well as choice of method for the treatment of NQE. This review has attempted to emphasize that even after choosing a software package with implemented state-of-the-art enzyme methods, there are still plenty of areas that require the attention of the researcher.

Table 1 Some Common Software Packages with Multiscale Capabilities

MM Software	Force Fields[a]	Built-In QM Capability	Interface to QM Software	QM/MM Boundary Method	NQE[b]
AMBER	AMBER, CHARMM, GAFF, GLYCAM, AMOEBA	SE, EVB	ADF, GAMESS-US, GAUSSIAN, NWCHEM, ORCA, Q-CHEM, TERACHEM	LINK ATOM	Yes
BOSS	OPLS	AM1, PM3, PDDG/PM3	GAUSSIAN	LINK ATOM	No
CHARMM	AMBER, CHARMM, CGENFF, DRUDE, OPLS	SE	CADPAC, GAMESS, GAMESS-UK, GAUSSIAN, Q-CHEM, TURBOMOLE	GHO, LINK ATOM, LSCF	Yes
CHEMSHELL (USES DL_POLY, GULP, CHARMM, GROMOS MM)	SAME AS MM SOFTWARE	NA[c]	DALTON, DMOL3, FHI-AIMS, GAMESS-UK, GAUSSIAN, NWCHEM, MNDO, MOLPRO, ORCA, Q-CHEM, TURBOMOLE	LINK ATOM	Yes
DISCOVERY STUDIO	CHARMm	NA[d]	DMOL3, VAMP	LINK ATOM	No
GAUSSIAN (ONIOM)	AMBER, DREIDING, UFF	AI, DFT, SE		LINK ATOM	No
GROMACS	AMBER, CHARMM, GROMOS, OPLS		CPMD, GAUSSIAN, GAMESS-US, MOPAC7, ORCA	LINK ATOM	No

Continued

Table 1 Some Common Software Packages with Multiscale Capabilities—cont'd

MM Software	Force Fields	Built-In QM Capability	Interface to QM Software	QM/MM Boundary Method	NQE
LICHEM (MM BY TINKER)	AMBER, AMOEBA, CHARMM, DANG, MM2/3, MMFF, OPLS	—[c]	GAUSSIAN, NWCHEM, PSI4	PSEUDO-BOND POTENTIALS, BOUNDARY ATOMS	No
MOLARIS-XG	MOLARIS	EVB	GAMESS, GAUSSIAN, MOPAC, MOLARIS QCFFPI, Q-CHEM	LINK ATOM	No
NWCHEM	AMBER, CHARMM	AI, DFT	—[c]	LINK ATOM	No
pDYNAMO	AMBER, CHARMM, OPLS	DFT, SE	ORCA	LINK ATOM	No
SCHRÖDINGER (QSITE)	OPLS	AI, DFT, SE (JAGUAR)	JAGUAR	LINK ATOM, FROZEN LOCALIZED ORBITALS	No
Q-CHEM	AMBER, CHARMM, OPLS	AI, DFT	—[c]	BOUNDARY (YING-YANG) ATOM	No
QMMM (MM BY TINKER)	AMBER, AMOEBA, CHARMM, DANG, MM2/3, MMFF, OPLS	—[c]	GAMESS, GAUSSIAN, ORCA	LINK ATOM	No

[a] Many different versions of each force field exist. Here, we only state which force field is included in one of these many versions.
[b] Only NQE methods accounting for vibrational effects, as well as tunneling, are considered as having NQE capabilities. Also, cases where the QM code has stand-alone NQE capabilities (ie, Q-CHEM has PI capabilities) are not included.
[c] Not applicable.
[d] Discovery Studio is based on the CHARMM code, so in principle all CHARMM capabilities are included. However, many CHARMM functionalities are inaccessible from the graphical user interface.

ACKNOWLEDGMENTS

This work has been supported by the Israel Science Foundation (Grant 1560/14) and the United States–Israel Binational Science Foundation (Grant #2012340).

REFERENCES

Alder, B. J., & Wainwright, T. E. (1957). Phase transition for a hard sphere system. *The Journal of Chemical Physics, 27*(5), 1208–1209.
Alder, B. J., & Wainwright, T. E. (1962). Phase transition in elastic disks. *Physical Review*, 359–361.
Alexov, E., Mehler, E. L., Baker, N., Baptista, A. M., Huang, Y., Milletti, F., et al. (2011). Progress in the prediction of pKa values in proteins. *Proteins: Structure, Function, and Bioinformatics, 79*, 3260–3275.
Alhambra, C., Corchado, J., Sanchez, M. L., Garcia-Viloca, M., Gao, J., & Truhlar, D. G. (2001). Canonical variational theory for enzyme kinetics with the protein mean force and multidimensional quantum mechanical tunneling dynamics. Theory and application to liver alcohol dehydrogenase. *Journal of Physical Chemistry B, 105*(45), 11326–11340.
Allen, M. P., & Tildesley, D. J. (1989). *Computer simulation of liquids.* Oxford: Oxford University Press.
Antes, I., & Thiel, W. (1999). Adjusted connection atoms for combined quantum mechanical and molecular mechanical methods. *The Journal of Physical Chemistry. A, 103*(46), 9290–9295.
Antony, J., Grimme, S., Liakos, D. G., & Neese, F. (2011). Protein–ligand interaction energies with dispersion corrected density functional theory and high-level wave function based methods. *The Journal of Physical Chemistry. A, 115*, 11210–11220.
Ashworth, J., Havranek, J. J., Duarte, C. M., Sussman, D., Jr., Monnat, R. J., Stoddard, B. L., et al. (2006). Computational redesign of endonuclease DNA binding and cleavage specificity. *Nature, 441*, 656–659.
Azuri, A., Engel, H., Doron, D., & Major, D. T. (2011). Path-integral calculations of nuclear quantum effects in model systems, small molecules, and enzymes via gradient-based forward corrector algorithms. *Journal of Chemical Theory and Computation, 7*(5), 1273–1286.
Bartels, C., & Karplus, M. (1997). Multidimensional adaptive umbrella sampling: Applications to main chain and side chain peptide conformations. *Journal of Computational Chemistry, 18*, 1450–1462.
Bartels, C., & Karplus, M. (1998). Probability distributions for complex systems: Adaptive umbrella sampling of the potential energy. *The Journal of Physical Chemistry. B, 102*, 865–880.
Beauchamp, K. A., Lin, Y. S., Das, R., & Pande, V. S. (2012). Are protein force fields getting better? A systematic benchmark on 524 diverse NMR measurements. *Journal of Chemical Theory and Computation, 8*, 1409–1414.
Beglov, D., & Roux, B. (1994). Finite representation of an infinite bulk system: Solvent boundary potential for computer simulations. *The Journal of Chemical Physics, 100*, 9050.
Benighaus, T., & Thiel, W. (2009). A general boundary potential for hybrid QM/MM simulations of solvated biomolecular systems. *Journal of Chemical Theory and Computation, 5*, 3114–3128.
Benighaus, T., & Thiel, W. (2010). Long-range electrostatic effects in QM/MM studies of enzymatic reactions: Application of the solvated macromolecule boundary potential. *Journal of Chemical Theory and Computation, 7*, 238–249.
Berendsen, H. J. C., Postma, J. P. M., van Gunsteren, W. F., & Hermans, J. (1981). Interaction models for water in relation to protein hydration. In B. Pullman (Ed.), *Intermolecular forces*. Dordrecht: Reidel.

Berne, B. J., & Thirumalai, D. (1986). On the simulation of quantum-systems—Path integral methods. *Annual Review of Physical Chemistry, 37*, 401–424.

Best, R. B., Mittal, J., Feig, M., & MacKerell, A. D., Jr. (2012). Inclusion of many-body effects in the additive CHARMM protein CMAP potential results in enhanced cooperativity of a-helix and b-hairpin formation. *Biophysical Journal, 103*(5), 1045–1051.

Best, R. B., Zhu, X., Shim, J., Lopes, P. E. M., Mittal, J., Feig, M., et al. (2012). Optimization of the additive CHARMM all-atom protein force field targeting improved sampling of the backbone ϕ, ψ and side-chain χ_1 and χ_2 dihedral angles. *Journal of Chemical Theory and Computation, 8*, 3257–3273.

Billeter, S. R., Webb, S. P., Agarwal, P. K., Iordanov, T., & Hammes-Schiffer, S. (2001a). Hydride transfer in liver alcohol dehydrogenase: Quantum dynamics, kinetic isotope effects, and role of enzyme motion. *Journal of the American Chemical Society, 123*, 11262–11272.

Billeter, S. R., Webb, S. P., Iordanov, T., Agarwal, Pratul K., & Hammes-Schiffer, S. (2001b). Hybrid approach for including electronic and nuclear quantum effects in molecular dynamics simulations of hydrogen transfer reactions in enzymes. *Journal of Chemical Physics, 114*, 6925–6936.

Bjelic, S., & Aqvist, J. (2004). Prediction of structure, substrate binding mode, mechanism, and rate for a malaria protease with a novel type of active site. *Biochemistry, 43*, 14521–14528.

Boehr, D. D., McElheny, D., & Dyson, H. J. (2006). The dynamic energy landscape of dihydrofolate reductase catalysis. *Science, 313*, 1638–1642.

Bolhuis, P. G., Dellago, C., Chandler, D., & Geissler, P. (2002). Transition path sampling, throwing ropes over rough mountain passes, in the dark. *Annual Review of Physical Chemistry, 53*, 291–318.

Boulanger, E., & Thiel, W. (2012). Solvent boundary potentials for hybrid QM/MM computations using classical Drude oscillators: A fully polarizable model. *Journal of Chemical Theory and Computation, 8*, 4527–4538.

Boulanger, E., & Thiel, W. (2014). Toward QM/MM simulation of enzymatic reactions with the drude oscillator polarizable force field. *Journal of Chemical Theory and Computation, 10*, 1795–1809.

Brooks, C. L., Brünger, A., & Karplus, M. (1985). Active site dynamics in protein molecules: A stochastic boundary molecular-dynamics approach. *Biopolymers, 24*(5), 843–865.

Brünger, A., Brooks, C. L., III, & Karplus, M. (1984). Stochastic boundary conditions for molecular dynamics simulations of ST2 water. *Chemical Physics Letters, 105*(5), 495–500.

Brunger, A. T., & Karplus, M. (1988). Polar hydrogen positions in proteins: Empirical energy placement and neutron diffraction comparison. *Proteins, 4*, 148–156.

Case, D. A., Cheatham, T. E. I., Darden, T., Gohlke, H., Luo, R., Merz, K. M. J., et al. (2005). The Amber biomolecular simulation programs. *Journal of Computational Chemistry, 26*, 1668–1688.

Cavanagh, J., Fairbrother, W. J., Palmer, A. G., III, Rance, M., & Skelton, N. J. (2006). *Protein NMR spectroscopy*. USA: Academic Press.

Cerutti, D. S., Freddolino, P. L., Duke, R. E., Jr., & Case, D. A. (2010). Simulations of a protein crystal with a high resolution X-ray structure: Evaluation of force fields and water models. *The Journal of Physical Chemistry. B, 114*, 12811–12824.

Corchado, J. C., Chuang, Y.-Y., Fast, P. L., Hu, W.-P., Liu, Y.-P., Lynch, G. C., et al. (2005). *POLYRATE 9.3.1*. Minneapolis: University of Minnesota.

Cornell, W. D., Cieplak, P., Bayly, C. I., Gould, I. R., Merz, K. M. J., Ferguson, D. M., et al. (1995). A second generation force field for the simulation of proteins, nucleic acids, and organic molecules. *Journal of the American Chemical Society, 117*, 5179–5197.

Cui, Q., Guo, H., & Karplus, M. (2002). Combining ab initio and density functional theories with semiempirical methods. *The Journal of Chemical Physics, 117*(12), 5617–5631.

Darden, T., York, D., & Pedersen, L. (1993). Particle mesh Ewald: An N · log(N) method for Ewald sums in large systems. *The Journal of Chemical Physics, 98*(12), 10089–10092.

Das, D., Eurenius, K. P., Billings, E. M., Sherwood, P., Chatfield, D. C., Hodoscek, M., et al. (2002). Optimization of quantum mechanical molecular mechanical partitioning schemes: Gaussian delocalization of molecular mechanical charges and the double link atom method. *The Journal of Chemical Physics, 117*, 10534–10547.

Davis, A. M., Teague, S. J., & Kleywegt, G. J. (2003). Application and limitations of x-ray crystallographic data in structure-based ligand and drug design. *Angewandte Chemie (International ed. in English), 42*, 2718–2736.

Dellago, C., Bolhuis, P. G., & Chandler, D. (1999). On the calculation of reaction rate constants in the transition path ensemble. *The Journal of Chemical Physics, 110*, 6617–6625.

Dellago, C., Bolhuis, P. G., Csajka, F. S., & Chandler, D. (1998). Transition path sampling and the calculation of rate constants. *The Journal of Chemical Physics, 108*, 1964–1977.

Doron, D., Kohen, A., & Major, D. T. (2012). Collective reaction coordinate for hybrid quantum and molecular mechanics simulations: A case study of the hydride transfer in dihydrofolate reductase. *Journal of Chemical Theory and Computation, 8*(7), 2484–2496.

Doron, D., Kohen, A., Nam, K., & Major, D. T. (2014). How accurate are transition states from simulations of enzymatic reactions? *Journal of Chemical Theory and Computation, 10*(5), 1863–1871.

Doron, D., Major, D. T., Kohen, A., Thiel, W., & Wu, X. (2011). Hybrid quantum and classical simulations of the dihydrofolate reductase catalyzed hydride transfer reaction on an accurate semi-empirical potential energy surface. *Journal of Chemical Theory and Computation, 7*(10), 3420–3437.

Doron, D., Stojkovic, V., Gakhar, L., Kohen, A., & Major, D. T. (2015). Free energy simulations of active-site mutants of dihydrofolate reductase. *The Journal of Physical Chemistry. B, 119*, 906–916.

Eyring, H. (1935). The activated complex in chemical reactions. *The Journal of Chemical Physics, 3*, 107–115.

Feierberg, I., Luzhkov, V., & Aqvist, J. (2000). Computer simulation of primary kinetic isotope effects in the proposed rate limiting step of the glyoxalase I reaction. *Journal of Biological Chemistry, 275*, 22657–22662.

Fernandez-Ramos, A., Ellingson, B. A., Garret, B. C., & Truhlar, D. G. (2007). Variational transition state theory with multidimensional tunneling. In K. B. Lipkowitz & T. R. Cundari (Eds.), *Reviews in computational chemistry: Vol. 23* (pp. 125–232). Hoboken, NJ: Wiley-VCH.

Feynman, R. P., & Hibbs, A. R. (1965). *Quantum mechanics and path integrals*. New York: McGraw-Hill.

Field, M. J., Bash, P. A., & Karplus, M. (1990). A combined quantum mechanical and molecular mechanical potential for molecular dynamics simulations. *Journal of Computational Chemistry, 11*(6), 700–733.

Fischer, M., Shoichet, B. K., & Fraser, J. S. (2015). One crystal, two temperatures—Cryocooling penalties alter ligand binding to transient protein sites. *ChemBioChem, 16*, 1560–1564.

Florian, J., & Warshel, A. (1999). Calculations of hydration entropies of hydrophobic, polar, and ionic solutes in the framework of the Langevin dipoles solvation model. *The Journal of Physical Chemistry. B, 103*, 10282–10288.

Frey, P. A., & Hegeman, A. D. (2007). *Enzymatic reaction mechanisms*. New York: Oxford University Press.

Friesner, R. A., & Guallar, V. (2005). Ab initio quantum chemical and mixed quantum mechanics/molecular mechanics (QM/MM) methods for studying enzymatic catalysis. *Annual Review of Physical Chemistry, 56*(1), 389–427.

Gao, J. (1994). Computation of intermolecular interactions with a combined quantum mechanical and classical approach. In D. A. Smith (Ed.), *Modeling the hydrogen bond* (pp. 8–21). Washington, DC: ACS.

Gao, J. (1995). Methods and applications of combined quantum mechanical and molecular mechanical potentials. In K. B. Lipkowitz & D. B. Boyd (Eds.), *Reviews in computational chemistry: Vol. 7.* (pp. 119–185). New York: VCH Publishers.

Gao, J., Amara, P., Alhambra, C., & Field, M. J. (1998). A generalized hybrid orbital (GHO) method for the treatment of boundary atoms in combined QM/MM calculations. *The Journal of Physical Chemistry A, 102*(24), 4714–4721.

Gao, J., Ma, S., Major, D. T., Nam, K., Pu, J., & Truhlar, D. G. (2006). Mechanisms and free energies of enzymatic reactions. *Chemical Reviews, 106*(8), 3188–3209.

Gao, J., & Xia, X. (1992). A priori evaluation of aqueous polarization effects through Monte Carlo QM-MM simulations. *Science, 258*(5082), 631–635.

Garcia-Viloca, M., Poulsen, T. D., Truhlar, D. G., & Gao, J. (2004). Sensitivity of molecular dynamics simulations to the choice of the X-ray structure used to model an enzymatic reaction. *Protein Science, 13*, 2341–2354.

Govender, K., Gao, J., & Naidoo, K. J. (2014). AM1/d-CB1: A semiempirical model for QM/MM simulations of chemical glycobiology systems. *Journal of Chemical Theory and Computation, 10*, 4694–4707.

Guillot, B. (2002). A reappraisal of what we have learnt during three decades of computer simulations on water. *Journal of Molecular Liquids, 101*, 219–260.

Halgren, T. A., & Damm, W. (2001). Polarizable force fields. *Current Opinion in Structural Biology, 11*(2), 236–242.

Hammes, G. G. (2002). Multiple conformational changes in enzyme catalysis. *Biochemistry, 41*, 8221–8228.

Hammes-Schiffer, S. (2004). Quantum-classical simulation methods for hydrogen transfer in enzymes: A case study of dihydrofolate reductase. *Current Opinion in Structural Biology, 2004*, 192–201.

Henzler-Wildman, K. A., Thai, V., Lei, M., Ott, M., Wolf-Watz, M., Fenn, T., et al. (2007). Intrinsic motions along an enzymatic reaction trajectory. *Nature, 450*, 838–844.

Hinsen, K., & Roux, B. (1997). Potential of mean force and reaction rates for proton transfer in acetylacetone. *The Journal of Chemical Physics, 106*, 3567–3577.

Honig, B., & Karplus, M. (1971). Implications of torsional potential of retinal isomers for visual excitation. *Nature, 229*(5286), 558–560.

Hooft, R. W., van Eijck, B. P., & Kroon, J. (1992). An adaptive umbrella sampling procedure in conformational analysis using molecular dynamics and its application to glycol. *The Journal of Chemical Physics, 97*, 6690–6694.

Hu, H., Boone, A., & Yang, W. (2008). Mechanism of OMP decarboxylation in orotidine 5′-monophosphate decarboxylase. *Journal of the American Chemical Society, 130*(44), 14493–14503.

Hu, H., & Yang, W. (2008). Free energies of chemical reactions in solution and in enzymes with ab initio quantum mechanics/molecular mechanics methods. *Annual Review of Physical Chemistry, 59*, 573–601.

Hub, J. S., de Groot, B. L., Grubmüller, H., & Groenhof, G. (2013). Quantifying artifacts in Ewald simulations of inhomogeneous systems with a net charge. *Journal of Chemical Theory and Computation, 10*(1), 381–390.

Hwang, J. K., Chu, Z. T., Yadav, A., & Warshel, A. (1991). Simulations of quantum mechanical corrections for rate constants of hydride-transfer reactions in enzymes and solutions. *Journal of Physical Chemistry, 95*(22), 8445–8448.

Hwang, J. K., & Warshel, A. (1993). A quantized classical path approach for calculations of quantum mechanical rate constants. *Journal of Physical Chemistry, 97*, 10053–10058.

Hwang, J.-K., & Warshel, A. (1996). How important are quantum mechanical nuclear motions in enzyme catalysis? *Journal of the American Chemical Society, 118*(47), 11745–11751.
Ishida, T. (2010). Effects of point mutation on enzyme activity: Correlation between protein electronic structure and motion in chorismate mutase reaction. *Journal of the American Chemical Society, 132*, 7104–7118.
Isom, D. G., Castaneda, C. A., Cannon, B. R., & Garcia-Moreno, B. (2011). Large shifts in pKa values of lysine residues buried inside a protein. *Proceedings of the National Academy of Sciences of the United States of America, 108*, 5260–5265.
Jiang, L., Althoff, E. A., Clemente, F. R., Doyle, L., Röthlisberger, D., Zanghellini, A., et al. (2008). De novo computational design of retro-aldol enzymes. *Science, 319*, 1387–1391.
Jorgensen, W. L., Chandrasekhar, J., Madura, J. D., Impey, R. W., & Klein, M. L. (1983). Comparison of simple potential functions for simulating liquid water. *The Journal of Chemical Physics, 79*(2), 926–935.
Jorgensen, W. L., & Madura, J. D. (1985). Temperature and size dependence for Monte Carlo simulations of TIP4P water. *Molecular Physics, 56*, 1381–1392.
Jorgensen, W. L., Maxwell, D. S., & Tirado-Rives, J. (1996). Development and testing of the OPLS all-atom force field on conformational energetics and properties of organic liquids. *Journal of the American Chemical Society, 118*(45), 11225–11236.
Jorgensen, W. L., & Tirado-Rives, J. (1988). The OPLS force field for proteins and energy minimizations for crystals of cyclic peptides and crambin. *Journal of the American Chemical Society, 110*, 1657–1666.
Jorgensen, W. L., & Tirado-Rives, J. (2005). Potential energy functions for atomic-level simulations of water and organic and biomolecular systems. *Proceedings of the National Academy of Sciences of the United States of America, 102*, 6665–6670.
Kamerlin, S. C. L., & Warshel, A. (2011). The empirical valence bond model: Theory and applications. *WIREs Computational Molecular Science, 1*, 30–45.
Kastenholz, M. A., & Hünenberger, P. H. (2003). Influence of artificial periodicity and ionic strength in molecular dynamics simulations of charged biomolecules employing lattice-sum methods. *The Journal of Physical Chemistry. B, 108*(2), 774–788.
Keedy, D. A., van den Bedem, H., Sivak, D. A., Petsko, G. A., Ringe, D., Wilson, M. A., et al. (2014). Crystal cryocooling distorts conformational heterogeneity in a model Michaelis complex of DHFR. *Structure, 22*, 899–910.
Kulik, H. J., Luehr, N., Ufimtsev, I. S., & Martinez, T. J. (2012). Ab initio quantum chemistry for protein structures. *The Journal of Physical Chemistry B, 116*(41), 12501–12509.
Kulik, H. J., Zhang, J., Klinman, J. P., & Martinez, T. J. (2015). *How large should the QM region be in QM/MM calculations? The case of catechol O-methyltransferase.* arXiv:1505.05730v1 [q-bio.BM].
Kumar, S., Rosenberg, J. M., Bouzida, D., Swendsen, R. H., & Kollman, P. A. (1992). The weighted histogram analysis method for free-energy calculations on biomolecules. I. The method. *Journal of Computational Chemistry, 13*(8), 1011–1021.
Kuwajima, S., & Warshel, A. (1988). The extended Ewald method: A general treatment of long-range electrostatic interactions in microscopic simulations. *The Journal of Chemical Physics, 89*(6), 3751–3759.
Layfield, J. P., & Hammes-Schiffer, S. (2014). Hydrogen tunneling in enzymes and biomimetic models. *Chemical Reviews, 114*, 3466–3494.
Lee, F. S., Chu, Z. T., & Warshel, A. (1993). Microscopic and semimicroscopic calculations of electrostatic energies in proteins by the POLARIS and ENZYMIX programs. *Journal of Computational Chemistry, 14*, 161–185.
Lever, G., Cole, D. J., Lonsdale, R., Ranaghan, K. E., Wales, D. J., Mulholland, A. J., et al. (2014). Large-scale density functional theory transition state searching in enzymes. *The Journal of Physical Chemistry Letters, 5*, 3614–3619.

Liao, R.-Z., & Thiel, W. (2013). Convergence in the QM-only and QM/MM modeling of enzymatic reactions: A case study for acetylene hydratase. *Journal of Computational Chemistry, 34*, 2389–2397.

Lin, Y.-l., & Gao, J. (2011). Kinetic isotope effects of L-dopa decarboxylase. *Journal of the American Chemical Society, 133*(12), 4398–4403.

Lindorff-Larsen, K., Maragakis, P., Piana, S., Eastwood, M. P., Dror, R. O., & Shaw, D. E. (2012). Systematic validation of protein force fields against experimental data. *PloS One, 7*, e32131–e32136.

Liu, C. T., Francis, K., Layfield, J. P., Huang, X., Hammes-Schiffer, S., Kohen, A., et al. (2014). Escherichia coli dihydrofolate reductase catalyzed proton and hydride transfers: Temporal order and the roles of Asp27 and Tyr100. *Proceedings of the National Academy of Sciences of the United States of America, 111*, 18231–18236.

Liu, H., & Warshel, A. (2007). Origin of the temperature dependence of isotope effects in enzymatic reactions: The case of dihydrofolate reductase. *Journal of Physical Chemistry B, 111*, 7852–7861.

Liu, H., & Warshel, A. (2009). Tunnelling does not contribute significantly to enzyme catalysis, but studying temperature dependence of isotope effects is useful. In R. K. Allemann & N. S. Scrutton (Eds.), *Quantum tunnelling in enzyme-catalysed reactions* (pp. 242–267): Cambridge, UK: RSC Publishing.

Lonsdale, R., Harvey, J. N., & Mulholland, A. J. (2012). Effects of dispersion in density functional based quantum mechanical/molecular mechanical calculations on cytochrome P450 catalyzed reactions. *Journal of Chemical Theory and Computation, 8*, 4637–4645.

Luk, L. Y. P., Ruiz-Pernía, J. J., Dawson, W. M., Roca, M., Loveridge, E. J., Glowacki, D. R., et al. (2013). Unraveling the role of protein dynamics in dihydrofolate reductase catalysis. *Proceedings of the National Academy of Sciences of the United States of America, 110*, 16344–16349.

MacKerell, A. D., Jr., Feig, M., & Brooks, C. L., III. (2004). Extending the treatment of backbone energetics in protein force fields: Limitations of gas-phase quantum mechanics in reproducing protein conformational distributions in molecular dynamics simulations. *Journal of Computational Chemistry, 25*(11), 1400–1415.

Mahoney, M. W., & Jorgensen, W. L. (2000). A five-site model for liquid water and the reproduction of the density anomaly by rigid, nonpolarizable potential functions. *The Journal of Chemical Physics, 112*, 8910–8922.

Major, D. T., & Gao, J. L. (2005). Implementation of the bisection sampling method in path integral simulations. *Journal of Molecular Graphics and Modelling, 24*(2), 121–127.

Major, D. T., & Gao, J. L. (2006). A combined quantum mechanical and molecular mechanical study of the reaction mechanism and α-amino acidity in alanine racemase. *Journal of the American Chemical Society, 128*(50), 16345–16357.

Major, D. T., & Gao, J. L. (2007). An integrated path integral and free-energy perturbation-umbrella sampling method for computing kinetic isotope effects of chemical reactions in solution and in enzymes. *Journal of Chemical Theory and Computation, 3*(3), 949–960.

Major, D. T., Garcia-Viloca, M., & Gao, J. L. (2006). Path integral simulations of proton transfer reactions in aqueous solution using combined QM/MM potentials. *Journal of Chemical Theory and Computation, 2*(2), 236–245.

Major, D. T., Heroux, A., Orville, A. M., Valley, M. P., Fitzpatrick, P. F., & Gao, J. (2009). Differential quantum tunneling contributions in nitroalkane oxidase catalyzed and the uncatalyzed proton transfer reaction. *Proceedings of the National Academy of Sciences of the United States of America, 106*(49), 20734–20739.

Major, D. T., Nam, K., & Gao, J. (2006). Transition state stabilization and alpha-amino carbon acidity in alanine racemase. *Journal of the American Chemical Society, 128*(25), 8114–8115.

Major, D. T., & Weitman, M. (2012). Electrostatically guided dynamics—The root of fidelity in a promiscuous terpene synthase? *Journal of the American Chemical Society, 134*(47), 19454–19462.

Major, D. T., York, D. M., & Gao, J. (2005). Solvent polarization and kinetic isotope effects in nitroethane deprotonation and implications to the nitroalkane oxidase reaction. *Journal of the American Chemical Society, 127*(47), 16374–16375.

Mavri, J., Liu, H., Olsson, M. H. M., & Warshel, A. (2008). Simulation of tunneling in enzyme catalysis by combining a biased propagation approach and the quantum classical path method: Application to lipoxygenase. *Journal of Physical Chemistry B, 112*(19), 5950–5954.

McCammon, J. A., & Karplus, M. (1979). Dynamics of activated processes in globular proteins. *Proceedings of the National Academy of Sciences of the United States of America, 76*, 3585–3589.

Mezei, M. (1987). Adaptive umbrella sampling: Self-consistent determination of the non-Boltzmann bias. *Journal of Computational Physics, 68*, 237–248.

Monari, A., Rivail, J.-L., & Assfeld, X. (2013). Theoretical modeling of large molecular systems. Advances in the local self consistent field method for mixed quantum mechanics/molecular mechanics calculations. *Accounts of Chemical Research, 46*(2), 596–603.

Morris, G. M., & Lim-Wilby, M. (2008). Molecular docking. *Methods in Molecular Biology, 443*, 365–382.

Nam, K., Gao, J., & York, D. M. (2004). An efficient linear-scaling Ewald method for long-range electrostatic interactions in combined QM/MM calculations. *Journal of Chemical Theory and Computation, 1*(1), 2–13.

Nilsson, L., Clore, G. M., Gronenborn, A. M., Brunger, A. T., & Karplus, M. (1986). Structure refinement of oligonucleotides by molecular dynamics with nuclear Overhauser effect interproton distance restraints: Application to 5' d(C-G-T-A-C-G)$_2$. *Journal of Molecular Biology, 188*(3), 455–475.

Northrup, S. H., Pear, M. R., Lee, C. Y., McCammon, J. A., & Karplus, M. (1982). Dynamical theory of activated processes in globular proteins. *Proceedings of the National Academy of Sciences of the United States of America, 79*(13), 4035–4039.

Olsson, M. H. M., Siegbahn, P. E. M., & Warshel, A. (2004a). Simulating large nuclear quantum mechanical corrections in hydrogen atom transfer reactions in metalloenzymes. *Journal of Biological Inorganic Chemistry, 9*(1), 96–99.

Olsson, M. H. M., Siegbahn, P. E. M., & Warshel, A. (2004b). Simulations of the large kinetic isotope effect and the temperature dependence of the hydrogen atom transfer in lipoxygenase. *Journal of the American Chemical Society, 126*(9), 2820–2828.

Oostenbrink, C., Villa, A., Mark, A. E., & Gunsteren, W. F. V. (2005). A biomolecular force field based on the free enthalpy of hydration and solvation: The GROMOS force-field parameter sets 53A5 and 53A6. *Journal of Computational Chemistry, 25*, 1656–1676.

Piana, S., Klepeis, J. L., & Shaw, D. E. (2014). Assessing the accuracy of physical models used in protein-folding simulations: Quantitative evidence from long molecular dynamics simulations. *Current Opinion in Structural Biology, 24*, 98–105.

Pu, J., Gao, J., & Truhlar, D. G. (2005). Generalized hybrid-orbital method for combining density functional theory with molecular mechanicals. *Chemphyschem, 6*, 1853–1865.

Pu, J., Gao, J., & Truhlar, D. G. (2006). Multidimensional tunneling, recrossing, and the transmission coefficient for enzymatic reactions. *Chemical Reviews, 106*(8), 3140–3169.

Ren, P., & Ponder, J. W. (2003). Polarizable atomic multipole water model for molecular mechanics simulation. *The Journal of Physical Chemistry. B, 107*, 5933–5947.

Reuter, N., Dejaegere, A., Maigret, B., & Karplus, M. (2000). Frontier bonds in QM/MM methods: A comparison of different approaches. *The Journal of Physical Chemistry A, 104*, 1720–1735.

Rohrdanz, M. A., Zheng, W., & Clementi, C. (2013). Discovering mountain passes via torchlight: Methods for the definition of reaction coordinates and pathways in complex macromolecular reactions. *Annual Review of Physical Chemistry*, *64*, 295–316.

Ross, G. A., Bodnarchuk, M. S., & Essex, J. W. (2015). Water sites, networks, and free energies with grand canonical Monte Carlo. *Journal of the American Chemical Society*, *137*, 14930–14943.

Rossi, I., & Truhlar, D. G. (1995). Parameterization of NDDO wavefunctions using genetic algorithms. An evolutionary approach to parameterizing potential energy surfaces and direct dynamics calculations for organic reactions. *Chemical Physics Letters*, *233*(3), 231–236.

Rubinstein, A., & Major, D. T. (2009). Catalyzing racemizations in the absence of a cofactor: The reaction mechanism in proline racemase. *Journal of the American Chemical Society*, *131*, 8513–8521.

Rubinstein, A., & Major, D. T. (2010). Understanding catalytic specificity in alanine racemase from quantum mechanical and molecular mechanical simulations of the arginine 219 mutant. *Biochemistry*, *49*(18), 3957–3964.

Sadeghian, K., Flaig, D., Blank, I. D., Schneider, S., Strasser, R., Stathis, D., et al. (2014). Ribose-protonated DNA base excision repair: A combined theoretical and experimental study. *Angewandte Chemie (International ed. in English)*, *53*, 10044–10048.

Schaefer, M., & Karplus, M. (1996). A comprehensive analytical treatment of continuum electrostatics. *The Journal of Physical Chemistry*, *100*(5), 1578–1599.

Schenter, G. K., Garret, B. C., & Truhlar, D. G. (2003). Generalized transition state theory in terms of the potential of mean force. *The Journal of Chemical Physics*, *119*, 5828–5833.

Schramm, V. L. (2011). Enzymatic transition states, transition-state analogues, dynamics, and lifetimes. *Annual Review of Biochemistry*, *80*, 703–732.

Schrepfer, P., Buettner, A., Goerner, C., Hertel, M., van Rijn, J., Wallrapp, F., et al. (2016). Identification of amino acid networks governing catalysis in the closed complex of class I terpene synthases. *Proceedings of the National Academy of Sciences of the United States of America*, *113*, E958–E967.

Schwartz, S. D., & Schramm, V. L. (2009). Enzymatic transition states and dynamic motion in barrier crossing. *Nature Chemical Biology*, *5*(8), 551–558.

Senn, H. M., & Thiel, W. (2009). QM/MM methods for biomolecular systems. *Angewandte Chemie (International Ed. in English)*, *48*(7), 1198–1229.

Sherwood, D., & Cooper, J. (2015). *Crystals, X-rays and proteins. Comprehensive protein crystallography*. Oxford, UK: Oxford University Press.

Stillinger, F. H., & Rahman, A. (1974). Improved simulation of liquid water by molecular dynamics. *The Journal of Chemical Physics*, *60*, 1545–1557.

Sun, Q., & Chan, G. K.-L. (2014). Exact and optimal quantum mechanics/molecular mechanics boundaries. *Journal of Chemical Theory and Computation*, *10*, 3784–3790.

ten Brink, T., & Exner, T. (2009). Influence of protonation, tautomeric, and stereoisomeric states on protein-ligand docking results. *Journal of Chemical Information and Modeling*, *49*, 1535–1546.

Torrie, G. M., & Valleau, J. P. (1977). Nonphysical sampling distributions in Monte Carlo free-energy estimation: Umbrella sampling. *Journal of Computational Physics*, *23*(2), 187–199.

van den Bedem, H., & Fraser, J. S. (2015). Integrative, dynamic structural biology at atomic resolution—It's about time. *Nature Methods*, *12*, 307–318.

van der Kamp, M. W., Chaudret, R. F., & Mulholland, A. J. (2013). QM/MM modelling of ketosteroid isomerase reactivity indicates that active site closure is integral to catalysis. *FEBS Journal*, *280*(13), 3120–3131.

van der Kamp, M. W., & Mulholland, A. J. (2013). Combined quantum mechanics/molecular mechanics (QM/MM) methods in computational enzymology. *Biochemistry, 52*(16), 2708–2728.

Vardi-Kilshtain, A., Doron, D., & Major, D. T. (2013). Quantum and classical simulations of orotidine monophosphate decarboxylase: Support for a direct decarboxylation mechanism. *Biochemistry, 52*(25), 4382–4390.

Vardi-Kilshtain, A., Major, D. T., Kohen, A., Engel, H., & Doron, D. (2012). Hybrid quantum and classical simulations of the formate dehydrogenase catalyzed hydride transfer reaction on an accurate semiempirical potential energy surface. *Journal of Chemical Theory and Computation, 8*(11), 4786–4796.

Vardi-Kilshtain, A., Nitoker, N., & Major, D. T. (2015). Nuclear quantum effects and kinetic isotope effects in enzyme reactions. *Archives of Biochemistry and Biophysics, 582*, 18–27.

Vega, C., Abascal, J. L. F., Conde, M. M., & Aragones, J. L. (2009). What ice can teach us about water interactions: A critical comparison of the performance of different water models. *Faraday Discussions, 141*, 251–276.

Wall, M. E., Benschoten, A. H. V., Sauter, N. K., Adams, P. D., Fraser, J. S., & Terwilliger, T. C. (2014). Conformational dynamics of a crystalline protein from microsecond-scale molecular dynamics simulations and diffuse X-ray scattering. *Proceedings of the National Academy of Sciences of the United States of America, 111*, 17887–17892.

Wang, Y., & Gao, J. (2015). Projected hybrid orbitals: A general QM/MM method. *The Journal of Physical Chemistry. B, 119*, 1213–1224.

Wang, Q., & Hammes-Schiffer, S. (2006). Hybrid quantum/classical path integral approach for simulation of hydrogen transfer reactions in enzymes. *Journal of Chemical Physics, 125*, 184102–184111.

Ward, A. B., Sali, A., & Wilson, I. A. (2013). Integrative structural biology. *Science, 339*, 913–915.

Warshel, A. (1991). *Computer modeling of chemical reactions in enzymes and solutions*. New York: John Wiley & Sons.

Warshel, A., & Bromberg, A. (1970). Oxidation of 4a,4b-dihydrophenanthrenes. III. A theoretical study of the large kinetic isotope effect of deuterium in the initiation step of the thermal reaction with oxygen. *The Journal of Chemical Physics, 52*, 1262–1269.

Warshel, A., & Karplus, M. (1972). Calculation of ground and excited state potential surfaces of conjugated molecules. I. Formulation and parametrization. *Journal of the American Chemical Society, 94*(16), 5612–5625.

Warshel, A., Kato, M., & Pisliakov, A. V. (2007). Polarizable force fields: History, test cases, and prospects. *Journal of Chemical Theory and Computation, 3*(6), 2034–2045.

Warshel, A., & Levitt, M. (1976). Theoretical studies of enzymic reactions: Dielectric, electrostatic and steric stabilization of the carbonium ion in the reaction of lysozyme. *Journal of Molecular Biology, 103*(2), 227–249.

Warshel, A., & Weiss, R. M. (1980). An empirical valence bond approach for comparing reactions in solutions and in enzymes. *Journal of the American Chemical Society, 102*, 6218–6226.

Watt, E. D., Shimada, H., Kovrigin, E. L., & Loria, J. P. (2007). The mechanism of rate-limiting motions in enzyme function. *Proceedings of the National Academy of Sciences of the United States of America, 104*, 11981–11986.

Weitman, M., & Major, D. T. (2010). Challenges posed to bornyl diphosphate synthase: Diverging reaction mechanisms in monoterpenes. *Journal of the American Chemical Society, 132*(18), 6349–6360.

Wolfenden, R., & Snider, M. J. (2001). The depth of chemical time and the power of enzymes as catalysts. *Accounts of Chemical Research, 34*, 938–945.

Word, J. M., Lovell, S. C., Richardson, J. S., & Richardson, D. C. (1999). Asparagine and glutamine: Using hydrogen atom contacts in the choice of side-chain amide orientation. *Journal of Molecular Biology, 285*, 1735–1747.

Xie, W., Orozco, M., Truhlar, D. G., & Gao, J. (2009). X-Pol potential: An electronic structure-based force field for molecular dynamics simulation of a solvated protein in water. *Journal of Chemical Theory and Computation, 5*(3), 459–467.

Zhang, J., Kulik, H. J., Martinez, T. J., & Klinman, J. P. (2015). Mediation of donor-acceptor distance in an enzymatic methyl transfer reaction. *Proceedings of the National Academy of Sciences of the United States of America, 112*, 7954–7959.

Zhang, Y., Lee, T.-S., & Yang, W. (1999). A pseudobond approach to combining quantum mechanical and molecular mechanical methods. *The Journal of Chemical Physics, 110*(1), 46–54.

Zhao, Y., & Truhlar, D. G. (2008). The M06 suite of density functionals for main group thermochemistry, thermochemical kinetics, noncovalent interactions, excited states, and transition elements: Two new functionals and systematic testing of four M06-class functionals and 12 other functionals. *Theoretical Chemistry Accounts, 120*(1–3), 215–241.

Zienau, J., & Cui, Q. (2012). Implementation of the solvent macromolecule boundary potential and application to model and realistic enzyme systems. *The Journal of Physical Chemistry. B, 116*, 12522–12534.

Zwanzig, R. W. (1954). High-temperature equation of state by a perturbation method. I. Nonpolar gases. *The Journal of Chemical Physics, 22*(8), 1420–1426.

CHAPTER ELEVEN

Examinations of the Chemical Step in Enzyme Catalysis

P. Singh[1], Z. Islam[1], A. Kohen[2]

University of Iowa, Iowa City, IA, United States
[2]Corresponding author: e-mail address: amnon-kohen@uiowa.edu

Contents

1. Introduction	288
2. KIE_{obs} vs KIE_{int}	290
3. The Northrop Method	292
4. Case Study 1: TSase	296
4.1 Experimental Section for TSase	301
5. Case Study 2: DHFR	305
5.1 Computational vs Experimental Studies	306
5.2 Experimental Section for DHFR	311
6. Summary	312
Acknowledgments	313
References	313

Abstract

Advances in computational and experimental methods in enzymology have aided comprehension of enzyme-catalyzed chemical reactions. The main difficulty in comparing computational findings to rate measurements is that the first examines a single energy barrier, while the second frequently reflects a combination of many microscopic barriers. We present here intrinsic kinetic isotope effects and their temperature dependence as a useful experimental probe of a single chemical step in a complex kinetic cascade. Computational predictions are tested by this method for two model enzymes: dihydrofolate reductase and thymidylate synthase. The description highlights the significance of collaboration between experimentalists and theoreticians to develop a better understanding of enzyme-catalyzed chemical conversions.

---

[1] These authors contributed equally to this work.

1. INTRODUCTION

Empirical studies are invaluable in understanding many kinetic, mechanistic, and structural features of enzyme catalysis. However, many microscopic aspects of catalysis, such as transition state (TS) structures, reactive intermediate constituents, and dynamic effects, should also be examined by computational methods, which complement experiments and can provide atomistic insights. One of the main challenges in comparing experimental studies to calculations is ensuring that the two approaches actually examine the same phenomenon. When the experiments involve enzyme kinetics, this task is not trivial: most kinetic measurements reflect a complex combination of microscopic events, while most computations address a single event, such as a single energy barrier crossing.

Enzymes such as dihydrofolate reductase (DHFR) and thymidylate synthase (TSase) catalyze the breaking and formation of carbon–hydrogen bonds; several computational methods, including molecular dynamics (MD) and hybrid quantum mechanics/molecular mechanics (QM/MM), have been used in the study of these catalytic events. A partial list of such studies includes Dametto, Antoniou, and Schwartz (2012), Hammes-Schiffer and Benkovic (2006), Klinman and Kohen (2013), Liu and Warshel (2007a), Luk et al. (2013), Rod, Radkiewicz, and Brooks (2003), Thorpe and Brooks (2004), Wang, Ferrer, Moliner, and Kohen (2013), and Wong, Selzer, Benkovic, and Hammes-Schiffer (2005). Each of these computational studies examined barrier crossing for a specific H-transfer reaction and predicted the involvement of several protein residues in facilitating C–H bond cleavage. A means of experimental examination of those predictions is the focus of this chapter. We describe several experimental probes of specific reactions' barriers including measurements of rate constants, their kinetic isotope effects (KIEs), and their temperature dependence. The comparison of experimental observations with computational predictions of individual catalytic events is also discussed.

A KIE is the ratio of reaction rates for two substrates that differ only in their isotopic composition (isotopologues). Please note that this chapter refers only to primary (1°) KIEs, ie, isotopic effects in which the isotope under study is part of the bond being cleaved or formed. When only the hydrogen on the C–H bond under study is isotopically labeled, KIEs offer an important means of probing activation of that bond. Isotopic labeling of the C–H that is cleaved in the reaction has little or no effect on other physical and chemical events along the catalytic cycle. This assumption follows that the C–H bond is not polarized and thus contributes little to the binding of large biological molecules, such as DHFR's and TSase's reactants.

While experimental measurements always yield an observed KIE (KIE_{obs}), the KIE for the bond cleavage per se is called the intrinsic KIE (KIE_{int}). By this definition KIE_{int} is free from kinetic complexities arising from the inclusion of other kinetic steps such as substrate binding, product release, or other chemical steps along the catalytic cascade that affect KIE_{obs}. Consequently, experimental evaluation of KIE_{int} (but not KIE_{obs}) can parallel the computational method in the sense that each examines a single chemical step.

We find that it is possible to use KIE_{int} to examine predictions of high-level computations by exploiting the temperature dependence of KIE_{int}. This can be used to probe the nature of C–H bond cleavage and subnanosecond protein motions coupled to the C–H bond cleavage (Klinman & Kohen, 2013; Kohen, 2015; Roston, Islam, & Kohen, 2014). An empirical interpretation of the temperature dependence of KIE_{int} is offered by "Activated Tunneling" models as described in the literature (Fan, Cembran, Ma, & Gao, 2013; Klinman, 2015; Klinman & Kohen, 2013; Kohen, 2015; Layfield & Hammes-Schiffer, 2013; Liu & Warshel, 2007a,2007b; Marcus, 2007; Nagel & Klinman, 2010; Pu, Gao, & Truhlar, 2006; Pudney et al., 2009; Roston, Islam, & Kohen, 2013). In the framework of these models small or no temperature dependence of KIE arise primarily from high frequency H-donor and H-acceptor distance (DAD) sampling, ie, a narrow distribution of DADs. Temperature-dependent KIE, on the other hand, stems from low-frequency DAD sampling (a broad distribution of DADs) at the TS. A high DAD sampling frequency indicates a well-organized TS, while a low DAD sampling frequency demonstrates a loose TS. A perturbation of the TS (for instance, through the mutation of a residue that affects the TS) alters the frequency of DAD sampling, and this manifests in an altered KIE temperature dependence. The temperature dependence of KIE_{int}, in other words, reflects the nature of a single bond cleavage step (within the complex kinetic cascade of an enzymatic reaction), which allows experimental comparisons to relevant computations.

That said, extracting the value of KIE_{int} from measured KIE_{obs} is a considerable challenge. While the Northrop method described below can be used with any KIE_{obs} (from single turnover, burst experiments, steady-state parameters, etc.), we recommend measuring KIE_{obs} on the second-order rate constant V/K competitively. The main advantage here is that only trace labeling with radioactive tritium (T) is needed in competitive experiments in which two isotopically labeled substrates compete for the enzyme at the same pot. Most other methods would require close to 100% tritium in the bond to be cleaved, and close to no tritium in any other positions on the substrate, which is experimentally challenging.

In the sections below we describe the process of extracting KIE_{int} from the observed value (KIE_{obs}) using the Northrop method then turn to examples of computational studies carried out on TSase and DHFR, followed by tests of these computational predictions with studies of temperature dependence of KIE_{int}. Finally, each case study includes the experimental procedure used to study a specific C–H bond activation.

2. KIE_{obs} VS KIE_{int}

Every experimental measurement of a KIE reflects a kinetically complex cascade rather than a single microscopic step. This kinetically complex chain of events may include binding and release of substrates and products, conformational changes associated with ligand binding or release, reprotonation of many enzymatic residues and the substrates, multiple chemical conversions within the reactive complex, etc. This problem is illustrated in Fig. 1 and shows that steps other than the chemical step of interest (eg, the C–H→C hydride transfer in the cases presented below) are partly rate-limiting for the different measurable kinetic parameters. This demonstrates how difficult it is to determine the rate constant or KIE_{int} on the intended chemical step per se, either from steady-state kinetics (eg, k_{cat} or k_{cat}/K_M) or from presteady-state kinetics (eg, single turnover or burst studies).

The KIE_{int} in Fig. 1 results from differences between the zero point energies of the ground state and the TS, and can be also affected by nuclear-quantum-mechanical tunneling (a phenomenon in which the atom is transferred under the classical energy barrier via its wave-like properties). These considerations are a matter of computational studies, but experimental information reflecting the cleavage of a single step is needed in order to compare the two. The discrepancy between calculated and measured rates has led to some misunderstandings in the past: for example, the catalytic turnover rate, k_{cat}, is the unimolecular rate constant representing the reciprocal of the time for a single enzymatic turnover (Cook & Cleland, 2007). However, since most enzymes did not evolve under saturating substrate concentrations, k_{cat} is often rate-limited by product release; thus comparing k_{cat}'s rates

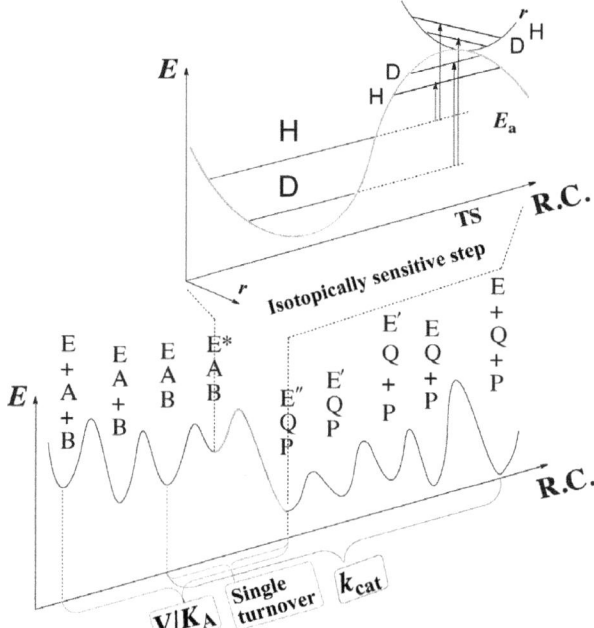

Fig. 1 Relations between the chemical step (*red (gray in the print version)*) and measurable rate constants. Both steady-state parameters (eg, k_{cat}/K_M (V/K) and k_{cat}) and presteady-state rates (eg, single-turnover rate) involve several microscopic rate constants. The KIE_{int} reflects the nature of the chemical step, but assessing KIE_{int} from observed values is quite challenging (see text). The "asterisk" represents a reactive ternary complex. *Reproduced from Kohen, A. (2015). Dihydrofolate reductase as a model for studies of enzyme dynamics and catalysis. F1000Research, 4, 1464, with permission from Faculty 1000.*

or KIEs with calculations examining merely the chemical step is not always meaningful.

The relations between the KIE_{int} of interest and the relevant measured KIE_{obs} and computational studies can be calculated per system. The mathematical expression for this relation always has a similar general form, although the specific expressions vary depending on the details of each system. For example, consider the following simple kinetic cascade, with only one irreversible chemical step and a single substrate:

$$E + S \underset{k_2}{\overset{k_1}{\rightleftharpoons}} ES \overset{k_3}{\rightarrow} EP \overset{k_4}{\rightarrow} E + P$$

Here the chemical step is the conversion of ES to EP, represented by the rate constant k_3. $KIE_{int} = k_{3\ light}/k_{3\ heavy}$ or, in the case of H/D KIE,

$^D k_3 = k_{3H}/k_{3D}$. As expressed in Eq. (1), the observed 1° KIE will be smaller than KIE_{int} by an amount that depends on commitment to catalysis. The KIE_{obs} on any steady-state rate constant is affected by commitments as follows (Cook & Cleland, 2007):

$$KIE_{obs} = \frac{KIE_{int} + C_f + C_r \cdot EIE}{1 + C_f + C_r} \quad (1)$$

where EIE is the equilibrium isotope effect. The forward commitment (C_f) is defined as the ratio of the isotopically sensitive rate constant (k_3 in this case) in the forward direction to the net rate constant for the breakdown of the reactive ES complex into E and S (k_2 in this case). The reverse commitment (C_r) is the ratio of the isotopically sensitive rate constant in the reverse direction (from EP to ES, zero in the current example) to the isotopically insensitive net rate constant of the decomposition of the EP complex to E and P (k_4). In cases where the chemistry is irreversible or cases where EIE = 1, Eq. (1) simplifies to:

$$KIE_{obs} = \frac{KIE_{int} + C}{1 + C} \quad (2)$$

where $C = C_f$ for irreversible reactions or $C = C_f + C_r$ when EIE = 1. In cases where C is much smaller than KIE_{int}, KIE_{obs} is similar to KIE_{int}. KIE_{int} values can then be directly compared to a computed barrier crossing event, the assessment of which is the focus of this chapter.

3. THE NORTHROP METHOD

There are several experimental methods that attempt to assess the nature of individual chemical steps: for instance, measurement of single-turnover rates as a function of pH (Fierke, Johnson, & Benkovic, 1987), or measurement of KIE_{obs} as function of substrate concentration (Hong, Maley, & Kohen, 2007). Measuring single-turnover rates can dramatically reduce the kinetic complexity by reducing the number of microscopic steps involved in the measurement, but as the initiation of the reaction involves substrate binding, and there is still a $C > 0$ that results from protonation changes and the conformational steps that follow that substrate binding, and of course from the chemical steps other than the one under study, in systems catalyzing sequential chemical conversions (eg, see Section 4).

The Northrop method eliminates C by making use of measurements of KIE_{obs} for all three hydrogen isotopes (Northrop, 1991). The KIE_{obs}

acquired for this method can be measured for single-turnover rates, steady-state rate constants, or other kinetically well-defined parameters. For example, for the simplified kinetic scheme presented earlier the expression for the second-order rate constant (k_{cat}/K_M, denoted as V/K) is given by:

$$(V/K)_i = \frac{k_1 k_{3i}}{k_2 + k_{3i}} \quad (3)$$

where i represents a given isotopologue, ie, a substrate with hydrogen (H), deuterium (D), or tritium (T). For H-transfer reactions, since there are only three possible isotopologues, every KIE_{obs} has an isotope in common with one other, eg, H/D KIE and H/T KIE share H as common isotope; H/D KIE and D/T KIE share D, and H/T KIE and D/T KIE share T as common isotope. All three combinations can be used, and each has different advantages and disadvantages as described in Sen, Yahashiri, and Kohen (2011). For each case, one must derive the Northrop equation so the commitment (C) in the KIE_{obs} expression is the C for the shared isotope in each case. For H/T KIE and D/T KIE, where T is the shared isotope, thus C for T must be the only commitment in each expression. In this case, one derives the reciprocal of KIE_{obs} on V/K for H/T, $^H(V/K)_{T,obs}$:

$$^H(V/K)_{T,obs} = \frac{(V/K)_T}{(V/K)_H} = \frac{\dfrac{k_1 k_{3T}}{k_1 + k_{3T}}}{\dfrac{k_1 k_{3H}}{k_2 + k_{3H}}} \quad (4)$$

where $(V/K)_i$ is the rate for isotope i, and $1/KIE_{obs}$ for H/T is $^H(V/K)_{T,obs}$.

$$^H(V/K)_{T,obs} = \frac{k_{3T} k_2 + k_{3T} k_{3H}}{k_{3H} k_2 + k_{3H} k_{3T}} \quad (5)$$

Dividing the numerator and denominator by $k_{3H} k_2$ gives:

$$^H(V/K)_{T,obs} = \frac{\left(\dfrac{k_{3T}}{k_{3H}} + \dfrac{k_{3T}}{k_2}\right)}{\left(1 + \dfrac{k_{3T}}{k_2}\right)} \quad (6)$$

$$^H(V/K)_{T,obs} = \frac{^H(k_3)_T + C_T}{1 + C_T} \quad (7)$$

where $C_T = k_{3T}/k_2$ is the commitment for tritium (k_{3T} is the C–T bond cleavage rate in this simple example), $^H(k_3)_T = k_T/k_H$ is reciprocal of the intrinsic tritium KIE for isotopes H vs T.

Similarly, the reciprocal of the KIE$_{obs}$ on V/K for D vs T is given by:

$$^D(V/K)_{T,obs} = \frac{^D(k_3)_T + C_T}{1 + C_T} \tag{8}$$

where $^D(k_3)_T = k_T/k_D$ is the reciprocal of the KIE$_{int}$ for D/T. Subtracting 1 from both sides in Eqs. (7) and (8) gives:

$$^H(V/K)_{T,obs} - 1 = \frac{^H(k_3)_T + C_T}{1 + C_T} - 1 = \frac{^H(k_3)_T - 1}{1 + C_T} \tag{9}$$

$$^D(V/K)_{T,obs} - 1 = \frac{^D(k_3)_T + C_T}{1 + C_T} - 1 = \frac{^D(k_3)_T - 1}{1 + C_T} \tag{10}$$

Dividing Eq. (9) by Eq. (10) results in an expression with no commitment at all, no matter how complex the expression for C:

$$\frac{^H(V/K)_{T,obs} - 1}{^D(V/K)_{T,obs} - 1} = \frac{^H(k_3)_T - 1}{^D(k_3)_T - 1} \tag{11}$$

Eq. (11) has two unknowns (the KIE$_{int}$ for H/T and D/T), but no commitment (C). The Swain–Schaad relationship for the pairs of hydrogen isotope (between H/T and D/T) is given by (Swain, Stivers, Reuwer, & Schaad, 1958):

$$\frac{k_H}{k_T} = \left(\frac{k_D}{k_T}\right)^{3.3} \tag{12}$$

where the value at the exponent (3.3) can be assessed by QM/MM calculations (like in the case of ecDHFR presented later) or calculated from basic principals as discussed later. Using the reciprocal of this relationship, the $^H(k_3)_T$ can be numerically determined from a pair of KIE$_{obs}$s:

$$\frac{^H(V/K)_{T,obs} - 1}{^D(V/K)_{T,obs} - 1} = \frac{^H(k_3)_T - 1}{^H(k_3)_T^{\frac{1}{3.3}} - 1} \tag{13}$$

Further rearrangement of Eq. (13) leads to:

$$\frac{^T(V/K)_{H,obs}^{-1} - 1}{^T(V/K)_{D,obs}^{-1} - 1} = \frac{^T(k_3)_H^{-1} - 1}{^T(k_3)_H^{\frac{-1}{3.3}} - 1} \tag{14}$$

where $^T(V/K)_{H,obs}$, $^T(V/K)_{D,obs}$, and $^T(k_3)_H$ are observed H/T, observed D/T, and intrinsic H/T KIEs, respectively. This equation only has one unknown, which is the KIE$_{int}$ that represents KIE on *only the bond cleavage event* (k_3).

Thus KIE$_{int}$ is extracted from its respective observed values using a numerical solution of Eq. (14). The equations were numerically solved for using a program that is freely available on our website under tools: https://chem.uiowa.edu/kohen-research-group/calculation-intrinsic-isotope-effects.

Like any other experimental method, the Northrop method makes assumptions, and its associated assumptions have to be carefully tested for each system. Three assumptions are embedded in the Northrop method: (1) KIEs on binding or other kinetic steps (except the one under study) are insignificant; (2) C_r in Eq. (1) is close to zero or EIE is close to unity; and (3) the two KIE$_{int}$ in Eq. (12) are related to each other, for example, via the Swain–Schaad exponent (SSE). The SSE can be calculated using QM/MM or assessed from $SSE = \left(\mu_{C-H}^{-0.5} - \mu_{C-T}^{-0.5}\right) / \left(\mu_{C-D}^{-0.5} - \mu_{C-T}^{-0.5}\right)$ (Swain et al., 1958), where μ_{C-i} is the reduced mass for isotope i and carbon 12. At the high temperature limit (ie, above 273 K), 1° SSE falls between 3.34 and 3.26, as the carbon's mass in the C–H/D/T bond can vary between 12 and infinity, respectively (depending on its coupling to the rest of the heavy atoms). While at temperatures below 270 K SSE can vary, SSE is mostly temperature independent in the range in which most biochemical experiments are conducted (Shelton, Hrovat, & Borden, 2007; Smedarchina & Siebrand, 2005). The value of SSE for specific systems can be calculated from QM/MM calculations with a tunneling correction (Pu, Ma, Gao, & Truhlar, 2005; Pu, Ma, Garcia-Viloca, et al., 2005), eg, 3.3 for the enzyme DHFR.

A recent suggestion that there is another, "hidden" assumption in the Northrop method resulted from a procedure in which C_T was calculated indirectly from $^T(V/K)_H$ and $^T(V/K)_D$ (Wang, Antoniou, Schwartz, & Schramm, 2016), rather than from their reciprocals as used in the Northrop method (Eq. 14). The correct procedure described in Eqs. (3)–(14) can be used for the complex kinetic scheme used in Wang et al. (2016) to calculate C_T directly, without the use of any further assumptions.

4. CASE STUDY 1: TSase

TSase catalyzes the last committed step of the de novo synthesis of the DNA building block thymidylate (2′-deoxythymidine-5′-monophosphate, dTMP). It plays an essential role in regulating the balance of nucleotide pools and the replication of DNA, and thus is a major target for chemotherapeutic drugs (Carosati et al., 2012; Wilson, Danenberg, Johnston, Lenz, & Ladner, 2014). In the TSase-catalyzed production of thymidylate, the cofactor methylene tetrahydrofolate (CH_2H_4folate) sequentially provides a methylene and a hydride to the substrate dUMP (2′-deoxyuridine-5′-monophosphate). Following the formation of the ternary complex of TSase and its substrates, the catalyzed chemistry involves a series of bond cleavages and formations, including two C–H activations: a proton abstraction and a hydride transfer. In the conventionally proposed mechanism of *E. coli* TSase (Carreras & Santi, 1995; Finer-Moore, Santi, & Stroud, 2003; McMurry & Begley, 2016), a highly conserved cysteine (C146) initiates the reaction by Michael addition at C6 of dUMP (step 1 in Scheme 1), forming an enolate. This enolate forms a methylene bridge with the preactivated CH_2H_4folate iminium cation (step 2 and 3). Two H-transfers follow: (i) a proton abstraction from the C5 of dUMP (step 4), leading to the elimination of tetrahydrofolate (H_4folate) and the formation of an exocyclic methylene intermediate (step 4) and (ii) a hydride transfer from the C6 of the H_4folate to the C7 of the exocyclic methylene intermediate to produce the product dTMP (step 5). The traditionally proposed mechanism (left column in Scheme 1) was constructed from numerous experimental studies including X-ray crystallography (Stroud & Finer-Moore, 2003), KIEs (Agrawal, Hong, Mihai, & Kohen, 2004; Spencer, Villafranca, & Appleman, 1997), steady-state kinetics, and mutagenesis (Carreras & Santi, 1995; Finer-Moore et al., 2003).

The mechanism of TSase has also been investigated by hybrid QM/MM calculations, which paint a slightly different picture (Kanaan et al., 2011; Kanaan, Marti, Moliner, & Kohen, 2007, 2009; Kanaan, Roca, Tunon, Marti, & Moliner, 2010; Wang, Ferrer, et al., 2013) (right column in Scheme 1). One of the main differences between the two proposed mechanisms is in the hydride transfer step (step 5 in Scheme 1), which the traditional view (left column) presents as a step-wise mechanism in which the

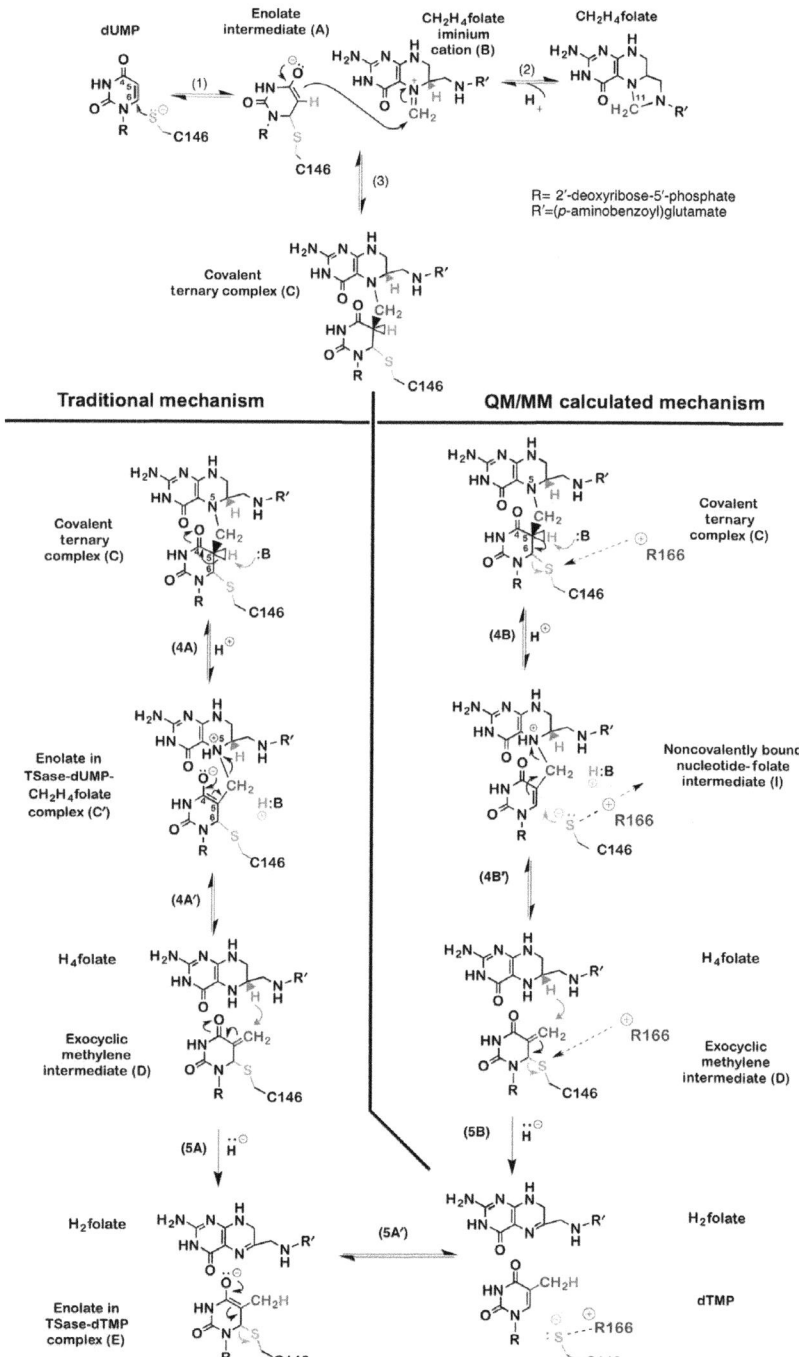

Scheme 1 Chemical mechanism of TSase. Traditional (*left panel*) and QM/MM calculated (*right panel*) mechanisms for proton abstraction and hydride transfer.

hydride transfer precedes the elimination of enzymatic nucleophile C146 (step 5A and 5A′). The QM/MM calculations (right column), on the other hand, suggest a concerted process in which the transfer of hydride and the cleavage of the thioether bond happen together (step 5B). The difference in hydride transfer mechanisms is significant: the concerted mechanism involves the formation of a charged species (compound E in Scheme 1) while the other seems to bypass it, an important consideration for designing a TS analogue. Key to the concerted mechanism's favorability, according to calculations, is a highly conserved, positively charged residue R166 that moves closer to the thioether bond and promotes its cleavage at the TS of the hydride transfer.

The QM/MM calculations also examined the mechanism of the proton abstraction from the C5 of dUMP in the ternary complex (compound C in Scheme 1) (Kanaan et al., 2007, 2009; Wang, Ferrer, et al., 2013). The calculations suggest the formation of an intermediate that had not been proposed in the traditional mechanism. It was calculated that the removal of the C5 proton from intermediate C causes dissociation of the C146, leaving an intermediate in which the folate and the nucleotide are bound via a methylene bridge, but are not covalently bound to the enzyme (intermediate in Scheme 1)! If confirmed experimentally, this previously unrecognized intermediate might serve as a new target for antibiotic and chemotherapeutic drugs. In contrast to current drugs, which are either derivatives of pyrimidine or folate, this hypothetical new class of inhibitors could comprise part of the nucleotide and part of the folate.

These theoretical predictions were examined experimentally: The calculations predicted that, as in the hydride transfer step, R166 also plays a vital role in polarizing and cleaving the thioether bond during proton abstraction (Wang, Ferrer, et al., 2013). A mutation at R166 should have altered the temperature dependence of KIE_{int} only if it was involved in TSs of H-transfers. KIE_{obs}s for H/T and D/T were measured over a range of temperatures for the both the hydride transfer and the proton abstraction in WT and in a variant of R166 (R166K, the only active mutant of R166) (Agrawal, Hong, et al., 2004; Islam, Strutzenberg, Ghosh, & Kohen, 2015; Islam, Strutzenberg, Gurevic, & Kohen, 2014; Wang & Kohen, 2010). KIE_{int}s were extracted from the KIE_{obs}s using the Northrop method described earlier (Section 3) and plotted against the inverse of temperature (Fig. 2).

As is apparent in Fig. 2, the mutation caused an increase in temperature dependence of KIE_{int} for both H-transfers. The KIE_{int} for the hydride transfer, which had been temperature independent in the WT, became

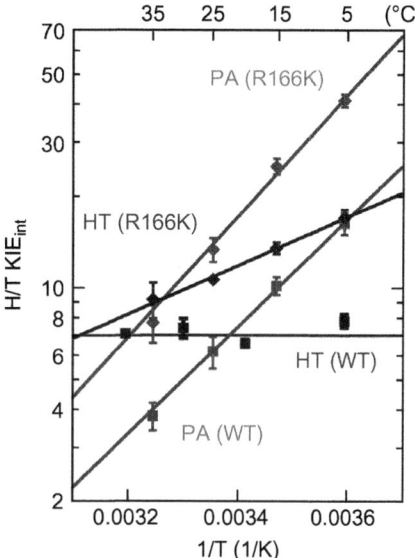

Fig. 2 KIE$_{int}$s on the hydride transfer (*blue (dark gray in the print version)*) and the proton abstraction (*red (gray in the print version)*) for the WT (cubes) and R166K (diamonds). *HT*, hydride transfer, *PA*, proton abstraction. *Reproduced from Islam, Z., Strutzenberg, T. S., Ghosh, A. K., & Kohen, A. (2015). Activation of two sequential H transfers in the thymidylate synthase catalyzed reaction. ACS Catalysis, 5, 6061–6068, with permission from ACS.*

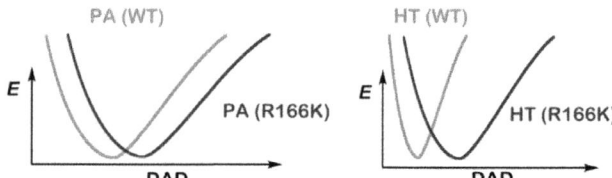

Fig. 3 Schematic representations of PESs for the H-transfers along the DAD coordinate. *Reproduced from Islam, Z., Strutzenberg, T. S., Ghosh, A. K., & Kohen, A. (2015). Activation of two sequential H transfers in the thymidylate synthase catalyzed reaction. ACS Catalysis, 5, 6061–6068, with permission from ACS.*

temperature-dependent in R166K, while the temperature dependence of the proton abstraction KIE increased in R166K (relative to WT). In addition, the mutation also increased the absolute magnitude of KIE$_{int}$ for both H-transfers. Seen in the light of the activated tunneling model, the larger magnitudes and the steeper temperature dependence of KIEs in R166K suggest lower-frequency DAD sampling and a higher average DAD at the TS. Fig. 3 illustrates the DAD sampling potential along the DAD coordinate.

A stiff potential energy surface (PES) represents a high DAD sampling frequency, suggesting a well-coordinated TS—as is seen in the hydride transfer in WT ecTSase—while a broad PES indicates a low DAD sampling frequency, demonstrating a loose TS, as can be seen in the proton abstraction in the wild type. The mutation at R166 caused the PES for the DAD to become wider for the both H-transfers, implying perturbation of the TSs. The mutation also caused a shift in the PESs' minima, yielding larger DADs for both H-transfers. These effects were attributed to R166's being an indispensable part of the reaction's TS for both the hydride transfer and the proton abstraction, in good agreement with the QM/MM calculations. Additionally, an observed normal secondary (2°) KIE on C6 of dUMP supported the concerted mechanism for the hydride transfer (Islam et al., 2014). As a control, mutation to several other residues closer to the reaction center was found to produce no or very minor changes in the temperature dependence of KIE_{int} (Abeysinghe & Kohen, 2015; Agrawal, Hong, et al., 2004; Wang, Abeysinghe, Finer-Moore, Stroud, & Kohen, 2012).

In summary, examining chemical steps such as C–H bond activations in enzyme-catalyzed reactions is of contemporary interest. While computational approaches have advantages and the virtue of examining a single chemical step, experimental tools for such examinations are few. The temperature dependence of KIE_{int} probes the TS of a given chemical transfer, thereby offering a test of computational predictions. As an example, TSase has conventionally been thought to catalyze two sequential H-transfers in a complex cascade of physical and chemical steps; QM/MM calculations predicted new mechanisms for the both hydride transfer and the proton abstraction. In both cases, the experimental findings, using KIE temperature dependence as a probe, offer strong support for the predictions made by QM/MM calculations. Not only was the role of a particular residue, R166, confirmed in both steps, but secondary KIEs (Islam et al., 2014) confirmed the concerted nature of the hydride transfer step proposed by QM/MM calculations.

The existence of the newly predicted intermediate (Wang, Ferrer, et al., 2013) has not yet been directly demonstrated because the involvement of R166 in steps before and after that proposed intermediate is quite indirect. More experiments and computational studies are needed to further examine these proposals, such as QM/MM calculations and temperature-dependence KIE_{int} measurements with TSase from other organisms, and other studies that would further expose the nature of TSase catalysis.

4.1 Experimental Section for TSase

[2-^{14}C]dUMP (specific radioactivity of 52 Ci/mol) was purchased from Moravek Biochemicals (Brea, CA, USA). [5-^{3}H]dUMP (15–30 Ci/mmol) and [^{3}H]NaBH$_4$ (15 Ci/mmol) were purchased from American Radiolabeled Chemicals (St. Louis, MO, USA). [2-^{3}H]isopropanol was synthesized by reducing acetone using [^{3}H]NaBH$_4$. Isopropanol-d8 was purchased from Sigma-Aldrich (St. Louis, MO, USA).

4.1.1 Measurements of KIEs on the Hydride Transfer Step

The upper panel in Scheme 2 shows the isotopic labeling pattern for the hydride-transfer KIE measurements, where the hydrogen at C6 of CH$_2$H$_4$folate (to be transferred at that step) was labeled as deuterium or tritium. The labeled (R)-[6-^{x}H]CH$_2$H$_4$folate (^{x}H=D or T) was synthesized through a chemoenzymatic process described elsewhere (Agrawal, Mihai, & Kohen, 2004). The KIE measurements were performed in 100 mM Tris buffer (pH 7.5), 2 mM TCEP, 1 mM EDTA, and 7 mM HCHO. Both H/T and D/T KIE were measured competitively, yielding a KIE$_{obs}$ on the second-order rate constant (V/K) (Cook & Cleland, 2007).

In competitive KIE measurements, both the light and the heavy isotopologue are present in the same reaction mixture. Here, the ^{14}C-labeled substrate ([2-^{14}C]dUMP) was used to trace the light isotope (H or D). For each reaction, trace (R)-[6-^{3}H]CH$_2$H$_4$folate (about 2 million disintegrations per minute, Mdpm) in protiated or deuterated CH$_2$H$_4$folate for H/T or D/T KIE experiments, respectively, was mixed with ^{14}C-labeled

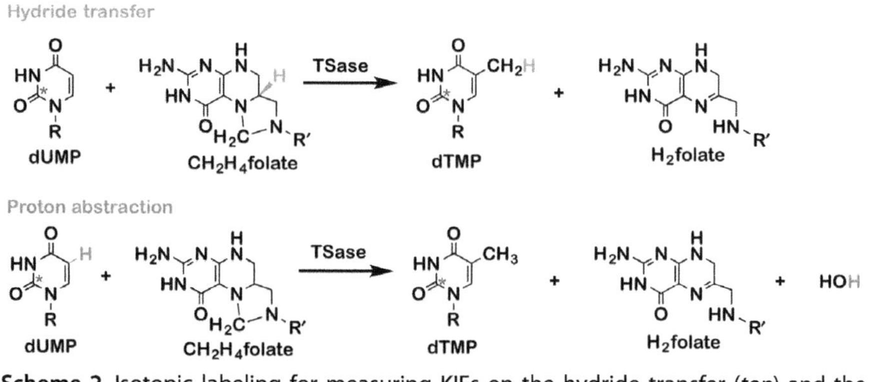

Scheme 2 Isotopic labeling for measuring KIEs on the hydride transfer (*top*) and the proton abstraction (*bottom*). The hydride (*red* (*gray* in the print version)) and the proton (*green* (*gray* in the print version)) are the sites of isotopic labeling. The *asterisks* on dUMP represent remote ^{14}C labeling to trace lighter isotopes.

substrate [2-^{14}C]dUMP (~0.3 Mdpm) in a ratio of 6/1 (^{3}H/^{14}C) or higher. Since tritium radiation is weaker than ^{14}C radiation, this ratio contributed to higher accuracies in the liquid scintillation counting (LSC) analyses. Using unlabeled dUMP, the concentration of [2-^{14}C]dUMP was adjusted to 20% in excess over the total CH$_2$H$_4$folate in order to enable measurement of fractional conversion of dUMP to dTMP (f). Since high CH$_2$H$_4$folate concentration (~500 μM) was previously found to be inhibitory (Agrawal, Hong, et al., 2004; Wang, Sapienza, et al., 2013), the chosen total concentration of dUMP was in the range of 100–400 μM so that with 20% in excess, CH$_2$H$_4$folate stayed below the inhibitory concentration. Aliquots of the reaction mixture were equilibrated to 5°C, 15°C, 25°C, and 35°C and adjusted to pH 7.5. Prior to initiating the reaction by adding enzyme, two aliquots (t_0) that served as controls were also taken and quenched (for consistency). At each temperature, the reaction was initiated with the addition of TSase, and six to eight aliquots of reaction mixture were removed and quenched at different time points (t) leading to fraction conversion between 20% and 80%. All aliquots were quenched with at least five times molar excess of 5-fluoro-dUMP, a nanomolar competitive inhibitor of TSase, over the substrate dUMP, and stored at −80°C for further HPLC analysis. A solution of concentrated enzyme was added to the remaining reaction mixture and incubated further to reach the completion of the reaction, then divided into three aliquots that were quenched (for consistency) and served as infinite time points (t_∞). All of the aliquots (t_0, t, and t_∞) were subjected to RP HPLC analyses in which reactants and products were separated from each other and their radioactivity measured by LSC. The KIE$_{obs}$ for each time point was calculated using Eq. (15):

$$\text{KIE} = \frac{\ln(1-f)}{\ln[1-f(R_t/R_\infty)]} \quad (15)$$

where f, R_t, and R_∞ are the fractional conversion, the ratio of ^{3}H/^{14}C in the product, and the ratio of ^{3}H/^{14}C in the product at the infinity time points, respectively. R_∞ was averaged over the infinite time points. The fractional conversion f for all time points was calculated using the following equation:

$$f = \frac{[^{14}\text{C}]\text{dUMP}}{(100 - \%\text{ excess of }[^{14}\text{C}]\text{dUMP})([^{14}\text{C}]\text{dUMP} + [^{14}\text{C}]\text{dTMP})} \quad (16)$$

The exact excess of [2-^{14}C]dUMP over CH$_2$H$_4$folate was calculated from the infinite time points using the following equation:

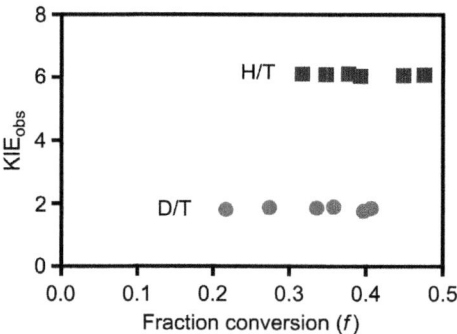

Fig. 4 Observed H/T and D/T KIEs on hydride transfer for R166K at 35°C.

$$\% \text{ excess of } \left[^{14}\text{C}\right]\text{dUMP} = \frac{\left[^{14}\text{C}\right]\text{dUMP}}{\left(\left[^{14}\text{C}\right]\text{dUMP} + \left[^{14}\text{C}\right]\text{dTMP}\right)} \quad (17)$$

Fig. 4 shows a typical plot of KIE_{obs} vs fraction conversion (f). The fact that the KIE_{obs}s are independent of fraction conversion serves as an important quality control, because most artifacts would lead to f-dependent KIE_{obs}. Each of the KIE_{obs}s was averaged over at least five time points at each temperature. The KIE_{int}s were calculated using Eq. (14).

4.1.2 Measurements of KIEs on the Proton Abstraction

The lower panel of Scheme 2 shows the isotopic labeling pattern for the proton-abstraction KIE measurements. Apart from different isotopic labeling, the experimental processes for the proton-abstraction KIEs are mostly identical to those of the hydride-transfer KIE measurements. Therefore, in this section, only the experimental procedures unique to hydride transfer are highlighted.

[2-^{14}C]dUMP and [5-^{3}H]dUMP were purchased from Moravek Biochemicals and American Radiolabeled Chemicals, respectively. [2-^{14}C, 5-^{2}H]dUMP was synthesized following a published procedure (Wataya & Hayatsu, 1972).

To measure the isotope effects in the proton abstraction, trace [5-^{3}H] dUMP was mixed with [2-^{14}C]dUMP or [2-^{14}C, 5-^{2}H]dUMP for H/T or D/T KIE measurements, respectively. The concentration of CH_2H_4folate was taken in excess to total dUMP concentration in order to obtain complete conversion of the labeled substrates.

It should be noted that the KIE_{obs} for proton abstraction is dependent on the concentration of CH_2H_4folate, as can be seen in Eqs. (2) and (18)

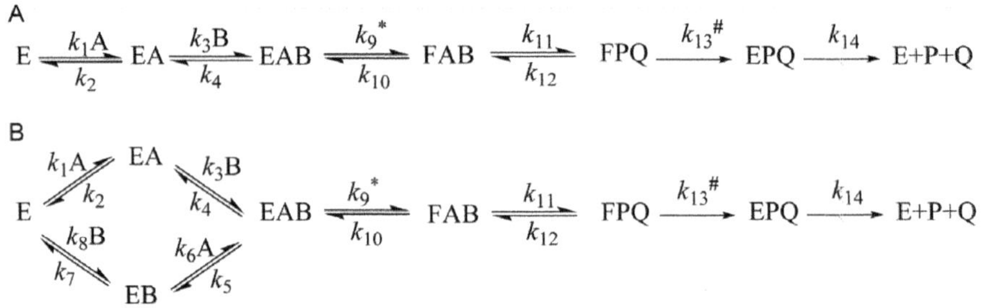

A

$$E \underset{k_2}{\overset{k_1 A}{\rightleftarrows}} EA \underset{k_4}{\overset{k_3 B}{\rightleftarrows}} EAB \underset{k_{10}}{\overset{k_9^*}{\rightleftarrows}} FAB \underset{k_{12}}{\overset{k_{11}}{\rightleftarrows}} FPQ \xrightarrow{k_{13}^\#} EPQ \xrightarrow{k_{14}} E+P+Q$$

B

$$E \begin{array}{c} \overset{k_1 A}{\underset{k_2}{\rightleftarrows}} EA \overset{k_3 B}{\underset{k_4}{\searrow}} \\ \underset{k_7}{\overset{k_8 B}{\rightleftarrows}} EB \overset{k_6 A}{\underset{k_5}{\nearrow}} \end{array} EAB \underset{k_{10}}{\overset{k_9^*}{\rightleftarrows}} FAB \underset{k_{12}}{\overset{k_{11}}{\rightleftarrows}} FPQ \xrightarrow{k_{13}^\#} EPQ \xrightarrow{k_{14}} E+P+Q$$

Scheme 3 Ordered (A) and random (B) binding mechanisms for substrates dUMP (A) and CH$_2$H$_4$folate (B) for TSase. *Asterisk* (*) and *number sign* (#) represent the proton abstraction (k_9) and the hydride transfer (k_{13}), respectively.

(Ghosh, Islam, Krueger, Abeysinghe, & Kohen, 2015; Hong et al., 2007; Wang & Kohen, 2010):

$$C_f = \frac{k_9}{k_5 + \frac{k_2 k_4}{k_2 + k_3[\text{CH}_2\text{H}_4\text{folate}]}} \quad (18)$$

where C_f is the forward commitment, and k_9 denotes the proton abstraction step (Scheme 3).

In some cases, at high concentration of CH$_2$H$_4$folate KIE$_{obs}$ goes to unity, which indicates a strictly ordered binding mechanism with dUMP binding first (panel A in Scheme 3), as can be seen in Eq. (19) (Ghosh et al., 2015; Hong et al., 2007; Wang & Kohen, 2010):

$$C_f = \frac{k_9}{k_4} + \frac{k_3[\text{CH}_2\text{H}_4\text{folate}]}{k_2 k_4} \quad (19)$$

In Eq. (19) C_f goes to infinity at high concentrations of CH$_2$H$_4$folate, so KIE$_{obs}$ goes to one (Eq. 1). For WT ecTSase an optimal concentration of CH$_2$H$_4$folate (3 μM) was chosen to yield large H/T and D/T KIE$_{obs}$ with low errors (Wang & Kohen, 2010). However, for many mutants (Abeysinghe & Kohen, 2015) including R166K (Islam et al., 2015) that follow a random binding mechanism, high concentrations of CH$_2$H$_4$folate yield large KIE$_{obs}$. The reaction mixture for the proton-abstraction KIE measurements contains 50 mM MgCl$_2$, which was added to reduce the commitment factor on this step (Islam et al., 2015; Wang & Kohen, 2013). Note that the presence of MgCl$_2$ does not affect the KIE$_{int}$s for the hydride transfer (Wang, Sapienza, et al., 2013). The KIE$_{obs}$ was calculated using Eq. (15), while the fraction conversion was determined using the following equation:

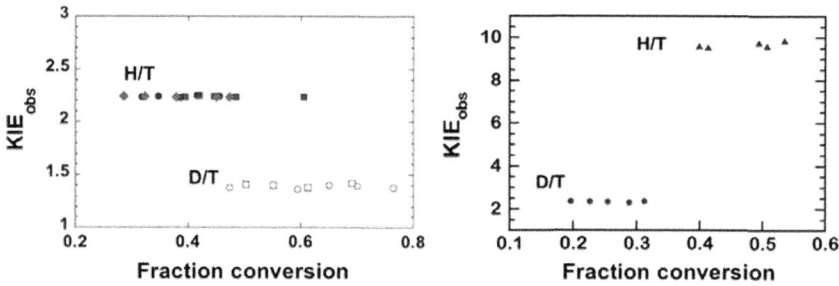

Fig. 5 Examples of KIE$_{obs}$ for the proton abstraction step for the WT ecTSase (*left*) (Wang & Kohen, 2010) and its R166K mutant (*right*) (Islam et al., 2015). Reproduced from Islam, Z., Strutzenberg, T. S., Ghosh, A. K., & Kohen, A. (2015). Activation of two sequential H transfers in the thymidylate synthase catalyzed reaction. ACS Catalysis, 5, 6061–6068, with permission from ACS.

$$f = \frac{[^{14}C]dTMP}{\left([^{14}C]dUMP + [^{14}C]dTMP\right)} \qquad (20)$$

As in the case of hydride transfer, the absence of any trends in KIE$_{obs}$s with fraction conversion (Fig. 5) indicates that no significant artifact such as protium contamination distorts the D/T KIEs. Though the proton abstraction is reversible, Eq. (14) for extracting KIE$_{int}$s is still valid since the EIE was found to be close to unity for this step (Hong et al., 2007). Furthermore, the reverse commitment on the tritium is essentially zero since once released, the tritium is diluted into 110 M protiated water. Given all these assumptions, quality control (QC) is most important in this case, and our experience is that the most sensitive QC is f-independent KIE$_{int}$ (Fig. 5). Almost every deviation from the above assumptions leads to an f-dependent KIE$_{int}$.

5. CASE STUDY 2: DHFR

DHFR catalyzes the NADPH-dependent reduction of 7,8-dihydrofolate (H$_2$F) to 5,6,7,8-tetrahydrofolate (H$_4$F) through a stereospecific transfer of the pro-*R* hydrogen from NADPH to C6 of the *si* face of the pterin ring (Scheme 4) (Hammes-Schiffer & Benkovic, 2006). DHFR from *E. coli* (ecDHFR) is a common model system for various studies due to its small size, flexibility, and the simple chemical transformation it catalyzes. The enzyme has served as a platform for many calculations and experimental studies, and the relationship between protein dynamics and its function has been

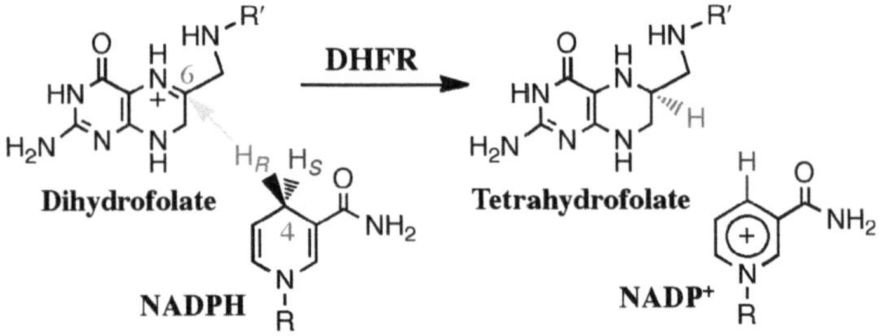

Scheme 4 DHFR catalyzes the stereospecific transfer of the pro-R hydride of C4 of NADPH to C6 of protonated N5-DHF, producing THF and the oxidized cofactor NADP$^+$. It is important for both measurements and computations that the hydride transfer (*green* (*light gray* in the print version) *arrow*) takes place after DHF is already protonated (Liu et al., 2014).

examined by both theoreticians and experimentalists. A partial list of references includes Arora and Brooks (2009), Bystroff and Kraut (1991), Dametto et al. (2012), Fan et al. (2013), Fierke et al. (1987), Klinman and Kohen (2013), Liu and Warshel (2007b), Loveridge, Behiry, Guo, and Allemann (2012), McElheny, Schnell, Lansing, Dyson, and Wright (2005), Pauling (1948), Singh, Abeysinghe, and Kohen (2015), Singh, Francis, and Kohen (2015), Singh, Morris, Tivanski, and Kohen (2015c, 2015d), Wang, Singh, Czekster, Kohen, and Schramm (2014), Wong, Watney, and Hammes-Schiffer (2004). The focus of this section is to demonstrate a combination of different approaches to studying the DHFR-catalyzed reaction emphasizing the hydride transfer step, primarily through KIE experiments and hybrid QM/MM/MD simulations. The current section emphasizes the interplay between experimental and computational approaches.

5.1 Computational vs Experimental Studies
5.1.1 Network of Coupled Motions in DHFR

The concept of remote residues influencing events at the active site has been debated for several enzymes (Fraser et al., 2009; Ghanem, Li, Wing, & Schramm, 2008; Henzler-Wildman & Kern, 2007; Loria, Berlow, & Watt, 2008; Meadows, Tsang, & Klinman, 2014). Calculations by Brooks and Hammes-Schiffer independently predicted that a "network of dynamically coupled motions" was part of the reaction coordinate of the C–H→C hydride transfer in ecDHFR (Hammes-Schiffer & Benkovic, 2006; Radkiewicz & Brooks, 2000; Rod et al., 2003; Wong et al., 2005). These coupled motions denote thermal equilibrium; ie, thermally averaged conformational changes along the collective reaction-coordinate, producing

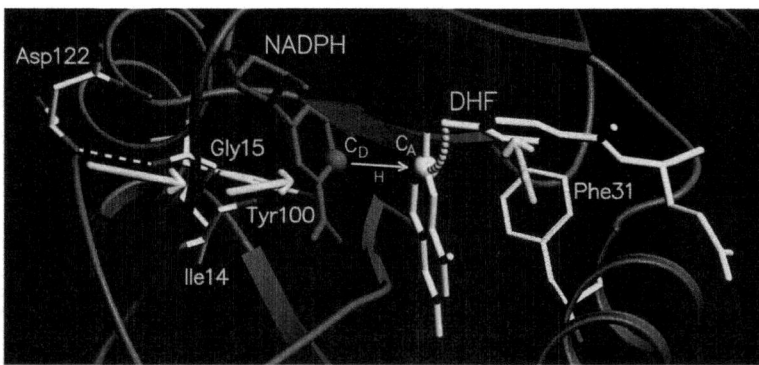

Fig. 6 Diagram of a portion of the network of coupled promoting motions in ecDHFR. The *yellow arrows* and *arc* indicate the coupled promoting motions. *From Agarwal, P. K., Billeter, S. R., Rajagopalan, P. T., Benkovic, S. J., & Hammes-Schiffer, S. (2002). Network of coupled promoting motions in enzyme catalysis. Proceedings of the National Academy of Sciences of the United States of America, 99, 2794–2799, with permission from Proceedings of the National Academy of Sciences of the United States of America.* (See the color plate.)

configurations conducive to the hydride transfer reaction with the preprotonated DHF (Scheme 4). A portion of this network of coupled motions in DHFR is illustrated in Fig. 6.

The equilibrium motions produce configurations that promote hydride transfer through short DADs, a proper electrostatic environment for charge transfer, and correct orientation of the substrate and cofactor. Further insights into this network of coupled motion have been provided by calculations performed on mutants of DHFR. For example, G121V, one of the most intensively studied DHFR mutants, results in a 40-fold decrease in NADPH-binding affinity and a 200-fold decrease in the single-turnover rates, despite the fact that the mutation is located ∼15 Å away from the active site (Boehr et al., 2013; Fan et al., 2013; Liu & Warshel, 2007a, 2007b; Mauldin, Sapienza, Petit, & Lee, 2012; Ohmae, Iriyama, Ichihara, & Gekko, 1998; Rajagopalan, Stefan, & Benkovic, 2002; Rod et al., 2003; Singh, Abeysinghe, et al., 2015; Singh, Sen, Francis, & Kohen, 2014; Swanwick, Shrimpton, & Allemann, 2004; Thorpe & Brooks, 2003, 2004; Wang, Tharp, Selzer, Benkovic, & Kohen, 2006). Simulations of G121V have resulted in a rate decrease that is consistent with experimental rate measurements and suggested that mutations may alter the network of coupled motions through nonlocal structural perturbations, raising the free energy barrier and reducing the reaction rate

(Watney, Agarwal, & Hammes-Schiffer, 2003). These calculations have suggested that remote mutations far from the active site could bring subtle structural perturbations that affect the catalytic rate by altering the conformational sampling of the entire enzyme, leading to alterations in the network of coupled motions of the wild-type enzyme. This concept provides an explanation for the experimentally observed nonadditive rates (Rajagopalan et al., 2002): the introduction of a mutation modifies the thermal motions of the entire enzyme because the remote regions of the enzyme are coupled to each other through long-range electrostatics and extended hydrogen bonding networks (Wong et al., 2004); in turn, altering the thermal motions of the enzyme affects the probability of the occurrence of sampling conformations conducive to the catalyzed chemical reaction, thereby influencing the free energy barrier and the rate. Other theoretical studies by Moliner, Allemann, and coworkers on WT-DHFR and G121V also imply that mutation causes nonlocal structural effects that may lead to perturbation of the network of coupled motions (Luk et al., 2013).

The QM/MM/MD calculations predicted that distal residues M42, G121, and F125, as well as active-site residue I14, are part of the network in question (Hammes-Schiffer & Benkovic, 2006; Hammes-Schiffer & Watney, 2006; Rod et al., 2003; Wong et al., 2005; Wong et al., 2004). Single-turnover rates examined by Benkovic and coworkers on G121, M42, and their double mutants provided some support for these predictions (Rajagopalan et al., 2002). KIE_{int}s and their temperature dependence were examined experimentally to test for the presence of the predicted network and effects of mutants on that network. As is true of most wild-type enzymes, WT ecDHFR has a KIE_{int} that is temperature independent (Sikorski et al., 2004), indicating a narrow DAD distribution and an active site that is perfect for hydride transfer. The temperature dependence of KIE_{int} was measured for I14A and M42W, G121V, F125M, and W133F, as well as for their double mutants (Singh et al., 2014; Wang, Goodey, Benkovic, & Kohen, 2006a, 2006b; Wang, Tharp, et al., 2006). Synergy between single mutations was used to indicate that the tested residues "work together," ie, are part of a network of motions coupled or correlated with the hydride transfer step. "Synergy" in this context means that the sum of changes in single-turnover rates or temperature dependence of KIE_{int} caused by single mutants is smaller than that caused by the relevant double mutant.

Fig. 7 summarizes the temperature dependences of the KIE_{int}s for remote single mutants, their double mutants, the single active site mutant, and a

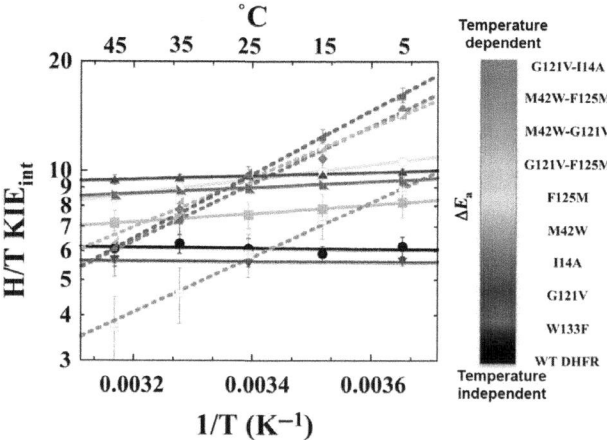

Fig. 7 Comparison of Arrhenius plots of intrinsic H/T KIEs of WT DHFR and its mutants. The *lines* represent the nonlinear regression to the Arrhenius equation, and the error bars represent standard deviation.

remote-active site double mutant. The synergistic effect of double-mutant KIEs is apparent in the nonadditive values of the temperature dependence, ie, ΔE_a, which is the difference between the energy of activations (E_a) of the two isotopes used per study. (Please note that for KIEs $\Delta E_a = \Delta H^{\ddagger}$ because $E_a = H^{\ddagger} - RT$, and RT is the same for both isotopes.) Fig. 8 summarizes the isotope effect on activation energy (ΔE_a) and the Arrhenius preexponential factors (A_{light}/A_{heavy}) for WT-ecDHFR and its mutants. Fig. 8 also reveals that the effect of single distal mutation is similar to the effect of active site mutations I14V and I14A, while the effect of double distal mutation is similar to the effect of active site mutation I14G, suggesting similarities in the way those mutants affect DAD distribution at the active site.

In summary, the influence of remote residues participating in a network of coupled motions on the catalyzed chemistry was validated for all the experimentally verified residues proposed by QM/MM simulations. Furthermore, W133F, the residue that had not been predicted by these calculations to be part of that network, did not seem to have an effect on hydride transfer. Because both the calculations and KIE_{int} examined the same single chemical step, the experimental findings can be compared directly to the computations that predicted that network. The findings thus validate calculations proposing that both remote and active-site residues constitute a network of coupled promoting motions correlated to the bond activation step (Singh, Francis, et al., 2015).

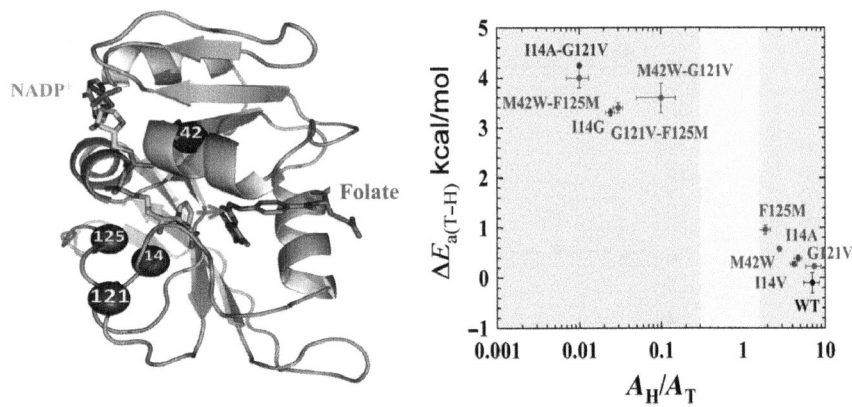

Fig. 8 Roles of active site and distal residues on the DHFR-catalyzed reaction. *Left panel*: Structure of WT-ecDHFR (PDB Code 1RX2), with folate in *magenta* and NADP⁺ in *light blue*. A *yellow arrow* marks the hydride's path from C4 of the nicotinamide to C6 of the folate, and the residues studied in Singh et al. (2014) are marked as *purple spheres*. *Right panel*: Presentation of the temperature-dependence parameters of KIE_{int} for WT (*black*), distal (*red*), and active site I14 (*green*) mutants of DHFR, where error bars represent standard deviation. For each point, the ordinate is the isotope effect on activation energy (ΔE_a), and the abscissa is the isotope effect on the Arrhenius preexponential factor (A_H/A_T), both of which were determined from a nonlinear regression of KIE_{int} to the Arrhenius equation. The *yellow block* represents the semiclassical range of the Arrhenius preexponential factor (0.3–1.7) (Kohen, 2006). Reprinted from Singh, P., Francis, K., & Kohen, A. (2015). Network of remote and local protein dynamics in dihydrofolate reductase catalysis. ACS Catalysis, 5, 3067–3073, with permission from the American Chemical Society. (See the color plate.)

After the experimental findings for G121V and M42W were published (Wang et al., 2006a), Warshel and coworkers calculated the temperature dependences of KIE_{int} for these mutants and for the wild type using the empirical valence bond–quantum classical path method (Liu & Warshel, 2007b). Those studies specifically explored the relationship between DAD and KIE_{int} at different temperatures. In good agreement with the empirical–activated tunneling models mentioned earlier (Section 4), Warshel's calculations found a broader DAD distribution for steeper temperature dependence of KIEs (ΔE_a).

A very different computational study of ecDHFR used transition-path sampling (TPS) and examined the role of protein promoting vibrations (PPVs) in ecDHFR catalysis (Dametto et al., 2012). This study found no such PPVs, which could be taken as contradictory to all the computational and experimental studies discussed earlier. However, the "network of coupled motions" along the reaction coordinate simulated by Hammes-Schiffer and others includes no nonstatistical elements. The calculated PPVs, on the other hand, are nonstatistical dynamic motions, so the apparent contradictions between the two methods likely reflect the fact that they examine very different phenomena. It seems that the experimental studies using KIE_{int} agree well with the first but did not test the second theoretical approach. Additionally, TPS calculations of the human DHFR (hsDHFR) predicted significant PPVs, but KIE_{int} for both light and heavy hsDHFR were identical throughout the temperature range (Francis, Sapienza, Lee, & Kohen, 2016), falling to support the theoretical prediction. Similarly, in a different TPS study of liver alcohol dehydrogenase, it was predicted that the valine residues at position 203 and 207 facilitate enzyme catalysis as part of the PPV (Caratzoulas, Mincer, & Schwartz, 2002). However, no experimental evidence was found to support a role of V207 in the dynamics of catalysis (Yahashiri, Rubach, & Plapp, 2014).

5.2 Experimental Section for DHFR

The cofactors $[Ad-^{14}C]$-NADPH (50 mCi/mmol), (R)-$[Ad-^{14}C, 4-^2H]$-NADPH (50 mCi/mmol), and (R)-$[4-^3H]$-NADPH (680 mCi/mmol) were synthesized according to previously published procedures (Agrawal & Kohen, 2003; Markham, Sikorski, & Kohen, 2003, 2004; McCracken, Wang, & Kohen, 2003; Sen et al., 2011).

5.2.1 Competitive KIE Measurements

For H/T 1° KIE, (R)-$[4-^3H]$-NADPH and $[Ad-^{14}C]$-NADPH were combined in a radioactivity ratio close to 5:1. Likewise, for D/T 1° KIE, (R)-$[4-^3H]$-NADPH and (R)-$[Ad-^{14}C, 4-^2H]$-NADPH were combined in the same manner. Each mixture was copurified on an analytical reverse phase HPLC column. The purified material was divided into aliquots containing 300,000 dpm of ^{14}C and stored at $-80°C$ for short-term storage.

All measurements were done in MTEN (50 mM MES, 25 mM Tris, 25 mM ethanolamine, and 100 mM NaCl) buffer at pH 9.0 over the temperature range of 5–45°C. NADPH was added to the reaction mixture so that the final concentration was 4 μM, while dihydrofolate was added to a final concentration of 0.85 mM (~200-fold excess over NADPH). A pH electrode was calibrated at each temperature using standard buffers at the respective temperatures and used to check the pH of all reaction mixtures. For DHFR, the largest KIE_{obs} can be measured at pH 9.0, though

KIE_{obs} at pH 8.0 and 7.0 have also been measured with good outcomes (Liu et al., 2014). The reaction was initiated by addition of DHFR and allowed to proceed until a fraction conversion of between 20% and 80% was reached, followed by quenching by adding an excess of methotrexate to a final concentration of 1.7 mM before storing in dry ice. Fraction conversions of 20–80% have been shown to produce minimal error when determining KIE_{obs} (Northrop, 1991). Samples were thawed and bubbled with oxygen before HPLC–LSC analysis to ensure complete oxidation of the product tetrahydrofolate. The samples were then separated by reverse phase HPLC and analyzed on a LSC. The KIE_{obs} was calculated using the following equation:

$$KIE = \frac{\ln(1-f)}{\ln[1-f(R_t/R_\infty)]} \quad (15)$$

where the fraction conversion (f) was determined from the ratio of ^{14}C in the product and reactant, and R_t and R_∞ are the ratios of $^3H/^{14}C$ at a particular time point and at infinite time, respectively. Subsequently, KIE_{int} was extracted from its respective observed values using the Northrop method as explained in an earlier (Section 3). All quality controls were performed for DHFR as for TSase, as described in Section 4.1 for TSase.

6. SUMMARY

In this chapter, we underscore the importance of examining the same chemical step, or microscopic event, in both calculations and experiments, and illustrate the necessity of such precise, parallel computational and experimental studies of DHFR and TSase. The utility of KIE_{int} as a probe for TS structures and their distribution is described. In the two case studies discussed earlier (Sections 4 and 5), the objective of the calculations is not only to explain experimental results but also to create experimentally testable predictions that permit the examination of the theory's underlying rationale. The objective of the experimental work is to conduct measurements that afford the examination of the same phenomenon studied by the relevant calculations. In that context, we suggest that assessment of KIE_{int} from measured KIE_{obs} is an important tool, and that in some cases the Northrop method can assist in that task. The two examples presented earlier (Sections 4 and 5) started from predictions made by QM/MM calculations (a newly proposed intermediate in TSase and a network of coupled motions in ecDHFR) and demonstrated how measurement of the temperature dependence of KIE_{int} supported these predictions, then suggested further need for calculations and experimental studies.

We suggest that a collaborative blend of theoretical and experimental approaches is beneficial to resolving quandaries of theoretical models and experimental results. Collaborative approaches are likely to deepen our knowledge of enzyme mechanisms in general, as well as our knowledge of the role of protein motions in catalysis, allowing the development of more detailed and comprehensive models of enzyme catalysis. The expected gain is most significant: experimental findings that agree with computations lend reassurance to both methods, and exposing "hidden" features in enzyme mechanisms may lead to the identification of new targets for inhibitors and potential drugs.

ACKNOWLEDGMENTS
This work was supported by NIH (R01GM65368) and NSF (CHE-1149023) and the Iowa Center of Biocatalysis and Bioprocessing associated with NIH T32 GM008365 to Z.I.

REFERENCES
Abeysinghe, T., & Kohen, A. (2015). Role of long-range protein dynamics in different thymidylate synthase catalyzed reactions. *International Journal of Molecular Sciences, 16*, 7304–7319.

Agrawal, N., Hong, B., Mihai, C., & Kohen, A. (2004). Vibrationally enhanced hydrogen tunneling in the Escherichia coli thymidylate synthase catalyzed reaction. *Biochemistry, 43*, 1998–2006.

Agrawal, N., & Kohen, A. (2003). Microscale synthesis of 2-tritiated isopropanol and 4R-tritiated reduced nicotinamide adenine dinucleotide phosphate. *Analytical Biochemistry, 322*, 179–184.

Agrawal, N., Mihai, C., & Kohen, A. (2004). Microscale synthesis of isotopically labeled R-[6-xH]N5, N10-methylene-5,6,7,8-tetrahydrofolate as a cofactor for thymidylate synthase. *Analytical Biochemistry, 328*, 44–50.

Arora, K., & Brooks, C. L., III. (2009). Functionally important conformations of the Met20 loop in dihydrofolate reductase are populated by rapid thermal fluctuations. *Journal of the American Chemical Society, 131*, 5642–5647.

Boehr, D. D., Schnell, J. R., McElheny, D., Bae, S.-H., Duggan, B. M., Benkovic, S. J., et al. (2013). A distal mutation perturbs dynamic amino acid networks in dihydrofolate reductase. *Biochemistry, 52*, 4605–4619.

Bystroff, C., & Kraut, J. (1991). Crystal structure of unliganded Escherichia coli dihydrofolate reductase. Ligand-induced conformational changes and cooperativity in binding. *Biochemistry, 30*, 2227–2239.

Caratzoulas, S., Mincer, J. S., & Schwartz, S. D. (2002). Identification of a protein-promoting vibration in the reaction catalyzed by horse liver alcohol dehydrogenase. *Journal of the American Chemical Society, 124*, 3270–3276.

Carosati, E., Tochowicz, A., Marverti, G., Guaitoli, G., Benedetti, P., Ferrari, S., et al. (2012). Inhibitor of ovarian cancer cells growth by virtual screening: A new thiazole derivative targeting human thymidylate synthase. *Journal of Medicinal Chemistry, 55*, 10272–10276.

Carreras, C. W., & Santi, D. V. (1995). The catalytic mechanism and structure of thymidylate synthase. *Annual Review of Biochemistry, 64*, 721–762.

Cook, P. F., & Cleland, W. W. (2007). *Enzyme kinetics and mechanism*: (pp. 253–324). New York, NY: Taylor and Francis Group LLC.

Dametto, M., Antoniou, D., & Schwartz, S. D. (2012). Barrier crossing in dihydrofolate reductase does not involve a rate-promoting vibration. *Molecular Physics, 110*, 531–536.

Fan, Y., Cembran, A., Ma, S., & Gao, J. (2013). Connecting protein conformational dynamics with catalytic function as illustrated in dihydrofolate reductase. *Biochemistry, 52*, 2036–2049.

Fierke, C. A., Johnson, K. A., & Benkovic, S. J. (1987). Construction and evaluation of the kinetic scheme associated with dihydrofolate reductase from Escherichia coli. *Biochemistry, 26*, 4085–4092.

Finer-Moore, J. S., Santi, D. V., & Stroud, R. M. (2003). Lessons and conclusions from dissecting the mechanism of a bisubstrate enzyme: Thymidylate synthase mutagenesis, function, and structure. *Biochemistry, 42*, 248–256.

Francis, K., Sapienza, P. J., Lee, A. L., & Kohen, A. (2016). The effect of protein mass modulation on human dihydrofolate reductase. *Biochemistry, 55*, 1100–1106.

Fraser, J. S., Clarkson, M. W., Degnan, S. C., Erion, R., Kern, D., & Alber, T. (2009). Hidden alternative structures of proline isomerase essential for catalysis. *Nature, 462*, 669–673.

Ghanem, M., Li, L., Wing, C., & Schramm, V. L. (2008). Altered thermodynamics from remote mutations altering human toward bovine purine nucleoside phosphorylase†. *Biochemistry, 47*, 2559–2564.

Ghosh, A. K., Islam, Z., Krueger, J., Abeysinghe, T., & Kohen, A. (2015). The general base in the thymidylate synthase catalyzed proton abstraction. *Physical Chemistry Chemical Physics, 17*, 30867–30875.

Hammes-Schiffer, S., & Benkovic, S. J. (2006). Relating protein motion to catalysis. *Annual Review of Biochemistry, 75*, 519–541.

Hammes-Schiffer, S., & Watney, J. B. (2006). Hydride transfer catalysed by Escherichia coli and Bacillus subtilis dihydrofolate reductase: Coupled motions and distal mutations. *Philosophical Transactions of the Royal Society B, 361*, 1365–1373.

Henzler-Wildman, K., & Kern, D. (2007). Dynamic personalities of proteins. *Nature, 450*, 964–972.

Hong, B., Maley, F., & Kohen, A. (2007). Role of Y94 in proton and hydride transfers catalyzed by thymidylate synthase. *Biochemistry, 46*, 14188–14197.

Islam, Z., Strutzenberg, T. S., Ghosh, A. K., & Kohen, A. (2015). Activation of two sequential H transfers in the thymidylate synthase catalyzed reaction. *ACS Catalysis, 5*, 6061–6068.

Islam, Z., Strutzenberg, T. S., Gurevic, I., & Kohen, A. (2014). Concerted versus stepwise mechanism in thymidylate synthase. *Journal of the American Chemical Society, 136*, 9850–9853.

Kanaan, N., Ferrer, S., Marti, S., Garcia-Viloca, M., Kohen, A., & Moliner, V. (2011). Temperature dependence of the kinetic isotope effects in thymidylate synthase. A theoretical study. *Journal of American Chemical Society, 133*, 6692–6702.

Kanaan, N., Marti, S., Moliner, V., & Kohen, A. (2007). A quantum mechanics/molecular mechanics study of the catalytic mechanism of the thymidylate synthase. *Biochemistry, 46*, 3704–3713.

Kanaan, N., Marti, S., Moliner, V., & Kohen, A. (2009). QM/MM study of thymidylate synthase: Enzymatic motions and the temperature dependence of the rate limiting step. *The Journal of Physical Chemistry. A, 113*, 2176–2182.

Kanaan, N., Roca, M., Tunon, I., Marti, S., & Moliner, V. (2010). Theoretical study of the temperature dependence of dynamic effects in thymidylate synthase. *Physical Chemistry Chemical Physics, 12*, 11657–11664.

Klinman, J. P. (2015). Dynamically achieved active site precision in enzyme catalysis. *Accounts of Chemical Research, 48*, 449–456.

Klinman, J. P., & Kohen, A. (2013). Hydrogen tunneling links protein dynamics to enzyme catalysis. *Annual Review of Biochemistry, 82*, 471–496.

Kohen, A. (2006). *Isotopes effects in chemistry and biology*: (pp. 743–764). Boca Raton, FL: Taylor and Francis.

Kohen, A. (2015). Role of dynamics in enzyme catalysis: Substantial versus semantic controversies. *Accounts of Chemical Research, 48*, 466–473.

Layfield, J. P., & Hammes-Schiffer, S. (2013). Hydrogen tunneling in enzymes and biomimetic models. *Chemical Reviews, 114*, 3466–3494.

Liu, C. T., Francis, K., Layfield, J., Huang, X., Hammes-Schiffer, S., Kohen, A., et al. (2014). The Escherichia coli dihydrofolate reductase catalyzed proton and hydride transfers: Order and the roles of Asp27 and Tyr100. *Proceedings of the National Academy of Sciences of the United States of America, 111*, 18231–18236.

Liu, H., & Warshel, A. (2007a). The catalytic effect of dihydrofolate reductase and its mutants is determined by reorganization energies†. *Biochemistry, 46*, 6011–6025.

Liu, H., & Warshel, A. (2007b). Origin of the temperature dependence of isotope effects in enzymatic reactions: The case of dihydrofolate reductase. *The Journal of Physical Chemistry. B, 111*, 7852–7861.

Loria, J. P., Berlow, R. B., & Watt, E. D. (2008). Characterization of enzyme motions by solution NMR relaxation dispersion. *Accounts of Chemical Research, 41*, 214–221.

Loveridge, E. J., Behiry, E. M., Guo, J., & Allemann, R. K. (2012). Evidence that a 'dynamic knockout' in Escherichia coli dihydrofolate reductase does not affect the chemical step of catalysis. *Nature Chemistry, 4*, 292–297.

Luk, L. Y. P., Javier Ruiz-Pernía, J., Dawson, W. M., Roca, M., Loveridge, E. J., Glowacki, D. R., et al. (2013). Unraveling the role of protein dynamics in dihydrofolate reductase catalysis. *Proceedings of the National Academy of Sciences of the United States of America, 110*, 16344–16349.

Marcus, R. A. (2007). H and other transfers in enzymes and in solution: Theory and computations, a unified view. 2. Applications to experiment and computations. *The Journal of Physical Chemistry. B, 111*, 6643–6654.

Markham, K. A., Sikorski, R. S., & Kohen, A. (2003). Purification, analysis, and preservation of reduced nicotinamide adenine dinucleotide 2′-phosphate. *Analytical Biochemistry, 322*, 26–32.

Markham, K. A., Sikorski, R. S., & Kohen, A. (2004). Synthesis and utility of 14C-labeled nicotinamide cofactors. *Analytical Biochemistry, 325*, 62–67.

Mauldin, R. V., Sapienza, P. J., Petit, C. M., & Lee, A. L. (2012). Structure and dynamics of the G121V dihydrofolate reductase mutant: Lessons from a transition-state inhibitor complex. *PloS One, 7*, e33252.

McCracken, J. A., Wang, L., & Kohen, A. (2003). Synthesis of R and S tritiated reduced beta-nicotinamide adenine dinucleotide 2' phosphate. *Analytical Biochemistry, 324*, 131–136.

McElheny, D., Schnell, J. R., Lansing, J. C., Dyson, H. J., & Wright, P. E. (2005). Defining the role of active-site loop fluctuations in dihydrofolate reductase catalysis. *Proceedings of the National Academy of Sciences of the United States of America, 102*, 5032–5037.

McMurry, J. E., & Begley, T. P. (2016). Nucleotide metabolism. In J. Murdzek (Ed.), *The organic chemistry of biological pathways* (p. 333). Greenwood Village, CO: Roberts and Company.

Meadows, C. W., Tsang, J. E., & Klinman, J. P. (2014). Picosecond-resolved fluorescence studies of substrate and cofactor-binding domain mutants in a thermophilic alcohol dehydrogenase uncover an extended network of communication. *Journal of the American Chemical Society, 136*, 14821–14833.

Nagel, Z. D., & Klinman, J. P. (2010). Update 1 of: Tunneling and dynamics in enzymatic hydride transfer. *Chemical Reviews, 110*, PR41–PR67.

Northrop, D. B. (1991). *Enzyme mechanism from isotope effects*: (pp. 181–202). Boca Raton, FL: CRC Press.

Ohmae, E., Iriyama, K., Ichihara, S., & Gekko, K. (1998). Nonadditive effects of double mutations at the flexible loops, glycine-67 and glycine-121, of Escherichia coli dihydrofolate reductase on its stability and function. *Journal of Biochemistry, 123*, 33–41.

Pauling, L. (1948). Chemical achievement and hope for the future. *American Scientist, 36*, 51–58.

Pu, J., Gao, J., & Truhlar, D. G. (2006). Multidimensional tunneling, recrossing, and the transmission coefficient for enzymatic reactions. *Chemical Reviews, 106*, 3140–3169.

Pu, J., Ma, S., Gao, J., & Truhlar, D. G. (2005a). Small temperature dependence of the kinetic isotope effect for the hydride transfer reaction catalyzed by Escherichia coli dihydrofolate reductase. *The Journal of Physical Chemistry. B, 109*, 8551–8556.

Pu, J., Ma, S., Garcia-Viloca, M., Gao, J., Truhlar, D. J., & Kohen, A. (2005b). Nonperfect synchronization of reaction center rehybridization in the transition state of the hydride transfer catalyzed by dihydrofolate reductase. *Journal of the American Chemical Society, 127*, 14879–14886.

Pudney, C. R., Hay, S., Levy, C., Pang, J., Sutcliffe, M. J., Leys, D., et al. (2009). Evidence to support the hypothesis that promoting vibrations enhance the rate of an enzyme catalyzed H-tunneling reaction. *Journal of the American Chemical Society, 131*, 17072–17073.

Radkiewicz, J. L., & Brooks, C. L. (2000). Protein dynamics in enzymatic catalysis: Exploration of dihydrofolate reductase. *Journal of the American Chemical Society, 122*, 225–231.

Rajagopalan, P. T. R., Stefan, L., & Benkovic, S. J. (2002). Coupling interactions of distal residues enhance dihydrofolate reductase catalysis: Mutaional effects on hydride transfer rates. *Biochemistry, 41*, 12618–12628.

Rod, T. H., Radkiewicz, J. L., & Brooks, C. L. (2003). Correlated motion and the effect of distal mutations in dihydrofolate reductase. *Proceedings of the National Academy of Sciences of the United States of America, 100*, 6980–6985.

Roston, D., Islam, Z., & Kohen, A. (2013). Isotope effects as probes for enzyme catalyzed hydrogen-transfer reactions. *Molecules, 18*, 5543–5567.

Roston, D., Islam, Z., & Kohen, A. (2014). Kinetic isotope effects as a probe of hydrogen transfers to and from common enzymatic cofactors. *Archives of Biochemistry and Biophysics, 544*, 96–104.

Sen, A., Yahashiri, A., & Kohen, A. (2011). Triple isotopic labeling and kinetic isotope effects: Exposing H-transfer steps in enzymatic systems. *Biochemistry, 50*, 6462–6468.

Shelton, G. R., Hrovat, D. A., & Borden, W. T. (2007). Calculations of the effect of tunneling on the Swain–Schaad exponents (SSEs) for the 1,5-hydrogen shift in 5-methyl-1,3-cyclopentadiene. Can SSEs be used to diagnose the occurrence of tunneling? *Journal of the American Chemical Society, 129*, 16115–16118.

Sikorski, R. S., Wang, L., Markham, K. A., Rajagopalan, P. T., Benkovic, S. J., & Kohen, A. (2004). Tunneling and coupled motion in the Escherichia coli dihydrofolate reductase catalysis. *Journal of the American Chemical Society, 126*, 4778–4779.

Singh, P., Abeysinghe, T., & Kohen, A. (2015). Linking protein motion to enzyme catalysis. *Molecules, 20*, 1192.

Singh, P., Francis, K., & Kohen, A. (2015). Network of remote and local protein dynamics in dihydrofolate reductase catalysis. *ACS Catalysis, 5*, 3067–3073.

Singh, P., Morris, H., Tivanski, A. V., & Kohen, A. (2015a). A calibration curve for immobilized dihydrofolate reductase activity assay. *Data in Brief, 4*, 19–21.
Singh, P., Morris, H., Tivanski, A. V., & Kohen, A. (2015b). Determination of concentration and activity of immobilized enzymes. *Analytical Biochemistry, 484*, 169–172.
Singh, P., Sen, A., Francis, K., & Kohen, A. (2014). Extension and limits of the network of coupled motions correlated to hydride transfer in dihydrofolate reductase. *Journal of the American Chemical Society, 136*, 2575–2582.
Smedarchina, Z., & Siebrand, W. (2005). Generalized Swain–Schaad relations including tunneling and temperature dependence. *Chemical Physics Letters, 410*, 370–376.
Spencer, H. T., Villafranca, J. E., & Appleman, J. R. (1997). Kinetic scheme for thymidylate synthase from Escherichia coli: Determination from measurements of ligand binding, primary and secondary isotope effects, and pre-steady-state catalysis. *Biochemistry, 36*, 4212–4222.
Stroud, R. M., & Finer-Moore, J. S. (2003). Conformational dynamics along an enzymatic reaction pathway: Thymidylate synthase, "the Movie" *Biochemistry, 42*, 239–247.
Swain, C. G., Stivers, E. C., Reuwer, J. F., & Schaad, L. J. (1958). Use of hydrogen isotope effects to identify the attacking nucleophile in the enolization of ketones catalyzed by acetic acid$^{1-3}$. *Journal of the American Chemical Society, 80*, 5885–5893.
Swanwick, R. S., Shrimpton, P. J., & Allemann, R. K. (2004). Pivotal role of Gly 121 in dihydrofolate reductase from Escherichia coli: The altered structure of a mutant enzyme may form the basis of its diminished catalytic performance†. *Biochemistry, 43*, 4119–4127.
Thorpe, I. F., & Brooks, C. L. (2003). Barriers to hydride transfer in wild type and mutant dihydrofolate reductase from E. coli. *The Journal of Physical Chemistry. B, 107*, 14042–14051.
Thorpe, I. F., & Brooks, C. L. (2004). The coupling of structural fluctuations to hydride transfer in dihydrofolate reductase. *Proteins: Structure, Function, and Bioinformatics, 57*, 444–457.
Wang, Z., Abeysinghe, T., Finer-Moore, J. S., Stroud, R. M., & Kohen, A. (2012). A remote mutation affects the hydride transfer by disrupting concerted protein motions in thymidylate synthase. *Journal of the American Chemical Society, 134*, 17722–17730.
Wang, Z., Antoniou, D., Schwartz, S. D., & Schramm, V. L. (2016). Hydride transfer in DHFR by transition path sampling, kinetic isotope effects, and heavy enzyme studies. *Biochemistry, 55*, 157–166.
Wang, Z., Ferrer, S., Moliner, V., & Kohen, A. (2013). QM/MM calculations suggest a novel intermediate following the proton abstraction catalyzed by thymidylate synthase. *Biochemistry, 52*, 2348–2358.
Wang, L., Goodey, N. M., Benkovic, S. J., & Kohen, A. (2006a). Coordinated effects of distal mutations on environmentally coupled tunneling in dihydrofolate reductase. *Proceedings of the National Academy of Sciences of the United States of America, 103*, 15753–15758.
Wang, L., Goodey, N. M., Benkovic, S. J., & Kohen, A. (2006b). The role of enzyme dynamics and tunnelling in catalysing hydride transfer: Studies of distal mutants of dihydrofolate reductase. *Philosophical Transactions of the Royal Society B, 361*, 1307–1315.
Wang, Z., & Kohen, A. (2010). Thymidylate synthase catalyzed H-transfers: Two chapters in one tale. *Journal of the American Chemical Society, 132*, 9820–9825.
Wang, Z., Sapienza, P. J., Abeysinghe, T., Luzum, C., Lee, A. L., Finer-Moore, J. S., et al. (2013). Mg2+ binds to the surface of thymidylate synthase and affects hydride transfer at the interior active site. *Journal of the American Chemical Society, 135*, 7583–7592.

Wang, Z., Singh, P., Czekster, C. M., Kohen, A., & Schramm, V. L. (2014). Protein mass-modulated effects in the catalytic mechanism of dihydrofolate reductase: Beyond promoting vibrations. *Journal of the American Chemical Society, 136*, 8333–8341.

Wang, L., Tharp, S., Selzer, T., Benkovic, S. J., & Kohen, A. (2006). Effects of a distal mutation on active site chemistry. *Biochemistry, 45*, 1383–1392.

Wataya, Y., & Hayatsu, H. (1972). Cysteine-catalyzed hydrogen isotope exchange at the 5 position of uridylic acid. *Journal of the American Chemical Society, 94*, 8927–8928.

Watney, J. B., Agarwal, P. K., & Hammes-Schiffer, S. (2003). Effect of mutation on enzyme motion in dihydrofolate reductase. *Journal of the American Chemical Society, 125*, 3745–3750.

Wilson, P. M., Danenberg, P. V., Johnston, P. G., Lenz, H.-J., & Ladner, R. D. (2014). Standing the test of time: Targeting thymidylate biosynthesis in cancer therapy. *Nature Reviews. Clinical Oncology, 11*, 282–298.

Wong, K. F., Selzer, T., Benkovic, S. J., & Hammes-Schiffer, S. (2005). Impact of distal mutations on the network of coupled motions correlated to hydride transfer in dihydrofolate reductase. *Proceedings of the National Academy of Sciences of the United States of America, 102*, 6807–6812.

Wong, K. F., Watney, J. B., & Hammes-Schiffer, S. (2004). Analysis of electrostatics and correlated motions for hydride transfer in dihydrofolate reductase. *The Journal of Physical Chemistry. B, 108*, 12231–12241.

Yahashiri, A., Rubach, J. K., & Plapp, B. V. (2014). Effects of cavities at the nicotinamide binding site of liver alcohol dehydrogenase on structure, dynamics and catalysis. *Biochemistry, 53*, 881–894.

CHAPTER TWELVE

Use of QM/DMD as a Multiscale Approach to Modeling Metalloenzymes

N.M. Gallup*, A.N. Alexandrova*,†,1
*University of California, Los Angeles, Los Angeles, CA, United States
†California NanoSystems Institute, Los Angeles, CA, United States
[1]Corresponding author: e-mail address: ana@chem.ucla.edu

Contents

1. Introduction 320
2. Overview of QM/DMD 321
   2.1 Advantages of QM/DMD over Current QM/MM Schemes 321
   2.2 QM/DMD Method 323
3. Setting Up QM/DMD 324
   3.1 Acquiring the Protein Crystal Structure 325
   3.2 Setting Up the DMD Region 326
   3.3 Setting Up the QM–DMD Region 330
4. Running QM/DMD and Details of the Procedure 331
   4.1 The Blueprint of the Algorithm 331
   4.2 Initial Iteration (Iteration 0) 333
   4.3 Full Iterations (Iterations 1 Through N) 333
   4.4 A Simple Analysis of Convergence 335
   4.5 Tasks for Which QM/DMD Is Best Suited 337
   4.6 Permissions, Copyrights, and Utilities 337
5. Conclusions 338
References 338

Abstract

Enzymes are complex biomolecules capable of performing unique catalysis under physiological conditions at neutral temperature and pH. However, the architecture of enzymatic catalysis is often a combination of the quantum influence of the immediate active site, as well as the electrostatic and configurational influences of amino acids surrounding the active site. As a result of this cooperation between baseline chemical reactivity and electrostatic assistance, it has become important to model enzymes using multiscale methods that take advantage of treating the active site with quantum mechanical methods, while approximately treating the surrounding protein using cheaper, classically driven force-field molecular mechanics methods. Here we describe the use of a multiscale engine which utilizes a combination of density functional theory with discrete molecular dynamics (dubbed QM/DMD) to aid in the characterization of metalloenzymes.

1. INTRODUCTION

Enzymes have become particular targets of chemical interest in recent decades due to their marked importance in biological function and the potential for use in industrial applications. However, describing enzyme catalysis, particularly metalloenzymes, possesses unique challenges over typical small-molecule systems (Chung, Li, & Morokuma, 2010; Warshel, 1997). Since the introduction of Warshel and Levitt's simplified protein models (1976), hybrid quantum mechanical and force-field-driven molecular mechanical methods (collectively known as hybrid QM/MM methods) have become an accepted means in describing the catalytic action that takes place at enzyme active sites and has expanded to become a versatile tool in examining solvated small-molecular complexes (Senn & Thiel, 2009). These methods typically rely on partitioning a system into QM and MM subcomponents, and iterating between a QM integration scheme and an MM integration scheme. The QM region usually includes species participating directly in chemical reactions of interest and may expand to include atoms, species, or residues that are thought to exert significant influence on the chemical reactivity that would unlikely be captured via electrostatic or polarizable embedding schemes. The boundary between QM and MM components often bisects covalent bonds. When this occurs, it is common practice in start-to-finish, fixed region schemes to cap these dangling bonds with frozen hydrogen atoms. Dynamic region definition schemes also exist and allow for covering an overall larger QM region, though exhibit more convergence problems. During the QM integration step, the region that surrounds the QM region is usually represented as fixed, polarizable, or nonpolarizable point charges. The MM region is normally the region around the active site that one would wish to be described by classical force fields, ie, harmonic bonding potentials, van der Waals, and Coulombic nonbonded interactions. Occasionally, the scheme may contain an intermediate boundary region, which is influenced by both QM and MM. In some methods, such as ONIOM, the MM calculation is done on both MM and QM regions.

This chapter will focus on the use of a hybrid QM and discrete molecular dynamics (QM/DMD) engine developed with particular focus on metalloenzymes (Sparta, Shirvanyants, Ding, Dokholyan, & Alexandrova, 2012). The QM/DMD engine was designed to adequately handle the additional challenges that embody metalloenzyme catalysis, and the substantial Coulombic influence and configurational preference that active site metals exert on their surroundings. Additionally it was also designed with scalability and speed in mind. QM/DMD takes advantage of the established motif of partitioning a system into QM and MM subcomponents. The QM region is treated with density functional theory (DFT), although other ab initio methods could easily be used. The MM region is treated with DMD which was developed by the Dokholyan group (Dokholyan, Buldyrev, Stanley, & Shakhnovich, 1998). Later, we describe in detail the qualities of QM/DMD that we believe are advantageous over other QM/MM methods. We then provide a thorough description of the method, setup, analysis of the results, and best known uses.

2. OVERVIEW OF QM/DMD

2.1 Advantages of QM/DMD over Current QM/MM Schemes

The hybrid QM/DMD engine attempts to overcome some of the issues relating to the use of traditional QM/MM schemes with enzymes and metalloenzymes. One of the most significant shortcomings of general QM/MM schemes is the difficulty in providing adequate sampling of the active site for a sufficient number of QM configurations. This becomes particularly expensive for QM/MM schemes that utilize Born–Oppenheimer molecular dynamics (BO-MD) for their QM methods. BO-MD seeks to integrate atomic motions of the QM region on the QM potential energy surface while damping the velocity of those motions with a thermostat set to a given temperature, eg, 300 K. While BO-MD provides a highly accurate description of the dynamics of the active site, the large number of calculations required often renders it prohibitively computationally expensive. More traditional QM/MM engines make use of ab initio geometry optimizations instead of QM dynamics. This is a more economical approach, but lacks sampling. In QM/DMD, we take advantage of DMD's ability to provide extensive sampling, but with a frequent reference to QM for the active site geometry. The only region that is never sampled dynamically is the immediate vicinity of the metal.

QM/MM schemes have employed a large variety of QM methods. Hartree–Fock (HF) and post-HF methods are often used, and post-HF description is essentially the only reliable tool when excited state QM/MM simulations are performed. For ground state simulations, DFT became much more popular over HF and post-HF methods, due the increased accuracy and improved scaling with respect to the number of electrons in the system. Semiempirical methods are also used, and are cheap and allow for extensive sampling, but accuracy is questionable. In addition, parameterization, particularly for metals, is extremely limited. The original aim of QM/DMD is to accurately and efficiently sample the ground state ensemble of metalloproteins, and therefore, DFT is the most natural choice for the QM method in QM/DMD.

DMD is particularly exciting part of QM/DMD, as it allows for very rapid sampling, and overcome some of the traditional problems with MM schemes. For example, it is generally acknowledged that solvation can play a critical role not only in providing the electrostatics necessary to facilitate a given reaction but also in the conformation of biomolecules. Generally, QM/MM schemes elect to account for this effect either via the inclusion of explicit water molecules or by freezing the initial backbone configuration of the protein for the duration of the QM/MM simulation.

The inclusion of explicit water molecules is an excellent way to model solvation. For large enzymes and biomolecules, however, a sufficiently solvated solute may require hundreds or thousands of water molecules. Under periodic boundary conditions that try to replicate the effects of a solvent continuum, this number is particularly large, since the solvent box must be large enough to prevent interactions of solutes from neighboring cells. Sometimes, this model is replaced with a finite solvent cap. In either implementation, this massive number of water molecules can drastically increase the computational expense of both the QM and MM portions, and sampling may become intractable, especially if a quality QM method is in use. In QM/DMD, solvation is implicit, apart from a few water molecules that might be directly bound to the active site and/or play a role in the catalyzed reaction. This is more granular, but our extensive tests show that the performance is adequate for a variety of tasks of interest, which will be discussed later. However, one can solvate the DMD-sampled enzyme in another, non-DMD simulation, if explicit placement of water molecules is essential for the project at hand.

Simulations start with a high-quality initial configuration of the protein typically obtained at the Protein Data Bank (PDB) (Berman et al., 2000).

The structures at the PDB typically are X-ray structures, crystallized at nonphysiological, and exceptionally low temperatures to allow for accurate imaging, and often in the presence of unusual solutes or ions that help induce crystallization. It is possible that the protein configuration in the PDB is not the same as at equilibrium at room temperature and normal conditions. Often, however, the protein backbone is kept frozen in QM/MM simulations, because backbone sampling, especially in explicit solvent, adds to computational expense, and equilibration becomes hard to reach. When this is done, it is implicitly asserted that the free-energy surface of the crystallized protein at its low temperature and in its current minima is approximate to neutral, physiological conditions. This assumption may not be valid, since entropy is as large as enthalpy for flexible systems in solution at finite temperatures. Thus, efficient sampling of the protein backbone is essential. DMD is a highly efficient and rapid sampling technique, and with it, we seek to overcome the potential intractability of explicit water, as well as the issues with frozen backbone approximation.

Another pitfall of QM/MM MD is the discontinuities of the potential at the QM–MM boundary. When forces acting on atoms need to be calculated in MD, these discontinuities create cusps. Various force-matching algorithms exist to overcome the problem. In QM/DMD, this problem does not occur, because of DMD's formulation which relies on ballistic equations of motion over Newtonian.

2.2 QM/DMD Method

QM/DMD relies on iterating between a QM integration scheme and an MM integration scheme, like most QM/MM methods. The QM region is defined at the beginning and is fixed start-to-finish. Where the QM region bisects covalent bonds, hydrogen atoms are used to cap the QM portion of the protein. The QM integration scheme consists of a geometry optimization via gradients with DFT to allow for accessible scaling, but the implementation of other ab initio methods is, of course, possible. The protein environment surrounding the QM region during the QM step is modeled using the conductor-like screening continuum solvation model, COSMO (Klamt & Schüürmann, 1993), to increase speed with mild penalty to accuracy. The QM region (in QM/DMD we call it QM–DMD region, because QM and DMD regions overlap in the method) should consist of all meaningful elements of chemical interest and expand outward to up to two or

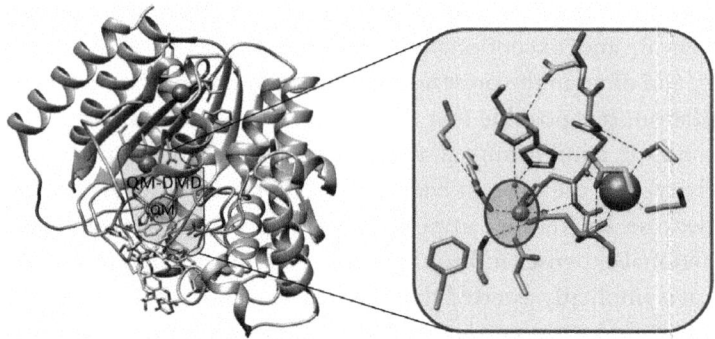

Fig. 1 The *purple* and *red* regions represent the QM–DMD boundary, QM–DMD region, and QM region. The rest of the protein constitutes the DMD-only region. The *inset* on the *right* is the cutout of the QM–DMD region. (See the color plate.)

even three coordination spheres if necessary. A representation of the cooperative quantum and classical regions can be found in Fig. 1.

The DMD region should consist of the entire protein, except for the active site metals and atoms directly coordinating to them. DMD provides for extensive sampling of protein conformations. It relies on an atomistic, event-driven integration of nuclei along spatial, not temporal, coordinates. The DMD method makes use of a slightly coarse-grained force field containing implicit solvation of amino acids in proteins, enabling exclusion of explicit water solvation and subsequently drastically reducing computational costs, while still including the effect of solvation on protein configuration.

The QM and DMD regions have a large overlap, allowing for structural information to transmit across boundaries between each integration step. We find that this approach saves many troubles having to do with the accuracy of treating of the QM–MM boundary. However, the energy of the shared QM–DMD region is double counted in QM/DMD due to this boundary overlap.

3. SETTING UP QM/DMD

Let us walk through a typical setup of QM/DMD. We will discuss the intuition one must utilize when creating the necessary constraints on the system, and then examine some of the potential results and insight that QM/DMD could provide. While much of the information contained herein pertains primarily to QM/DMD, a fair amount of the intuition that goes into selection of QM and DMD regions, respectively, is widely applicable to other QM/MM engines. We will use the metalloenzyme histone deacetylase 8, HDAC8, as an example. The expression of this protein is associated with a variety of diseases, and whose targeted inhibition is an FDA-approved therapy for some cancers (Marks, 2007).

3.1 Acquiring the Protein Crystal Structure

All protein crystal structures at the PDB are assigned unique accession codes. There is often a number of crystal structures for any given enzyme, whose structure could differ based on crystallization environment, degree of success in crystallizing with substrates or important metals, or even mutants that facilitated crystallization of important states. Here, we are going to select the PDB structure 2V5W, which was published in 2007 (Vannini et al., 2007) and has a resolution of 2.0 Å. When selecting a PDB structure, it is important to weight the resolution heavily in your consideration because this represents the experimental confidence in the location of all atoms. If the resolution is extremely low, the orientations of entire residues, or even their identities, may be obfuscated. Most PDB structures have too low a resolution to resolve the position of hydrogen atoms, thus we will need to manually add them later. Besides the resolution, 2V5W contains many other essential components that will allow it to be studied, including the bound substrate and as additional metal ion in the vicinity of the active site. More often than not, it is not possible to find a crystal structure with bound substrate, in which case the substrate can be manually docked to the protein during the QM/DMD setup.

The 2V5W structure is unique in that it is a noncatalytic variant of wild-type HDAC8. In order to facilitate the crystallization of HDAC8 with bound substrate, the catalysis had to be inhibited, which was done by mutating Y306 into F306, whose original identity had been implicated in being essential for catalysis. Since we are interested in describing this catalysis, we need to reverse this mutation. To start, we need to clean up the PDB file of several extraneous components. As seen in Fig. 2, our PDB contains several HDAC8 images. This represents the repeat unit cell of the crystal structure, and the orientation of each enzyme upon crystallization. Differences in these enzymes can sometimes reveal interesting properties about the enzymes themselves or their substrates, but in this case they are redundant and only

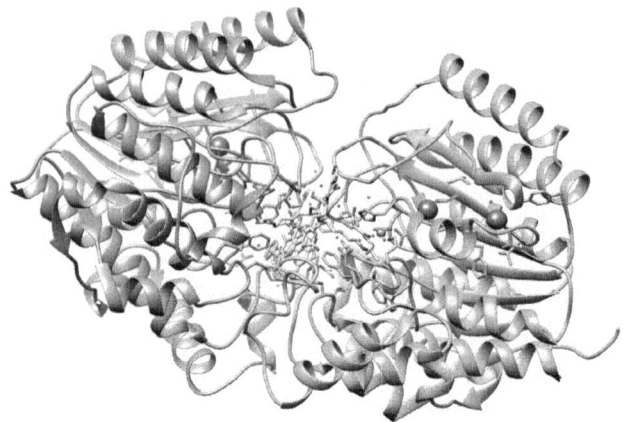

Fig. 2 The two crystallized enzymes are colored in *blue* (*gray* in the print version) and *orange* (*light gray* in the print version). The interface contains a substrate analog and substantial water oxygens.

one is necessary. We kept only one chain (chain A) for QM/DMD simulations.

However, our structure still has a substantial amount of unnecessary components: solvent oxygens and excess substrate; other solutes that were part of the crystallization process but not critical to the functioning of the enzyme may also be present. These excess components are more visible in Fig. 3 and should be manually removed by editing the PDB text file or using a program such as *PyMol* or *UCSF Chimera*. Now our structure has a single enzyme unit, no extraneous solvent, and a viable substrate is already bound (Fig. 4). It will be receiving a few final modifications to make it a high-quality structure for use in QM/DMD.

3.2 Setting Up the DMD Region

The DMD force field includes only polar hydrogens, eg, the εH of a His residue whose coordinates need to be provided. However, particular attention must be given to the active site and substrate, as they may exhibit nontypical protonation states. Since the substrate is nonnative to the DMD force field, it will need to be manually protonated. Since DMD makes a distinction between polar (labeled H in the PDB) and nonpolar hydrogens (Eh in the PDB), one must exercise his/her chemical intuition as to what each hydrogen of the substrate is. It is also important to choose correct protonation states for metal-bound amino acids. His residues often switch their protonation from the εN to the δN, if the former is bound to a metal.

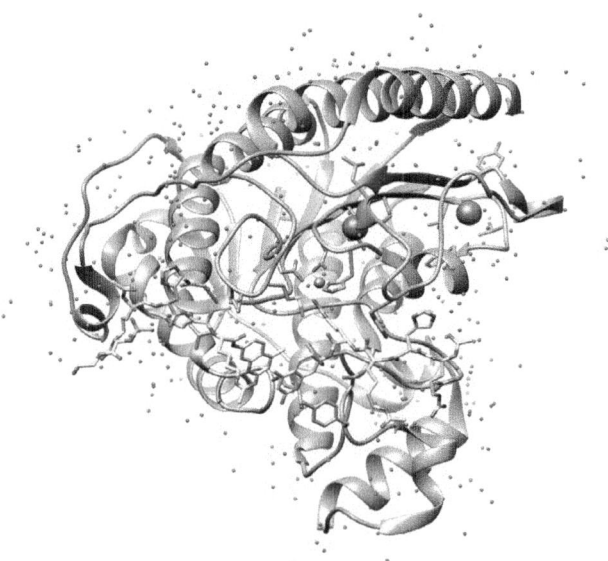

Fig. 3 The cocrystallized solvent is more apparent. Each water oxygen is represented by a *red* (*gray* in the print version) *sphere*. The substrate analogs bound to the remaining enzyme, and the former enzyme can also be observed (tan).

Asp, Glu, and Cys are deprotonated. QM/DMD has built-in ways to override the protonation states of amino acids for use in the active site. Additionally, DMD only has built-in parameters for the Zn ion, thus it is necessary to change the identity of our catalytic metal at the active site to that of Zn, but only for the DMD integration step. The metal's identity will be reversed for the QM step. This is likely to be a reasonable general approximation as, outside of the local bonding environment, any divalent metal ion is likely to exert similar influences to its second coordination sphere, and furthermore, the immediate metal coordination is not sampled by DMD in QM/DMD. The geometry of the local bonding environment can be changed by QM, and QM will see the metal with its true identity.

While DMD offers significant advantages in terms of speed and subsequently sampling, it does come with caveats whose effects should be minimized. In particular, because the substrate is unlikely to be natively included in DMD's force field, DMD will generate parameters for it on-the-fly. These parameters, however, are unlikely to be as high quality as native parameters, and if left unchecked, may facilitate unusual geometries with its coordinating neighbors. Therefore, it might be necessary to adjust the force-field parameters, or apply constraints to the system to prevent it from

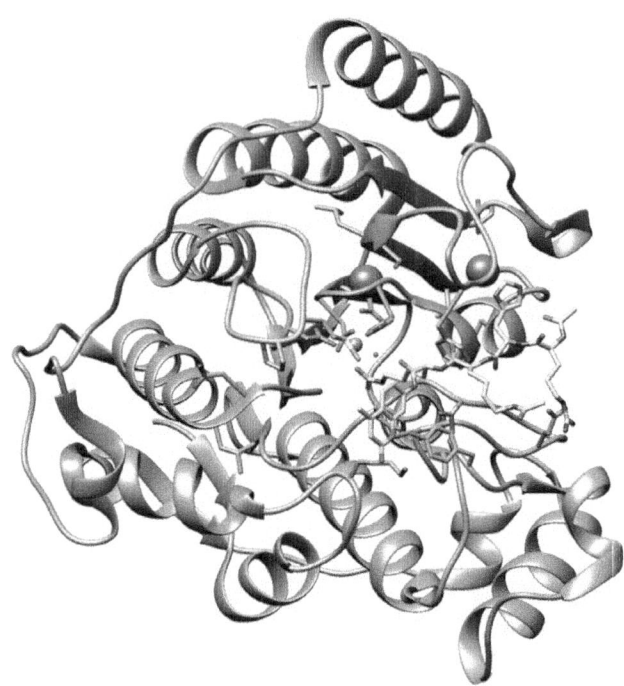

Fig. 4 The final PDB structure. Few modifications remain to make this functional for QM/DMD.

taking on unphysical configurations. Constraints are also always applied to the metal–ligand interactions. Fig. 5 highlights some example considerations at the active site of 2V5W.

Generally speaking, the entire substrate is frozen for the DMD integration step to prevent DMD from contorting it into unnatural configurations. Nonbonded interactions of the substrate with other protein parts are included in DMD, however. Likewise, coordinating ligand bonds to the substrate are also good targets for constraints. As an example, one potential candidate for a constraint is a hydrogen bond highlighted in red in Fig. 5. A small variety of constraints are also available within DMD, such as frozen nuclei positions in Cartesian space or frozen atom–atom distances. DMD's atom–atom constraints are analogous to the common harmonic constraint that can be applied in MD simulations to a pair of atoms; DMD's atom–atom constraint is modeled as a square well, in line with DMD's use of step-wise and discontinuous potentials. It is generally good practice to constrain important substrate–ligand bonds, such as hydrogen bonds, using atom–atom constraints to allow for some sampling of these bonds. Bond length

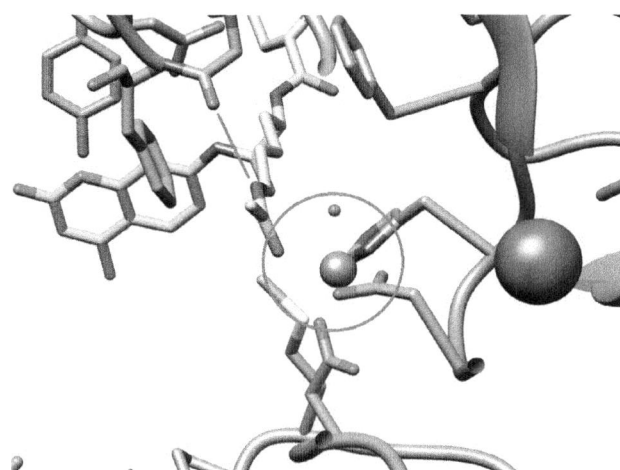

Fig. 5 *Red marks* indicate potential targets for DMD constraints. The *red line* connects a hydrogen bond between the substrate and the backbone of the protein, while the *red circle* highlights the coordinating ligands around the active site metal. (See the color plate.)

constraints on the order of ±0.1 to 1.0 Å is generally adequate in providing a compromise between sampling of reasonable bond lengths and disallowing unusual configurations. Note again that all constraints introduced for DMD are released during QM, during every Iteration. Therefore, sampling in these areas does take place, but only at the QM level.

The active site metal, circled red in Fig. 5, also requires constraints, to prevent DMD from controlling its coordination, and instead follow only the physically relevant QM gradients. It is best practice to apply freezes to all the Cartesian coordinates of the metal and all its coordinating atoms in DMD. Then, DMD would sample the backbone and side chain atoms of the metal ligands, excluding only the metal-coordinating atoms.

Sometimes, it is also important to constrain parts of the system in the second or third coordination spheres, if they exhibit unphysical and systematic fluctuations when other parts of the protein are absent or present. This is often the case for His–His dyads, whose conservation is often important in catalysis, while configuration can vary dramatically if allowed. It is also generally accepted that in the case of a His–His dyad, these two residues will exhibit significant parallel planarity relative to each other. His–His dyads are unique in that it is probably best practice to apply two atom–atom constraints between the two residues to both reduce ligand–ligand distance sampling, but also reduce the degree to which the plane of these two residues can become orthonormal. These constraints can generally be loose (± 1.0 Å), but may need to become more strict depending on the extent of the charge relay network such as in the case of a triad and so on, if the orientation of these residues could have significant impact on the QM region energy. In general, with all these considerations, the number of constraints should be as small as possible, to avoid biases in sampling.

3.3 Setting Up the QM–DMD Region

The QM–DMD region is shared by QM and DMD in QM/DMD. In other QM/MM methods, it would be the analog of a pure-QM region. QM/DMD has been designed to work with *Turbomole* using DFT with the resolution of identity approximation for acceleration, and COSMO implicit solvation scheme. However, QM/DMD has been coded in such a way that the use of any QM software package or method could be feasible. It is also possible to utilize explicit point charge embedding, if desired. Explicit solvation, apart from a few QM water molecules present at the active site and participating in the chemistry, should, again, not be used. This is because DMD is parameterized to include solvation implicitly, and additional solvent would create double counting.

The choice of the shared QM–DMD region should be heavily guided by chemical intuition. It will only consist of a small fraction of the overall protein. In the case of 2V5W, the QM–DMD region should encompass the active site metal, its coordinating ligands, the substrate, the coordinated water molecule, and any other ligands that may play an important role in catalysis. There are no hard rules for what should be included in the QM–DMD region. In 2V5W, the substrate is quite large, and it may be reasonable to truncate it at some point in the chain so that computational resources are not wasted on extraneous parts of the substrate that do not participate in catalysis. In general, QM/DMD allows for significantly expanded QM–DMD regions, but chemical accuracy must be appropriately balanced with the available computational resources. QM–DMD regions on the order of 100–200 atoms are common and tractable.

It is very likely that the choice of QM–DMD region will bisect some covalent bonds. If left unmodified, covalent bond cleavage could lead to an excessive buildup of charge around the dangling bonds and lead to artifacts. There are a variety of ways to ameliorate this issue; however, QM/DMD caps dangling bonds with hydrogen atoms. Both atoms along the dangling bond are frozen in Cartesian space to reflect the attachment to the rest of the protein structure, which remains untreated with QM.

Before the QM step, the DMD ensemble of structures is clustered, as described in detail later, and a single structure is chosen to undergo a QM relaxation. During the QM step, the selected structure undergoes a geometric relaxation, either to convergence or to a predefined maximum number of optimization steps. The relaxed QM geometry is then reinstalled back into the protein. It is important to note that the QM–DMD region may consist of a significant number of atoms that are not frozen in the DMD scheme. This flexibility is important as it allows QM/DMD to relay structural information in this boundary region between the QM and DMD integration steps. QM bond lengths are also used to reparameterize the DMD force field for the QM–DMD region on the fly, by recentering the potential wells.

4. RUNNING QM/DMD AND DETAILS OF THE PROCEDURE

This section will cover, in more technical detail, how QM/DMD actually interfaces the QM and DMD integrations steps. It will describe the manner in which QM/DMD selects appropriate configurations with which to continue additional QM/DMD simulations, as well as some of the metrics that can be used to probe convergence of geometries. It will also highlight the actual processes that take place during the QM and DMD steps. Almost all of the operations described here are automated and take place without user oversight. We will discuss how to use the produced sets of configurations for further studies.

4.1 The Blueprint of the Algorithm

Once QM/DMD has been properly setup, a user needs to define only a few additional parameters, such as the starting annealing temperature, duration of the annealing step in DMD, and duration of the equilibration stage. After that, QM/DMD software is ready to run. The QM/DMD engine will perform the necessary steps in line with its original design philosophy (Fig. 6). First, a QM optimization is performed to relax any poor contacts at the active site. After, this new active site configuration is installed back into the PDB, and a DMD integration step is performed. The enzyme structure begins sampling at an elevated temperature and cooled until it reaches the desired temperature specified by the user at the beginning of the QM/DMD simulation, similar to simulated annealing. Once the target temperature is reached, equilibrium DMD is carried out for a length of time also specified by the user and data collected. Once DMD has concluded, the ensemble is

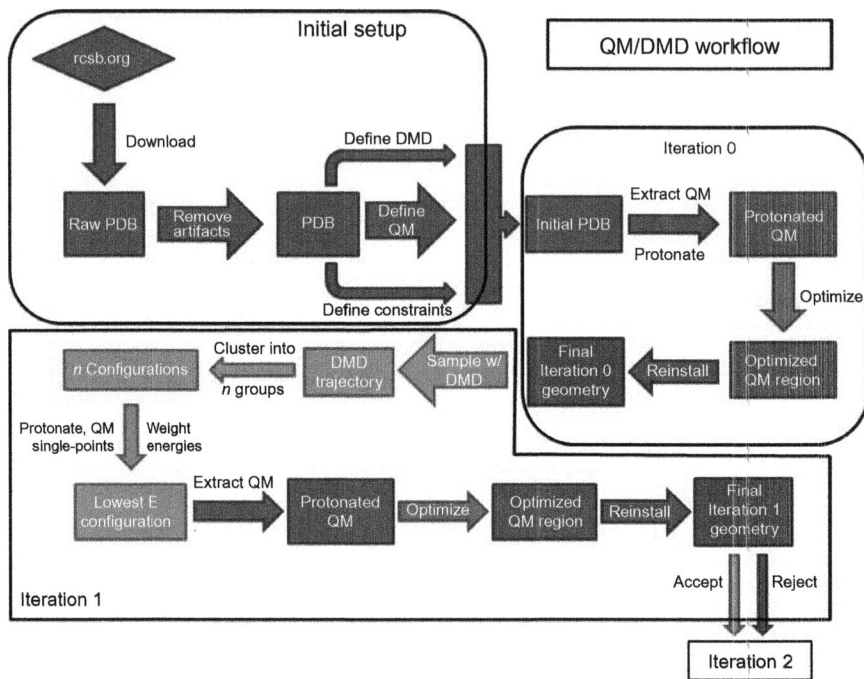

Fig. 6 Schematic overview of the steps QM/DMD goes through in generating publishable configurations for mechanistic analysis. Final configurations are highlighted in *purple* (*dark gray* in the print version). QM optimizations are represented in *light blue* (*gray* in the print version), and steps present only in Iterations >0 are represented in *orange* (*light gray* in the print version) (DMD sampling and handling steps).

clustered, several configurations are selected, their QM–DMD regions are extracted and capped with hydrogens, and single-point QM energies are calculated. The most suitable configuration is then chosen based on both QM and DMD energies, and its geometry is optimized using QM. The optimized QM–DMD region is installed back in the protein and the cycle begins anew. Each step will be described in more detail as we proceed through this section.

QM/DMD subdivides its progress into what have been dubbed Iterations. Each Iteration consists of a DMD integration step, followed by a QM integration step, yields the energy, and coordinates of a DMD-sampled, QM- protein configuration. As QM/DMD progresses, these iterations will, on average, decrease in total energy and eventually converge toward a minimum geometry. Looser constraints during the setup will produce greater fluctuations in energy and will likely require a greater number of Iterations to reach an appropriate minimum. Overly rigid constraints may produce less energetic fluctuations at the expense of potentially preventing the protein from reaching a lower energy structure where catalysis may be more favorable. A balance must be struck between constraint rigidity and desired sampling. It is also generally recommended to increase the temperature at the annealing, and the duration of the simulation, if the starting configuration is suspected to be far from equilibrium. However, convergence can generally be reached within 20–100 Iterations.

4.2 Initial Iteration (Iteration 0)

The QM/DMD engine's initial Iteration is not a complete Iteration, in the sense that it does not consist of both a DMD and QM steps. At the beginning of the simulation, bonding contacts have not been relaxed in either the QM–DMD or DMD regions, and so it is important to relax this configuration with priority given to the active site geometry. For this reason, the initial Iteration (Iteration 0) consists only of a QM gradient-following relaxation on the QM–DMD region. Once an optimized QM geometry is acquired, it is reinstalled back into the enzyme structure, and subsequent Iterations can take place using this as an initial reference geometry.

4.3 Full Iterations (Iterations 1 Through N)

All subsequent Iterations after Iteration 0 utilize both a DMD and QM integration step, in that order. For a given set of Iterations, the lowest energy structure is used as the reference structure. This means that, if the simulation takes the system toward a lower energy conformation, this new conformation will be used to start a new iteration; if, however, a particular Iteration produces a higher energy species, this species is retained for the statistics of the ensemble, but the next Iteration will not start from it, and will use the lowest energy structure instead. This is implemented for equilibrium sampling near a minimum. This option can be turned off, if desired, with a single keyword in the input file. For example, if one might wish to cross low free-energy barriers and access new minima, the option must be turned off.

DMD begins with slowly cooling the structure from an elevated to the target temperature, and then sampling takes place for a specified period of time. The DMD step within each Iteration produces approximately 10,000 configurations, of which every 10th configuration is saved by default, producing 1000 saved snap-shots. To produce the most rigorous results, QM single points would be calculated for each of the configurations along the DMD trajectory. However, due to the amount of sampling DMD performs, this quickly becomes intractable. Instead, a special selection scheme is utilized to sample likely candidates and produce reliable convergence among Iterations. For each configuration saved from the DMD trajectory, the Kabsch RMSD (Kabsch, 1976) is calculated. Configurations with significant similarities are then clustered together into bins of configurations. The maximum number of clusters can be defined prior to executing QM/DMD, allowing the user to specify the degree of granularity implemented for the QM/DMD engine. A typical recommendation would be to use three to five such bins. For each bin, either the structure closest to the centroid, or the lowest energy structure in the bin, or both, can be chosen as the representative geometries for QM single-point calculations to follow. A representative image of this process can be found in Fig. 7.

Representative geometries' QM–DMD regions are then extracted, capped with hydrogens, and evaluated with single-point QM calculations. The DMD energy and QM energy for each structure are combined, to choose the best candidate structure. In traditional QM/MM schemes, complex coupling terms are often employed to provide balance between these

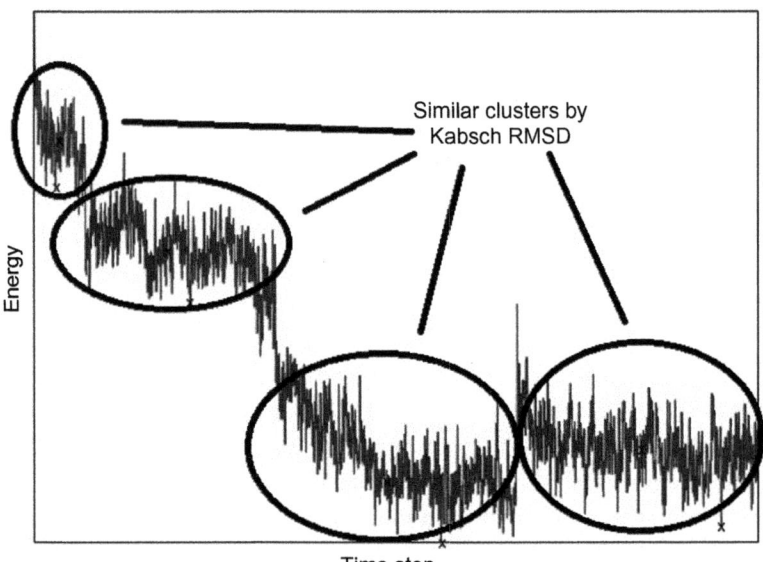

Fig. 7 *Circled regions* represent geometrically similar clusters based on Kabsch RMSD. The configurations marked by a "×" represent the chosen representative configurations based on centroid proximity or lowest energy structure.

two terms, but in QM/DMD these two energies are weighted equally. Thus the total QM/DMD energy becomes

$$E_{\text{total}} = \frac{1}{2}\left(E_{\text{DMD}} + E_{\text{QM}}\right)$$

where E_{total} is the total energy, E_{DMD} is the energetic contributions from DMD, and E_{QM} is the energetic contributions from the QM routine. The total energy is the means by which QM/DMD evaluates which configuration to proceed with for subsequent treatment. Note that it is not the true total energy of the system, due to double counting in the shared QM–DMD region.

The lowest energy structure chosen from cluster single points undergoes a geometric optimization at the QM level, with the constraints representing its attachment to the rest of the protein being observed. After this, the QM–DMD region is reinstalled into the protein, the QM–DMD boundary shrinks back to the QM-only region, and the wells in the DMD potential are recentered in accord with QM-optimized bond length. The system is ready for the DMD step of the next Iteration, when QM information can propagate to the rest of the protein.

4.4 A Simple Analysis of Convergence

An essential goal for QM/DMD is to quickly converge to a reasonable set of geometries that would exist at room temperature and neutral conditions. The simplest means to evaluate convergence over a set of Iterations is to calculate the RMSD between them and compare fluctuations in the total energy. All-atom RMSD of the active site (QM–DMD region), and backbone-only RMSD of the entire protein are monitored. Additionally, the convergence in terms of DMD and QM energies is evaluated. Note, however, that the DMD energy fluctuations can be very large (ca. 20 kcal/mol) and should not be considered on par with the QM energies, which are much more narrowly distributed. This arises because of the discontinuous form of the DMD potential: upon a small change in the protein structure, some interactions might be turned on or completely off, in accord with the square-well potentials. This produces large fluctuations in energy. This behavior of DMD is normal. Fig. 8 shows an example of potential RMSD and energy curves that could be expected from a QM/DMD run.

Once a converged set of geometries has been reached, one can then use a set of these geometries, or the lowest energy conformer for further QM analysis (ie, a mechanistic study). The spread in computed values of interest,

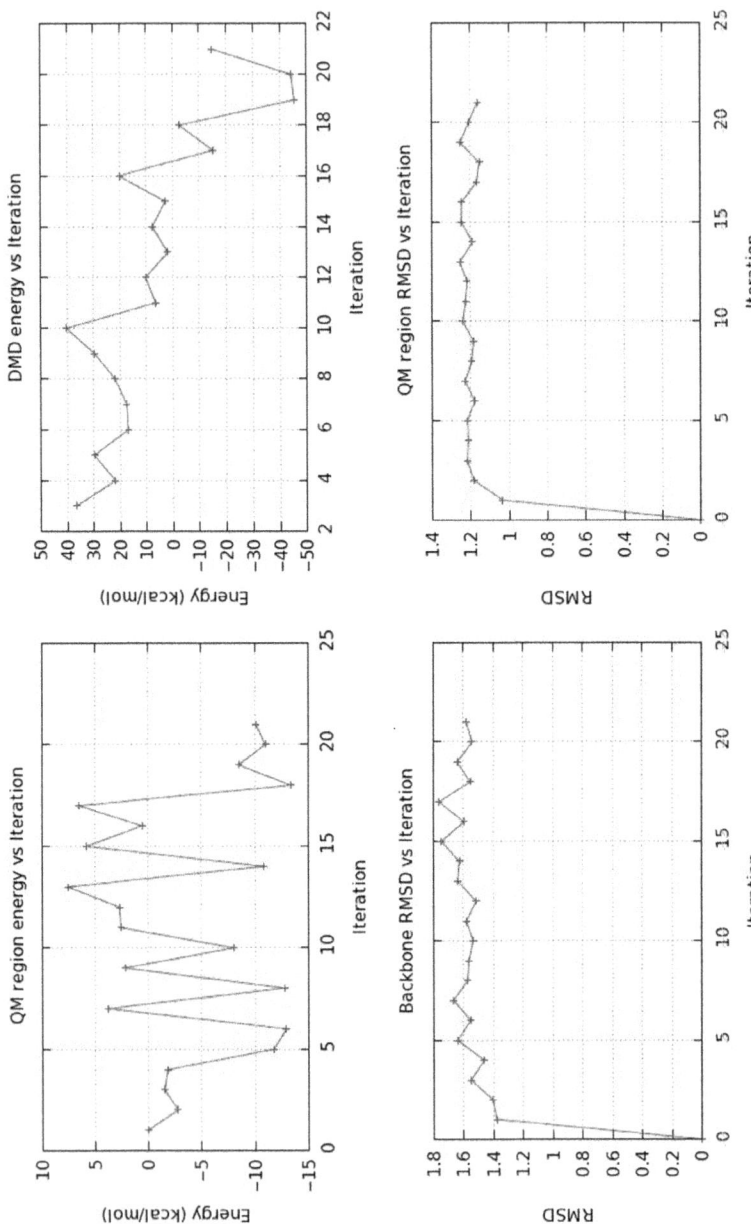

Fig. 8 Comparison of QM and DMD energies (*top*) and RMSDs ×*bottom*) plotted against Iteration number. The combination of a flat RMSD and total energy curve seen here is indicative of geometric convergence.

such as atomic charges or reaction barriers, will represent the ensemble. Further averaging could be done using Boltzmann statistics.

4.5 Tasks for Which QM/DMD Is Best Suited

QM/DMD has been tested and shown to perform well for such tasks as recapitulation of natural protein structures (Sparta et al., 2012), recovery of these structures back to the equilibrium after moderate distortions such as temporary removal and reinstallation of the metal (Sparta et al., 2012), prediction of structural changes upon mutagenesis (Sparta et al., 2012), flexible docking of substrates to metalloproteins that involve motion of large protein parts such as loops (Valdez, Sparta, & Alexandrova, 2013), predictions of protein structure after replacement of the metal or change of its oxidation state and accompanying changes in the coordination sphere geometry and number of ligands (Nedd, Redler, Proctor, & Dokholyan, 2014; Sparta et al., 2012; Sparta, Valdez, & Alexandrova, 2013; Valdez, Gallup, & Alexandrova, 2014), and predictions of protein rearrangements after removal of the metal in cases where the protein does not completely unfold, or the use of a short sequence within 20–30 amino acids (Nedd et al., 2014). Structures produced by QM/DMD are in good agreement with quality X-ray structures, if available (Sparta et al., 2012). For the sets of structures produced by QM/DMD, calculated parameters, such as reaction barriers, are found in good agreement with the experiment (Sparta et al., 2012, 2013; Valdez et al., 2013).

There are also tasks for which QM/DMD is not adequate. Predictions of the absolute values of the reduction potential of the metal cannot be satisfactorily computed with QM/DMD. For these types of calculations, one can use QM/DMD structures and improve the level of QM theory by increasing the basis set, and including the electrostatic embedding. Likewise, small (within 50 mV) changes in the reduction potential, for example, upon mutations around the active site, are not well captured (Sparta et al., 2012). Additionally, the use of QM/DMD for the full metalloprotein folding is not recommended. For small metalloproteins, folding possibly could be done, but this has not been fully tested yet.

4.6 Permissions, Copyrights, and Utilities

QM/DMD is available free of charge. However, in the current implementation it uses Turbomole for the QM calculations, and also DMD, both of which are proprietary software. The user is responsible for obtaining individual licenses for these programs. If one desires to use QM/DMD in research, a request with a short justification must be submitted to the developers (Anastassia Alexandrova, ana@chem.ucla.edu, and Nikolay Dokholyan, nikolay_dokholyan@med.unc.edu). Every publication using QM/DMD must cite *Biophysical Journal,* 2012, 103, 767–776. In addition, in order to make QM/DMD easy to use, we provide a script for setting up a simulation for any metalloprotein starting from a PDB file. It takes care of troubleshooting, or points at possible problems with the provided PDB file, such as presence of unresolvable geometric clashes or the absence of a metal site. We also accompany the QM/DMD distribution with the user manual and are willing to provide reasonable amount of advice for setting up the runs and interpreting the results.

5. CONCLUSIONS

QM/DMD is a hybrid engine designed specifically with biomolecules in mind. It combines DMD with DFT, although other ab initio methods can feasibly be employed, to provide for fast sampling of a protein. The use of DMD over more traditional MD methods overcomes many of the computationally costly obstacles presented by the need for explicit solvation, and algorithmic challenges notorious for QM/MM, such as boundary force matching. QM/DMD provides a means to acquire converged ensembles of geometries for biomolecules at a finite temperature, in a short period of time, for a subsequent more focused study by the user.

REFERENCES

Berman, H. M., Westbrook, J., Feng, Z., Gilliland, G., Bhat, T. N., Weissig, H., et al. (2000). The protein data bank (www.rcsb.org). *Nucleic Acids Research, 28,* 235–242.

Chung, L. W., Li, X., & Morokuma, K. (2010). Modeling enzymatic reactions in metalloenzymes and photobiology by quantum mechanics (QM) and quantum mechanics/molecular mechanics (QM/MM) calculations. *Quantum Biochemistry, 1,* 85–130.

Dokholyan, N. V., Buldyrev, S. V., Stanley, H. E., & Shakhnovich, E. I. (1998). Discrete molecular dynamics studies of the folding of a protein-like model. *Folding & Design, 3,* 577–587.

Kabsch, W. (1976). A solution of the best rotation to relate two sets of vectors. *Acta Crystallographica. Section A, 32,* 922–923.

Klamt, A., & Schüürmann, G. (1993). COSMO: A new approach to dielectric screening in solvents with explicit expressions for the screening energy and its gradient. *Journal of the Chemical Society, Perkin Transactions 2, 2,* 799–805.

Marks, P. A. (2007). Discovery and development of SAHA as an anticancer agent. *Oncogene, 26,* 1351–1356.

Nedd, S., Redler, R. L., Proctor, E. A., & Dokholyan, N. V. (2014). Cu, Zn-superoxide dismutase without Zn is folded but catalytically inactive. *Journal of Molecular Biology, 426*, 4112–4124.

Senn, H. M., & Thiel, W. (2009). QM/MM methods for biomolecular systems. *Angewandte Chemie International Edition, 48*, 1198–1229.

Sparta, M., & Alexandrova, A. N. (2012). How metal substitution affects the enzymatic activity of COMT. *PloS One, 7*, e47172.

Sparta, M., Shirvanyants, D., Ding, F., Dokholyan, N. V., & Alexandrova, A. N. (2012). Hybrid dynamics simulation engine for metalloproteins. *Biophysical Journal, 103*, 767–776.

Sparta, M., Valdez, C. E., & Alexandrova, A. N. (2013). Metal-dependent activity of Fe and Ni acireductone dioxygenases: How two electrons reroute the catalytic pathway. *Journal of Molecular Biology, 245*, 3007–3018.

Valdez, C. E., Gallup, N. M., & Alexandrova, A. N. (2014). Co^{2+} acireductone dioxygenase: Fe^{2+} mechanism, Ni^{2+} mechanism, or something else? *Chemical Physics Letters, 604*, 77–82.

Valdez, C. E., Sparta, M., & Alexandrova, A. N. (2013). The role of the flexible L43-S54 protein loop of the CcrA metallo-beta-lactamase in binding structurally dissimilar beta-lactam antibiotics. *Journal of Chemical Theory and Computation, 9*, 730–737.

Vannini, A., Volpari, C., Gallinari, P., Jones, P., Mattu, M., Carfi, A., et al. (2007). Substrate binding to histone deacetylases as revealed by crystal structure of Hdac8-substrate complex. *EMBO Reports, 8*, 879.

Warshel, A. (1997). Computer modelling of chemical reactions in enzymes and solutions. *Trends in Biotechnology, 15*, 439.

Warshel, A., & Levitt, M. (1976). Theoretical studies of enzymic reactions: Dielectric, electrostatic and steric stabilization of the carbonium ion in the reaction of lysozyme. *Journal of Molecular Biology, 103*, 227–249.

CHAPTER THIRTEEN

Adaptive Partitioning QM/MM Dynamics Simulations for Substrate Uptake, Product Release, and Solvent Exchange

A. Duster, C. Garza, H. Lin[1]
University of Colorado Denver, Denver, CO, United States
[1]Corresponding author: e-mail address: hai.lin@ucdenver.edu

Contents

1. Introduction 342
2. Methodology 344
   2.1 Defining the Buffer Zone 344
   2.2 Interpolating Energy and Gradients 345
   2.3 Setting Zeros of Energy 348
3. Implementation 349
4. Applications of AP Schemes 351
   4.1 Choosing Parameters for AP Calculations 351
   4.2 Parallel Scalability 352
   4.3 Application Example: Solvent Exchange of Protein Active Site 352
5. Summary 353
Acknowledgments 354
References 354

Abstract

Combined quantum mechanics/molecular mechanics (QM/MM) plays an important role in multiscale simulations of biological systems including enzymes. The adaptive-partitioning (AP) schemes surpass the conventional QM/MM methods in that they allow the on-the-fly, smooth exchange of particles between QM and MM subsystems in molecular dynamics simulations, leading to a seamless and dynamic integration of the QM and MM realms. Originally developed for simulating ion solvation in bulk solutions, the AP schemes have recently been extended to the treatment of proteins, fostering applications in the simulations of enzymes. The present contribution provides a detailed account of the AP schemes. We delineate the background of the algorithms and their parallel implementation, as well as offer practical advice and examples for their applications in the simulations of biological systems.

1. INTRODUCTION

Combined quantum mechanics/molecular mechanics (QM/MM) (Chung, Hirao, Li, & Morokuma, 2012; Duarte, Amrein, Blaha-Nelson, & Kamerlin, 2015; Ferrer et al., 2011; Field, Bash, & Karplus, 1990; Gao, 1996; Gao & Truhlar, 2002; Hammes-Schiffer, 2000; Hu & Yang, 2008; Lin & Truhlar, 2007; Lonsdale, Harvey, & Mulholland, 2012; Meier et al., 2013; Menikarachchi & Gascon, 2010; Mennucci, 2013; Monard & Merz, 1999; Monari, Rivail, & Assfeld, 2012; Pezeshki & Lin, 2014a, 2014b; Riccardi et al., 2006; Sabin & Brändas, 2010; Senn & Thiel, 2007; Sherwood, 2000; Sherwood, Brooks, & Sansom, 2008; Singh & Kollmann, 1986; Wallrapp & Guallar, 2011; Warshel & Levitt, 1976; Woodcock et al., 2011; Wu, Cao, & Zhang, 2012) has a long history in the computational modeling of enzymes. In a QM/MM system, a primary subsystem (PS) is treated by an accurate QM level of theory while the rest of the system, which forms the secondary subsystem (SS), is modeled by computationally efficient MM force fields. The PS is also called the QM subsystem, or active zone, and the SS is known as the MM subsystem, or environmental zone. The QM/MM potential energy can be formally expressed as the sum of the energy of the PS, the energy of the SS, and the interaction energy between them (Lin & Truhlar, 2007).

$$E(\text{QM/MM;ES}) = E(\text{QM;PS}) + E(\text{MM;SS}) + E(\text{QM/MM;PS|SS}), \quad (1)$$

The interaction energy usually comprises of both bonded interactions (bond stretching, bond bending, and internal rotation interactions) and nonbonded interactions (electrostatic and van der Waals interactions). While van der Waals interactions are normally included at the MM level, electrostatic interactions are handled differently in various QM/MM schemes. This hybrid delineation allows for realistic descriptions of the PS in the context of its surrounding environment.

Conventionally, all atoms are assigned to either the QM or MM subsystems at the beginning of a QM/MM dynamics simulation and do not change their identities throughout the simulation. Such conventional QM/MM schemes are the standard implementation in all QM/MM software packages. However, the inability to change an atom's identity between QM and MM during simulations precludes the use of conventional QM/MM schemes from wider applications where these identity changes are necessary or preferred. An example is the modeling of an enzyme active site that is in proximity to bulk solvent, for which exchange of solvent molecules with the bulk solvent may occur during the simulations. As the trajectory propagates, a solvent molecule originally from the MM zone may enter the active site and participate in the catalysis. It would be prudent here to update the identity of this solvent molecule from MM to QM, but this cannot be done within the framework of conventional QM/MM. Similar situations are conceivable for substrate uptake, product release, as well as protein conformational changes that lead to a catalytic side-chain moving into or out of the active site. Usually, the possibility of seeing these dynamical changes increases as the simulation time is extended.

One straightforward solution to minimize the impact of the relocation of particles across the QM/MM boundary is to include as many atoms as possible inside the QM subsystem. However, doing so can dramatically increase the cost of the QM calculations, limiting the length and scope of the simulations. The problem may also be circumvented by imposing a restraining potential (or other special treatments) on the QM/MM boundary to prevent QM molecules from diffusing into the MM zone and vice versa; this strategy forms the basis of the flexible inner region ensemble separator (Lev, Roux, & Noskov, 2013; Rowley & Roux, 2012) and boundary-based-on-exchange-symmetry-theory (BEST) (Shiga & Masia, 2013a, 2013b, 2014). However, the use of a restraining potential method perturbs the dynamical properties obtained from these simulations. The third solution is to adopt an adaptive algorithm that allows on-the-fly, smooth exchange of particles between QM and MM subsystems in simulations. This approach has the promise of achieving seamless and dynamical integration of the QM and MM realms.

A number of adaptive QM/MM algorithms have been proposed, including the hot spot scheme (Hofer, Hitzenberger, & Randolf, 2012; Kerdcharoen, Liedl, & Rode, 1996), the "our own n-layered integrated molecular orbital and molecular mechanics-exchange of solvent" scheme (Kerdcharoen & Morokuma, 2002, 2003), the permuted adaptive-partitioning (PAP) scheme (Heyden, Lin, & Truhlar, 2007; Pezeshki, Davis, Heyden, & Lin, 2014; Pezeshki & Lin, 2011, 2015a, 2015b), the sorted adaptive-partitioning (SAP) scheme (Heyden et al., 2007; Pezeshki et al., 2014; Pezeshki & Lin, 2015a, 2015b), the difference-based adaptive solvation scheme (Bulo, Ensing, Sikkema, & Visscher, 2009; Bulo, Michel, Fleurat-Lessard, & Sautet, 2013; Nielsen, Bulo, Moore, & Ensing, 2010), the buffered-force (BF) scheme (Bernstein et al., 2012; Mones et al., 2015; Peguiron, Colombi Ciacchi, De Vita, Kermode, & Moras, 2015; Várnai, Bernstein, Mones, & Csányi, 2013), the number-adaptive scheme

(Takenaka, Kitamura, Koyano, & Nagaoka, 2012a, 2012b), the size-consistent multipartitioning scheme (Watanabe, Kubař, & Elstner, 2014), the density-based adaptive scheme (Waller, Kumbhar, & Yang, 2014), and the time-adaptive switching interaction potential scheme (Böckmann, Doltsinis, & Marx, 2015). Each scheme has its own merits and pitfalls. A number of excellent reviews of these methods are available in literature (Nielsen et al., 2010; Pezeshki et al., 2014; Pezeshki & Lin, 2015a, 2015b).

In the present contribution, we offer a detailed account of the adaptive-partitioning (AP) schemes, which were originally introduced by Heyden et al. (2007) for the simulations of solute in bulk solvent and have since been extended by us to treat fragmental buffer groups of biopolymers (Pezeshki & Lin, 2011), solvent molecules entering and leaving protein active sites (Pezeshki et al., 2014), and proton hopping (Pezeshki & Lin, 2015a, 2015b) via the Grotthuss mechanism (Agmon, 1995). The background of the AP algorithms and their parallel implementation will be delineated. In addition, we will give practical advice concerning parameter selections for the simulations of biological systems.

2. METHODOLOGY
2.1 Defining the Buffer Zone

The AP schemes are an improvement upon conventional QM/MM schemes in that they permit on-the-fly reclassification of atoms as QM or MM throughout simulations. To accomplish smooth transitions, a narrow buffer zone is inserted between the active (QM) and environmental (MM) zones. Although not a requirement, all three zones are usually defined as a series of concentric, spherical shells (Fig. 1). In the AP schemes, we adopt a group-based prescription, where an atom, an ion including polyatomic ion, an entire molecule like water, or a molecular fragment like a CH_3 functional group is treated as a single entity in the partitioning procedure. If the distance r_i from a group to the active-zone center falls between r_{min} and r_{max}, the group can be designated as a buffer group and has dual QM and MM characteristics.

The measurement of r_i requires two reference points: the active-zone center and a reference point for the group. For convenience, both can be taken to be the coordinates of a specific atom (eg, the O in a water molecule) or an imaginary particle, the location of which depends on the coordinates of a number of preselected atoms (eg, the center of mass of a residue side-chain). When monitoring particle exchanges between an enzyme

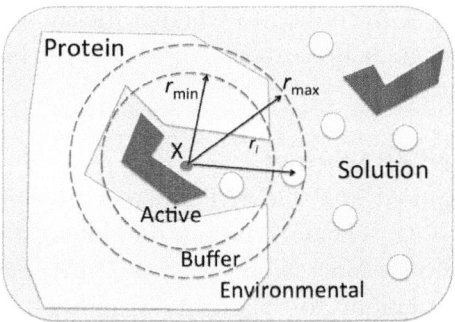

Fig. 1 Illustration of adaptive-partitioning setups. The protein is displayed as a *light-green polygon*, substrate molecules as *dark-green polygons*, and solvent molecules as *yellow spheres*. The active zone is defined as a sphere of radius r_{min} (*dashed red circle*) centered at the active-site center X (*small red circle*). The environmental zone is outside the sphere of radius r_{max} (*dashed green circle*). The thin shell between r_{min} and r_{max} represents the buffer zone. The distance r_i is measured between the active-site center and the *i*th group, which may be a solvent molecule, a substrate molecule, a cofactor molecule, or a residue side-chain of the protein. (See the color plate.)

active site and its surroundings, the active-zone center can be set to a pseudoatom, X, placed in the active site (Pezeshki et al., 2014). Because X does not interact, its location is, in principle, fixed and will not be changed by the diffusion of substrate and solvent molecules or by the swinging of a side chain, making X a convenient choice to represent the active zone.

In some cases (such as proton hopping in water), it is not trivial to decide which atoms should be selected for the determination of the imaginary particle (proton indicator), and the relations between the coordinates of the imaginary particle and of the selected atoms can be very complicated (Pezeshki & Lin, 2015a, 2015b). Finally, we note that the location of the imaginary particle can be selected based on atomic properties (eg, partial charges) other than coordinates.

2.2 Interpolating Energy and Gradients

The interpolations of energy and gradients play a key role in almost all adaptive schemes (one notable exception is the BF scheme). Smoothing functions are applied in order to interpolate the energy and/or gradients of the system during the transition of a group between the active and environmental zones. Otherwise, abrupt changes can occur in energy and gradients, leading to numerical instability and simulation artifacts

(Heyden et al., 2007). According to whether the smoothing functions are applied to energy or gradients, an adaptive scheme can be classified as either energy- or force-based.

2.2.1 Energy-Based AP Schemes

The originally proposed PAP and SAP schemes are *energy* based (Heyden et al., 2007). The PAP scheme expresses the potential of the system in a many-body expansion manner, where the QM/MM energies of all possible partitioning configurations for a given "snapshot" of the system are weighted and summed as follows (Heyden et al., 2007):

$$V = V^A + \sum_{i=1}^{N} P_i(V_i^A - V^A) + \sum_{i=1}^{N-1}\sum_{j=i+1}^{N} P_i P_j \left(V_{i,j}^A - \left[V^A + \sum_{r=i,j}^{N}(V_r^A - V^A) \right] \right)$$

$$\sum_{i=1}^{N-2}\sum_{j=i+1}^{N-1}\sum_{k=j+1}^{N} P_i P_j P_k \left(V_{i,j,k}^A - \left(V^A + \sum_{r=i,j,k}^{N}(V_r^A - V^A) \right) \right)$$

$$+ \sum_{(p,q)=(i,j),(i,k),(j,k)}^{N-1,N} \left(V_{p,q}^A - \left(V^A + \sum_{r=i,j}^{N}(V_r^A - V^A) \right) \right) + \cdots$$

(2)

or more succinctly as (Pezeshki & Lin, 2011):

$$V = V^A \prod_{i=1}^{N}(1-P_i) + \sum_{i=1}^{N} P_i V_i^A \prod_{j\neq i}^{N}(1-P_j) + \sum_{i=1}^{N-1}\sum_{j=i+1}^{N} P_i P_j V_{i,j}^A \prod_{k\neq j\neq i}^{N}(1-P_k) + \cdots$$

(3)

Here, V^A is the QM/MM energy of the configuration with no buffer group treated at the QM level, V_i^A with the ith buffer group treated at the QM level, $V_{i,j}^A$ with the ith and jth buffer groups treated at the QM level, ... $V_{1,2,\ldots,N}^A$ with all N buffer groups at the QM level, and P_i is the smoothing function for the ith buffer group, the value of which varies between 0 and 1 depending on the location of the buffer group. A buffer group is therefore classified as QM in some configurations and as MM in others, in line with its dual QM–MM characteristics. This contrasts with the groups in the active and environmental zones, which remain as QM and MM, respectively, in all possible configurations.

In total, 2^N configurations must be evaluated in order to obtain the *full* PAP potential. However to reduce computational expense, the expansion can be truncated because higher-order terms typically contribute much less to the full potential than lower-order terms. The product of the weights, P_i, further scales down the higher-order contributions. In test calculations, we found negligible discontinuities in energy and gradients, if the expansion was truncated at the fifth order (Heyden et al., 2007; Pezeshki & Lin, 2011).

The smoothing function P_i is required to vary continuously and smoothly between 0 (when the buffer group is at the buffer-environmental boundary) and 1 (when the buffer group is at the active-buffer boundary). All derivatives of the potential energy with respect to the coordinates will vary smoothly up to the same order for which P_i varies continuously. We have examined a variety of smoothing functions and recommended a fifth-order spline (Heyden et al., 2007):

$$P_i(\alpha_i) = -6\alpha_i^5 + 15\alpha_i^4 - 10\alpha_i^3 + 1, \qquad (4)$$

where α_i is a metric for the position of the buffer group:

$$\alpha_i = \frac{r_i - r_{min}}{r_{max} - r_{min}} \quad \text{for } r_{min} \leq r_i \leq r_{max} \qquad (5)$$

The SAP scheme can be regarded as a simplified PAP scheme with greatly reduced computational costs. In SAP, the buffer groups are sorted in a canonical order according to their distances to the active-zone center: the first buffer group is the closest to the active-zone center, and the Nth buffer group is the farthest. Only $N+1$ configurations are considered, where the buffer groups are added to the QM treatment one at a time according to their decreasing ranks. Therefore, the first configuration has no buffer group treated at the QM level, the second configuration includes the first buffer group at the QM level, and the last configuration includes all buffer groups at the QM level. The SAP potential is given by

$$V = V^A \prod_{j=1}^{N}(1-\Phi_j) + \Phi_1 V_1^A \prod_{j=2}^{N}(1-\Phi_j) + \Phi_2 V_{1,2}^A \prod_{j=3}^{N}(1-\Phi_j) + \cdots \qquad (6)$$

where Φ_i is the smoothing function that ensures not only the smooth changes of the energy and gradients of one group moving in the buffer zone but also constant energy and gradients when two buffer groups switch ranks (Heyden et al., 2007). The additional requirement for rank switching has led to far more complicated smoothing functions being implemented (and smaller time step sizes to be used for retaining numerical stability).

One important benefit of using SAP or PAP schemes is that there is a well-defined energy for the entire system that is conserved throughout the simulation. Moreover, the forces are computed exactly as the negative of the gradients of the potential. As a result, the forces experienced by *all* atoms are smoothed, and the momentums of the system are strictly conserved. The rigorousness of the algorithms is demonstrated by producing the same thermodynamic properties (eg, radial distribution functions) in both NVE and NVT simulations of an argon liquid, in agreement with what one will expect from the thermodynamics limit (Heyden et al., 2007).

2.2.2 Force-Based AP Schemes

The energy-based AP schemes, however, suffer from a major drawback: the existence of "extra forces" due to the gradients of the smoothing functions, which can lead to artifacts if they are not negligible compared with the "true" forces due to the interatomic interactions. In principle, these extra forces can be eliminated or minimized by carefully aligning the QM and MM potentials in the boundary region, as was done in the simulations of argon liquid. However, this can be difficult to achieve, especially for polar molecules, which have anisotropic potentials. Often, a simple alignment is performed such that the QM and MM potentials agree when the interacting groups are separated indefinitely. A general and straightforward alternative solution to handle the artificial extra forces is to simply delete them, as was done in the modified AP schemes (Pezeshki et al., 2014).

The removal of the extra forces in the modified AP schemes is equivalent to adding on the system external forces that exactly cancel the extra forces. The system is no longer isolated, as external forces do work on the system. Consequently, the energy is no longer conserved. On the other hand, because the extra forces exist in pairs, their removal does not violate Newton's Third Law of Motion, and the momentums of the system are strictly conserved. The modified AP schemes describe non-Hamiltonian systems, which can be coupled to thermostats so that NVT simulations can be carried out. The modified AP schemes are classified as *force*-based schemes, because the quantities of central interest are the interpolated forces, although they are no longer exactly the negative of the gradients of the energy.

2.3 Setting Zeros of Energy

In the AP schemes, the zero of QM or MM energy of a molecule is set to the QM or MM energy of the isolated molecule of QM- or MM-optimized

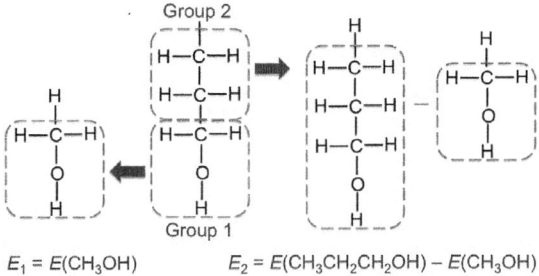

Fig. 2 Example of zeros of energy for molecular fragments as buffer groups. A super-group $CH_2CH_2CH_2OH$ is formed by merging Group 1 (CH_2OH) and Group 2 (CH_2CH_2). The energy of Group 1 is set to that of the H-capped fragment (CH_3OH). The energy of Group 2 is set to the energy difference between the capped super-group ($CH_3CH_2CH_2OH$) and Group 1.

geometry. Complications arise for a molecular fragment, which can be a functional group of a large substrate or a residue side-chain of the protein. These complications are due to the additions of link atoms (usually H atoms) to saturate the dangling bonds of the molecular fragment. The link-atom approach is straightforward and popular, though more sophisticated treatments of the QM/MM boundary are available (Antes & Thiel, 1999; Assfeld & Rivail, 1996; Gao, Amara, Alhambra, & Field, 1998; Lin & Truhlar, 2005; Pezeshki & Lin, 2014a, 2014b; Zhang, Lee, & Yang, 1999; Zhang & Lin, 2008, 2010; Zhang, Lin & Truhlar, 2007). As illustrated in Fig. 2, two fragmental groups will merge into a super-group, if they both are treated at the QM level of theory. To be self-consistent, the sum of the energies of both groups should equal the energy of the super-group. An automated procedure is therefore implemented to merge the groups, place the link atoms, and compute the energies based on the molecular topology (Pezeshki & Lin, 2011).

3. IMPLEMENTATION

The PAP and SAP algorithms have been implemented in the QMMM program (Lin, Zhang, Pezeshki, Duster, & Truhlar, 2015). The QMMM program is an interfacing program, which calls an MM package (NAMD (Phillips et al., 2005) or Tinker (Ponder, 2010)) for MM calculations and a QM package (MNDO (Thiel, 2005), Gaussian (Frisch et al., 2010), or ORCA (Neese, 2011)) for the QM calculations. The QMMM program then synthesizes the QM and MM gradients and propagates the trajectory in molecular dynamics simulations.

An important advantage of the AP algorithms is that the calculations of the partitioning configurations are parallel in nature, making code parallelization straightforward. In principle, all configurations can be evaluated at the same time, and the needed wall-clock time will be dominated by the calculation of the configuration with the largest number of buffer groups treated at the QM level. In practice, the efficiency, or the scaling behavior, will depend on the number of CPU cores available and on the number of buffer groups. The number of buffer groups varies during a simulation due to the constant motion of the atoms, requiring special care in the implementation.

Fig. 3 shows the flowchart of the PAP method implemented in QMMM. After obtaining an initial set of coordinates and user-defined groups, the program calculates the gradients and energy of the entire system at the MM level. Next, the code compiles/updates the lists of groups in the buffer and active zones, respectively. For each group in the buffer zone, the value of P_i is calculated. The program then assembles a list of all partitioning configurations through permutations in the many-body expansion up to the

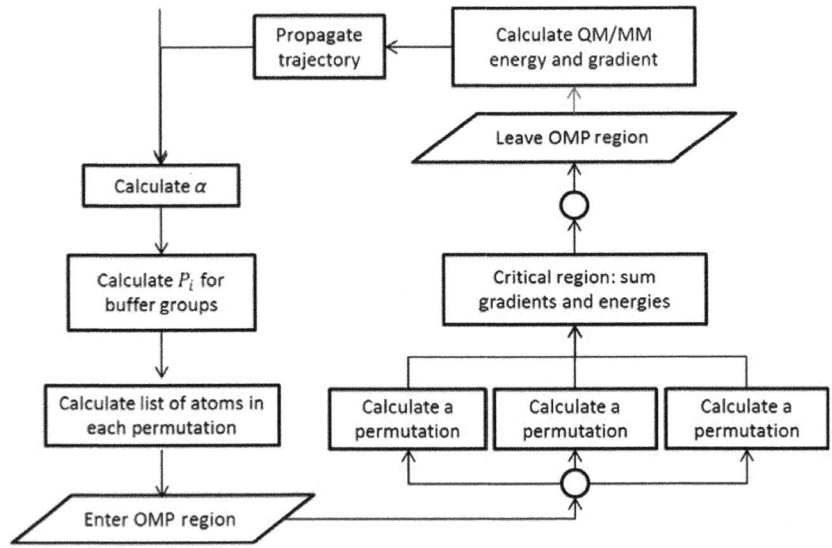

Fig. 3 Flowchart of PAP implementation in QMMM.

truncated order. Using the OpenMP protocol, the code spawns a number of threads. The configurations are assigned to the threads using dynamic scheduling. For each configuration, only the active and buffer groups need to be calculated owing to the subtractive QM/MM-energy definition (Sherwood, 2000). After all configurations are calculated, the threads enter an OMP critical region where the components are summed to yield the total QM/MM energy and gradients, based on which the trajectory is propagated. The SAP can be implemented similarly.

4. APPLICATIONS OF AP SCHEMES

4.1 Choosing Parameters for AP Calculations

The on-the-fly update of the contents of the QM and MM subsystems makes the finite QM subsystems "infinite" and, in principle, can sustain simulations of arbitrary length. During simulations of substrate uptake, product release, and solvent exchange for an enzyme, one usually concentrates on the active site, where the active zone is typically located. As mentioned earlier, the active-zone center can be located at a specific atom of the protein in the active site, the center of mass of a few preselected protein atoms, or a pseudoatom placed at the geometric center of the active site. If one is more interested in following the substrate or product wherever it goes, the active-zone center can also be placed at the substrate or product. In either case, the active zone should incorporate important QM interactions of the reaction being catalyzed, eg, cofactors and key catalytic residues. Of course, one can always designate a specific residue and/or a specific type of molecule to be QM or MM, regardless its distance to the active-zone center.

The buffer zone should be sufficiently large to ensure a smooth transition between QM and MM potentials. On the other hand, a larger buffer zone can lead to more buffer groups and higher computational costs. Fortunately, this is less an issue for an enzyme active site *partially* open to the bulk solvent than for a solute *fully* solvated in bulk solution. Based on our experience, a buffer thickness of at least 1 Å is recommended, if it is computationally feasible. The use of small time step size such as 0.5 fs instead of the common choice 1.0 fs also helps to smooth the transition, allowing a smaller buffer zone to be used.

4.2 Parallel Scalability

The scalability of the AP schemes depends strongly on n, the number of configurations to be evaluated, which varies from one model system to another and from time to time in a given simulation. It should be noted that in PAP, n does not scale linearly with N, the number of buffer groups, unless the potential is truncated at the first order. Speedup from parallel processing reaches maximum once the available number of CPU cores catches up with n. In a test calculation of proton transfer in bulk water, with a small active zone ($r_{min} = 4$ Å) and a small buffer zone ($r_{max} = 4.5$ Å), the average number of configurations is 6 per step. The scalability almost reaches the theoretical limit of 6 when 6 CPU cores are used.

Our code is currently parallelized through the OpenMP protocol, and only the cores in one node can be used. In the future, we will attempt a hybrid OpenMP/MPI approach such that multiple nodes can be utilized. The improvement will make the AP schemes very attractive for long-time scale QM/MM simulations of enzymes employing high-performance computers, which have massive numbers of nodes.

4.3 Application Example: Solvent Exchange of Protein Active Site

The AP schemes have been applied to the monitoring of solvent molecules entering and leaving the binding site of the *Escherichia coli* CLC (EcCLC) transport protein (Pezeshki et al., 2014). Although the transport protein is not an enzyme, the simulations are illuminating, because the diffusive nature of solvent exchange between the active site and the bulk solvent for this protein is the same as for an enzyme. Two EcCLC-binding sites (S_{int} and S_{ext}) were simulated using the modified PAP (mPAP) scheme, each for 50 ps. For each binding site, the active-zone sphere of $r_{min} = 6.0$ Å is centered at a pseudoatom that is placed at the initial position of the bound Cl^- ion, and the buffer is 0.5 Å thick. The PM3 semiempirical Hamiltonian was selected for the QM calculations, and the CHARMM22 force fields and TIP3P water model were employed for the MM descriptions. Fig. 4 shows the simulations of the S_{int} binding site. The results were found in satisfactory agreement with conventional QM/MM simulations on the same model system but using *larger* QM subsystems. For example, the minimum, average, and maximum numbers of water molecules in the active zone were 3, 4.7, and 6 in the mPAP simulations, respectively, which are compared favorably with 4, 5.4, and 7 in the conventional QM/MM simulations. For the buffer

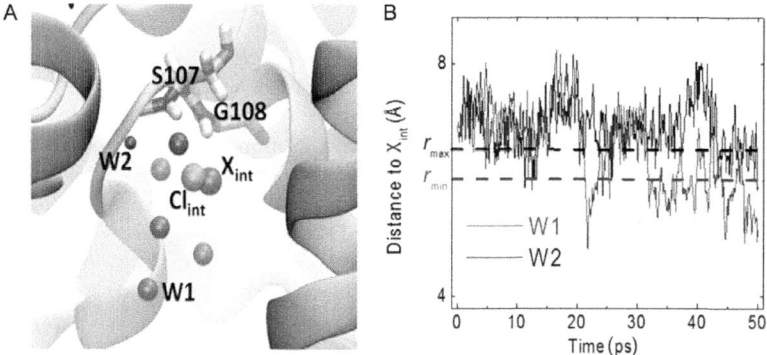

Fig. 4 (A) Snapshot from the mPAP simulation. The protein is shown as cartoon in *yellow*, with two key residues S107 and G108 in *sticks* (N: *blue*, C: *cyan*, O: *red*, and H: *white*). The bound Cl⁻ (Cl_{int}) and pseudoatom (X_{int}) are indicated as *green* and *purple* spheres, respectively, and water in the active and buffer zones as *large and small red spheres*, respectively. W1 and W2 are two water molecules, whose distances to X_{int} are plotted in (B) over simulation time, where the *green* and *black dashed lines* indicate the active-buffer and buffer-environmental boundaries, respectively. *Reprinted with permission from Pezeshki, S., Davis, C., Heyden, A., & Lin, H. (2014). Adaptive-partitioning QM/MM dynamics simulations: 3. Solvent molecules entering and leaving protein binding sites. Journal of Chemical Theory and Computation 10, 4765–4776. Copyright 2014 American Chemical Society.* (See the color plate.)

region, the corresponding numbers were 0, 0.6, and 4 in mPAP and 0, 0.7, and 3 in conventional QM/MM, respectively. The results suggest that mPAP provides a realistic description of the frequent exchange of solvent molecules between the binding site and bulk solvent for the EcCLC transport protein.

5. SUMMARY

In this contribution, we presented a detailed account of the AP QM/MM algorithms. The theoretical background, code implementation, parameter selection, and parallel scalability were discussed. The capability of on-the-fly smooth exchanging particles between the QM and MM regions in dynamics simulations makes the AP methods very suitable in the simulations of substrate uptake, product release, and solvent exchange for enzymes, especially when long simulation times are sought. We envision that the AP schemes will continue to be refined and applied in the future.

ACKNOWLEDGMENTS

This work is supported by National Science Foundation (CHE-0952337 and CHE-1564349). This work used the Extreme Science and Engineering Discovery Environment (XSEDE) under Grant CHE-140070, which is supported by National Science Foundation Grant number ACI-1053575, and the National Energy Research Scientific Computing Center (NERSC), a DOE Office of Science User Facility supported by the Office of Science of the U.S. Department of Energy under Contract No. DE-AC02-05CH11231. H.L. thanks the Camille & Henry Dreyfus Foundation for support (TH-14-028). C.G. is supported by the Undergraduate Research Opportunity Program of the University of Colorado Denver. Helps from Nara Chon and Mackenzie Zarecki are greatly appreciated.

Conflict of interest: The authors declare no competing financial interest.

REFERENCES

Agmon, N. (1995). The Grotthuss mechanism. *Chemical Physics Letters, 244*, 456–462.
Antes, I., & Thiel, W. (1999). Adjusted connection atoms for combined quantum mechanical and molecular mechanical methods. *Journal of Physical Chemistry A, 103*, 9290–9295.
Assfeld, X., & Rivail, J.-L. (1996). Quantum chemical computations on parts of large molecules: The ab initio local self consistent field method. *Chemical Physics Letters, 263*, 100–106.
Bernstein, N., Varnai, C., Solt, I., Winfield, S. A., Payne, M. C., Simon, I., et al. (2012). QM/MM simulation of liquid water with an adaptive quantum region. *Physical Chemistry Chemical Physics, 14*, 646–656.
Böckmann, M., Doltsinis, N. L., & Marx, D. (2015). Adaptive switching of interaction potentials in the time domain: An extended Lagrangian approach tailored to transmute force field to QM/MM simulations and back. *Journal of Chemical Theory and Computation, 11*, 2429–2439.
Bulo, R. E., Ensing, B., Sikkema, J., & Visscher, L. (2009). Toward a practical method for adaptive QM/MM simulations. *Journal of Chemical Theory and Computation, 5*, 2212–2221.
Bulo, R. E., Michel, C., Fleurat-Lessard, P., & Sautet, P. (2013). Multiscale modeling of chemistry in water: Are we there yet? *Journal of Chemical Theory and Computation, 9*, 5567–5577.
Chung, L. W., Hirao, H., Li, X., & Morokuma, K. (2012). The ONIOM method: Its foundation and applications to metalloenzymes and photobiology. *Wiley Interdisciplinary Reviews: Computational Molecular Science, 2*, 327–350.
Duarte, F., Amrein, B. A., Blaha-Nelson, D., & Kamerlin, S. C. L. (2015). Recent advances in QM/MM free energy calculations using reference potentials. *Biochimica et Biophysica Acta (BBA) - General Subjects, 1850*, 954–965.
Ferrer, S., Ruiz-Pernia, J., Marti, S., Moliner, V., Tunon, I., Bertran, J., et al. (2011). Hybrid schemes based on quantum mechanics/molecular mechanics simulations: Goals to success, problems, and perspectives. In C. Christov (Ed.), *Computational chemistry methods in structural biology: Vol. 85. Advances in protein chemistry and structural biology* (pp. 81–142.) San Diego, CA: Elsevier Academic Press Inc.
Field, M. J., Bash, P. A., & Karplus, M. (1990). A combined quantum mechanical and molecular mechanical potential for molecular dynamics simulations. *Journal of Computational Chemistry, 11*, 700–733.
Frisch, M. J., Trucks, G. W., Schlegel, H. B., Scuseria, G. E., Robb, M. A., Cheeseman, J. R., et al. (2010). *Guassian09*. Wallingford: CT Gaussian, Inc.

Gao, J. (1996). Methods and applications of combined quantum mechanical and molecular mechanical potentials. *Reviews in Computational Chemistry, 7*, 119–185.

Gao, J., Amara, P., Alhambra, C., & Field, M. J. (1998). A generalized hybrid orbital (GHO) method for the treatment of boundary atoms in combined QM/MM calculations. *Journal of Physical Chemistry A, 102*, 4714–4721.

Gao, J., & Truhlar, D. G. (2002). Quantum mechanical methods for enzyme kinetics. *Annual Review of Physical Chemistry, 53*, 467–505.

Hammes-Schiffer, S. (2000). Theoretical perspectives on proton-coupled electron transfer reactions. *Accounts of Chemical Research, 34*, 273–281.

Heyden, A., Lin, H., & Truhlar, D. G. (2007). Adaptive partitioning in combined quantum mechanical and molecular mechanical calculations of potential energy functions for multiscale simulations. *Journal of Physical Chemistry B, 111*, 2231–2241.

Hofer, T. S., Hitzenberger, M., & Randolf, B. R. (2012). Combining a dissociative water model with a hybrid QM/MM approach—A simulation strategy for the study of proton transfer reactions in solution. *Journal of Chemical Theory and Computation, 8*, 3586–3595.

Hu, H., & Yang, W. (2008). Free energies of chemical reactions in solution and in enzymes with ab initio quantum mechanics/molecular mechanics methods. *Annual Review of Physical Chemistry, 59*, 573–601.

Kerdcharoen, T., Liedl, K. R., & Rode, B. M. (1996). A QM/MM simulation method applied to the solution of Li^+ in liquid ammonia. *Chemical Physics, 211*, 313–323.

Kerdcharoen, T., & Morokuma, K. (2002). ONIOM-XS: An extension of the ONIOM method for molecular simulation in condensed phase. *Chemical Physics Letters, 355*, 257–262.

Kerdcharoen, T., & Morokuma, K. (2003). Combined quantum mechanics and molecular mechanics simulation of Ca^{2+}/ammonia solution based on the ONIOM-XS method: Octahedral coordination and implication to biology. *Journal of Chemical Physics, 118*, 8856–8862.

Lev, B., Roux, B., & Noskov, S. Y. (2013). Relative free energies for hydration of monovalent ions from QM and QM/MM simulations. *Journal of Chemical Theory and Computation, 9*, 4165–4175.

Lin, H., & Truhlar, D. G. (2005). Redistributed charge and dipole schemes for combined quantum mechanical and molecular mechanical calculations. *Journal of Physical Chemistry A, 109*, 3991–4004.

Lin, H., & Truhlar, D. G. (2007). QM/MM: What have we learned, where are we, and where do we go from here? *Theoretical Chemistry Accounts, 117*, 185–199.

Lin, H., Zhang, Y., Pezeshki, S., Duster, A., & Truhlar, D. G. (2015). *QMMM*. Minneapolis, MN: University of Minnesota.

Lonsdale, R., Harvey, J. N., & Mulholland, A. J. (2012). A practical guide to modelling enzyme-catalysed reactions. *Chemical Society Reviews, 41*, 3025–3038.

Meier, K., Choutko, A., Dolenc, J., Eichenberger, A. P., Riniker, S., & van Gunsteren, W. F. (2013). Multi-resolution simulation of biomolecular systems: A review of methodological issues. *Angewandte Chemie, International Edition, 52*, 2820–2834.

Menikarachchi, L. C., & Gascon, J. A. (2010). QM/MM approaches in medicinal chemistry research. *Current Topics in Medicinal Chemistry, 10*, 46–54.

Mennucci, B. (2013). Modeling environment effects on spectroscopies through QM/classical models. *Physical Chemistry Chemical Physics, 15*, 6583–6594.

Monard, G., & Merz, K. M., Jr. (1999). Combined quantum mechanical/molecular mechanical methodologies applied to biomolecular systems. *Accounts of Chemical Research, 32*, 904–911.

Monari, A., Rivail, J.-L., & Assfeld, X. (2012). Theoretical modeling of large molecular systems. Advances in the local self consistent field method for mixed quantum mechanics/molecular mechanics calculations. *Accounts of Chemical Research, 46*, 596–603.

Mones, L., Jones, A., Götz, A. W., Laino, T., Walker, R. C., Leimkuhler, B., et al. (2015). The adaptive buffered force QM/MM method in the CP2K and AMBER software packages. *Journal of Computational Chemistry, 36*, 633–648.
Neese, F. (2011). *ORCA*. Bonn, Germany: University of Bonn.
Nielsen, S. O., Bulo, R. E., Moore, P. B., & Ensing, B. (2010). Recent progress in adaptive multiscale molecular dynamics simulations of soft matter. *Physical Chemistry Chemical Physics, 12*, 12401–12414.
Peguiron, A., Colombi Ciacchi, L., De Vita, A., Kermode, J. R., & Moras, G. (2015). Accuracy of buffered-force QM/MM simulations of silica. *Journal of Chemical Physics, 142*, 064116/1-13.
Pezeshki, S., Davis, C., Heyden, A., & Lin, H. (2014). Adaptive-partitioning QM/MM dynamics simulations: 3. Solvent molecules entering and leaving protein binding sites. *Journal of Chemical Theory and Computation, 10*, 4765–4776.
Pezeshki, S., & Lin, H. (2011). Adaptive-partitioning redistributed charge and dipole schemes for QM/MM dynamics simulations: On-the-fly relocation of boundaries that pass through covalent bonds. *Journal of Chemical Theory and Computation, 7*, 3625–3634.
Pezeshki, S., & Lin, H. (2014a). Molecular dynamics simulations of ion solvation by flexible-boundary QM/MM: On-the-fly partial charge transfer between QM and MM subsystems. *Journal of Computational Chemistry, 35*, 1778–1788.
Pezeshki, S., & Lin, H. (2014b). Recent developments in QM/MM methods towards open-boundary multi-scale simulations. *Molecular Simulation, 41*, 168–189.
Pezeshki, S., & Lin, H. (2015a). Adaptive-partitioning QM/MM for molecular dynamics simulations: 4. Proton hopping in bulk water. *Journal of Chemical Theory and Computation, 11*, 2398–2411.
Pezeshki, S., & Lin, H. (2015b). Recent developments in adaptive QM/MM. In J.-L. Rivail, M. Ruiz-Lopez, & X. Assfeld (Eds.), *Quantum modeling of complex molecular systems* (pp. 93–114). New York: Springer.
Phillips, J. C., Braun, R., Wang, W., Gumbart, J., Tajkhorshid, E., Villa, E., et al. (2005). Scalable molecular dynamics with NAMD. *Journal of Computational Chemistry, 26*, 1781–1802.
Ponder, J. W. (2010). *TINKER*. St. Louis, MO: Washington University.
Riccardi, D., Schaefer, P., Yang, Y., Yu, H., Ghosh, N., Prat-Resina, X., et al. (2006). Development of effective quantum mechanical/molecular mechanical (QM/MM) methods for complex biological process. *Journal of Physical Chemistry B, 110*, 6458–6469.
Rowley, C. N., & Roux, B. (2012). The solvation structure of Na^+ and K^+ in liquid water determined from high level ab Initio molecular dynamics simulations. *Journal of Chemical Theory and Computation, 8*, 3526–3535.
Sabin, J. R., & Brändas, E. (Eds.), (2010). *Advances in quantum chemistry: Vol. 59. Combining quantum mechanics and molecular mechanics. Some recent progresses in QM/MM methods:* (pp. 1–416). Cambridge, MA: Academic Press.
Senn, H. M., & Thiel, W. (2007). QM/MM methods for biological systems. *Topics in Current Chemistry, 268*, 173–290.
Sherwood, P. (2000). Hybrid quantum mechanics/molecular mechanics approaches. In J. Grotendorst (Ed.), *Modern methods and algorithms of quantum chemistry: Vol. 3* (pp. 285–305). Jülich: John von Neumann-Institute.
Sherwood, P., Brooks, B. R., & Sansom, M. S. P. (2008). Multiscale methods for macromolecular simulations. *Current Opinion in Structural Biology, 18*, 630–640.
Shiga, M., & Masia, M. (2013a). Boundary based on exchange symmetry theory for multilevel simulations. I. Basic theory. *Journal of Chemical Physics, 139*, 044120/1-8.
Shiga, M., & Masia, M. (2013b). Boundary based on exchange symmetry theory for multilevel simulations. II. Multiple time scale approach. *Journal of Chemical Physics, 139*, 144103/1-14.

Shiga, M., & Masia, M. (2014). Quasi-boundary based on exchange symmetry theory for multilevel simulations. *Molecular Simulation*, *41*, 827–831.

Singh, U. C., & Kollmann, P. A. (1986). A combined ab initio quantum mechanical and molecular mechanical method for carrying out simulations on complex molecular systems: Applications to the $CH_3Cl + Cl^-$ exchange reaction and gas phase protonation of polyethers. *Journal of Computational Chemistry*, *7*, 718–730.

Takenaka, N., Kitamura, Y., Koyano, Y., & Nagaoka, M. (2012a). An improvement in quantum mechanical description of solute-solvent interactions in condensed systems via the number-adaptive multiscale quantum mechanical/molecular mechanical-molecular dynamics method: Application to zwitterionic glycine in aqueous solution. *Journal of Chemical Physics*, *137*, 024501/1-11.

Takenaka, N., Kitamura, Y., Koyano, Y., & Nagaoka, M. (2012b). The number-adaptive multiscale QM/MM molecular dynamics simulation: Application to liquid water. *Chemical Physics Letters*, *524*, 56–61.

Thiel, W. (2005). *MNDO2005*. Mülheim an der Ruhr, Germany: Max-Planck-Institut für Kohlenforschung.

Várnai, C., Bernstein, N., Mones, L., & Csányi, G. (2013). Tests of an adaptive QM/MM calculation on free energy profiles of chemical reactions in solution. *Journal of Physical Chemistry B*, *117*, 12202–12211.

Waller, M. P., Kumbhar, S., & Yang, J. (2014). A density-based adaptive quantum mechanical/molecular mechanical method. *ChemPhysChem*, *15*, 3218–3225.

Wallrapp, F. H., & Guallar, V. (2011). Mixed quantum mechanics and molecular mechanics methods: Looking inside proteins. *Wiley Interdisciplinary Reviews: Computational Molecular Science*, *1*, 315–322.

Warshel, A., & Levitt, M. (1976). Theoretical studies of enzymic reactions: Dielectric, electrostatic and steric stabilization of the carbonium ion in the reaction of lysozyme. *Journal of Molecular Biology*, *103*, 227–249.

Watanabe, H. C., Kubař, T., & Elstner, M. (2014). Size-consistent multipartitioning QM/MM: A stable and efficient adaptive QM/MM method. *Journal of Chemical Theory and Computation*, *10*, 4242–4252.

Woodcock, H. L., Miller, B. T., Hodoscek, M., Okur, A., Larkin, J. D., Ponder, J. W., et al. (2011). MSCALE: A general utility for multiscale modeling. *Journal of Chemical Theory and Computation*, *7*, 1208–1219.

Wu, R., Cao, Z., & Zhang, Y. (2012). Computational simulations of zinc enzyme: Challenges and recent advances. *Progress in Chemistry*, *24*, 1175–1184.

Zhang, Y., Lee, T.-S., & Yang, W. (1999). A pseudobond approach to combining quantum mechanical and molecular mechanical methods. *Journal of Chemical Physics*, *110*, 46–54.

Zhang, Y., & Lin, H. (2008). Flexible-boundary quantum-mechanical/molecular-mechanical calculations: Partial charge transfer between the quantum-mechanical and molecular-mechanical subsystems. *Journal of Chemical Theory and Computation*, *4*, 414–425.

Zhang, Y., & Lin, H. (2010). Flexible-boundary QM/MM calculations: II. Partial charge transfer across the QM/MM boundary that passes through a covalent bond. *Theoretical Chemistry Accounts*, *126*, 315–322.

Zhang, Y., Lin, H., & Truhlar, D. G. (2007). Self-consistent polarization of the boundary in the redistributed charge and dipole scheme for combined quantum-mechanical and molecular-mechanical calculations. *Journal of Chemical Theory and Computation*, *3*, 1378–1398.

CHAPTER FOURTEEN

Enzymatic Kinetic Isotope Effects from Path-Integral Free Energy Perturbation Theory

J. Gao[1]

Theoretical Chemistry Institute, Jilin University, Changchun, Jilin Province, PR China
University of Minnesota, Minneapolis, MN, United States
[1]Corresponding author: e-mail address: gao@jialigao.org

Contents

1. Introduction — 360
2. Methods — 362
   2.1 Potential Energy Surface — 363
   2.2 Potential of Mean Force — 366
   2.3 KIEs and the PI-FEP Method — 368
3. Illustrative Examples — 371
   3.1 Primary and Secondary KIE for Non-Hydrogen Atoms — 371
   3.2 The RGM and Swain–Schaad Exponent — 374
   3.3 NAO: An Enzyme Enhancing Nuclear Quantum Effects — 377
   3.4 Protein Dynamics on Enzymatic KIEs — 378
4. Concluding Remarks — 382
Acknowledgment — 382
References — 383

Abstract

Path-integral free energy perturbation (PI-FEP) theory is presented to directly determine the ratio of quantum mechanical partition functions of different isotopologs in a single simulation. Furthermore, a double averaging strategy is used to carry out the practical simulation, separating the quantum mechanical path integral exactly into two separate calculations, one corresponding to a classical molecular dynamics simulation of the centroid coordinates, and another involving free-particle path-integral sampling over the classical, centroid positions. An integrated centroid path-integral free energy perturbation and umbrella sampling (PI-FEP/UM, or simply, PI-FEP) method along with bisection sampling was summarized, which provides an accurate and fast convergent method for computing kinetic isotope effects for chemical reactions in solution and in enzymes. The PI-FEP method is illustrated by a number of applications, to highlight the computational precision and accuracy, the rule of geometrical mean in kinetic isotope effects, enhanced nuclear quantum effects in enzyme catalysis, and protein dynamics on temperature dependence of kinetic isotope effects.

1. INTRODUCTION

Although the dominant factor in enzyme catalysis is the lowering of the free energy barrier of a catalyzed reaction (Garcia-Viloca, Gao, Karplus, & Truhlar, 2004; Schowen, 1978), it is of great interest to understand the detailed mechanism of transition state stabilization and the connection between protein dynamics and rate acceleration. Kinetic isotope effects (KIEs) provide a direct probe of the transition state, and a comparison of experimental KIEs for the enzymatic and uncatalyzed reactions in water can reveal surprising differences. KIEs are not only significant for proton, hydride, and hydrogen-atom transfer reactions due to their small mass, but also they are important for reactions involving heavy atoms as illustrated by the work of Schramm (2003), who used the experimental KIE to develop transition state models for inhibitor design. KIEs are of quantum mechanical origin, including contributions from zero-point energy and tunneling (Pu, Gao, & Truhlar, 2006). A challenging question is whether enzymes have evolved to enhance nuclear quantum effects in rate acceleration (Major et al., 2009), and another intriguing question is by what means amino acid mutations that alter protein dynamics affect the measured KIE (Klinman & Kohen, 2014). To rationalize the experimental findings (Gao et al., 2006; Garcia-Viloca et al., 2004; Kohen & Limbach, 2005; Layfield & Hammes-Schiffer, 2014; Wang, Fried, Boxer, & Markland, 2014; Wong, Xu, & Xu, 2015), computational studies are often necessary, and quantum mechanics is essential for modeling enzymatic mechanism and kinetics (Gao et al., 2006; Gao & Truhlar, 2002). In this chapter, we present methods that have been developed in our group for studying enzymatic reactions, with a particular emphasis on accurate computation of primary and secondary KIEs.

Quantum mechanics is essential in enzyme kinetics modeling, and there are two major challenges to accurately determine the rate constant and KIEs (Gao & Truhlar, 2002). First, it is necessary to construct an accurate potential energy surface (PES) to model enzymatic reactions in aqueous solution (Gao et al., 2006). To this end, combined quantum mechanical and molecular mechanical (QM/MM) methods (Liu, Wang, Chen, Field, & Gao, 2014), in which the active site is modeled by an electronic structural theory and the surrounding solvent and protein environment is represented by a force field, provide an effective procedure to treat the bond-forming and bond-breaking processes (Gao, 1995a, 1996; Senn & Thiel, 2009). Such a combined QM/MM approach offers both accuracy and computational efficiency (Gao, 1996), and methods that can treat the entire protein–solvent system with a quantum mechanical representation are being actively pursued by a number of research groups (Gao, Truhlar, Wang, et al., 2014; Gao, Zhang, & Houk, 2014; Merz, 2014; Xie, Orozco, Truhlar, & Gao, 2009).

The second challenge to theory is to explicitly incorporate nuclear quantum effects, including quantum mechanical treatment of vibrational motions and tunneling (Pu et al., 2006). A variety of methods are available to treat nuclear quantum effects for gas-phase reactions (Fernandez-Ramos, Miller, Klippenstein, & Truhlar, 2006). In principle, these techniques can be directly extended to condensed-phase systems; however, the size and complexity of enzyme systems make it intractable computationally. One approach successfully applied to enzymatic reactions is the ensemble-averaged variational transition state theory with QM/MM sampling (EA-VTST-QM/MM) (Gao & Truhlar, 2002; Pu et al., 2006). This technique extends the VTST method developed for gas-phase reactions into condensed-phase reactions through ensemble averaging over configurations sampled through dynamics simulations (Alhambra et al., 2001). In another work, a grid-based hybrid approach was used to model nuclear quantum effects by numerically solving the vibrational wave function of the quantized nucleus (Layfield & Hammes-Schiffer, 2014). Recently, transition path sampling was used with normal model centroid molecular dynamics to determine KIE for enzymatic reactions (Dzierlenga, Antoniou, & Schwartz, 2015).

The theoretical framework in our study of enzymatic KIE is discrete path-integral simulation, in which the ratio of the quantum mechanical partition functions between different isotopic systems is determined in a *single simulation*, providing the necessary precision required for computing KIEs (Major & Gao, 2007). In particular, our method makes use of free energy perturbation (FEP) theory (Zwanzig, 1954) by carrying out path-integral simulation of one isotopolog, eg, the light (L) isotope, and by treating the difference in another isotope distribution, eg, the heavy (H) isotope, as a perturbation. Thus,

$$\frac{Q_H^{qm}}{Q_L^{qm}} = \left\langle e^{-\beta \Delta V_{\text{eff}}^{H \leftarrow L}} \right\rangle_L \qquad (1)$$

where $\beta = 1/k_B T$, $\Delta V_{\text{eff}}^{H \leftarrow L}$ is the difference in the quantum mechanical (qm) effective potential (Feynman & Hibbs, 1965) between the light and heavy isotopes, and the average $\langle\cdots\rangle_L$ is carried out based on the effective potential for the light particle. In principle, the average of Eq. (1) can be obtained from path-integral simulations directly, or using various other procedures such as centroid molecular dynamics (Cao & Voth, 1996; Hernandez, Cao, & Voth, 1995) and ring-polymer dynamics. Importantly, an even more efficient method is to separate the direct quantum mechanical average into two separate steps (Major & Gao, 2007). In particular, using a free-particle reference state, the quantum mechanical average of a property, A, can be rigorously obtained through path-integral free-particle sampling over classical configurations from molecular dynamics or Monte Carlo simulations:

$$\langle A \rangle = \langle \langle A \rangle_{FP} \rangle_{cm} \tag{2}$$

where the inner average $\langle\cdots\rangle_{FP}$ represents free-particle sampling over classical atomic coordinates as the centroids of the discrete paths, and the outer average $\langle\cdots\rangle_{cm}$ represents a classical mechanical ensemble average.

The double averaging strategy was given by Feynman (Feynman & Hibbs, 1965) and used by Sprik et al. in Monte Carlo simulations for a system consisting of one electron embedded in random hard spheres (Sprik, Klein, & Chandler, 1985). It was also called quantized classical path and pointed out that the expression of Eq. (2) is particularly useful (Hwang & Warshel, 1993, 1996; Major & Gao, 2007) because the quantum free energy of the system (here, in Eq. (1), the difference between two isotopic systems) can be obtained through two separate simulations: first by carrying out a Newtonian simulation to determine generate the classical ensemble, and then, by evaluating quantum contributions through free-particle sampling in a path-integral simulation.

The rest of the chapter focuses on discussion of treatment of the PES $U(\bar{r})$, and the technical details of using Eqs. (1) and (2) to compute free energies of activation and nuclear quantum effects in enzymatic reactions.

2. METHODS

Almost all enzyme reactions can be well described by the Born–Oppenheimer approximation (exceptions include, for example, photochemical processes), in which the sum of the electronic energy and the nuclear repulsion provides a potential energy function, or PES, governing the interatomic motions. Therefore the molecular modeling problem breaks into two parts: the PES and the dynamics simulations.

2.1 Potential Energy Surface

The potential energy function describes the energetic changes as a function of the variations in atomic coordinates, including thermal fluctuations and rearrangements of the chemical bonds. The accuracy of the potential energy function used to carry out molecular dynamics simulation directly affects the reliability of the computational results (Truhlar et al., 2002). Although it is possible to treat an entire enzyme–solvent system by QM methods (Car & Parrinello, 1985; Gao, Zhang, et al., 2014; Rohrig, Guidoni, & Rothlisberger, 2005; Stewart, 1996; Titmuss, Cummins, Rendell, Bliznyuk, & Gready, 2002), the computational costs are still too large to be practical for free energy simulations of enzymatic reactions. The most effective approach to model chemical reactions both in condensed phases and in enzymes is QM/MM methods (Liu et al., 2014), which offer the advantage of both computational efficiency and accuracy (Gao, 1995a).

In QM/MM methods, a system is divided into a QM region and an MM region (Field, Bash, & Karplus, 1990; Gao, 1992, 1995a; Gao & Xia, 1992; Liu et al., 2014; Singh & Kollman, 1986; Warshel & Levitt, 1976). The QM region typically includes atoms that are directly involved in the chemical step and they are treated explicitly by an electronic structure method, whereas the MM region consists of the rest of the system that is approximated by an MM force field. The QM/MM potential is given by Field et al. (1990) and Gao (1995a):

$$U = \langle \Psi(S) | H^o_{qm}(S) + H_{qm/mm}(S) | \Psi(S) \rangle + U_{mm} \quad (3)$$

where $H^o_{qm}(S)$ is the Hamiltonian of the QM-subsystem (the substrate and key amino acid residues), U_{mm} is the classical (MM) potential energy of the remainder of the system, and $H_{qm/mm}(S)$ represents the interactions between the two regions. $\Psi(S)$ is the molecular wave function of the QM-subsystem optimized for $H^o_{qm}(S) + H_{qm/mm}(S)$.

It is most convenient to rewrite Eq. (3) as follows (Gao, 1992; Gao & Xia, 1992):

$$U = E^o_{qm}(S) + \Delta E_{qm/mm}(S) + U_{mm} \quad (4)$$

where $E^o_{qm}(S)$ is the energy of the QM-subsystem in the gas phase, and $\Delta E_{qm/mm}(S)$ is the energy change of transfer to bring the QM-subsystem from the gas phase into the enzyme. The latter, which is also called QM–MM interaction energy, is defined by

$$\Delta E_{\text{qm/mm}}(S) = \langle \Psi(S)|H_{\text{qm}}^{\circ}(S) + H_{\text{qm/mm}}(S)|\Psi(S)\rangle - E_{\text{qm}}^{\circ}(S) \quad (5)$$

Eq. (4) is especially useful in that the total energy of a hybrid QM and MM system is separated into two "*independent*" terms—the gas-phase energy and the interaction energy—which can now be evaluated using different QM methods. Here, we have identified the energy terms involving electronic degrees of freedom by E and those purely empirical functions by U. First, the high accuracy needed to describe the intrinsic chemical reactivity of the QM-subsystem must be obtained by using an accurate quantum chemical model. This typically prevents its use in statistical mechanical simulations, especially for macromolecular systems, because of the high computation cost. Fortunately, the required accuracy is primarily associated with the first term in Eq. (4), $E_{\text{qm}}^{\circ}(S)$, in the gas phase, and it typically does not demand for extensive configurational sampling. Thus, it is possible to use methods such as coupled-cluster theory CCSD(T), complete active space self-consistent field with perturbation theory CASPT2, or density functional theory with a well-tested functional along with large basis functions. In other cases, it is also possible to reparameterize semiempirical methods to fit experimental data or high-level QM results. To emphasize the use of a high-level (HL) theory to substitute the $E_{\text{qm}}^{\circ}(S)$ term, which could be different from the QM/MM interaction term (Eq. 5), we have

$$E_{\text{qm}}^{\text{HL}}(S) = E_{\text{qm}}^{\circ}(S). \quad (6)$$

This substitution of QM models can be made on the fly during a dynamics simulation, or post priori.

For condensed-phase systems such as enzymatic reactions, the most critical issue is adequate configuration sampling, and this affects the accuracy in the calculation of the $\Delta E_{\text{qm/mm}}(S)$ term. It was recognized early on, when QM/MM simulations were first carried out that combined QM/MM potential is an empirical model, which contains empirical parameters and should be optimized to describe QM/MM interactions (Freindorf & Gao, 1996; Gao, 1994a; Gao & Xia, 1992). By systematically optimizing the associated van der Waals parameters for the "QM-atoms," both semiempirical and ab initio (Hartree–Fock) QM/MM potentials can yield excellent results for hydrogen-bonding and dispersion interactions in comparison with experimental data. The use of semiempirical methods (Cui, Elstner,

Kaxiras, Frauenheim, & Karplus, 2001; Dewar, Zoebisch, Healy, & Stewart, 1985; Gaus, Cui, & Elstner, 2011; Stewart, 1989) in QM/MM simulations has been validated through extensive studies of a variety of properties and molecular systems, including computations of free energies of solvation and polarization energies of organic compounds (Gao & Xia, 1992; Orozco, Luque, Habibollahzadeh, & Gao, 1995), the free energy profiles for organic reactions (Gao, 1995b, 1996; Major, York, & Gao, 2005), and the effects of solvation on molecular structures and on electronic transitions (Gao, 1994b; Gao & Byun, 1997; Lin & Gao, 2007). Importantly, these relatively lower-level (LL) methods for QM/MM interactions (but with a high degree of accuracy) offer an opportunity to perform adequate configuration sampling essential for understanding the contributions of protein dynamics to catalysis.

$$\Delta E_{qm/mm}^{LL}(S) = \Delta E_{qm/mm}(S) \qquad (7)$$

It should be noted that unlike the use of a preoptimized reaction path and the associated empirical parameters including partial charges in subsequent MM simulations, the prescription of Eqs. (5) and (7) includes the instantaneous electronic polarization of the QM-subsystem due to the thermal fluctuations of the enzyme and solvent environment (Garcia-Viloca, Truhlar, & Gao, 2003). The latter is essential for enzymes.

The approach described above represents a highly accurate dual-level (DL) total energy for enzymatic systems (Byun, Mo, & Gao, 2001):

$$U^{DL} = E_{qm}^{HL}(S) + \Delta E_{qm/mm}^{LL}(S) + U_{mm} \qquad (8)$$

The DL QM/MM approach is akin to the ONIOM model developed by Maseras and Morokuma (1995), but physically and conceptually different in that it is not an energy subtraction–addition scheme, and it is a gas-phase and condensed-phase separation algorithm.

The QM/MM PES combines the generality of quantum mechanical methods for treating chemical processes with the computational efficiency of molecular mechanics for large molecular systems. The use of an explicit electronic structure method to describe the enzyme active site is important because understanding the changes in electronic structure along the reaction path can help design inhibitors and novel catalysts. It is also important because the dynamic fluctuations of the enzyme and aqueous solvent system have a major impact on the polarization of the species involved in the chemical reaction, which, in turn, affects the chemical reactivity (Gao et al., 2006; Gao & Xia, 1992).

2.2 Potential of Mean Force

The rate constant in our study of enzymatic reactions is determined using path-integral quantum transition state theory (QTST) (Voth, Chandler, & Miller, 1989a, 1989b), which was derived by writing the rate expression analogous to classical transition state theory (TST) (Fernandez-Ramos et al., 2006).

$$k_{\text{QTST}} = \frac{1}{\beta h} e^{-\beta \Delta F_{\text{qm}}^{\neq}} \qquad (9)$$

where h is Planck's constant, and $\Delta F_{\text{qm}}^{\neq}$ is the free energy of activation, which can be obtained from path-integral or quantum mechanical potential of mean force (PMF), $w_{\text{qm}}(\bar{z})$, as a function of the centroid reaction coordinate denoted as $\bar{z}$. As in classical TST, the quantum free energy of activation can then be obtained from the PMF:

$$\Delta F_{\text{qm}}^{\neq} = w_{\text{qm}}(\bar{z}^{\neq}) - w_{\text{qm}}(\bar{z}^R) \qquad (10)$$

where the symbol $\bar{z}^{\neq}$ specifies the value of the centroid reaction coordinate, at which $w_{\text{qm}}(\bar{z})$ has the maximum value, and $\bar{z}^R$ is the coordinate at the reactant state.

Feynman path-integral simulation provides a convenient procedure to determine $w_{\text{qm}}(\bar{z})$, incorporating quantum effects on tunneling and nuclear vibrations (Feynman & Hibbs, 1965). In this approach, ensemble averages for the quantum system are obtained from a classical simulation in which the quantized particles are represented by ring polymers of classical particles. To determine the PMF, the average positions, or centroids, of the quantized particles are used as classical variable to define a corresponding reaction coordinate $\bar{z}$ (Cao & Voth, 1994; Gillan, 1988; Messina, Schenter, & Garrett, 1993; Voth et al., 1989a). Practical methods include centroid molecular dynamics (Jang & Voth, 2000; Makarov & Topaler, 1995; Messina, Schenter, & Garrett, 1995; Mills, Schenter, Makarov, & Jonsson, 1997) and ring-polymer molecular dynamics (Habershon, Manolopoulos, Markland, & Miller, 2013). We employ an alternative approach, in which the quantum mechanical PMF $w_{\text{qm}}(\bar{z})$ is determined through a double averaging procedure. This approach has been used in simple applications and has been applied to enzymatic reactions, and it has been called quantized classical path (Hwang & Warshel, 1993, 1996; Major & Gao, 2007; Villa & Warshel, 2001).

In the discrete path-integral method, each quantized nucleus is represented by a ring of P quasi-particles called beads. The coordinates for the M-quantized particles are collectively denoted as $\mathbf{r} = \{\mathbf{r}_i^q; i = 1, \cdots, P; q = 1, \cdots, M\}$, with the definition of $\mathbf{r}_{P+1} = \mathbf{r}_1$. Each bead associated with a given quantized particle is connected to its two neighbors via harmonic springs and is subjected to a fraction, $1/P$, of the full classical potential, $U(\mathbf{r}_i^q, \mathbf{S})$, where $\mathbf{S}$ represents all classical protein–solvent coordinates. In centroid path integral, the centroid positions $\{\bar{\mathbf{r}}\}$ are used as the principle variable and the canonical QM partition function of the hybrid system can be written as follows:

$$Q_P^{qm} = \int d\mathbf{S} \int d\bar{\mathbf{r}} \left(\frac{P}{2\pi\lambda_q^2}\right)^{3P/2} \int d\mathbf{R}\, e^{-\beta V_{\text{eff}}(\mathbf{r}, \mathbf{S})} \qquad (11)$$

where $V_{\text{eff}}(\mathbf{r}, \mathbf{S})$ is the effective quantum mechanical potential, $\int d\mathbf{R} = \int d\mathbf{r}_1 \cdots \int d\mathbf{r}_P \delta(\bar{\mathbf{r}})$, the centroid coordinates $\{\bar{\mathbf{r}}\}$ of the quasi-particles is defined as $\bar{\mathbf{r}}^q = 1/P \sum_{i=1}^{P} \mathbf{r}_i^q$, and the de Broglie thermal wavelength λ_q of a particle of mass m_q is $\lambda_q^2 = \beta\hbar^2/m_q$.

The quantum partition function of Eq. (11) can be exactly rewritten as a double average (Hwang & Warshel, 1993, 1996; Major & Gao, 2005, 2007; Major, Garcia-Viloca, & Gao, 2006; Sprik et al., 1985):

$$Q_P^{qm} = Q_P^{cm} \left\langle \left\langle e^{-\beta \Delta U(\bar{\mathbf{r}}, \mathbf{S})} \right\rangle_{\text{FP},\bar{\mathbf{r}}} \right\rangle_U \qquad (12)$$

where Q_P^{cm} is the classical partition function (Feynman & Hibbs, 1965). As noted above, the average $\langle \cdots \rangle_U$ is obtained according the potential $U(\bar{\mathbf{r}}, \mathbf{S})$, which is of QM/MM type, but purely classical in nuclear degrees of freedom, and the inner average $\langle \cdots \rangle_{\text{FP},\bar{\mathbf{r}}}$ represents free-particle sampling, constrained to match the path-integral centroid coordinates with the classical atomic positions. The average differential potential energy in Eq. (12) is given as follows:

$$\Delta \bar{U}(\bar{z}) = \frac{1}{P}\sum_{q}^{M}\sum_{i}^{P}\{U(\mathbf{r}_i^q, \mathbf{S}) - U(\bar{\mathbf{r}}^q, \mathbf{S})\} \quad (13)$$

and the free-particle sampling is carried out according to the distribution $\exp\left[-\beta(P/2\beta\lambda_m^2)\sum_i^P(\Delta\mathbf{r}_i)^2\right]$, or

$$\langle\cdots\rangle_{FP,\bar{\mathbf{r}}} = \frac{\int d\mathbf{r}_P\{\cdots\}\delta(\bar{\mathbf{r}})e^{-(P/2\lambda^2)\sum_i^P(\Delta\mathbf{r}_i)^2}}{\int d\mathbf{r}_P\,\delta(\bar{\mathbf{r}})e^{-(P/2\lambda_m^2)\sum_i^P(\Delta\mathbf{r}_i)^2}} \quad (14)$$

where $\Delta\mathbf{r}_i = \mathbf{r}_i - \mathbf{r}_{i+1}$.

The double averaging procedure yields the exact path-integral centroid density, thereby, the centroid quantum mechanical PMF:

$$w_{qm}(\bar{z}) = w_{cm}(z) - RT\ln\left\langle\delta(\bar{z})\langle e^{-\beta\Delta\bar{U}(\bar{z})}\rangle_{FP,\bar{\mathbf{r}}}\right\rangle_U \quad (15)$$

where $w_{qm}(\bar{z})$ and $w_{cm}(z=\bar{z})$ are the centroid quantum mechanical and classical mechanical PMF, respectively.

2.3 KIEs and the PI-FEP Method

KIE is defined as the ratio of the rate constants between light and heavy isotopic reactions, which can be computed simply by applying QTST as follows:

$$\text{KIE} = \frac{k^L}{k^H} = \frac{e^{-\beta\{w_{qm}^L(\bar{z}_L^{\neq}) - w_{qm}^L(\bar{z}_L^R)\}}}{e^{-\beta\{w_{qm}^H(\bar{z}_H^{\neq}) - w_{qm}^H(\bar{z}_H^R)\}}} \quad (16)$$

where $w_{qm}^L(\bar{z})$ and $w_{qm}^H(\bar{z})$ are the centroid PMF for the reactions with light and heavy isotopologs, respectively. Eq. (16) indicates that KIEs may be determined by computing the potentials of mean force separately for the L and H isotopes (whether or not using path-integral simulations) (Hwang & Warshel, 1993, 1996; Major & Gao, 2007; Major et al., 2006), and this indeed is typically the procedure used in many applications. However, statistical fluctuations in the actual simulations for computing the potentials of mean force are typically of similar, or even greater, magnitude as the free energy difference in isotope effects (Major & Gao, 2007). Thus, it is extremely difficult to obtain meaningful results for heavy atom and especially for secondary KIE using this strategy.

KIEs can also be expressed in terms of the ratio of partial partition functions at the centroid reactant and transition state:

$$\text{KIE} = \frac{k^{\text{L}}}{k^{\text{H}}} = \left[\frac{Q_{\text{qm}}^{\text{L}}(\bar{z}_{\text{L}}^{\neq})}{Q_{\text{qm}}^{\text{H}}(\bar{z}_{\text{H}}^{\neq})}\right]\left[\frac{Q_{\text{qm}}^{\text{H}}(\bar{z}_{\text{H}}^{\text{R}})}{Q_{\text{qm}}^{\text{L}}(\bar{z}_{\text{L}}^{\text{R}})}\right] \qquad (17)$$

This expression is particularly useful because it transforms two separate calculations of the PMF for different isotopic reactions in Eq. (16) into evaluation of the relative probabilities of the light and heavy isotopes at the transition state and in the reactant state region. As highlighted in Eq. (1), the ratio of the quantum partition functions for different isotopes can be determined directly through FEP theory by perturbing the mass from one isotope into the other (Ceriotti & Markland, 2013; Major & Gao, 2007; Marsalek et al., 2014). Therefore, only one path-integral simulation of a given isotopic reaction is needed to yield the free energy difference, ie, the ratio to the partition function of anther isotope. This method is called path-integral free energy perturbation (PI-FEP) theory (it seems that the SC-FEP scheme in Ceriotti and Markland (2013) is the same as the PI-FEP method, but the authors were apparently not aware of the earlier work), which results in a major improvement in precision for KIE calculations such that not only H/D primary KIE can be computed, but also heavy atom and secondary KIEs can be determined accurately.

In PI-FEP, we further separate the PI quantum average into a hybrid classical molecular dynamics and path-integral simulation with free-particle sampling. The theory is exact, but the separate procedures of classical MD and PI-free-particle sampling are especially advantageous because it allows an extensive sampling of the protein–solvent (bath) conformational space, and the free-particle sampling scheme avoids the discrete-bead distribution being trapped in a local minimum. Specifically, the ratio of the distribution functions for a system with light and heavy isotope substitutions is given as follows:

$$\frac{Q_{\text{H}}^{\text{qm}}(\bar{z})}{Q_{\text{L}}^{\text{qm}}(\bar{z})} = \frac{\left\langle \delta(\bar{z}) \left\langle e^{-\beta \Delta \Delta \bar{U}_{\text{H} \leftarrow \text{L}}} e^{-\beta \Delta \bar{U}_{\text{L}}} \right\rangle_{\text{FP,L}} \right\rangle_{\text{U}}}{\left\langle \delta(\bar{z}) \left\langle e^{-\beta \Delta \bar{U}_{\text{L}}} \right\rangle_{\text{FP,L}} \right\rangle_{\text{U}}} \qquad (18)$$

where the subscripts L specifies that the ensemble averages are done using the light isotope, $\Delta \bar{U}_L$ is the difference (Eq. 13) between the average potential of the P-discrete beads of the light particle and that of the centroid $\{\bar{\mathbf{r}}\}$, and $\Delta\Delta U_{H\leftarrow L} = \Delta\overline{U}_H - \Delta\overline{U}_L$. One may recall the reweighting expression of umbrella sampling (UM) of Valleau and Torrie (1977); thus, our method really is an integrated PI-FEP and UM approach.

A practical difficulty in path-integral simulations is to obtain converged results for the computed PMF within a tolerable amount of simulation time, especially at a precision needed for computing KIE. To this end, we have developed a bisection sampling technique, called BQCP (Major & Gao, 2005; Major et al., 2006). The method was based on the approach developed by Ceperley for free-particle sampling in which the initial and final beads are not connected (Ceperley, 1995; Pollock & Ceperley, 1984). In our implementation, the bisection sampling was made as originally proposed (Ceperley, 1995; Pollock & Ceperley, 1984), but we enforce the first and last beads to be identical to enclose the ring particles. We then impose the condition that the centroid position matches the corresponding atomic coordinates by rigid-body translocation (Major & Gao, 2005; Major et al., 2006). Since free-particle distribution is known exactly at a given temperature, each discrete-bead distribution is generated according to this distribution and thus 100% accepted. Furthermore, in this construction, each new configuration is created independently, starting from a single initial bead position, allowing the ring-polymer configurations to move into completely different regions of configurational space, independent of previous beads distribution. This latter point is especially important to enhance convergence by avoiding being trapped in a local minimum of the classical potential in the path-integral sampling. Using this sampling technique, the PI-FEP method has been applied to a range of condensed-phase and enzymatic systems (Fan, Cembran, Ma, & Gao, 2013; Gao, Wong, & Major, 2008; Lin, Gao, Rubinstein, & Major, 2011; Major & Gao, 2005, 2007; Major et al., 2006, 2009, 2005).

In the bisection sampling scheme, the perturbed heavy isotope positions are related to the lighter ones by

$$\frac{\mathbf{r}^q_{i,L}}{\mathbf{r}^q_{i,H}} = \frac{\lambda_{q_L}\boldsymbol{\theta}^q_i}{\lambda_{q_H}\boldsymbol{\theta}^q_i} = \sqrt{\frac{m_{q,H}}{m_{q,L}}}; \quad i=1,2,\ldots,P; q=1,2,\ldots,M \qquad (19)$$

where $\mathbf{r}^q_{i,L}$ and $\mathbf{r}^q_{i,H}$ are the coordinates for bead i of the corresponding light and heavy isotopes of the classical atom q, $m_{q,L}$, and $m_{q,H}$ are the masses for the light and heavy nuclei, and $\boldsymbol{\theta}^q_i$ is a position vector with respect to the position of the centroid, or classical atomic coordinates in the bisection sampling scheme (Major & Gao, 2005). In our PI-FEP (or PI-FEP/UM) simulation scheme, the position vectors for the perturbed beads, eg, the heavy isotope, are made identical to the reference (ie, light) isotope distribution. As a result, beads positions of the perturbed mass are solely determined by the square root of mass ratio (Eq. 19).

3. ILLUSTRATIVE EXAMPLES

In the following, I have carefully selected a few illustrative examples to highlight applications of the PI-FEP method. These include (1) the computation of primary and secondary KIEs for heavy elements, (2) the rule of geometric mean (RGM) and Swain–Schaad exponents, (3) an enzyme that has been optimized to enhance nuclear quantum effects over the uncatalyzed process, and (4) the effect of protein dynamics on temperature dependence of KIE.

3.1 Primary and Secondary KIE for Non-Hydrogen Atoms

It is straightforward to compute H/D primary KIE, which typically have reasonably large values that can be determined from separate PMF calculations of the hydrogen and deuterium systems. However, the calculation of primary KIE for heavy isotopologs as well as secondary KIEs (hydrogen and non-hydrogen) presents a major challenge because the relative free energies of activation associated with these isotope changes are too small to be adequately treated by separate simulations. To illustrate the computational precision of the PI-FEP method, we first present a study of the primary and secondary KIEs involving heavy atoms in the decarboxylation reaction of N-methyl picolinate in water (Major & Gao, 2007; Rishavy & Cleland, 2000). We first performed Newtonian molecular dynamics simulations to obtain the (classical) PMF along the $C_2 - C_{O_2}$ distance reaction coordinate for N-methyl picolinate in a box of 888 water molecules (Major & Gao, 2007). Combined QM/MM potential was used, in which the N-methyl picolinate ion was treated by the semiempirical AM1 Hamiltonian and water by the TIP3P model. Then, centroid path-integral free-particle sampling was carried out on the classical ensemble of configurations to yield the ratio of the quantum mechanical partition functions for different isotopes (^{12}C, ^{13}C, ^{14}N, and ^{15}N) associated with the decarboxylation reaction. Note that the classical PMF is independent of atomic masses. Also, in addition to the isotopically substituted nucleus, the immediately adjacent atoms were also quantized, and each quantized atom was described by 32 beads.

Solvent effects are significant, increasing the free energy barrier by 15.2 kcal/mol to a value of 26.8 kcal/mol. The large solvent effect is due to the presence of a positive charge on the pyridine nitrogen, which is annihilated in the decarboxylation reaction. There is only about 2 kcal/mol of CO_2 recombination barrier, which would post difficulty for computing KIEs using only a single transition structure.

The computed $^{12}C/^{13}C$ primary KIE at the carboxyl carbon position and the $^{14}N/^{15}N$ secondary KIE are listed in Table 1. Figs. 1 and 2 depict the difference in computing KIE by (1) using two separate PMF simulations for different isotopes (QTST) and (2) using the single-simulation PI-FEP method. In the first approach (denoted by QTST, although KIE calculations using PI-FEP are also based on QTST), one needs to take difference of two quantities with large fluctuations illustrated by the error bars along the PMFs (Fig. 1), resulting in large standard errors in the computed KIE (Table 1). On the other hand, the standard error for KIE from the PI-FEP method is simply that of a single simulation (Fig. 2). The computed intrinsic ^{13}C primary KIE is 1.0345 ± 0.0028 from PI-FEP at 25°C for the decarboxylation of N-methyl picolinate in water. Although the standard deviation is greater than the accuracy of experimental measurement (Table 1), the confidence level is much higher than that from separate PMF calculations, even though the same beads configurations were used in PI simulations. To emphasize the sensitivity of the computational result, the computed KIE is equivalent to a free energy difference of merely 0.0187 kcal/mol. For the secondary ^{15}N KIE, the PI-FEP simulation yields an average value of 1.0083 ± 0.0016,

Table 1 Computed and Experimental Primary $^{12}C/^{13}C$ and Secondary $^{14}N/^{15}N$ Kinetic Isotope Effects for the Decarboxylation Reaction of N-Methyl Picolinate in Water at 25°C

	$^{12}k/^{13}k$	$^{14}k/^{15}k$
Exp (25°C)	1.0281 ± 0.0003	1.0070 ± 0.0003
PI-QTST	1.0346 ± 0.8773	1.0067 ± 0.8862
PI-FEP	1.0345 ± 0.0028	1.0083 ± 0.0016

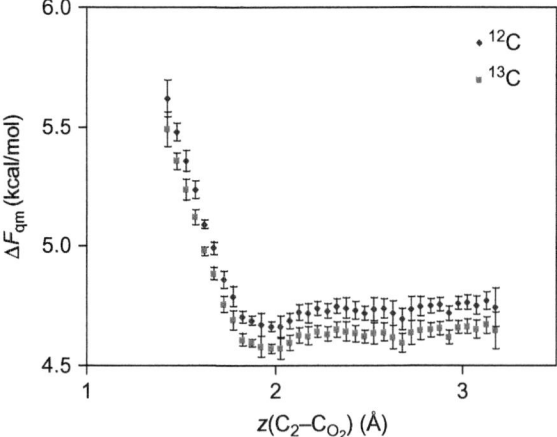

Fig. 1 Computed nuclear quantum effects along the potential of mean force for the decarboxylation reaction of N-methyl picolinate ion in water. ^{13}C denotes isotope substitution at the carboxyl carbon position.

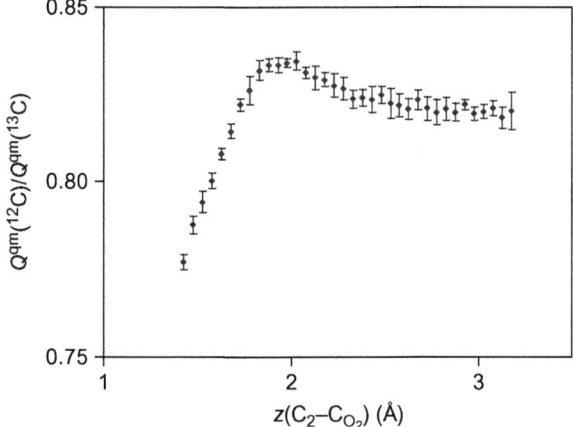

Fig. 2 Computed ratio of quantum partition function between ^{12}C and ^{13}C isotope substitutions at the carboxyl carbon position along the centroid reaction coordinate for the decarboxylation of N-methyl picolinate in water.

which may be compared with experiment (1.0070 ± 0.0003) (Rishavy & Cleland, 2000).

This example shows that both the accuracy and the precision of the computed KIE using the PI-FEP method are of similar magnitude in comparison with experiment. This cannot be said with alternative simulation approaches applied to enzyme systems.

3.2 The RGM and Swain–Schaad Exponent

Tunneling in chemical reactions is of great interest both for chemical processes and for enzyme catalysis. However, it is not easy to separate tunneling from other nuclear quantum effects such as zero-point energy. The effect of a second isotope substitution on KIE, also called the rule of geometric mean (RGM), is often used to provide insights on tunneling contributions to a chemical process:

$$g_{HD}^{HD} = \frac{(k_H^H/k_D^H)}{(k_H^D/k_D^D)} \tag{20}$$

where the superscripts indicate secondary sites, and the subscripts denote primary positions. It was originally derived at high temperature limit with small quantum tunneling contribution (Bigeleisen, 1955 #6990), and it has been shown to have negligible deviations on model systems using semiclassical TST (Saunders, 1985). If $g_{HD}^{HD} = 1$, it implies that there is no isotope effect from a second site on the measured KIE at the first site, and Eq. (20) can be rearranged as follows

$$\frac{k_H^H}{k_D^H} = \frac{k_H^D}{k_D^D} = \frac{(k_H^H/k_D^D)}{(k_H^D/k_H^D)} \tag{21}$$

or equivalently,

$$\frac{k_H^H}{k_D^D} = \left(\frac{k_H^H}{k_D^H}\right)\left(\frac{k_H^H}{k_H^D}\right) \tag{22}$$

Eq. (22) shows that if RGM holds, the total KIE will be the product of the individual primary and secondary KIEs, or the change in activation free energy is additive. Deviation or the observation of RGM breakdown is often used as a measure of the extent of tunneling in a given system (Huskey, 2006).

Another widely used indicator for tunneling is the so-called Swain–Schaad exponent (Swain, Stivers, Reuwer, & Schaad, 1958), which is expressed below using the notation of Huskey (2006),

$$n_{HD} = \frac{\ln(k_H^H/k_T^H)}{\ln(k_H^H/k_D^H)} \tag{23}$$

$$n_{DT} = \frac{\ln(k_H^H/k_T^H)}{\ln(k_D^H/k_T^H)} \tag{24}$$

These equations assume that the isotope effects are determined solely by the change in the zero-point energy from the mode corresponding to the reaction coordinate without tunneling. Studies have shown that the value of n_{HD} for primary KIEs is typically in the range of 1.43–1.45 (Huskey, 2006). Deviations from these values are thought to be indicative of presence of tunneling (Cha, Murray, & Klinman, 1989).

The Swain–Schaad exponent in Eqs. (23) and (24) are restricted to isotope substitutions on a single site, primary KIEs as indicated (but certainly possible using secondary KIEs). It turns out that mixed Swain–Schaad exponents from both primary and secondary isotope substitutes are more sensitive which have been generally favored for assessing tunneling (Swain et al., 1958).

$$n_{DT}^{DD} = \frac{\ln(k_H^H/k_T^H)}{\ln(k_D^D/k_T^D)} \tag{25}$$

$$n_{DD}^{DT} = \frac{\ln(k_H^H/k_H^T)}{\ln(k_D^D/k_D^T)} \tag{26}$$

The first equation is the primary Swain–Schaad exponent, which reflects the secondary isotope effects on primary KIEs. The use of different primary KIEs, H/T vs D/T is to increase the magnitude of the exponent. The second equation describes a similar relationship for the secondary KIEs. Semiclassical arguments dictate that values of the mixed Swain–Schaad exponents greater than 3.3 are strong indication that tunneling contribution is significant (Cha et al., 1989).

Here, we use the proton abstraction reaction of nitroalkane by acetate ion in aqueous solution to illustrate the above phenomena or tunneling indicators. Interestingly, the reaction itself is a classic example that shows an unusual Brønsted relationship in water, known as the nitroalkane anomaly (Bernasconi, 1992; Bordwell & Boyle, 1972; Kresge, 1974). This process is also catalyzed by the nitroalkane oxidase (NAO) in the initial step of the oxidation of nitroalkanes (Valley & Fitzpatrick, 2003), which will be addressed in Section 3.3.

Summarized in Table 2 are the computed primary and secondary KIEs for D and T substitutions (Gao et al., 2008; Major & Gao, 2007; Major et al., 2005). The computed H/D primary and secondary intrinsic KIEs are $k_H^H/k_D^H = 6.63 \pm 0.31$ and $k_H^H/k_H^D = 1.34 \pm 0.13$, respectively. Experimentally, only the total effects where both primary and secondary hydrogen

Table 2 Computed Primary (1 degree) and Secondary (2 degree) Kinetic Isotope Effects, and Computed and Experimental Total Deuterium Isotope Effects for the Proton Transfer from Nitroethane to Asp402 in NAO and to Acetate Ion in Water

Primary KIE	k_H^H/k_D^H	k_H^H/k_T^H	k_D^D/k_T^D
H$_2$O	6.63 ± 0.31	13.0 ± 1.0	2.17 ± 0.04
NAO	8.36 ± 0.58	18.1 ± 2.4	2.38 ± 0.05
Secondary KIE	k_H^H/k_H^D	k_H^H/k_H^T	k_D^D/k_D^T
H$_2$O	1.340 ± 0.132	1.375 ± 0.183	1.096 ± 0.048
NAO	1.213 ± 0.150	1.229 ± 0.209	1.050 ± 0.025
Total k_H^H/k_D^D	Calc.	Expt.	
H$_2$O	8.3 ± 1.1	7.8 ± 0.1	
NAO	10.1 ± 1.4	9.2 ± 0.4	

Subscripts and superscripts are used to specify the rate constant for isotope substitutions at the primary and secondary position, respectively.

atoms are replaced by deuterium were determined (7.8 ± 0.1) (Valley & Fitzpatrick, 2004); the PI-FEP result is $k_H^H/k_D^D = 8.3 \pm 1.1$ (Valley & Fitzpatrick, 2004). Interestingly, the RGM of Eq. (22) yields a value of 8.88 (6.63 × 1.34). This gives a ratio of $g_{HD}^{HD} = 1.07$ over the actual computed value (8.31). Another way of interpreting the results is that there is a secondary KIE of 1.07 on the primary KIEs according to Eq. (20). These values are reasonably close to unity, therefore, there is a rather small nonadditive effect (in free energy terms), or tunneling contribution, in the proton abstraction in solution.

Alternatively, single-site Swain–Schaad exponents can be calculated for primary KIE at a value of $n_{HD}^{(1)} = 1.35$, and for secondary KIE at $n_{HD}^{(2)} = 1.09$; these values are close to the semiclassical limits. It is interesting to note that the exponents, $n_{DT}^{(1)}$ and $n_{DT}^{(2)}$, for D/T ratios are rather large, at values of 3.8 and 12.3, respectively, which exhibit a significant deviation from the semiclassical limit, particularly on secondary KIEs. The latter tends to have greater computational errors because of the small free energy difference. Mixed-site Swain–Schaad exponents, which is favored by experimentalists, can also be determined using the data in Table 2, for a primary exponent of $n_{DT}^{DD} = 3.31$, and for a secondary exponent of $n_{DT}^{DD} = 3.47$. These results are again close to the semiclassical limit, suggesting also small tunneling contributions in the proton abstraction of nitroethane in aqueous solution.

3.3 NAO: An Enzyme Enhancing Nuclear Quantum Effects

The proton abstraction of nitroethane is catalyzed by the flavoenzyme NAO, which accelerates the proton transfer by a factor of 10^9 relative to the uncatalyzed reaction in water (Valley & Fitzpatrick, 2004). Valley and Fitzpatrick reported a KIE of 9.2 for the dideuterated substrate [1,1-^{2}H] nitroethane by NAO, noticeably greater than that in water (7.8) (Valley & Fitzpatrick, 2004; Valley, Tichy, & Fitzpatrick, 2005). Here, we provide an example, revealing that the enzyme NAO has been optimized to enhance nuclear quantum effects relative to that of the identical, uncatalyzed process in water.

The same combined QM/MM PES for the aqueous reaction described in Section 3.2 was used to model the NAO catalysis (Gao & Xia, 1992; Major et al., 2005), and the PI-FEP approach was applied to model the corresponding nuclear quantum effects. Thus, both the electronic structure of the reacting system and the nuclear dynamics are treated at an equal footing by quantum mechanics. It is interesting to note that a parallel computational and X-ray crystallographic study resulted in an identical Michaelis complex structure, which was used in the enzyme simulation (Major et al., 2009). The computed centroid PMFs from PI-FEP coupled with UM simulations in NAO and in water are shown in Fig. 3. The centroid

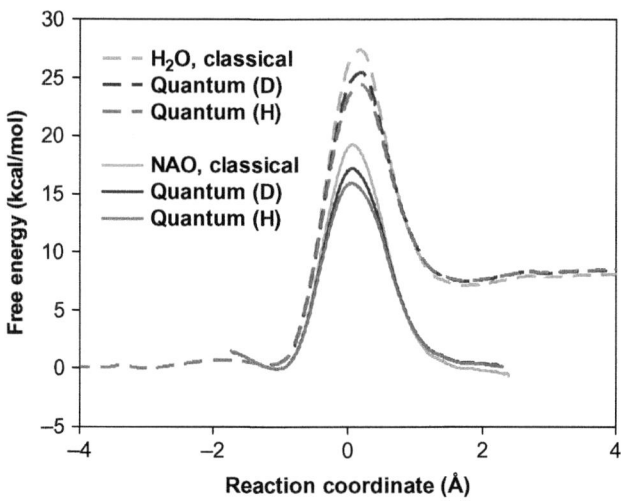

Fig. 3 Classical (*blue*) and quantum mechanical potential of mean force for the proton (*red*) and deuteron (*green*) transfer from nitroethane to acetate ion in water (*dashed curves*), and to Asp402 (*solid curves*) in the active site of nitroalkane oxidase. (See the color plate.)

reaction coordinate is defined as the difference between the breaking (C–H) and forming (H–O) bond distances. The computed free energies of activation are 15.9 and 24.4 kcal/mol for the enzymatic and the uncatalyzed reaction in water, respectively, in agreement with experimental results (14.0 and 24.8 kcal/mol) (Valley & Fitzpatrick, 2004).

The results in Table 2 and Fig. 3 reveal a number of findings, supporting the conclusion that the NAO-catalyzed process has greater nuclear quantum effects that the corresponding uncatalyzed reaction in water (Major et al., 2009). First, the double averaging strategy in the PI-FEP method allows us to determine both classical and quantum mechanical free energy barriers. Fig. 3 shows that the classical barrier in the enzyme active site is lowered more (-3.4 kcal/mol) than that of the uncatalyzed reaction in water (-3.0 kcal/mol), due to nuclear quantum effects (Major et al., 2009). Second, the difference in quantum effects between proton and deuteron transfer is more significant for the enzymatic reaction than the uncatalyzed one as indicated by the primary H/D KIE (8.4 vs 6.6). Third, both experiment (9.2 vs 7.8) and computation (10.1 vs 8.3) show enhanced total KIEs in the enzyme over that in water. Fourth, the mixed Swain–Schaad exponent for the enzymatic reaction is greater than the semiclassical limit without tunneling and it is increased for the enzyme process (4.3 vs 3.5 in water). Finally, the tunneling transmission coefficient was found to be about three times greater for the enzyme reaction than the model reaction in water from another analysis employing an entirely different approach, namely EA-VTST/MT (Alhambra et al., 2001; Gao & Truhlar, 2002; Pu et al., 2006) (ensemble-averaged variational TST with multidimensional tunneling).

Analyses of the tunneling paths in EA-VTST revealed that the origin for the difference is due to a narrowing effect in the effective potential for tunneling in the enzyme than that in aqueous solution (Major et al., 2009). These studies demonstrate that differential quantum tunneling contributions are utilized in certain enzymatic catalysis as illustrated by NAO.

3.4 Protein Dynamics on Enzymatic KIEs

The connection between protein dynamics and enzyme activity is an intriguing question that has been actively debated. Of course, it is clear that protein dynamic fluctuations are essential for their biological functions; it is of interest to understand the mechanism by which protein dynamics affects enzyme kinetics. In this example, we show the effect of amino acid mutations that quench protein dynamic fluctuations in going from the Michaelis complex to the transition state on enzyme kinetics.

Dihydrofolate reductase (DHFR) is a small, flexible protein with a relatively large cofactor, nicotinamide adenine dinucleotide phosphate (NADPH), and a rather large substrate, 7,8-dihydrofolate (DHF). This interesting combination results in a protein ternary complex that is dynamically flexible. For example, NMR experiments established that conformational changes of several flexible loops in the protein play critical roles in ligand specificity and catalytic turnover (Bhabha et al., 2011; Boehr, McElheny, Dyson, & Wright, 2006, 2010; Oyeyemi et al., 2011, 2010). The double amino acid mutation of M42W and G121V as well as other replacements, which are remote from the catalytic center, can have profound, nonadditive effects on reaction rate, activation parameters, and the temperature dependence of KIE (Carroll et al., 2012; Loveridge, Behiry, Guo, & Allemann, 2012; Mauldin & Lee, 2010; Mauldin, Sapienza, Petit, & Lee, 2012; Rajagopalan, Lutz, & Benkovic, 2002; Swanwick, Shrimpton, & Allemann, 2004; Wang, Goodey, Benkovic, & Kohen, 2006; Watney, Agarwal, & Hammes-Schiffer, 2003; Wong, Selzer, Benkovic, & Hammes-Schiffer, 2005).

Using a QM/MM methodology established for the DHFR reaction (Garcia-Viloca et al., 2004, 2003; Pu, Ma, Gao, & Truhlar, 2005; Truhlar et al., 2002), we obtained the potentials of mean force for the hydride transfer from NADPH to DHF catalyzed by wild-type (wt)-DHFR and the M42W/G121V double mutant (dm)-DHFR at 5°C, 25°C, and 45°C. We obtained a free energy of activation of 16.4 kcal/mol for the hydride transfer in wt-DHFR, in agreement with experiment (16.7 kcal/mol) (Wang et al., 2006), whereas the activation free energy is increased by 1.2 kcal/mol in the M42W/G121V mutant, somewhat smaller than the corresponding experimental value (2.3 kcal/mol). Decomposition of the activation free energies into enthalpy and entropy contributions reveals that the double mutation significantly lowers the entropy of activation by −18.2 cal/K mol from computation and −13.2 cal/K mol from experiment (Fan et al., 2013). Analyses of the dynamic fluctuations and structural changes, we attribute the decreased dynamic flexibility during the catalyzed process, as expressed in kinetic parameters, to the annihilation of the Met20 loop motions at the transition state and the product state in the double mutant (Fig. 4).

The altered Met20 loop dynamics, resulting in decreased entropy of activation in the G121V/M42W double mutation, is also reflected in the

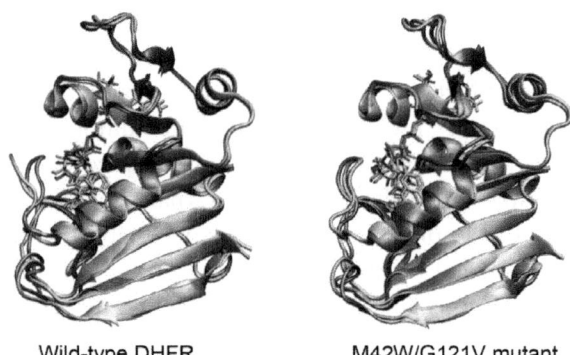

Wild-type DHFR   M42W/G121V mutant

Fig. 4 Superposition comparison of snapshots of the Michaelis complex (*blue*) and transition state (*brown*) structures for the wild-type enzyme dihydrofolate reductase (*left*) and the M42W/G121V double mutant (*right*). (See the color plate.)

measured and computed KIEs. Wild-type *Escherichia coli* DHFR exhibits a temperature-*independent* KIE for the hydride transfer step (Sikorski et al., 2004). However, it becomes temperature *dependent* for the hydride transfer in M42W/G121V dm-DHFR (Wang et al., 2006). In fact, a large number of enzymes have been found to show temperature independence of KIEs in their optimal operating temperature range, but the KIEs are changed to temperature-dependent outside this temperature range, or as a result of mutations (Nagel & Klinman, 2006, 2009). Several studies have proposed that the wild-type enzyme has evolved to optimize the average donor–acceptor distance for tunneling, resulting in mass-independent thermal activation (Nagel & Klinman, 2006, 2009; Roston, Cheatum, & Kohen, 2012; Stojkovic, Perissinotti, Willmer, Benkovic, & Kohen, 2012; Wang et al., 2006).

Our study showed that there are two major contributions to the overall nuclear quantum effects, responsible for the observed KIEs: the change in quantum vibrational free energy, predominantly zero-point effects, and tunneling (Gao & Truhlar, 2002; Pu et al., 2006, 2005; Truhlar et al., 2002). For the hydride transfer catalyzed by DHFR where the observed intrinsic primary KIE is only about 3 both in the wt-DHFR and in mutants (Loveridge et al., 2012; Sikorski et al., 2004; Wang et al., 2006), it would be important for a mechanism to also account for the change in zero-point vibrational energy. Using EA-VTST/MT (Alhambra et al., 2001; Gao & Truhlar, 2002), we identified two general features responsible for the observed temperature *independence* of KIEs in wt-DHFR (Pu et al.,

2005). First, we found that the location of the transition state for the hydride transfer coordinate is slightly shifted at different temperatures (Pu et al., 2005). Consequently, the difference in vibrational energy between H-transfer and D-transfer is increased at a higher temperature, resulting in a nearly temperature invariant Boltzmann factor. Second, the tunneling potential becomes narrower at higher temperature and a small increase in the tunneling transmission factor, also giving rise to a temperature insensitive effect.

The temperature dependence of the H/D primary KIE for the hydride transfer in the M42W/G121 dm-DHFR was examined using PI-FEP simulations (Fan et al., 2013). Unlike the EA-VTST approach (Pu et al., 2005), vibrational free energy and nuclear tunneling are not separable in path-integral simulations, but the total nuclear quantum effects as well as KIE are obtained directly from dynamics simulations. Arrhenius plots of the calculated and experimental H/D KIEs are shown in Fig. 5. Good accord was obtained for the wild-type DHFR both in the absolute value of KIEs and in the temperature-independent behavior. The agreement for the M42W/G121V double mutant is also good, although the slope of the temperature dependence plot is smaller than that measured experimentally. Since the

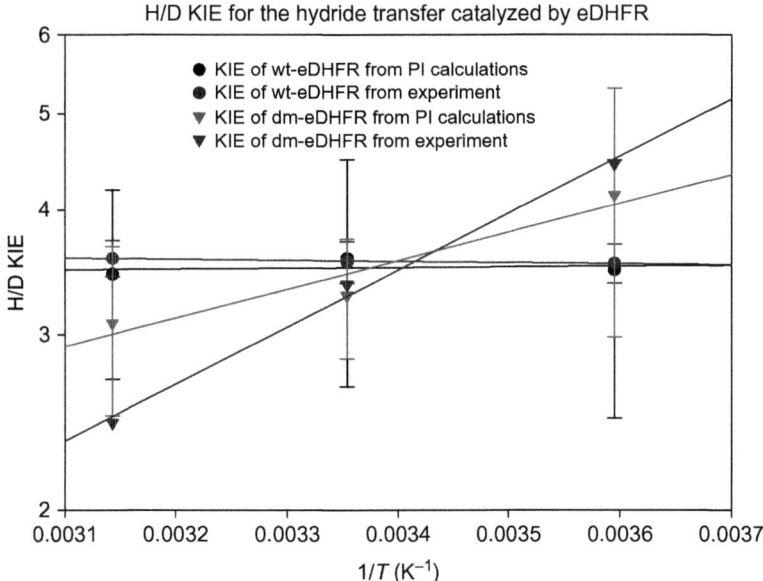

Fig. 5 Computed and experimental H/D primary kinetic isotope effects for the hydride transfer process catalyzed by wild-type and M42W/G121V mutant dihydrofolate reductase at different temperatures.

average donor–acceptor distances are not different between wt-DHFR and the M42W/G121V mutant, suggesting similar zero-point effects, we attribute the change is temperature dependence of the KIEs in the double mutant to altered PES for tunneling. This is consistent with structural analyses and computed change in entropy of activation for the wild-type and double mutant DHFR.

4. CONCLUDING REMARKS

We presented a method for incorporating quantum mechanics into enzyme kinetics modeling, in which (1) the PES is represented by combined QM/MM methods to describe bond forming and breaking processes and (2) PI-FEP theory is used to directly determine the ratio of quantum mechanical partition functions for different isotopologs in a single simulation. In the latter simulation approach, we further employ a double averaging strategy, separate the quantum mechanical path-integral simulation exactly into two separate calculations, one corresponding to classical molecular dynamics simulation of the centroid coordinates, and another involving free-particle path-integral sampling over the classical, centroid positions. An integrated centroid path-integral free energy perturbation and umbrella sampling (PI-FEP/UM, or simply, PI-FEP) method along with a bisection sampling procedure was summarized, which provides an accurate, fast convergent method for computing KIEs for chemical reactions in solution and in enzymes. These methods are illustrated by applications to the decarboxylation reaction of N-methyl picolinate in water to highlight the precision of the PI-FEP method contrasted to conventional approaches, the deprotonation process of nitroethane in water to highlight the RGM and Swain–Schaad exponent for assessing tunneling, the same deprotonation process catalyzed by NAO to reveal an enzyme that has been evolved to enhance nuclear quantum effects, and the effects of protein dynamics on temperature dependence of KIEs in DHFR. These examples show that the incorporation of quantum mechanical effects is essential for enzyme kinetics simulations, and our computational approaches provided insights on the importance of nuclear quantum effects.

ACKNOWLEDGMENT

This work has been supported by the National Institutes of Health. The author wishes to thank his coworkers whose name are shown in references cited.

REFERENCES

Alhambra, C., Corchado, J., Sanchez, M. L., Garcia-Viloca, M., Gao, J., & Truhlar, D. G. (2001). Canonical variational theory for enzyme kinetics with the protein mean force and multidimensional quantum mechanical tunneling dynamics. Theory and application to liver alcohol dehydrogenase. *The Journal of Physical Chemistry. B, 105,* 11326–11340.

Bernasconi, C. F. (1992). The principle of nonperfect synchronization: More than a qualitative concept? *Accounts of Chemical Research, 25,* 9–16.

Bhabha, G., Lee, J., Ekiert, D. C., Gam, J., Wilson, I. A., et al. (2011). A dynamic knockout reveals that conformational fluctuations influence the chemical step of enzyme catalysis. *Science, 332,* 234–238.

Bigeleisen, J. (1955). Statistical mechanics of isotopic systems with small quantum corrections. I. General considerations and the rule of the geometric mean.. *The Journal of Chemical Physics, 23,* 2264.

Boehr, D. D., McElheny, D., Dyson, H. J., & Wright, P. E. (2006). The dynamic energy landscape of dihydrofolate reductase catalysis. *Science, 313,* 1638–1642.

Boehr, D. D., McElheny, D., Dyson, H. J., & Wright, P. E. (2010). Millisecond timescale fluctuations in dihydrofolate reductase are exquisitely sensitive to the bound ligands. *Proceedings of the National Academy of Sciences of the United States of America, 107,* 1373–1378.

Bordwell, F. G., & Boyle, W. J., Jr. (1972). Acidities, Broensted coefficients, and transition state structures for 1-arylnitroalkanes. *Journal of the American Chemical Society, 94,* 3907–3911.

Byun, K., Mo, Y., & Gao, J. (2001). New insight on the origin of the unusual acidity of Meldrum's acid from ab initio and combined QM/MM simulation study. *Journal of the American Chemical Society, 123,* 3974–3979.

Cao, J., & Voth, G. A. (1994). The formulation of quantum statistical mechanics based on the Feynman path centroid density. V. Quantum instantaneous normal mode theory of liquids. *The Journal of Chemical Physics, 101,* 6184–6192.

Cao, J., & Voth, G. A. (1996). A unified framework for quantum activated rate processes. I. General theory. *The Journal of Chemical Physics, 105,* 6856–6870.

Car, R., & Parrinello, M. (1985). Unified approach for molecular dynamics and density-functional theory. *Physical Review Letters, 55,* 2471–2474.

Carroll, M. J., Mauldin, R. V., Gromova, A. V., Singleton, S. F., Collins, E. J., & Lee, A. L. (2012). Evidence for dynamics in proteins as a mechanism for ligand dissociation. *Nature Chemical Biology, 8,* 246–252.

Ceperley, D. M. (1995). Path integrals in the theory of condensed helium. *Reviews of Modern Physics, 67,* 279–355.

Ceriotti, M., & Markland, T. E. (2013). Efficient methods and practical guidelines for simulating isotope effects. *The Journal of Chemical Physics, 138,* 014112.

Cha, Y., Murray, C. J., & Klinman, J. P. (1989). Hydrogen tunneling in enzyme reactions. *Science, 243,* 1325–1330.

Cui, Q., Elstner, M., Kaxiras, E., Frauenheim, T., & Karplus, M. (2001). A QM/MM implementation of the self-consistent charge density functional tight binding (SCC-DFTB) method. *The Journal of Physical Chemistry. B, 105,* 569–585.

Dewar, M. J. S., Zoebisch, E. G., Healy, E. F., & Stewart, J. J. P. (1985). Development and use of quantum mechanical molecular models. 76. AM1: A new general purpose quantum mechanical molecular model. *Journal of the American Chemical Society, 107,* 3902–3909.

Dzierlenga, M. W., Antoniou, D., & Schwartz, S. D. (2015). Another look at the mechanisms of hydride transfer enzymes with quantum and classical transition path sampling. *Journal of Physical Chemistry Letters, 6,* 1177–1181.

Fan, Y., Cembran, A., Ma, S., & Gao, J. (2013). Connecting protein conformational dynamics with catalytic function as illustrated in dihydrofolate reductase. *Biochemistry*, *52*, 2036–2049.

Fernandez-Ramos, A., Miller, J. A., Klippenstein, S. J., & Truhlar, D. G. (2006). Modeling the kinetics of bimolecular reactions. *Chemical Reviews*, *106*, 4518–4584.

Feynman, R. P., & Hibbs, A. R. (1965). *Quantum mechanics and path integrals*. New York: McGraw-Hill.

Field, M. J., Bash, P. A., & Karplus, M. (1990). A combined quantum mechanical and molecular mechanical potential for molecular dynamics simulations. *Journal of Computational Chemistry*, *11*, 700–733.

Freindorf, M., & Gao, J. (1996). Optimization of the Lennard-Jones parameters for a combined ab initio quantum mechanical and molecular mechanical potential using the 3-21G basis set. *Journal of Computational Chemistry*, *17*, 386–395.

Gao, J. (1992). Absolute free energy of solvation from Monte Carlo simulations using combined quantum and molecular mechanical potentials. *The Journal of Physical Chemistry*, *96*, 537–540.

Gao, J. (1994a). Computation of intermolecular interactions with a combined quantum mechanical and classical approach. *ACS Symposium Series*, *569*, 8–21.

Gao, J. (1994b). Monte Carlo quantum mechanical-configuration interaction and molecular mechanics simulation of solvent effects on the n.fwdarw.pi.* blue shift of acetone. *Journal of the American Chemical Society*, *116*, 9324–9328.

Gao, J. (1995a). Methods and applications of combined quantum mechanical and molecular mechanical potentials. In K. B. Lipkowitz & D. B. Boyd (Eds.), *Reviews in computational chemistry* (pp. 119–185). New York: VCH.

Gao, J. (1995b). An automated procedure for simulating chemical reactions in solution. Application to the decarboxylation of 3-carboxybenzisoxazole in water. *Journal of the American Chemical Society*, *117*, 8600–8607.

Gao, J. (1996). Hybrid quantum mechanical/molecular mechanical simulations: An alternative avenue to solvent effects in organic chemistry. *Accounts of Chemical Research*, *29*, 298–305.

Gao, J., & Byun, K. (1997). Solvent effects on the n->pi* transition of pyrimidine in aqueous solution. *Theoretical Chemistry Accounts*, *96*, 151–156.

Gao, J., Ma, S., Major, D. T., Nam, K., Pu, J., & Truhlar, D. G. (2006). Mechanisms and free energies of enzymatic reactions. *Chemical Reviews*, *106*, 3188–3209.

Gao, J., & Truhlar, D. G. (2002). Quantum mechanical methods for enzyme kinetics. *Annual Review of Physical Chemistry*, *53*, 467–505.

Gao, J. L., Truhlar, D. G., Wang, Y. J., Mazack, M. J. M., Loffler, P., et al. (2014). Explicit polarization: A quantum mechanical framework for developing next generation force fields. *Accounts of Chemical Research*, *47*, 2837–2845.

Gao, J., Wong, K.-Y., & Major, D. T. (2008). Combined QM/MM and path integral simulations of kinetic isotope effects in the proton transfer reaction between nitroethane and acetate ion in water. *Journal of Computational Chemistry*, *29*, 514–522.

Gao, J., & Xia, X. (1992). A prior evaluation of aqueous polarization effects through Monte Carlo QM-MM simulations. *Science*, *258*, 631–635.

Gao, J. L., Zhang, J. Z. H., & Houk, K. N. (2014). Beyond QM/MM: Fragment quantum mechanical methods. *Accounts of Chemical Research*, *47*, 2711.

Garcia-Viloca, M., Gao, J., Karplus, M., & Truhlar, D. G. (2004). How enzymes work: Analysis by modern rate theory and computer simulations. *Science*, *303*, 186–195.

Garcia-Viloca, M., Truhlar, D. G., & Gao, J. (2003). Importance of substrate and cofactor polarization in the active site of dihydrofolate reductase. *Journal of Molecular Biology*, *327*, 549–560.

Gaus, M., Cui, Q. A., & Elstner, M. (2011). DFTB3: Extension of the self-consistent-charge density-functional tight-binding method (SCC-DFTB). *Journal of Chemical Theory and Computation, 7*, 931–948.

Gillan, M. J. (1988). The quantum simulation of hydrogen in metals. *Philosophical Magazine A, 58*, 257–283.

Habershon, S., Manolopoulos, D. E., Markland, T. E., & Miller, T. F. (2013). Ring-polymer molecular dynamics: Quantum effects in chemical dynamics from classical trajectories in an extended phase space. *Annual Review of Physical Chemistry, 64*, 387–413.

Hernandez, R., Cao, J., & Voth, G. A. (1995). On the Feynman path centroid density as a phase space distribution in quantum statistical mechanics. *The Journal of Chemical Physics, 103*, 5018–5026.

Huskey, W. P. (2006). Multiple-isotope probes of hydrogen tunneling. In J. T. Hynes, J. P. Klinman, H. H. Limbach, & R. L. Schowen (Eds.), *Hydrogen-transfer reactions* (p. 1285). Weinheim: Wiley-VCH Verlag GmbH & Co. KGaA.

Hwang, J. K., & Warshel, A. (1993). A quantized classical path approach for calculations of quantum mechanical rate constants. *The Journal of Physical Chemistry, 97*, 10053–10058.

Hwang, J.-K., & Warshel, A. (1996). How important are quantum mechanical nuclear motions in enzyme catalysis? *Journal of the American Chemical Society, 118*, 11745–11751.

Jang, S., & Voth, G. A. (2000). A relationship between centroid dynamics and path integral quantum transition state theory. *The Journal of Chemical Physics, 112*, 8747–8757. Erratum: 2001. 114: 1944.

Klinman, J. P., & Kohen, A. (2014). Evolutionary aspects of enzyme dynamics. *The Journal of Biological Chemistry, 289*, 30205–30212.

Kohen, A., & Limbach, H. H. (Eds.), (2005). *Isotope effects in chemistry and biology*. New York: Taylor & Francis Group, CRC Press.

Kresge, A. (1974). The nitroalkane anomaly. *Canadian Journal of Chemistry, 52*, 1897–1903.

Layfield, J. P., & Hammes-Schiffer, S. (2014). Hydrogen tunneling in enzymes and biomimetic models. *Chemical Reviews, 114*, 3466–3494.

Lin, Y.-L., & Gao, J. (2007). Solvatochromic shifts of the n->pi* transition of acetone from steam vapor to ambient aqueous solution: A combined configuration interaction QM/MM simulation study incorporating solvent polarization. *Journal of Chemical Theory and Computation, 3*, 1484–1493.

Lin, Y.-L., Gao, J., Rubinstein, A., & Major, D. T. (2011). Molecular dynamics simulations of the intramolecular proton transfer and carbanion stabilization in the pyridoxal 5′-phosphate dependent enzymes l-dopa decarboxylase and alanine racemase. *Biochimica et Biophysica Acta (BBA)—Proteins and Proteomics, 1814*, 1438–1446.

Liu, M. Y., Wang, Y. J., Chen, Y. K., Field, M. J., & Gao, J. L. (2014). QM/MM through the 1990s: The first twenty years of method development and applications. *Israel Journal of Chemistry, 54*, 1250–1263.

Loveridge, E. J., Behiry, E. M., Guo, J. N., & Allemann, R. K. (2012). Evidence that a 'dynamic knockout' in *Escherichia coli* dihydrofolate reductase does not affect the chemical step of catalysis. *Nature Chemistry, 4*, 292–297.

Major, D. T., & Gao, J. (2005). Implementation of the bisection sampling method in path integral simulations. *Journal of Molecular Graphics & Modelling, 24*, 121–127.

Major, D. T., & Gao, J. (2007). An integrated path integral and free-energy perturbation-umbrella sampling method for computing kinetic isotope effects of chemical reactions in solution and in enzymes. *Journal of Chemical Theory and Computation, 3*, 949–960.

Major, D. T., Garcia-Viloca, M., & Gao, J. (2006). Path integral simulations of proton transfer reactions in aqueous solution using combined QM/MM potentials. *Journal of Chemical Theory and Computation, 2*, 236–245.

Major, D. T., Heroux, A., Orville, A. M., Valley, M. P., Fitzpatrick, P. F., & Gao, J. (2009). Differential quantum tunneling contributions in nitroalkane oxidase catalyzed and the uncatalyzed proton transfer reaction. *Proceedings of the National Academy of Sciences of the United States of America, 106*, 20736–20739.

Major, D. T., York, D. M., & Gao, J. (2005). Solvent polarization and kinetic isotope effects in nitroethane deprotonation and implications to the nitroalkane oxidase reaction. *Journal of the American Chemical Society, 127*, 16374–16375.

Makarov, D. E., & Topaler, M. (1995). Quantum transition-state theory below the crossover temperature. *Physical Review E: Statistical Physics, Plasmas, Fluids, and Related Interdisciplinary Topics, 52*, 178–188.

Marsalek, O., Chen, P. Y., Dupuis, R., Benoit, M., Meheut, M., et al. (2014). Efficient calculation of free energy differences associated with isotopic substitution using path-integral molecular dynamics. *Journal of Chemical Theory and Computation, 10*, 1440–1453.

Maseras, F., & Morokuma, K. (1995). IMOMM: A new integrated ab initio + molecular mechanics geometry optimization scheme of equilibrium structures and transition states. *Journal of Computational Chemistry, 16*, 1170–1179.

Mauldin, R. V., & Lee, A. L. (2010). Nuclear magnetic resonance study of the role of M42 in the solution dynamics of *Escherichia coli* dihydrofolate reductase. *Biochemistry, 49*, 1606–1615.

Mauldin, R. V., Sapienza, P. J., Petit, C. M., & Lee, A. L. (2012). Structure and dynamics of the G121V dihydrofolate reductase mutant: Lessons from a transition-state inhibitor complex. *PloS One, 7*, e33252.

Merz, K. M. (2014). Using quantum mechanical approaches to study biological systems. *Accounts of Chemical Research, 47*, 2804–2811.

Messina, M., Schenter, G. K., & Garrett, B. C. (1993). Centroid-density, quantum rate theory: Variational optimization of the dividing surface. *The Journal of Chemical Physics, 98*, 8525–8536.

Messina, M., Schenter, G. K., & Garrett, B. C. (1995). A variational centroid density procedure for the calculation of transmission coefficients for asymmetric barriers at low temperature. *The Journal of Chemical Physics, 103*, 3430–3435.

Mills, G., Schenter, G. K., Makarov, D. E., & Jonsson, H. (1997). Generalized path integral based quantum transition state theory. *Chemical Physics Letters, 278*, 91–96.

Nagel, Z. D., & Klinman, J. P. (2006). Tunneling and dynamics in enzymatic hydride transfer. *Chemical Reviews, 106*, 3095–3118.

Nagel, Z. D., & Klinman, J. P. (2009). A 21(st) century revisionist's view at a turning point in enzymology. *Nature Chemical Biology, 5*, 543–550.

Orozco, M., Luque, F. J., Habibollahzadeh, D., & Gao, J. (1995). The polarization contribution to the free energy of hydration. *The Journal of Chemical Physics, 103*, 9112. Erratum to document cited in CA122:299891.

Oyeyemi, O. A., Sours, K. M., Lee, T., Kohen, A., Resing, K. A., et al. (2011). Comparative hydrogen-deuterium exchange for a mesophilic vs thermophilic dihydrofolate reductase at 25°C: Identification of a single active site region with enhanced flexibility in the mesophilic protein. *Biochemistry, 50*, 8251–8260.

Oyeyemi, O. A., Sours, K. M., Lee, T., Resing, K. A., Ahn, N. G., & Klinman, J. P. (2010). Temperature dependence of protein motions in a thermophilic dihydrofolate reductase and its relationship to catalytic efficiency. *Proceedings of the National Academy of Sciences of the United States of America, 107*, 10074–10079.

Pollock, E. L., & Ceperley, D. M. (1984). Simulation of quantum many-body systems by path-integral methods. *Physical Review B, 30*, 2555–2568.

Pu, J., Gao, J., & Truhlar, D. G. (2006). Multidimensional tunneling, recrossing, and the transmission coefficient for enzymatic reactions. *Chemical Reviews, 106*, 3140–3169.

Pu, J., Ma, S., Gao, J., & Truhlar, D. G. (2005). Small temperature dependence of the kinetic isotope effect for the hydride transfer reaction catalyzed by *Escherichia coli* dihydrofolate reductase. *Journal of Physical Chemistry B, 109*, 8551–8556.

Rajagopalan, P. T., Lutz, S., & Benkovic, S. J. (2002). Coupling interactions of distal residues enhance dihydrofolate reductase catalysis: Mutational effects on hydride transfer rates. *Biochemistry, 41*, 12618–12628.

Rishavy, M. A., & Cleland, W. W. (2000). Determination of the mechanism of orotidine 5′-monophosphate decarboxylase by isotope effects. *Biochemistry, 39*, 4569–4574.

Rohrig, U. F., Guidoni, L., & Rothlisberger, U. (2005). *ChemPhysChem, 6*, 1836.

Roston, D., Cheatum, C. M., & Kohen, A. (2012). Hydrogen donor-acceptor fluctuations from kinetic isotope effects: A phenomenological model. *Biochemistry, 51*, 6860–6870.

Saunders, W. H., Jr. (1985). Calculations of isotope effects in elimination reactions. New experimental criteria for tunneling in slow proton transfers. *Journal of the American Chemical Society, 107*, 164–169.

Schowen, R. L. (1978). Catalytic power and transition-state stabilization. In R. D. Gandour & R. L. Schowen (Eds.), *Transition states of biochemical processes* (pp. 77–114). New York: Plenum Press.

Schramm, V. L. (2003). Enzymatic transition state poise and transition state analogues. *Accounts of Chemical Research, 36*, 588–596.

Senn, H. M., & Thiel, W. (2009). QM/MM methods for biomolecular systems. *Angewandte Chemie International Edition, 48*, 1198–1229.

Sikorski, R. S., Wang, L., Markham, K. A., Rajagopalan, P. T. R., Benkovic, S. J., & Kohen, A. (2004). Tunneling and coupled motion in the *Escherichia coli* dihydrofolate reductase catalysis. *Journal of the American Chemical Society, 126*, 4778–4779.

Singh, U. C., & Kollman, P. A. (1986). A combined ab initio quantum mechanical and molecular mechanical method for carrying out simulations on complex molecular systems: Applications to the $CH_3Cl + Cl^-$ exchange reaction and gas phase protonation of polyenes. *Journal of Computational Chemistry, 7*, 718–730.

Sprik, M., Klein, M. L., & Chandler, D. (1985). Staging: A sampling technique for the Monte Carlo evaluation of path integrals. *Physical Review B, 31*, 4234–4244.

Stewart, J. J. P. (1989). Optimization of parameters for semiempirical methods I. Method. *Journal of Computational Chemistry, 10*, 209–220.

Stewart, J. J. P. (1996). Application of localized molecular orbitals to the solution of semiempirical self-consistent field equations. *International Journal of Quantum Chemistry, 58*, 133–146.

Stojkovic, V., Perissinotti, L. L., Willmer, D., Benkovic, S. J., & Kohen, A. (2012). Effects of the donor-acceptor distance and dynamics on hydride tunneling in the dihydrofolate reductase catalyzed reaction. *Journal of the American Chemical Society, 134*, 1738–1745.

Swain, C. G., Stivers, E. C., Reuwer, J. F., & Schaad, L. J. (1958). Use of hydrogen isotope effects to identify the attacking nucleophile in the enolization of ketones catalyzed by acetic acid. *Journal of the American Chemical Society, 80*, 5885–5893.

Swanwick, R. S., Shrimpton, P. J., & Allemann, R. K. (2004). Pivotal role of Gly 121 in dihydrofolate reductase from *Escherichia coli*: The altered structure of a mutant enzyme may form the basis of its diminished catalytic performance. *Biochemistry, 43*, 4119–4127.

Titmuss, S. J., Cummins, P. L., Rendell, A. P., Bliznyuk, A. A., & Gready, J. E. (2002). Comparison of linear-scaling semiempirical methods and combined quantum mechanical/molecular mechanical methods for enzymic reactions. II. An energy decomposition analysis. *Journal of Computational Chemistry, 23*, 1314–1322.

Truhlar, D. G., Gao, J., Alhambra, C., Garcia-Viloca, M., Corchado, J., et al. (2002). The incorporation of quantum effects in enzyme kinetics modeling. *Accounts of Chemical Research, 35*, 341–349.

Valleau, J. P., & Torrie, G. M. (1977). A guide to Monte Carlo for statistical mechanics: 2. Byways. In B. J. Berne (Ed.), *Modern theoretical chemistry* (pp. 169–194). New York: Plenum Press.

Valley, M. P., & Fitzpatrick, P. F. (2003). Reductive half-reaction of nitroalkane oxidase: Effect of mutation of the active site aspartate to glutamate. *Biochemistry*, *42*, 5850–5856.

Valley, M. P., & Fitzpatrick, P. F. (2004). Comparison of enzymatic and non-enzymatic nitroethane anion formation: Thermodynamics and contribution of tunneling. *Journal of the American Chemical Society*, *126*, 6244–6245.

Valley, M. P., Tichy, S. E., & Fitzpatrick, P. F. (2005). Establishing the kinetic competency of the cationic imine intermediate in nitroalkane oxidase. *Journal of the American Chemical Society*, *127*, 2062–2066.

Villa, J., & Warshel, A. (2001). Energetics and dynamics of enzymatic reactions. *The Journal of Physical Chemistry. B*, *105*, 7887–7907.

Voth, G. A., Chandler, D., & Miller, W. H. (1989a). Rigorous formulation of quantum transition state theory and its dynamical corrections. *The Journal of Chemical Physics*, *91*, 7749–7760.

Voth, G. A., Chandler, D., & Miller, W. H. (1989b). Time correlation function and path integral analysis of quantum rate constants. *The Journal of Physical Chemistry*, *93*, 7009–7015.

Wang, L., Fried, S. D., Boxer, S. G., & Markland, T. E. (2014). Quantum delocalization of protons in the hydrogen-bond network of an enzyme active site. *Proceedings of the National Academy of Sciences of the United States of America*, *111*, 18454–18459.

Wang, L., Goodey, N. M., Benkovic, S. J., & Kohen, A. (2006). Coordinated effects of distal mutations on environmentally coupled tunneling in dihydrofolate reductase. *Proceedings of the National Academy of Sciences of the United States of America*, *103*, 15753–15758.

Warshel, A., & Levitt, M. (1976). Theoretical studies of enzymic reactions: Dielectric, electrostatic and steric stabilization of the carbonium ion in the reaction of lysozyme. *Journal of Molecular Biology*, *103*, 227–249.

Watney, J. B., Agarwal, P. K., & Hammes-Schiffer, S. (2003). Effect of mutation on enzyme motion in dihydrofolate reductase. *Journal of the American Chemical Society*, *125*, 3745–3750.

Wong, K. F., Selzer, T., Benkovic, S. J., & Hammes-Schiffer, S. (2005). Impact of distal mutations on the network of coupled motions correlated to hydride transfer in dihydrofolate reductase. *Proceedings of the National Academy of Sciences of the United States of America*, *102*, 6807–6812.

Wong, K. Y., Xu, Y., & Xu, L. (2015). Review of computer simulations of isotope effects on biochemical reactions: From the Bigeleisen equation to Feynman's path integral. *Biochimica et Biophysica Acta*, *1854*, 1782–1794.

Xie, W., Orozco, M., Truhlar, D. G., & Gao, J. (2009). X-Pol potential: An electronic structure-based force field for molecular dynamics simulation of a solvated protein in water. *Journal of Chemical Theory and Computation*, *5*, 459–467.

Zwanzig, R. (1954). High-temperature equation of state by a perturbation method. I. Nonpolar gases. *The Journal of Chemical Physics*, *22*, 1420–1426.

CHAPTER FIFTEEN

Simulating Nuclear and Electronic Quantum Effects in Enzymes

L. Wang*, C.M. Isborn[†], T.E. Markland[‡,1]
*Rutgers University, Piscataway, NJ, United States
[†]School of Natural Sciences, University of California, Merced, CA, United States
[‡]Stanford University, Stanford, CA, United States
[1]Corresponding author: e-mail address: tmarkland@stanford.edu

Contents

1. Introduction	390
2. Electronic Quantum Effects in Biological Systems	392
2.1 Basis Sets	393
2.2 Density Functionals and the Role of Exact Exchange	395
2.3 QM Regions in QM/MM Calculations	398
3. Incorporating NQEs in AIMD Simulations	401
3.1 Path Integral Molecular Dynamics	401
3.2 Accelerating PIMD Convergence Using Generalized Langevin Equations	402
3.3 Extracting Thermodynamic Isotope Effects from PIMD Simulations	403
3.4 Electronic and Nuclear Quantum Fluctuations in Biological Hydrogen Bond Networks	406
4. Outlook	409
Acknowledgments	410
References	410

Abstract

An accurate treatment of the structures and dynamics that lead to enhanced chemical reactivity in enzymes requires explicit treatment of both electronic and nuclear quantum effects. The former can be captured in ab initio molecular dynamics (AIMD) simulations, while the latter can be included by performing ab initio path integral molecular dynamics (AI-PIMD) simulations. Both AIMD and AI-PIMD simulations have traditionally been computationally prohibitive for large enzymatic systems. Recent developments in streaming computer architectures and new algorithms to accelerate path integral simulations now make these simulations practical for biological systems, allowing elucidation of enzymatic reactions in unprecedented detail. In this chapter, we summarize these recent developments and discuss practical considerations for applying AIMD and AI-PIMD simulations to enzymes.

1. INTRODUCTION

Tremendous effort has been devoted to using molecular simulation to unravel how enzymes catalyze chemical reactions with such remarkable efficiency and selectivity. A large amount of this work has utilized classical molecular mechanical (MM) methods combined with fixed charge and, more recently, polarizable empirical force fields (Ponder & Case, 2003). These classical methods have been successful in predicting and elucidating important properties ranging from ligand binding free energies to pK_as of active-site residues (Benkovic & Hammes-Schiffer, 2003; Nielsen & McCammon, 2003; Simonson, Archontis, & Karplus, 2002). However, empirical force fields are unable to describe the full electronic reorganization in enzyme active sites arising from bond making/breaking and charge transfer. In addition, most force fields do not include parametrization for transition metals or nonstandard ligands. For these systems, single-point electronic structure calculations and geometry optimizations (ie, quenching to the local 0 K structure) have been highly useful in determining bonding topologies, identifying transition states and intermediates, as well as predicting reaction kinetics (Benkovic & Hammes-Schiffer, 2003; Gao & Truhlar, 2002; Siegbahn & Borowski, 2006). However, one must go beyond these snapshot-based methods to fully determine how the electron density redistributes as the enzyme's structure fluctuates; these changes in the electron density can lead to bond cleavage and proton movement.

Ab initio molecular dynamics (AIMD) simulations evolve the nuclei using forces generated from the instantaneous electronic structure obtained at each time step, which allows coupling between nuclear motion and electronic rearrangement. These simulations treat the electrons quantum mechanically and the nuclei classically. However, if light atoms are present, nuclear quantum effects (NQEs), such as tunneling and zero point energy (ZPE), can also play an important role. For example, kinetic isotope effects of over two orders of magnitude on the enzymes' catalytic rates have been experimentally observed upon substituting hydrogen (H) by deuterium (D) (Klinman & Kohen, 2013; Sutcliffe & Scrutton, 2002) and have been attributed to tunneling. Also, the ZPE in a typical oxygen–hydrogen (O–H) bond is ~5 kcal/mol. As a result, including the ZPE can significantly alter the structure and dynamics of systems containing short hydrogen bonds, which are commonly observed in biomolecules where the protein fold can position groups much closer than typically seen in solution

(Cleland, Frey, & Gerlt, 1998; Mildvan et al., 2002). To treat electronic quantum effects and NQEs one can perform ab initio path integral molecular dynamics (AI-PIMD) simulations, which exactly include the effect of NQEs on static equilibrium properties of a given electronic surface (Berne & Thirumalai, 1986; Marx & Parrinello, 1996; Chen, Ivanov, Klein, & Parrinello, 2003; Morrone & Car, 2008). The imaginary time path integral formalism on which PIMD simulations are founded also forms the basis for the approximate centroid molecular dynamics (CMD) (Cao & Voth, 1994; Jang & Voth, 1999) and ring-polymer molecular dynamics (RPMD) (Craig & Manolopoulos, 2004; Habershon, Manolopoulos, Markland, & Miller, 2013) approaches to obtain quantum dynamics.

Performing AIMD simulations for enzyme active sites has traditionally been a formidable computational task because electronic structure calculations must be performed at each time step. Due to the need to perform many thousands of electronic structure calculations for each picosecond evolved, these simulations are typically performed using density functional theory (DFT) to generate the wave function, although cheaper semiempirical methods such as density functional tight binding have also been employed when longer timescales are required, albeit at a loss in accuracy (Gaus, Cui, & Elstner, 2014; Riccardi et al., 2006). However, recent theoretical and algorithmic advances and novel computer processing architectures have greatly accelerated DFT simulations and allowed for trajectories of hundreds of picoseconds or even nanosecond timescales for systems with quantum regions on the order of 60–300 atoms. In particular, electronic structure calculations have been accelerated by the development of linear scaling (Bowler & Miyazaki, 2012; Bowler, Miyazaki, & Gillan, 2002; Goedecker, 1999; Ordejón, Artacho, & Soler, 1996; Skylaris, Haynes, Mostofi, & Payne, 2005) DFT methods and the advent of codes that use the massively parallel stream processing capabilities of graphical processing units (GPUs) (Ufimtsev & Martinez, 2008, 2009a). The former methods take advantage of the locality of the density matrix as well as the fact that both the Coulomb/Hartree and DFT exchange-correlation energies are functions of the local spin density, which allows for a linear growth in computational cost when the system size increases. An example of the latter, which will be the major focus of this chapter, is the TeraChem electronic structure program, which uses GPUs to accelerate the computation of the electronic wave function and has demonstrated speedups of over 100-fold compared to CPU-based codes (Isborn, Luehr, Ufimtsev, & Martnez, 2011; Ufimtsev & Martinez, 2008, 2009a, 2009b). This speedup has enabled ab initio energy calculations on both the ground and excited states of systems containing thousands of atoms, including polypeptides and proteins (Isborn et al., 2011; Kulik, Luehr, Ufimtsev, & Martinez, 2012; Ufimtsev & Martinez, 2008, 2009a, 2009b).

AI-PIMD simulations have typically required about two orders of magnitude more computational cost than the corresponding AIMD simulations. A large amount of this additional cost arises from the need to make many replicas of the system, each of which requires a separate electronic structure calculation. Hence a standard PIMD implementation requires 30–50 electronic structure calculations to evolve a single time step for a typical hydrogen-containing system at room temperature (Berne & Thirumalai, 1986; Markland & Manolopoulos, 2008a; Wallqvist & Berne, 1985). In addition, the PIMD Hamiltonian contains high-frequency motions that limit the time step that can be employed and do not efficiently sample the full phase space if used directly. However, recent developments have significantly alleviated these issues, reducing the computational overhead for including NQEs in simulations (Ceriotti & Manolopoulos, 2012; Ceriotti, Parrinello, Markland, & Manolopoulos, 2010; Markland & Manolopoulos, 2008a, 2008b) and making it more efficient to extract isotope effects on free energy changes from these simulations (Ceriotti & Markland, 2013; Marsalek et al., 2014; Vanicek & Miller, 2007).

Here, we briefly review some considerations when performing electronic structure calculations of large biological systems. We then outline how these electronic structure developments have recently been combined with the latest PIMD algorithms to allow AI-PIMD simulations, which include nuclear and electronic quantum effects, to be performed for enzyme active sites.

2. ELECTRONIC QUANTUM EFFECTS IN BIOLOGICAL SYSTEMS

Electronic structure codes that are written to take advantage of GPUs, such as TeraChem, have made it possible to perform DFT single-point energy calculations, geometry optimizations, and AIMD simulations on quantum mechanical (QM) regions of many hundreds to thousands of atoms. The ability to now perform such large calculations has provided a number of insights into biological processes. For example, large-scale geometry optimizations showed that changes in protein structure upon mutation are directly correlated with enzymatic methyl transfer efficiency (Zhang, Kulik, Martinez, & Klinman, 2015). In addition, AIMD simulations have been used to examine charge transfer and polarization in the BPTI protein (Ufimtsev, Luehr, & Martinez, 2011), discover new pathways for glycine synthesis from primitive compounds (Wang, Titov, et al., 2014), and determine amorphous indium phosphide nanostructures (Zhao, Xie, & Kulik, 2015).

However, calculations of large systems have also highlighted deficiencies in DFT. In particular, semilocal DFT methods lead to large size-dependent errors and nonlocal exact exchange is necessary to fix these errors. As nonlocal exact exchange is extremely difficult to include in linear-scaling DFT methods, parallel stream processing computer hardware such as GPUs and software advances like those implemented in TeraChem are essential for large-scale exact exchange DFT calculations. While TeraChem is not linear scaling with the number of basis functions, N, over a large range of practical system sizes it achieves scaling of around $N^{1.5}$, which is considerably better than the formal DFT scaling of N^3 (Luehr, Sisto, & Martinez, 2016). In this section we discuss the choice of basis set, DFT exchange-correlation functional, and QM region in the hybrid quantum mechanics/molecular mechanics (QM/MM) calculations for biological systems.

2.1 Basis Sets

The basis set describes the electronic wave functions and must balance the accuracy of a given property with the computational cost. For example, calculating gas-phase bond dissociation energies with a high level of theory to sub-kcal/mol convergence may require approaching the complete basis set limit (Haworth & Bacskay, 2002; Henry, Parkinson, & Radom, 2002; Ochterski, Petersson, & Montgomery, 1996). However, for very large QM calculations, using such a large basis set is often not necessary, and including diffuse functions may lead to difficulties converging the wave function. In many condensed phase applications, one is concerned with properties that are less sensitive to basis set effects, eg, transfer of a proton between similar chemical groups. For such geometry-based properties, smaller (eg, double-zeta) basis sets can often be sufficient to obtain the desired accuracy (Wang & Wilson, 2004). Fig. 1 shows the potential energy profile for moving a proton along the active-site hydrogen bond network in an enzyme KSI^{D40N} (see Section 3.4 for more details) (Wang, Fried, Boxer, & Markland, 2014). All basis sets larger than 6-31G produce potential energy curves in quantitative agreement with that predicted using the largest basis set aug-cc-pVDZ.

Fig. 2 shows results from Kulik et al., who performed DFT structural optimizations of 58 proteins using TeraChem and showed that small basis

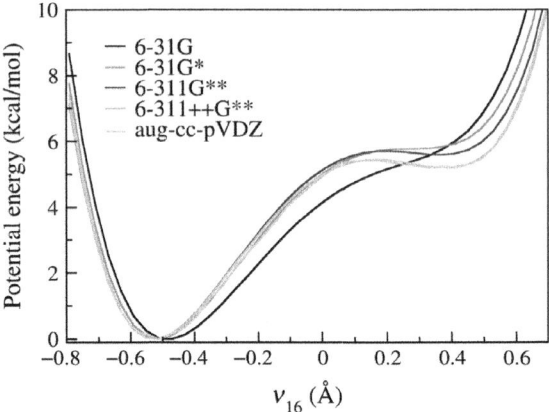

Fig. 1 Potential energy profile for proton transfer in the active-site hydrogen bond network of KSI^{D40N}, as calculated at the B3LYP-D3 level (Becke, 1993; Grimme et al., 2010) using five different basis sets. The proton transfer coordinate $\nu_{16} = d_{O16,H16} - d_{O57,H16}$ and $\nu_{16} \geq 0$ represents a proton transfer from residue Tyr16 to Tyr57 (see Section 3.4).

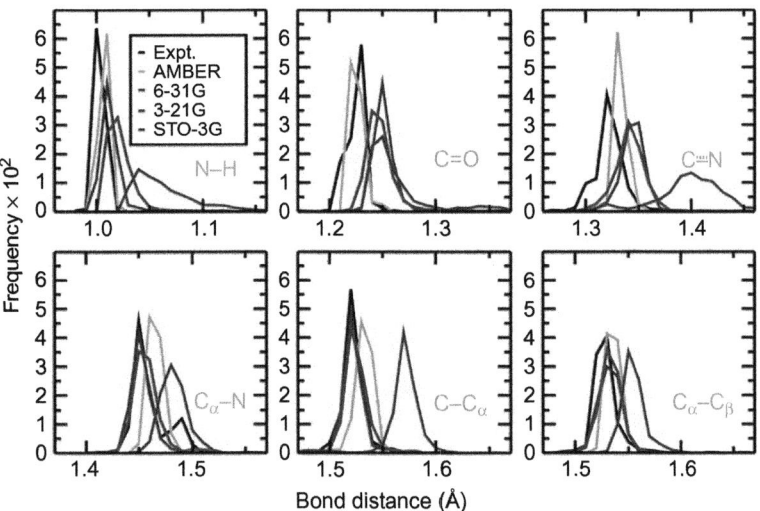

Fig. 2 Distributions of bond length for experimental structures (*black*) compared to those obtained from structural optimization with the classical AMBER force field (*orange*), ωPBEh/STO-3G (*red*), ωPBEh/3-21G (*green*), and ωPBEh/6-31G (*blue*) for a collection of 58 proteins. *Reproduced from Kulik, H. J., Luehr, N., Ufimtsev, I. S., & Martinez, T. J. (2012). Ab initio quantum chemistry for protein structures. The Journal of Physical Chemistry B, 116 (41), 12501–12509. doi: 10.1021/jp307741u with permission from American Chemical Society.* (See the color plate.)

sets such as MINI and STO-3G, while computationally efficient, lead to significant errors in predicted properties such as bond lengths and structural clashes as compared to the experimental crystal structures (Kulik et al., 2012). This problem can be alleviated by using a double-zeta basis set such as 3-21G, although this basis set is still relatively small and thus is likely subject to basis set superposition errors and may be too inflexible for modeling more subtle changes in polarization. For an accurate description of the electronic wave function and hydrogen bonding, the basis set should include polarization functions on all atoms. For systems with electrons far from the nucleus such as anions or excited states, diffuse functions should also be included.

2.2 Density Functionals and the Role of Exact Exchange

While DFT is formally exact in principle, owing to the Hohenberg–Kohn and Kohn–Sham theorems (Hohenberg & Kohn, 1964; Kohn & Sham, 1965), the functional form of the exchange-correlation energy is unknown and therefore it is necessary to make approximations for practical calculations (Burke, 2012; Perdew, Ruzsinszky, Constantin, Sun, & Csonka, 2009). DFT methods in practice often approximate the electron exchange and correlation interactions by considering only semilocal properties of the electron density. For example, the generalized gradient approximation (GGA) functionals use the local spin density and its gradient at a given point in space (Perdew & Yue, 1986). These semilocal functionals are often quite successful in regions of slowly varying electron density and can also benefit from cancellation of errors in the exchange and correlation functionals.

Dispersion is important for accurately modeling thermodynamic and substrate binding properties of enzymes. However, as dispersion is a long-range dynamical correlation effect, it is not accounted for in semilocal density functionals (Grimme, 2011; Johnson, Mackie, & DiLabio, 2009). A common technique to overcome this deficiency is to add an explicit, empirically parametrized attractive term to the DFT energy to represent interactions between atomic pairs, in which case "-D" is usually appended to the name of the functional (Elstner, Hobza, Frauenheim, Suhai, & Kaxiras, 2001; Grimme, 2006; Wodrich, Jana, von Ragu Schleyer, & Corminboeuf, 2008; Wu, Vargas, Nayak, Lotrich, & Scoles, 2001). These DFT-D approaches require little additional computational cost and often work quite well. Nevertheless, the accuracy of the DFT-D predictions depends on the size of the basis sets and is limited by the atom types included in the parametrization. An alternative, nonempirical approach is to compute the dispersion interaction between atoms based on the exchange-hole dipole model (XDM) (Becke & Johnson, 2005, 2007; Johnson & Becke, 2006). This model also adds negligible computational cost to a DFT calculation, but its implementation in geometry optimizations and dynamical simulations is currently not practical because forces obtained from the XDM method have not been widely available.

The approximate exchange term in semilocal DFT methods produces errors in both the electron density and energy, which leads to the derivative discontinuity (Kraisler & Kronik, 2014; Perdew, Parr, Levy, & Balduz, 1982; Schmidt, Kraisler, Makmal, Kronik, & Kuemmel, 2014; Tozer, 2003), the many-electron self-interaction error (Mori-Sanchez, Cohen, & Yang, 2006; Zhang & Yang, 1998), and the delocalization error (Heaton-Burgess & Yang, 2010; Kim, Sim, & Burke, 2013; Mori-Sánchez, Cohen, & Yang, 2008; Zheng, Liu, Johnson, Contreras-Garcia, & Yang, 2012). These semilocal density functionals underestimate the energy gap between the highest-occupied and lowest-unoccupied molecular orbitals (Cohen, Mori-Sánchez, & Yang, 2008a, 2008b; Mori-Sánchez et al., 2008; Stein, Autschbach, Govind, Kronik, & Baer, 2012), which leads to severe problems as the size of the QM region increases. We have shown that for molecules in aqueous solution, semilocal DFT methods predict that the band gap approaches zero and eventually the ground state self-consistent field (SCF) calculation can no longer converge if more and more solvent molecules are included in the QM region (Isborn, Mar, Curchod, Tavernelli, & Martnez, 2013). Similar difficulties were found by Kulik et al. when performing semilocal DFT calculations on entire proteins (Kulik et al., 2012). Additionally, we have recently shown that approximate semilocal exchange functionals generate a size-dependent error in the ionization potential, with the size of the error increasing as the size of the system grows (Sosa Vazquez & Isborn, 2015; Whittleton, Sosa Vazquez, Isborn, & Johnson, 2015).

Many of these inaccuracies can be improved by incorporating nonlocal exact exchange into the density functional via the generalized Kohn–Sham scheme (Seidl, Görling, Vogl, Majewski, & Levy, 1996). Exact exchange is often included in approximate semilocal exchange methods through either global hybrid or range-separated hybrid (RSH) techniques. For global hybrids, such as the B3LYP functional (Becke, 1993), a fixed fraction of exact exchange (typically 20–50%) is included for all interelectronic distances. RSH methods vary the amount of exact exchange based on the interelectronic distance $r_{12} \equiv |\mathbf{r}_{12}| = |\mathbf{r}_1 - \mathbf{r}_2|$, with the error function usually used to make the variation smooth for splitting the short-and long-range parts of the Coulomb operator

$$\frac{1}{r_{12}} = \frac{1 - erf(\omega r_{12})}{r_{12}} + \frac{erf(\omega r_{12})}{r_{12}}. \qquad (1)$$

For a long-range corrected (LC) hybrid, full exact exchange is used at long range (the $erf(\omega r_{12})$ term), while semilocal DFT exchange dominates at short range (the $1 - erf(\omega r_{12})$ term) to provide good balance with semilocal DFT correlation. The ratio of local DFT exchange to nonlocal exact exchange is determined by a range-separation parameter, ω, given in atomic units of inverse bohrs, a_0^{-1} (Baer & Neuhauser, 2005; Chai & Head-Gordon, 2008; Gill, Adamson, & Pople, 1996; Iikura, Tsuneda, Yanai, & Hirao, 2001; Tawada, Tsuneda, Yanagisawa, Yanai, & Hirao, 2004; Toulouse, Colonna, & Savin, 2004; Yanai, Tew, & Handy, 2004). The default value of ω is often between 0.2 and 0.5 a_0^{-1}, with smaller values leading to local DFT exchange having a longer range. A system-dependent "optimal" ω can also be determined by tuning ω to enforce Koopmans' theorem (Autschbach & Srebro, 2014; Baer, Livshits, & Salzner, 2010; Kronik, Stein, Refaely-Abramson, & Baer, 2012; Salzner & Baer, 2009; Stein, Eisenberg, Kronik, & Baer, 2010).

Long-range corrected hybrid functionals have the correct asymptotic limit for the exchange interaction, which often yields improved ionization potentials, band gaps, and excitation energies compared to experiment (Autschbach & Srebro, 2014; Kronik et al., 2012; Kuritz, Stein, Baer, & Kronik, 2011; Refaely-Abramson, Baer, & Kronik, 2011; Refaely-Abramson et al., 2012; Stein et al., 2012), as well as improved convergence of the SCF ground state for systems containing large QM regions (Isborn et al., 2013; Kulik et al., 2012). However, although all long-range corrected hybrid functionals have the correct asymptotic behavior of the exchange interaction, care must be taken to ensure that the range-separation parameter ω is large enough to fix the deficiencies of using approximate semilocal exchange at short range. We have recently shown that the choice of ω is key in determining the correct physical polarization in response to electron ionization. For example, for the ionization of a solvated ethene molecule, if ω is too small (below ~ 0.4 a_0^{-1}), the surrounding water molecules will unphysically contribute some of their own electron density to the ionization process, while for a larger ω they will correctly polarize in response to the electron being removed from the solute (Sosa Vazquez & Isborn, 2015). This can be seen in the density differences for the neutral and cation systems shown in Fig. 3. Thus the correct choice of functional, and particularly the treatment of exchange in the functional, is crucial for correctly describing processes that are key in enzymatic reactions such as polarization and charge transfer. Hence, in addition to including dispersion through methods mentioned earlier, we recommend a long-range corrected hybrid functional such as LC-BLYP, LC-ωPBE, and ωB97, with an ω value in the range $\omega = 0.4$–$0.6 \; a_0^{-1}$.

2.3 QM Regions in QM/MM Calculations

When simulating large biomolecules, it is desirable to accurately describe the region of interest, such as an enzyme's active site, using a high-level QM method with a large basis set. Such QM treatments are only feasible for a fraction of atoms in biomolecules due to the high computational cost of the QM approach. However, unlike an isolated system in vacuum, the electrostatic environment in the condensed phase strongly polarizes the QM system and therefore it is crucial to include the environmental effects for a correct description of the energy, geometry, and electron density distribution in the QM region. The hybrid QM/MM approach takes advantage of both the accuracy of the QM method and the speed of the MM method,

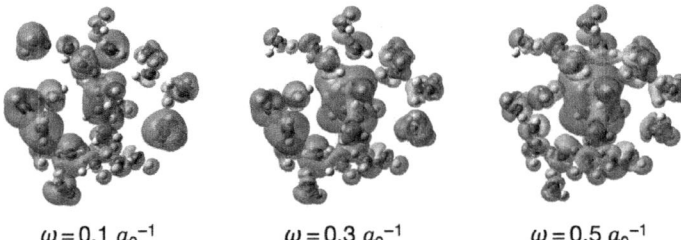

$\omega = 0.1 \; a_0^{-1}$ $\omega = 0.3 \; a_0^{-1}$ $\omega = 0.5 \; a_0^{-1}$

Fig. 3 The DFT error in size-dependent ionization potential is shown by analyzing the neutral and cation SCF density differences for ethene surrounded by 25 water molecules for the LC-BLYP functional with $\omega = 0.1$, 0.3, and 0.5 a_0^{-1}. The *purple* (*dark gray* in the print version) *isosurface* shows where the electron density has been removed, and the *blue* (*light gray* in the print version) *isosurface* shows where the electron density has been gained. The correct polarization response of the solvent is seen for $\omega = 0.5 \; a_0^{-1}$, where the water molecules polarize their electron density toward the cationic solute. Smaller ω values show the solvent being incorrectly partially ionized due to the DFT delocalization error present when using approximate exchange based on the local spin density.

allowing the MM charges to polarize the QM wave function (Friesner & Guallar, 2005; Warshel & Levitt, 1976). The coupling between the QM and MM regions has been extensively investigated (see Altun, Shaik, & Thiel, 2006; Lennartz, Schfer, Terstegen, & Thiel, 2002 for examples; and for recent reviews of the QM/MM method, see Lin & Truhlar, 2006; Ranaghan & Mulholland, 2010; Riccardi et al., 2006; Senn & Thiel, 2009).

To combine large-scale GPU-accelerated quantum chemistry calculations with MM force fields, we have created an extensive TeraChem interface for QM/MM calculations (Isborn, Goetz, Clark, Walker, & Martinez, 2012). In this approach, TeraChem receives the point charges of the MM atoms and performs a QM calculation with the electrostatic embedding method to account for the polarization of the QM region due to the electrostatic MM environment. TeraChem then outputs the QM forces on each atom for the propagation of the QM/MM system. Communication between the QM and MM regions is either through files or the message passing interface (MPI). Although the TeraChem QM/MM interface was originally implemented with the AMBER molecular dynamics program (Case et al., 2012), it is general enough to be used with any program that can accept QM forces for molecular dynamics simulations. Indeed, the AI-PIMD discussed in the following section were performed using the TeraChem MPI interface to connect to an in-house PIMD code to perform the dynamics.

Ideally, the QM region in the QM/MM calculation will be large enough that the results are independent of QM region size. To test how large the QM region should be to compute converged properties, we performed excited state QM/MM calculations on photoactive yellow protein (PYP) and computed its absorption spectrum using six different QM regions of increasing size, which range from the photoactive chromophore alone (22 QM atoms, 217 basis functions) to including the entire protein and counterions (1935 QM atoms, 16,827 basis functions) (Isborn et al., 2012). For all calculations the total system was the same, which included the entire protein, counterions, and a 32 Å solvation sphere, and any atoms not included in the QM region were modeled with MM point charges. We found that although the electronic excitation was for the most part localized on the chromophore in all QM/MM calculations, a large QM region (723 QM atoms) was required to reproduce the energy of the largest QM system (the entire protein included in the calculation). Furthermore, the predicted excitation energy did not have a monotonic trend as the size of the QM region increased, but instead jumped around substantially (up by 0.4 eV when going from the fourth to fifth largest QM region) as additional protein residues were included in the QM region (see Fig. 4). Overall, when sampling over many snapshots, this dependence of excitation energy on the size of the QM region led to significant differences in the final computed absorption spectrum (Isborn et al., 2012).

The slow convergence of a local property, such as an excitation energy, with the size of the QM region suggests that the electrostatic interactions at the QM/MM boundary can cause large changes in the QM electron density. Therefore the QM region should be as large as possible to push this boundary away from the active site to obtain an accurate QM density. Others have arrived at similar conclusions about the large size required for the QM region, citing problems with charges at the QM/MM junction (Flaig, Beer, & Ochsenfeld, 2012; Fox et al., 2011; Hu, Sderhjelm, & Ryde, 2011; Lpez-Canut, Mart, Bertrn, Moliner, & Tun, 2009; Sumowski & Ochsenfeld, 2009).

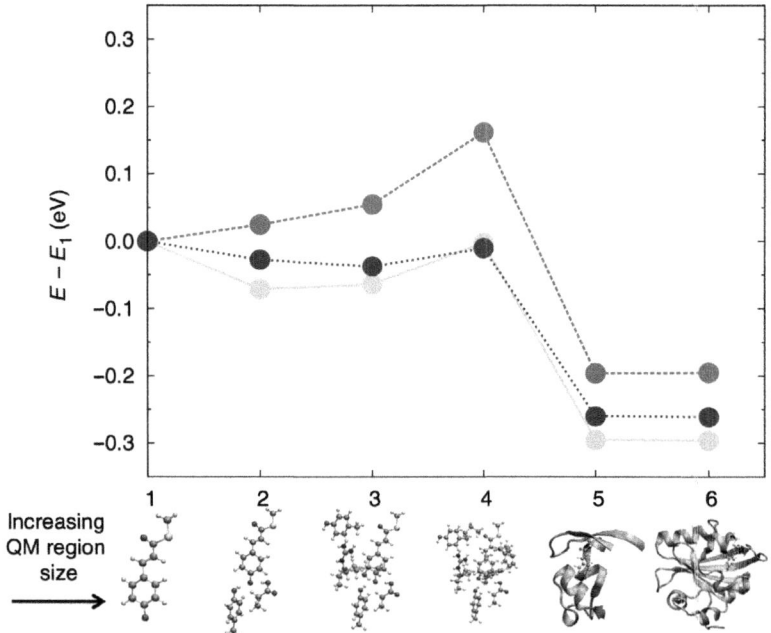

Fig. 4 Dependence of the PYP excited state energy on the size of the QM region, as obtained from QM/MM calculations for three different PYP configurations. Very large QM regions are required to generate the converged excitation energies. The y-axis is the difference in the computed excited state energy compared to that for QM region 1.

3. INCORPORATING NQES IN AIMD SIMULATIONS

The interplay between nuclear and electronic quantum effects can dramatically alter the structure and dynamics of hydrogen-bonded systems. It has been shown that the distance between the donor and acceptor heavy atoms, R, plays a crucial role in determining the proton behavior in a hydrogen bond (McKenzie, 2012; Perrin & Nielson, 1997). In particular, when R is around or below 2.7 Å, the influence of NQEs becomes significant. For example, in liquid water, which has an average O–O distance of around 2.8 Å, the quantum nature of the hydrogen bonds leads to transient proton excursion events in which a proton is closer to the hydrogen bond acceptor than donor oxygen (Ceriotti, Cuny, Parrinello, & Manolopoulos, 2013; Wang, Ceriotti, & Markland, 2014). In addition, the competition between quantum effects, which leads to NQEs weakening long hydrogen bonds and strengthening short ones, is heavily affected by R (Ceriotti et al., 2016; Habershon, Markland, & Manolopoulos, 2009; Li, Walker, & Michaelides, 2011; Markland & Berne, 2012; McKenzie, Bekker, Athokpam, & Ramesh, 2014).

Despite the importance of NQEs, most molecular simulations are performed treating the nuclei as classical particles. The path integral formalism of quantum mechanics provides an efficient way to incorporate NQEs in molecular simulations (Feynman & Hibbs, 1964). By combining the path integral methods and on-the-fly electronic structure calculations, one can perform AI-PIMD simulations to examine the nuclear and electronic quantum fluctuations in chemical and biological systems. Below we provide the basic concepts, implementation, and examples for AI-PIMD simulations.

3.1 Path Integral Molecular Dynamics

PIMD simulations allow the exact inclusion of NQEs on static equilibrium properties for a given electronic surface by exploiting the isomorphism between the quantum partition function of a QM system and the classical partition function of a "ring polymer," which is a cyclic structure containing multiple copies of the classical system with adjacent copies (beads) connected by harmonic springs (Berne & Thirumalai, 1986).

For a system of N classical particles of masses m_i the Hamiltonian is

$$H = \sum_{i=1}^{N} \frac{\mathbf{p}_i^2}{2m_i} + V(\mathbf{r}_1, \ldots, \mathbf{r}_N), \qquad (2)$$

where in AIMD the potential energy $V(\mathbf{r}_1, \ldots, \mathbf{r}_N)$ is obtained from an electronic structure calculation. The path integral expression for the partition function for this Hamiltonian is

$$Q_P = \frac{1}{(2\pi\hbar)^f} \int d^f\mathbf{p} \int d^f\mathbf{r}\, e^{-\beta_P H_P(\mathbf{p},\mathbf{r})}, \qquad (3)$$

where $f = 3NP$, P is the number of beads in the ring polymer, $\mathbf{p} \equiv \{\mathbf{p}_i^{(j)}\}_{i=1\ldots N}^{j=1\ldots P}$, $\mathbf{r} \equiv \{\mathbf{r}_i^{(j)}\}_{i=1\ldots N}^{j=1\ldots P}$, and $\beta_P = \frac{\beta}{P} = \frac{1}{k_B T_P}$. The effective temperature, $T_P = PT$, under which each bead evolves is thus P times higher than the physical temperature, allowing the copies of the system to sample areas of phase space that are not accessible to a classical particle. The PIMD Hamiltonian $H_P(\mathbf{p}, \mathbf{r})$ is,

$$H_P(\mathbf{p},\mathbf{r}) = \sum_{j=1}^{P} \left(\sum_{i=1}^{N} \frac{|\mathbf{p}_i^{(j)}|^2}{2m_i} + \frac{1}{2} m_i \omega_P^2 |\mathbf{r}_i^{(j)} - \mathbf{r}_i^{(j-1)}|^2 \right) + V(\mathbf{r}_1^{(j)}, \ldots, \mathbf{r}_N^{(j)}), \qquad (4)$$

where $\omega_P = 1/\beta_P \hbar$ and cyclic boundary conditions, $j + P \equiv j$, are implied. H_P corresponds to the Hamiltonian of a classical ring polymer, in which adjacent beads are linked by harmonic springs with a force constant of $m_i \omega_P^2$. In PIMD simulations the system is evolved using Hamilton's equations of motion from this Hamiltonian (Berne & Thirumalai, 1986). Note that PIMD simulations utilize the momentum term in Eq. (4) to sample phase space and hence the dynamics generated from this Hamiltonian are not exact quantum dynamics.

Static equilibrium properties of the QM system can be exactly calculated as

$$\langle A \rangle = \frac{\int d^f\mathbf{p} \int d^f\mathbf{r}\, A e^{-\beta_P H_P}}{\int d^f\mathbf{p} \int d^f\mathbf{r}\, e^{-\beta_P H_P}}. \qquad (5)$$

3.2 Accelerating PIMD Convergence Using Generalized Langevin Equations

In Eq. (3), Q_P is the exact quantum partition function when $P \to \infty$. In practice, a finite number of replicas P are used to converge the properties to the desired accuracy. A useful indication as to the number of replicas is $P > \beta\hbar\omega_{\max}$, where $\omega_{\max}$ is the highest frequency present in the system. These criteria can be understood as stating that the role of P is to increase the effective thermal energy of each replica (Pk_BT) to be larger than the energy level spacings in the system ($\hbar\omega_{max}$ for a harmonic oscillator), ie, allowing each replica to approach the classical limit (Markland & Manolopoulos, 2008a; Wallqvist & Berne, 1985). For example, the number of replicas used for convergence is typically $P \geq 32$ for a hydrogen-containing system at 300 K. Since each replica of the system requires an ab initio calculation to evaluate its energy and forces, the computational cost of PIMD simulations is therefore at least P times higher than the corresponding classical simulation.

Computational overhead due to the convergence requirement can be significantly reduced using a generalized Langevin equation (GLE) thermostat (Ceriotti, Bussi, & Parrinello, 2010; Ceriotti & Manolopoulos, 2012; Ceriotti, Manolopoulos, & Parrinello, 2011). By coupling the path integral system to a GLE thermostat (the PI + GLE approach), one can tune the correlated noise such that convergence of the potential energy, $\langle V \rangle$, to its quantum mechanical expectation value can be achieved with considerably fewer path integral beads (Ceriotti et al., 2010, 2011). However, the PI + GLE approach does not guarantee fast convergence of the quantum kinetic energy, $\langle T \rangle$, as the centroid virial estimator for the kinetic energy (Eq. 7) (Cao & Berne, 1989; Herman, Bruskin, & Berne, 1982) involves correlations between the beads and the ring-polymer centroid. To allow rapid convergence of the kinetic energy, Ceriotti and coworkers have recently introduced the PIGLET method, which enforces the necessary correlations to accelerate the convergence of the centroid virial kinetic energy estimator (Ceriotti & Manolopoulos, 2012). The PIGLET approach allows the convergence of thermodynamic properties and isotope effects that depend on the kinetic energy (see discussions in the following section) using $P = 6$ for hydrogen-containing systems at 300 K. This reduces the computational cost by over fivefold compared to a typical PIMD simulation (Ceriotti et al., 2013; Ceriotti & Manolopoulos, 2012). Parameters for the GLE thermostat can be conveniently obtained using the GLE4MD input generator (Ceriotti, 2010).

3.3 Extracting Thermodynamic Isotope Effects from PIMD Simulations

Isotope substitution techniques have been widely used in fields ranging from molecular biology to atmospheric chemistry (Kohen & Limbach, 2006). Computer simulations that accurately and efficiently predict isotope effects form an important complement to experiments and have enabled detailed analysis of the reaction mechanism and thermodynamic equilibrium in a wide variety of chemical and biological processes (Kohen & Limbach, 2006).

Since isotopes of the same element have almost identical electronic structures, the change in structural and thermodynamic properties due to isotope substitution arises entirely from the quantum mechanical nature of the nuclei. The central quantity for calculating these thermodynamic isotope effects is the free energy change upon isotope substitution, which can be computed exactly using PIMD simulations for a given electronic surface. For example, the free energy change upon exchanging D for H in a given system k, ΔA_k, is directly related to the quantum kinetic energies of the isotopes by the thermodynamic integration (Ceriotti & Markland, 2013)

$$\Delta A_k = \int_{m_H}^{m_D} d\mu \frac{\langle T_k(\mu) \rangle}{\mu}. \tag{6}$$

Here m_H and m_D are the masses of H and D, respectively. $T_k(\mu)$ is the quantum kinetic energy of a hydrogen isotope of mass μ and can be determined from AI-PIMD simulations using the centroid virial estimator (Cao & Berne, 1989; Herman et al., 1982),

$$\langle T_k(\mu) \rangle = \langle \frac{3}{2\beta} + \frac{1}{2P} \sum_{j=1}^{P} (\mathbf{r}^{(j)} - \bar{\mathbf{r}}) \cdot \frac{\partial V}{\partial \mathbf{r}^{(j)}} \rangle, \tag{7}$$

where V is the potential energy of the system and $\bar{\mathbf{r}} = \sum_{j=1}^{P} \mathbf{r}^{(j)}/P$ is the centroid position.

From Eq. (6), one can in principle perform multiple PIMD simulations with varying μ and obtain ΔA_k by carrying out the integration. This process can be greatly accelerated by using the free energy perturbation (FEP) method (Ceriotti & Markland, 2013), which allows one to extract $\langle T_k(\mu) \rangle$ from a single simulation of the most abundant isotope of hydrogen. From the FEP approach (Ceriotti & Markland, 2013),

$$\langle T_k(\mu) \rangle = \frac{\langle T_k(m_H) e^{-h(\mu/m_H; \mathbf{r})} \rangle_{m_H}}{\langle e^{-h(\mu/m_H; \mathbf{r})} \rangle_{m_H}} \tag{8}$$

with

$$h(\alpha; \mathbf{r}) = \frac{(\alpha - 1)\beta m_H \omega_P^2}{2P} \sum_{j=1}^{P} (\mathbf{r}^{(j)} - \mathbf{r}^{(j+1)})^2. \tag{9}$$

In the following, we use pK_a as an example and demonstrate how one can efficiently extract thermodynamic isotope effects from PIMD simulations. Let us consider the side-chain pK_a of an amino acid X in an enzyme upon H/D substitution. In H_2O, this pK_a corresponds to the chemical equilibrium where the neutral amino acid, XH_{EnzH}, dissociates into the anion X^-_{EnzH} and the proton H^+ (shown in the scheme below). In D_2O, we denote the neutral and ionized species of X as XD_{EnzD} and X^-_{EnzD}, respectively. Note that in D_2O all labile protons in the enzyme are exchanged to D, and thus X^-_{EnzH} and X^-_{EnzD} are not necessarily identical. These chemical reactions can be connected using the thermodynamic cycle:

$$
\begin{array}{ccccc}
XH_{EnzH} & \rightleftharpoons & X^-_{EnzH} & + & H^+ \qquad \Delta A_1 \\
\Delta A_{Enz} \uparrow & & \uparrow \Delta A_{Enz-} & & \uparrow \Delta A_H \\
XD_{EnzD} & \rightleftharpoons & X^-_{EnzD} & + & D^+ \qquad \Delta A_2
\end{array}
$$

Assuming the free energy changes of the aforementioned dissociation processes are ΔA_1 and ΔA_2, then the pK_a isotope effect is

$$\Delta pK_a^{Enz} \equiv pK_a^{EnzD} - pK_a^{EnzH} = \frac{\Delta A_2 - \Delta A_1}{2.303 k_B T} = \frac{\Delta A_{Enz} - \Delta A_{Enz-} - \Delta A_H}{2.303 k_B T}. \tag{10}$$

ΔA_{Enz} and ΔA_{Enz-} are the free energy changes upon converting XD_{EnzD} and X^-_{EnzD} to XH_{EnzH} and X^-_{EnzH}, respectively, and can be efficiently calculated from PIMD simulations using Eq. (6) along with the FEP method (Eq. 8).

To obtain the absolute value of ΔpK_a^{Enz}, one also needs ΔA_H, which is the free energy change upon converting D^+ to H^+ in aqueous solution. It can be computed from PIMD simulations of protons in liquid water using the same level of QM description as that for the enzyme. However, it is often computationally costly to properly sample the solvent configurations around the proton in order to determine an accurate ΔA_H. Alternatively, a useful metric is to compare ΔpK_a^{Enz} to a reference state such as the pK_a isotope effect of the amino acid X in aqueous solution, ΔpK_a^{Sol}. This results in an excess isotope effect, $\Delta\Delta pK_a \equiv \Delta pK_a^{Enz} - \Delta pK_a^{Sol}$, which represents the additional NQEs arising from the unique enzyme environment as compared to aqueous solution.

Similar to the enzyme case, ΔpK_a^{Sol} can be computed from the thermodynamic cycle.

$$XH_{Sol} \rightleftharpoons X^-_{Sol} + H^+ \qquad \Delta A_3$$
$$\Delta A_{Sol} \uparrow \qquad\qquad\qquad \uparrow \Delta A_H$$
$$XD_{Sol} \rightleftharpoons X^-_{Sol} + D^+ \qquad \Delta A_4$$

XH_{Sol} and XD_{Sol} denote the neutral amino acid X in H_2O and D_2O, respectively, while X^-_{Sol} represents the ionized species. Therefore,

$$\Delta pK_a^{Sol} = pK_a^{SolD} - pK_a^{SolH} = \frac{\Delta A_4 - \Delta A_3}{2.303 k_B T} = \frac{\Delta A_{Sol} - \Delta A_H}{2.303 k_B T}. \qquad (11)$$

Comparing ΔpK_a^{Enz} and ΔpK_a^{Sol} makes the ΔA_H term cancel, and thus the excess isotope effect is

$$\Delta\Delta pK_a = \Delta pK_a^{Enz} - \Delta pK_a^{Sol} = \frac{\Delta A_{Enz} - \Delta A_{Enz-} - \Delta A_{Sol}}{2.303 k_B T}. \qquad (12)$$

In this way, $\Delta\Delta pK_a$ can be extracted from PIMD simulations by computing the free energy changes upon isotope substitution in the enzyme and solution environment.

3.4 Electronic and Nuclear Quantum Fluctuations in Biological Hydrogen Bond Networks

Combing the technical advances introduced earlier, we now demonstrate how one can employ AI-PIMD simulations to elucidate the interplay between electronic and nuclear quantum effects in enzymes. We have recently studied the behavior of an active-site hydrogen bond network in an enzyme mutant KSI^{D40N} by interfacing with TeraChem and taking advantage of the PIGLET algorithm (Ceriotti & Manolopoulos, 2012) and the quantum FEP method (Ceriotti & Markland, 2013). As the majority of computational cost comes from ab initio force evaluations on the P ring-polymer beads, we have parallelized these calculations across GPUs with one GPU dedicated for each bead. These methods allow us to achieve a simulation speed that is three orders of magnitude higher than existing AI-PIMD approaches. In addition, in TeraChem the speed of electronic structure evaluations scales almost linearly with the number of GPUs. Therefore, one can further speed up the simulations by assigning one node for each bead and using multiple GPUs on each node to carry out electronic structure calculations.

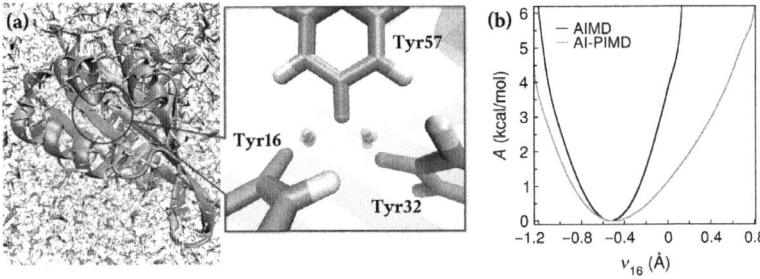

Fig. 5 (A) A snapshot of KSI^{D40N} from the AI-PIMD simulation. Its active-site hydrogen bond network is enlarged, in which *green, red*, and *white* represent carbon, oxygen, and hydrogen atoms, respectively. The *blue spheres* are the full ring-polymer representation of the protons. For clarity, all the other atoms are shown as their centroids. (B) The free energy surface of the proton movement along ν_{16}, as calculated from AIMD and AI-PIMD simulations. (See the color plate.)

In KSI^{D40N}, the side-chain groups of the active-site residues form a triad structural motif, in which residue Tyr57 sits at the center and forms two hydrogen bonds with the hydroxyl groups of residues Tyr16 and Tyr32 (Fig. 5A). In several high-resolution crystal structures (Ha, Kim, Lee, Choi, & Oh, 2000; Sigala et al., 2013) the hydrogen-bonded O57–O16 and O57–O32 distances are found to be around 2.6 Å, noticeably shorter than those commonly observed in hydrogen-bonded systems such as liquid water.

In order to choose the proper DFT functional and basis set for the QM region, we have calculated the potential energy profiles for proton transfer within the tyrosine triad. We define the proton transfer coordinate as $\nu_{16} = d_{O16,H16} - d_{O57,H16}$, where $d_{Oi,Hj}$ is the distance between the oxygen O of residue i and hydrogen H of residue j (Fig. 5A), and a quantity $\Delta E_{\nu=0}$ which represents the energy required to move the proton from its equilibrium position to the perfectly shared position, $\nu_{16} = 0$. As shown in Fig. 6, all density functionals tested generate results that qualitatively agree with each other, although GGA functionals BLYP and PBE underestimate $\Delta E_{\nu=0}$ by about 1.5 kcal/mol. Hybrid functionals B3LYP and PBE0 produce almost identical potential energy profiles, with the CAM-B3LYP RSH method in excellent agreement with the global hybrids. Dispersion corrections (D3 in this case) lead to very little change. Our basis set studies in Fig. 1 demonstrate that 6-31G* is the smallest basis set that is able to produce the correct energy profile for proton transfer. Based on these tests, we choose to treat the QM

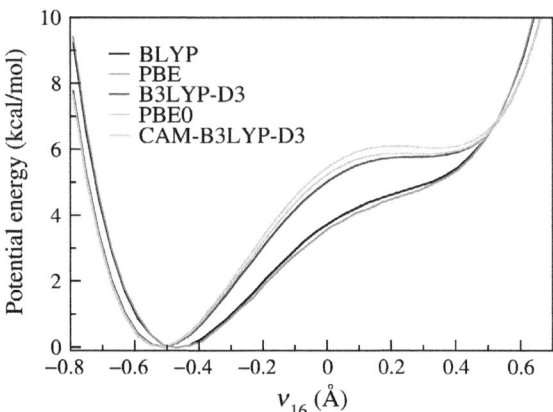

Fig. 6 Potential energy profile for proton transfer in KSI^{D40N}, as obtained using five density functionals and the 6-31G* basis set.

region at the B3LYP-D3 level (Becke, 1993; Grimme, Antony, Ehrlich, & Krieg, 2010) with the 6-31G* basis set.

In our AI-PIMD simulations, the QM region includes the tyrosine triad (47 atoms and 174 electrons) and is treated quantum mechanically in both the electronic and nuclear degrees of freedom. We have shown that increasing the size of the QM region by incorporating nearby residues does not significantly change the potential energy profile for proton transfer (Wang, Fried, et al., 2014). As environmental fluctuations are indispensable for a correct description of the QM region, especially the O–O distances, we include the rest of the protein as well as solvent molecules and counterions in the MM region (over 52,000 atoms), which is described using the AMBER03 (Duan et al., 2003) force field and the TIP3P water model (Jorgensen, Chandrasekhar, Madura, Impey, & Klein, 1983). All bonds across the QM/MM boundary are capped with hydrogen link atoms in the QM region, which are constrained along the bisecting bonds and do not interact with the MM region. The total interactions in the QM/MM system include (1) forces between QM atoms, (2) forces between MM atoms, and (3) electrostatic and van der Waals interactions between QM and MM atoms. Our simulation program obtains forces within the QM region and the electrostatic QM/MM interactions through an interface with TeraChem (Isborn et al., 2012; Ufimtsev & Martinez, 2009b), while the forces within the MM region and the Lennard–Jones QM/MM interactions are acquired via an MPI interface with the LAMMPS molecular dynamics package (Plimpton, 1995). A snapshot of the QM/MM system is shown in Fig. 5A.

The impact of NQEs can be understood by comparing AIMD and AI-PIMD simulations of KSI^{D40N}. Fig. 5B shows that NQEs significantly decrease the free energy required to move the proton from Tyr16 to Tyr57 ($\nu_{16} \geq 0$), thus allowing the proton to be quantum mechanically delocalized in the active-site hydrogen bond network. This is manifested by the wide spread of the protons' ring-polymer beads, which represents their position uncertainty (Fig. 5A). This quantum delocalization leads to a 10,000-fold increase in the acidity dissociation constant of Tyr57 and a $\Delta\Delta pK_a$ of 0.50 compared to tyrosine in solution, in excellent agreement with experiment (Wang, Fried, et al., 2014).

Due to the short R in the active-site hydrogen bond network, the ZPE possessed by the O–H bonds and $\Delta E_{\nu=0}$ are both about 5 kcal/mol (Fig. 6), which facilitates proton movements along ν_{16}. Hence the interplay between nuclear and electronic quantum effects qualitatively and quantitatively changes the proton behavior from classical hydrogen bonding to quantum delocalization.

4. OUTLOOK

With the development of new algorithms and their combination with novel computer streaming architectures, path integral methods and AIMD simulations can now be combined to elucidate the movements of electrons and light nuclei in unprecedented detail during biological processes. In addition, recent developments will accelerate methods such as CMD (Cao & Voth, 1994; Jang & Voth, 1999) and RPMD (Craig & Manolopoulos, 2004; Habershon et al., 2013) to obtain approximate quantum dynamics. These methods are not amenable to acceleration using the PIGLET approach. However, very recently it has been shown how the ring-polymer contraction approach (Fanourgakis, Markland, & Manolopoulos, 2009; Markland & Manolopoulos, 2008a), which exploits the separation of the forces acting on each copy of the system (Eq. 4) into rapidly and slowly varying parts, can be extended to AI-PIMD and AI-RPMD simulations (Kapil, VandeVondele, & Ceriotti, 2016; Marsalek & Markland, 2016).

As was illustrated in Section 3.4, AI-PIMD simulations can already provide novel insights into the structure and functional roles of short hydrogen bonds ($R \leq 2.6$ Å) in biological systems. Another area where AI-PIMD simulations may be able to offer important insights is the study of proton-coupled electron transfer reactions, which play a fundamental role in photosynthesis, respiration, and enzymatic catalysis (Chang, Chang, Damrauer, & Nocera, 2004; Cukier, 2004; Hammes-Schiffer & Stuchebrukhov, 2010). A comprehensive understanding of these biological processes requires simulation techniques that properly describe the QM nature of the electrons and the light nuclei. With the recent advent of efficient AI-PIMD simulations, these techniques will provide powerful tools to unravel the function of complex biological systems.

ACKNOWLEDGMENTS
C.M.I. and T.E.M. are supported by US Department of Energy, Office of Science, Office of Basic Energy Sciences under Award Number DE-SC0014437. L.W. acknowledges a postdoctoral fellowship from the Stanford Center for Molecular Analysis and Design. T.E.M. also acknowledges support from a Cottrell Scholarship from the Research Corporation for Science Advancement and an Alfred P. Sloan Research fellowship.

REFERENCES
Altun, A., Shaik, S., & Thiel, W. (2006). Systematic QM/MM investigation of factors that affect the cytochrome P450-catalyzed hydrogen abstraction of camphor. *Journal of Computational Chemistry*, 27(12), 1324–1337. http://dx.doi.org/10.1002/jcc.20398.
Autschbach, J., & Srebro, M. (2014). Delocalization error and functional tuning in Kohn–Sham calculations of molecular properties. *Accounts of Chemical Research*, 47(8), 2592–2602. http://dx.doi.org/10.1021/ar500171t.
Baer, R., Livshits, E., & Salzner, U. (2010). Tuned range-separated hybrids in density functional theory. S. R. Leone S. R., P. S. Cremer P. S., J. T. Grove J. T., M. A. Johnon M. A., G. Richmond G. (Ed.), *Annual review of physical chemistry*, Vol. 61, 85–109.
Baer, R., & Neuhauser, D. (2005). Density functional theory with correct long-range asymptotic behavior. *Physical Review Letters*, 94, 043002. http://dx.doi.org/10.1103/PhysRevLett.94.043002.
Becke, A. D. (1993). Density-functional thermochemistry. III. The role of exact exchange. *The Journal of Chemical Physics*, 98(7), 5648–5652. http://dx.doi.org/10.1063/1.464913.
Becke, A. D., & Johnson, E. R. (2005). Exchange-hole dipole moment and the dispersion interaction. *The Journal of Chemical Physics*, 122(15), 154104. http://dx.doi.org/10.1063/1.1884601.
Becke, A. D., & Johnson, E. R. (2007). Exchange-hole dipole moment and the dispersion interaction revisited. *The Journal of Chemical Physics*, 127(15), 154108. http://dx.doi.org/10.1063/1.2795701.
Benkovic, S. J., & Hammes-Schiffer, S. (2003). A perspective on enzyme catalysis. *Science*, 301(5637), 1196–1202. http://dx.doi.org/10.1126/science.1085515. Retrieved from, http://science.sciencemag.org/content/301/5637/1196.
Berne, B. J., & Thirumalai, D. (1986). On the simulation of quantum systems: Path integral methods. *Annual Review of Physical Chemistry*, 37(1), 401–424.
Bowler, D. R., & Miyazaki, T. (2012). O(N) methods in electronic structure calculations. *Reports on Progress in Physics*, 75(3), 036503. Retrieved from, http://stacks.iop.org/0034-4885/75/i=3/a=036503.

Bowler, D. R., Miyazaki, T., & Gillan, M. J. (2002). Recent progress in linear scaling ab initio electronic structure techniques. *Journal of Physics: Condensed Matter, 14*(11), 2781. Retrieved from, http://stacks.iop.org/0953-8984/14/i=11/a=303.

Burke, K. (2012). Perspective on density functional theory. *The Journal of Chemical Physics, 136*(15), 150901. http://dx.doi.org/10.1063/1.4704546.

Cao, J., & Berne, B. J. (1989). On energy estimators in path integral Monte Carlo simulations: Dependence of accuracy on algorithm. *The Journal of Chemical Physics, 91*(10), 6359.

Cao, J., & Voth, G. A. (1994). The formulation of quantum statistical mechanics based on the Feynman path centroid density. *IV. Algorithms for centroid molecular dynamics. The Journal of Chemical Physics, 101*(7), 6168–6183.

Case, D. A., Darden, T. A., Cheatham, T. E., Simmerling, C. L., Wang, J., Duke, R. E., ... Kollman, P. A. (2012). *AMBER 12*. San Francisco: University of California. Retrieved from, http://ambermd.org/.

Ceriotti, M. (2010). GLE4MD. epfl-cosmo.github.io/gle4md/.

Ceriotti, M., Bussi, G., & Parrinello, M. (2010). Colored-noise thermostats à la carte. *Journal of Chemical Theory and Computation, 6*(4), 1170–1180.

Ceriotti, M., Cuny, J., Parrinello, M., & Manolopoulos, D. E. (2013). Nuclear quantum effects and hydrogen bond fluctuations in water. *Proceedings of the National Academy of Sciences of the United States of America, 110*(39), 15591–15596.

Ceriotti, M., Fang, W., Kusalik, P. G., McKenzie, R. H., Morales, M. A., Michaelides, A., & Markland, T. E. (2016). Nuclear quantum effects in water and aqueous systems: Experiment, theory, and current challenges. *Chemical Reviews*. http://dx.doi.org/10.1021/acs.chemrev.5b00674.

Ceriotti, M., & Manolopoulos, D. E. (2012). Efficient first-principles calculation of the quantum kinetic energy and momentum distribution of nuclei. *Physical Review Letters, 109*(10), 100604.

Ceriotti, M., Manolopoulos, D. E., & Parrinello, M. (2011). Accelerating the convergence of path integral dynamics with a generalized Langevin equation. *The Journal of Chemical Physics, 134*(8), 84104.

Ceriotti, M., & Markland, T. E. (2013). Efficient methods and practical guidelines for simulating isotope effects. *The Journal of Chemical Physics, 138*(1), 014112.

Ceriotti, M., Parrinello, M., Markland, T. E., & Manolopoulos, D. E. (2010). Efficient stochastic thermostatting of path integral molecular dynamics. *The Journal of Chemical Physics, 133*(12), 124104.

Chai, J.-D., & Head-Gordon, M. (2008). Systematic optimization of long-range corrected hybrid density functionals. *The Journal of Chemical Physics, 128*, 084106.

Chang, C. J., Chang, M. C. Y., Damrauer, N. H., & Nocera, D. G. (2004). Proton-coupled electron transfer: A unifying mechanism for biological charge transport, amino acid radical initiation and propagation, and bond making/breaking reactions of water and oxygen. *Biochimica et Biophysica Acta, 1655*, 13–28.

Chen, B., Ivanov, I., Klein, M. L., & Parrinello, M. (2003). Hydrogen bonding in water. *Physical Review Letters, 91*, 215503.

Cleland, W. W., Frey, P. A., & Gerlt, J. A. (1998). The low barrier hydrogen bond in enzymatic catalysis. *The Journal of Biological Chemistry, 273*(40), 25529–25532. http://dx.doi.org/10.1074/jbc.273.40.25529.

Cohen, A. J., Mori-Sánchez, P., & Yang, W. (2008a). Fractional charge perspective on the band gap in density-functional theory. *Physical Review B, 77*, 115123.

Cohen, A. J., Mori-Sánchez, P., & Yang, W. (2008b). Insights into current limitations of density functional theory. *Science, 321*, 792–794.

Craig, I. R., & Manolopoulos, D. E. (2004). Quantum statistics and classical mechanics: Real time correlation functions from ring polymer molecular dynamics. *The Journal of Chemical Physics, 121*, 3368.

Cukier, R. I. (2004). Theory and simulation of proton-coupled electron transfer, hydrogen-atom transfer, and proton translocation in proteins. *Biochimica et Biophysica Acta, 1655*, 37–44.

Duan, Y., Wu, C., Chowdhury, S., Lee, M. C., Xiong, G., Zhang, W., ... Kollman, P. (2003). A point-charge force field for molecular mechanics simulations of proteins based on condensed-phase quantum mechanical calculations. *Journal of Computational Chemistry, 24*(16), 1999–2012. http://dx.doi.org/10.1002/jcc.10349.

Elstner, M., Hobza, P., Frauenheim, T., Suhai, S., & Kaxiras, E. (2001). Hydrogen bonding and stacking interactions of nucleic acid base pairs: A density-functional-theory based treatment. *The Journal of Chemical Physics, 114*(12), 5149–5155. http://dx.doi.org/10.1063/1.1329889.

Fanourgakis, G. S., Markland, T. E., & Manolopoulos, D. E. (2009). A fast path integral method for polarizable force fields. *The Journal of Chemical Physics, 131*(9), 094102. http://dx.doi.org/10.1063/1.3216520.

Feynman, R. P., & Hibbs, A. R. (1964). *Quantum mechanics and path integrals*. New York: McGraw-Hill.

Flaig, D., Beer, M., & Ochsenfeld, C. (2012). Convergence of electronic structure with the size of the QM region: Example of QM/MM NMR shieldings. *Journal of Chemical Theory and Computation, 8*(7), 2260–2271. http://dx.doi.org/10.1021/ct300036s.

Fox, S. J., Pittock, C., Fox, T., Tautermann, C. S., Malcolm, N., & Skylaris, C.-K. (2011). Electrostatic embedding in large-scale first principles quantum mechanical calculations on biomolecules. *The Journal of Chemical Physics, 135*(22), 224107. http://dx.doi.org/10.1063/1.3665893.

Friesner, R. A., & Guallar, V. (2005). Ab initio quantum chemical and mixed quantum mechanics/molecular mechanics (QM/MM) methods for studying enzymatic catalysis. *Annual Review of Physical Chemistry, 56*(1), 389–427. http://dx.doi.org/10.1146/annurev.physchem.55.091602.094410.

Gao, J., & Truhlar, D. G. (2002). Quantum mechanical methods for enzyme kinetics. *Annual Review of Physical Chemistry, 53*(1), 467–505. http://dx.doi.org/10.1146/annurev.physchem.53.091301.150114.

Gaus, M., Cui, Q., & Elstner, M. (2014). Density functional tight binding: application to organic and biological molecules. *Wiley Interdisciplinary Reviews: Computational Molecular Science, 4*(1), 49–61. http://dx.doi.org/10.1002/wcms.1156.

Gill, P. M. W., Adamson, R. D., & Pople, J. A. (1996). Coulomb-attenuated exchange energy density functionals. *Molecular Physics, 88*, 1005–1009.

Goedecker, S. (1999). Linear scaling electronic structure methods. *Reviews of Modern Physics, 71*(4), 1085–1123. http://dx.doi.org/10.1103/RevModPhys.71.1085.

Grimme, S. (2006). Semiempirical GGA-type density functional constructed with a long-range dispersion correction. *Journal of Computational Chemistry, 27*(15), 1787–1799. http://dx.doi.org/10.1002/jcc.20495.

Grimme, S. (2011). Density functional theory with London dispersion corrections. *Wiley Interdisciplinary Reviews: Computational Molecular Science, 1*(2), 211–228. http://dx.doi.org/10.1002/wcms.30.

Grimme, S., Antony, J., Ehrlich, S., & Krieg, H. (2010). A consistent and accurate ab initio parametrization of density functional dispersion correction (DFT-D) for the 94 elements H-Pu. *The Journal of Chemical Physics, 132*(15), 154104.

Ha, N.-C., Kim, M.-S., Lee, W., Choi, K. Y., & Oh, B.-H. (2000). Detection of large pKa perturbations of an inhibitor and a catalytic group at an enzyme active site, a mechanistic basis for catalytic power of many enzymes. *The Journal of Biological Chemistry, 275*(52), 41100–41106. http://dx.doi.org/10.1074/jbc.M007561200. Retrieved from, http://www.jbc.org/content/275/52/41100.abstract.

Habershon, S., Manolopoulos, D. E., Markland, T. E., & Miller, T. F., III. (2013). Ring-polymer molecular dynamics: Quantum effects in chemical dynamics from classical trajectories in an extended phase space. *Annual Review of Physical Chemistry, 64*(1), 387–413. http://dx.doi.org/10.1146/annurev-physchem-040412-110122.

Habershon, S., Markland, T. E., & Manolopoulos, D. E. (2009). Competing quantum effects in the dynamics of a flexible water model. *The Journal of Chemical Physics, 131*, 024501.

Hammes-Schiffer, S., & Stuchebrukhov, A. A. (2010). Theory of coupled electron and proton transfer reactions. *Chemical Reviews, 110*(12), 6939–6960. http://dx.doi.org/10.1021/cr1001436.

Haworth, N. L., & Bacskay, G. B. (2002). Heats of formation of phosphorus compounds determined by current methods of computational quantum chemistry. *The Journal of Chemical Physics, 117*(24), 11175–11187. http://dx.doi.org/10.1063/1.1521760.

Heaton-Burgess, T., & Yang, W. (2010). Structural manifestation of the delocalization error of density functional approximations: C4N+2 rings and C-20 bowl, cage, and ring isomers. *The Journal of Chemical Physics, 132*, 234113.

Henry, D. J., Parkinson, C. J., & Radom, L. (2002). An assessment of the performance of high-level theoretical procedures in the computation of the heats of formation of small open-shell molecules. The Journal of Physical Chemistry. *A, 106*(34), 7927–7936. http://dx.doi.org/10.1021/jp0260752.

Herman, M. F., Bruskin, E. J., & Berne, B. J. (1982). On path integral Monte Carlo simulations. *The Journal of Chemical Physics, 76*(10), 5150.

Hohenberg, P., & Kohn, W. (1964). Inhomogeneous electron gas. *Physics Review, 136*(3B), B864–B871. http://dx.doi.org/10.1103/PhysRev.136.B864.

Hu, L., Sderhjelm, P., & Ryde, U. (2011). On the convergence of QM/MM energies. *Journal of Chemical Theory and Computation, 7*(3), 761–777. http://dx.doi.org/10.1021/ct100530r.

Iikura, H., Tsuneda, T., Yanai, T., & Hirao, K. (2001). A long-range correction scheme for generalized-gradient-approximation exchange functionals. *The Journal of Chemical Physics, 115*, 3540–3544.

Isborn, C. M., Goetz, A. W., Clark, M. A., Walker, R. C., & Martinez, T. J. (2012). Electronic absorption spectra from MM and ab initio QM/MM molecular dynamics: Environmental effects on the absorption spectrum of photoactive yellow protein. *Journal of Chemical Theory and Computation, 8*(12), 5092–5106. http://dx.doi.org/10.1021/ct3006826.

Isborn, C. M., Luehr, N., Ufimtsev, I. S., & Martinez, T. J. (2011). Excited-state electronic structure with configuration interaction singles and Tamm–Dancoff time-dependent density functional theory on graphical processing units. *Journal of Chemical Theory and Computation, 7*(6), 1814–1823. http://dx.doi.org/10.1021/ct200030k.

Isborn, C. M., Mar, B. D., Curchod, B. F. E., Tavernelli, I., & Martnez, T. J. (2013). The charge transfer problem in density functional theory calculations of aqueously solvated molecules. *The Journal of Physical Chemistry B, 117*(40), 12189–12201. http://dx.doi.org/10.1021/jp4058274.

Jang, S., & Voth, G. A. (1999). A derivation of centroid molecular dynamics and other approximate time evolution methods for path integral centroid variables. *The Journal of Chemical Physics, 111*(6), 2371–2384. http://dx.doi.org/10.1063/1.479515.

Johnson, E. R., & Becke, A. D. (2006). A post-Hartree–Fock model of intermolecular interactions: Inclusion of higher-order corrections. *The Journal of Chemical Physics, 124*(17), 174104. http://dx.doi.org/10.1063/1.2190220.

Johnson, E. R., Mackie, I. D., & DiLabio, G. A. (2009). Dispersion interactions in density-functional theory. *Journal of Physical Organic Chemistry, 22*(12), 1127–1135. http://dx.doi.org/10.1002/poc.1606.

Jorgensen, W. L., Chandrasekhar, J., Madura, J. D., Impey, R. W., & Klein, M. L. (1983). Comparison of simple potential functions for simulating liquid water. *The Journal of Chemical Physics, 79*(2), 926.

Kapil, V., VandeVondele, J., & Ceriotti, M. (2016). Accurate molecular dynamics and nuclear quantum effects at low cost by multiple steps in real and imaginary time: Using density functional theory to accelerate wavefunction methods. *The Journal of Chemical Physics, 144*(5), 054111.

Kim, M.-C., Sim, E., & Burke, K. (2013). Understanding and reducing errors in density functional calculations. *Physical Review Letters, 111*, 073003.

Klinman, J. P., & Kohen, A. (2013). Hydrogen tunneling links protein dynamics to enzyme catalysis. *Annual Review of Biochemistry, 82*(1), 471–496. http://dx.doi.org/10.1146/annurev-biochem-051710-133623.

Kohen, A., & Limbach, H. (Eds.), (2006). *Isotope effects in chemistry and biology*. Florida: CRC Press.

Kohn, W., & Sham, L. J. (1965). Self-consistent equations including exchange and correlation effects. *Physics Review, 140*(4A), A1133–A1138. http://dx.doi.org/10.1103/PhysRev.140.A1133.

Kraisler, E., & Kronik, L. (2014). Fundamental gaps with approximate density functionals: The derivative discontinuity revealed from ensemble considerations. *The Journal of Chemical Physics, 140*, 18A540.

Kronik, L., Stein, T., Refaely-Abramson, S., & Baer, R. (2012). Excitation gaps of finite-sized systems from optimally tuned range-separated hybrid functionals. *Journal of Chemical Theory and Computation, 8*(5), 1515–1531. http://dx.doi.org/10.1021/ct2009363.

Kulik, H. J., Luehr, N., Ufimtsev, I. S., & Martinez, T. J. (2012). Ab initio quantum chemistry for protein structures. *The Journal of Physical Chemistry. B, 116*(41), 12501–12509. http://dx.doi.org/10.1021/jp307741u.

Kuritz, N., Stein, T., Baer, R., & Kronik, L. (2011). Charge-transfer-like pi($\rightarrow$)pi* excitations in time-dependent density functional theory: A conundrum and its solution. *Journal of Chemical Theory and Computation, 7*, 2408–2415.

Lennartz, C., Schfer, A., Terstegen, F., & Thiel, W. (2002). Enzymatic reactions of triosephosphate isomerase: A theoretical calibration study. *The Journal of Physical Chemistry B, 106*(7), 1758–1767. http://dx.doi.org/10.1021/jp012658k.

Li, X., Walker, B., & Michaelides, A. (2011). Quantum nature of the hydrogen bond. *Proceedings of the National Academy of Sciences of the United States of America, 108*, 6369.

Lin, H., & Truhlar, D. G. (2006). QM/MM: What have we learned, where are we, and where do we go from here? *Theoretical Chemistry Accounts, 117*(2), 185–199. http://dx.doi.org/10.1007/s00214-006-0143-z.

Lpez-Canut, V., Mart, S., Bertrn, J., Moliner, V., & Tun, I. (2009). Theoretical modeling of the reaction mechanism of phosphate monoester hydrolysis in alkaline phosphatase. *The Journal of Physical Chemistry B, 113*(22), 7816–7824. http://dx.doi.org/10.1021/jp901444g.

Luehr, N., Sisto, A., & Martinez, T. J. (2016). Gaussian basis set Hartree–Fock, density functional theory, and beyond on GPUs. In R. Walker & A. Goetz (Eds.), *Electronic structure calculations on graphics processing units: From quantum chemistry to condensed matter physics* (pp. 67–100). West Sussex: Wiley.

Markland, T. E., & Berne, B. J. (2012). Unraveling quantum mechanical effects in water using isotope fractionation. *Proceedings of the National Academy of Sciences of the United States of America, 109*, 7988–7991.

Markland, T. E., & Manolopoulos, D. E. (2008a). An efficient ring polymer contraction scheme for imaginary time path integral simulations. *The Journal of Chemical Physics, 129*(2), 024105.

Markland, T. E., & Manolopoulos, D. E. (2008b). A refined ring polymer contraction scheme for systems with electrostatic interactions. *Chemical Physics Letters, 464*(4–6), 256.

Marsalek, O., Chen, P.-Y., Dupuis, R., Benoit, M., Meheut, M., Bacic, Z., & Tuckerman, M. E. (2014). Efficient calculation of free energy differences associated with isotopic substitution using path-integral molecular dynamics. *Journal of Chemical Theory and Computation, 10*(4), 1440–1453. http://dx.doi.org/10.1021/ct400911m.

Marsalek, O., & Markland, T. E. (2016). Ab initio molecular dynamics with nuclear quantum effects at classical cost: Ring polymer contraction for density functional theory. *The Journal of Chemical Physics, 144*(5), 054112.

Marx, D., & Parrinello, M. (1996). Ab initio path integral molecular dynamics: Basic ideas. *The Journal of Chemical Physics, 104*(11), 4077.

McKenzie, R. H. (2012). A diabatic state model for donor-hydrogen vibrational frequency shifts in hydrogen bonded complexes. *Chemical Physics Letters, 535*, 196–200.

McKenzie, R. H., Bekker, C., Athokpam, B., & Ramesh, S. G. (2014). Effect of quantum nuclear motion on hydrogen bonding. *The Journal of Chemical Physics, 140*, 174508.

Mildvan, A. S., Massiah, M. A., Harris, T. K., Marks, G. T., Harrison, D. H. T., Viragh, C., ... Kovach, I. M. (2002). Short, strong hydrogen bonds on enzymes: NMR and mechanistic studies. *Journal of Molecular Structure, 615*(13), 163–175. http://dx.doi.org/10.1016/S0022-2860(02)00212-0. Retrieved from, http://www.sciencedirect.com/science/article/pii/S0022286002002120.

Mori-Sanchez, P., Cohen, A. J., & Yang, W. (2006). Many-electron self-interaction error in approximate density functionals. *The Journal of Chemical Physics, 125*, 201102.

Mori-Sánchez, P., Cohen, A. J., & Yang, W. (2008). Localization and delocalization errors in density functional theory and implications for band-gap prediction. *Physical Review Letters, 100*, 146401.

Morrone, J. A., & Car, R. (2008). Nuclear quantum effects in water. *Physical Review Letters, 101*(1), 17801.

Nielsen, J. E., & McCammon, J. A. (2003). Calculating pKa values in enzyme active sites. *Protein Science: A Publication of the Protein Society, 12*(9), 1894–1901. Retrieved from, http://www.ncbi.nlm.nih.gov/pmc/articles/PMC2323987/.

Ochterski, J. W., Petersson, G. A., & Montgomery, J. A. (1996). A complete basis set model chemistry. V. Extensions to six or more heavy atoms. *The Journal of Chemical Physics, 104*(7), 2598–2619. http://dx.doi.org/10.1063/1.470985.

Ordejón, P., Artacho, E., & Soler, J. M. (1996). Self-consistent order-n density-functional calculations for very large systems. *Physical Review B, 53*(16), R10441–R10444. http://dx.doi.org/10.1103/PhysRevB.53.R10441.

Perdew, J. P., Parr, R. G., Levy, M., & Balduz, J. L. (1982). Density-functional theory for fractional particle number—Derivative discontinuities of the energy. *Physical Review Letters, 49*, 1691–1694.

Perdew, J. P., Ruzsinszky, A., Constantin, L. A., Sun, J., & Csonka, G. I. (2009). Some fundamental issues in ground-state density functional theory: A guide for the perplexed. *Journal of Chemical Theory and Computation, 5*(4), 902–908. http://dx.doi.org/10.1021/ct800531s.

Perdew, J. P., & Yue, W. (1986). Accurate and simple density functional for the electronic exchange energy: Generalized gradient approximation. *Physical Review B, 33*(12), 8800–8802. http://dx.doi.org/10.1103/PhysRevB.33.8800.

Perrin, C. L., & Nielson, J. B. (1997). "Strong" hydrogen bonds in chemistry and biology. *Annual Review of Physical Chemistry, 48*(1), 511–544. http://dx.doi.org/10.1146/annurev.physchem.48.1.511.

Plimpton, S. (1995). Fast parallel algorithms for short-range molecular dynamics. *Journal of Computational Physics, 117*(1), 1–19.

Ponder, J. W., & Case, D. A. (2003). Force fields for protein simulations. *Advances in Protein Chemistry, 66*, 27–85.

Ranaghan, K. E., & Mulholland, A. J. (2010). Investigations of enzyme-catalysed reactions with combined quantum mechanics/molecular mechanics (QM/MM) methods.

International Reviews in Physical Chemistry, 29(1), 65–133. http://dx.doi.org/10.1080/01442350903495417.

Refaely-Abramson, S., Baer, R., & Kronik, L. (2011). Fundamental and excitation gaps in molecules of relevance for organic photovoltaics from an optimally tuned range-separated hybrid functional. Physical Review B, 84, 075144.

Refaely-Abramson, S., Sharifzadeh, S., Govind, N., Autschbach, J., Neaton, J. B., Baer, R., & Kronik, L. (2012). Quasiparticle spectra from a nonempirical optimally tuned range-separated hybrid density functional. Physical Review Letters, 109, 226405.

Riccardi, D., Schaefer, P., Yang, Y., Yu, H., Ghosh, N., Prat-Resina, X., ... Cui, Q. (2006). Development of effective quantum mechanical/molecular mechanical (QM/MM) methods for complex biological processes. The Journal of Physical Chemistry B, 110(13), 6458–6469. http://dx.doi.org/10.1021/jp056361o.

Salzner, U., & Baer, R. (2009). Koopmans' springs to life. The Journal of Chemical Physics, 131, 231101.

Schmidt, T., Kraisler, E., Makmal, A., Kronik, L., & Kuemmel, S. (2014). A self-interaction-free local hybrid functional: Accurate binding energies vis-à-vis accurate ionization potentials from Kohn–Sham eigenvalues. The Journal of Chemical Physics, 140, 18A510.

Seidl, A., Görling, A., Vogl, P., Majewski, J. A., & Levy, M. (1996). Generalized Kohn–Sham schemes and the band-gap problem. Physical Review B, 53(7), 3764–3774. http://dx.doi.org/10.1103/PhysRevB.53.3764.

Senn, H. M., & Thiel, W. (2009). QM/MM methods for biomolecular systems. Angewandte Chemie International Edition, 48(7), 1198–1229. http://dx.doi.org/10.1002/anie.200802019.

Siegbahn, P. E. M., & Borowski, T. (2006). Modeling enzymatic reactions involving transition metals. Accounts of Chemical Research, 39(10), 729–738. http://dx.doi.org/10.1021/ar050123u.

Sigala, P. A., Fafarman, A. T., Schwans, J. P., Fried, S. D., Fenn, T. D., Caaveiro, J. M. M., ... Herschlag, D. (2013). Quantitative dissection of hydrogen bond-mediated proton transfer in the ketosteroid isomerase active site. Proceedings of the National Academy of Sciences of the United States of America, 110(28), E2552–E2561. http://dx.doi.org/10.1073/pnas.1302191110. Retrieved from, http://www.pnas.org/content/110/28/E2552.abstract.

Simonson, T., Archontis, G., & Karplus, M. (2002). Free energy simulations come of age: Protein-ligand recognition. Accounts of Chemical Research, 35(6), 430–437. http://dx.doi.org/10.1021/ar010030m.

Skylaris, C.-K., Haynes, P. D., Mostofi, A. A., & Payne, M. C. (2005). Introducing ONETEP: Linear-scaling density functional simulations on parallel computers. The Journal of Chemical Physics, 122(8), 084119. http://dx.doi.org/10.1063/1.1839852.

Sosa Vazquez, X. A., & Isborn, C. M. (2015). Size-dependent error of the density functional theory ionization potential in vacuum and solution. The Journal of Chemical Physics, 143(24), 244105. http://dx.doi.org/10.1063/1.4937417.

Stein, T., Autschbach, J., Govind, N., Kronik, L., & Baer, R. (2012). Curvature and frontier orbital energies in density functional theory. Journal of Physical Chemistry Letters, 3, 3740–3744.

Stein, T., Eisenberg, H., Kronik, L., & Baer, R. (2010). Fundamental gaps in finite systems from eigenvalues of a generalized Kohn–Sham method. Physical Review Letters, 105, 266802.

Sumowski, C. V., & Ochsenfeld, C. (2009). A convergence study of QM/MM isomerization energies with the selected size of the QM region for peptidic systems. The Journal of Physical Chemistry A, 113(43), 11734–11741. http://dx.doi.org/10.1021/jp902876n.

Sutcliffe, M. J., & Scrutton, N. S. (2002). A new conceptual framework for enzyme catalysis. European Journal of Biochemistry, 269(13), 3096–3102. Retrieved from, http://dx.doi.org/10.1046/j.1432-1033.2002.03020.x.

Tawada, Y., Tsuneda, T., Yanagisawa, S., Yanai, T., & Hirao, K. (2004). A long-range-corrected time-dependent density functional theory. *The Journal of Chemical Physics, 120*, 8425–8433.

Toulouse, J., Colonna, F., & Savin, A. (2004). Long-range-short-range separation of the electron-electron interaction in density-functional theory. *Physical Review A, 70*, 062505.

Tozer, D. J. (2003). Relationship between long-range charge-transfer excitation energy error and integer discontinuity in Kohn–Sham theory. *The Journal of Chemical Physics, 119*(24), 12697–12699. http://dx.doi.org/10.1063/1.1633756.

Ufimtsev, I. S., Luehr, N., & Martinez, T. J. (2011). Charge transfer and polarization in solvated proteins from ab initio molecular dynamics. *Journal of Physical Chemistry Letters, 2*(14), 1789–1793. http://dx.doi.org/10.1021/jz200697c.

Ufimtsev, I. S., & Martinez, T. J. (2008). Quantum chemistry on graphical processing units. 1. Strategies for two-electron integral evaluation. *Journal of Chemical Theory and Computation, 4*(2), 222–231. http://dx.doi.org/10.1021/ct700268q.

Ufimtsev, I. S., & Martinez, T. J. (2009a). Quantum chemistry on graphical processing units. 2. Direct self-consistent-field implementation. *Journal of Chemical Theory and Computation, 5*(4), 1004–1015. http://dx.doi.org/10.1021/ct800526s.

Ufimtsev, I. S., & Martinez, T. J. (2009b). Quantum chemistry on graphical processing units. 3. Analytical energy gradients, geometry optimization, and first principles molecular dynamics. *Journal of Chemical Theory and Computation, 5*(10), 2619–2628. http://dx.doi.org/10.1021/ct9003004.

Vanicek, J., & Miller, W. H. (2007). Efficient estimators for quantum instanton evaluation of the kinetic isotope effects: Application to the intramolecular hydrogen transfer in pentadiene. *The Journal of Chemical Physics, 127*(11), 114309. http://dx.doi.org/10.1063/1.2768930.

Wallqvist, A., & Berne, B. (1985). Path-integral simulation of pure water. *Chemical Physics Letters, 117*(3), 214–219. http://dx.doi.org/10.1016/0009-2614(85)80206-2. Retrieved from, http://www.sciencedirect.com/science/article/pii/0009261485802062.

Wang, L., Ceriotti, M., & Markland, T. E. (2014). Quantum fluctuations and isotope effects in ab initio descriptions of water. *The Journal of Chemical Physics, 141*(10), 104502. http://dx.doi.org/10.1063/1.4894287.

Wang, L., Fried, S. D., Boxer, S. G., & Markland, T. E. (2014). Quantum delocalization of protons in the hydrogen-bond network of an enzyme active site. *Proceedings of the National Academy of Sciences of the United States of America, 111*(52), 18454–18459. http://dx.doi.org/10.1073/pnas.1417923111. Retrieved from, http://www.pnas.org/content/111/52/18454.

Wang, L.-P., Titov, A., McGibbon, R., Liu, F., Pande, V. S., & Martínez, T. J. (2014). Discovering chemistry with an ab initio nanoreactor. *Nature Chemistry, 6*(12), 1044–1048. Retrieved from, http://dx.doi.org/10.1038/nchem.2099.

Wang, N. X., & Wilson, A. (2004). The behavior of density functionals with respect to basis set. I. The correlation consistent basis sets. *The Journal of Chemical Physics, 121*(16), 7632–7646. http://dx.doi.org/10.1063/1.1792071.

Warshel, A., & Levitt, M. (1976). Theoretical studies of enzymic reactions: Dielectric, electrostatic and steric stabilization of the carbonium ion in the reaction of lysozyme. *Journal of Molecular Biology, 103*(2), 227–249. http://dx.doi.org/10.1016/0022-2836(76)90311-9.

Whittleton, S. R., Sosa Vazquez, X. A., Isborn, C. M., & Johnson, E. R. (2015). Density-functional errors in ionization potential with increasing system size. *The Journal of Chemical Physics, 142*(18), 184106. http://dx.doi.org/10.1063/1.4920947.

Wodrich, M. D., Jana, D. F., von Ragu Schleyer, P., & Corminboeuf, C. (2008). Empirical corrections to density functional theory highlight the importance of nonbonded intramolecular interactions in alkanes. *The Journal of Physical Chemistry A, 112*(45), 11495–11500. http://dx.doi.org/10.1021/jp806619z.

Wu, X., Vargas, M. C., Nayak, S., Lotrich, V., & Scoles, G. (2001). Towards extending the applicability of density functional theory to weakly bound systems. *The Journal of Chemical Physics, 115*(19), 8748–8757. http://dx.doi.org/10.1063/1.1412004.

Yanai, T., Tew, D. P., & Handy, N. C. (2004). A new hybrid exchange-correlation functional using the Coulomb-attenuating method (CAM-B3LYP). *Chemical Physics Letters, 393*, 51–57.

Zhang, J., Kulik, H. J., Martinez, T. J., & Klinman, J. P. (2015). Mediation of donoracceptor distance in an enzymatic methyl transfer reaction. *Proceedings of the National Academy of Sciences of the United States of America, 112*(26), 7954–7959.

Zhang, Y., & Yang, W. (1998). A challenge for density functionals: Self-interaction error increases for systems with a noninteger number of electrons. *The Journal of Chemical Physics, 109*, 2604–2608.

Zhao, Q., Xie, L., & Kulik, H. J. (2015). Discovering amorphous indium phosphide nanostructures with high-temperature ab initio molecular dynamics. *Journal of Physical Chemistry C, 119*(40), 23238–23249. http://dx.doi.org/10.1021/acs.jpcc.5b07264.

Zheng, X., Liu, M., Johnson, E. R., Contreras-Garcia, J., & Yang, W. (2012). Delocalization error of density-functional approximations: A distinct manifestation in hydrogen molecular chains. *The Journal of Chemical Physics, 137*, 214106.

CHAPTER SIXTEEN

Using Molecular Simulation to Study Biocatalysis in Ionic Liquids

K.G. Sprenger, J. Pfaendtner[1]
University of Washington, Seattle, WA, United States
[1]Corresponding author: e-mail address: jpfaendt@uw.edu

Contents

1. Introduction 420
   1.1 Motivation for Studying Biomolecules in Ionic Liquids 420
   1.2 Experiments of Biocatalysis in ILs 420
   1.3 Simulations of Biocatalysis in ILs 421
2. Methods for Simulating Biomolecules in ILs 425
   2.1 Choice of Force Field/Parameterization Process 425
   2.2 System Setup Example Using GAFF/Antechamber Tools 428
   2.3 Electrostatics and Charge Scaling Considerations 430
   2.4 IL-Specific Aspects of Enzyme MD Simulations 433
   2.5 Typical Analysis Approaches 434
3. Perspective: Challenges and Future Directions 436
References 437

Abstract

The practice of computational biocatalysis in ionic liquids (ILs) is still in its infancy, and thus best simulation practices are still developing. Herein, we examine the computational and experimental literature to date featuring systems of enzymes in aqueous and neat ILs. The many different approaches taken to parameterize ILs and set up simulations of enzymes in ILs are discussed, and common analysis techniques are reviewed. We also shed light on potential drawbacks and limitations to simulating enzymes in ILs, which include a lack of experimental data with which to validate computational models and inadequate sampling arising from the slow dynamics of many ILs that can lead to inaccurate descriptions of transport and equilibrium thermodynamic properties. A small case study illustrates the effects of scaling IL partial charges, which is a common practice in the field, on the conformational transitions of alanine dipeptide. The degree of charge scaling has a significant effect on the transition times between states of the biomolecule and highlights the importance of carefully setting up systems of enzymes in ILs. Finally, we discuss means to overcome these challenges and briefly consider possible new directions for the field.

1. INTRODUCTION

1.1 Motivation for Studying Biomolecules in Ionic Liquids

The study of biomolecules in room temperature ionic liquids (sometimes referred to as RTILs; here called simply ILs) has applications in biochemistry, biomedicine, and nanotechnology (Benedetto & Ballone, 2015). Early examples in the area of biocatalysis in ILs included new biomass pretreatment methods to facilitate the breakdown and conversion of cellulose to green fuels and chemicals (Fort et al., 2007), and the enzymatic breakdown of waste food oils (Guo & Xu, 2006; Ha, Lan, Lee, Hwang, & Koo, 2007). Compared to traditional organic solvents, ILs have many beneficial properties that could support novel uses of enzymes in industry (eg, negligible vapor pressures, low flammability, high recoverability, and extremely tunable solvent properties) (Moniruzzaman, Nakashima, Kamiya, & Goto, 2010). Although there are many examples of combining ILs with the full spectrum of various biomolecules, in this chapter, we are focusing on enzymes due to their unique ability to catalyze chemical transformations. Both computational and experimental efforts have provided valuable insights into the microscopic interactions of enzymes with ILs. However, this book chapter is focused on simulations of biocatalysis in ILs, and thus experimental studies will be highlighted only briefly. A very recent review article by Benedetto and Ballone paints a more detailed picture of the current status of experiments related to biomolecules in ILs (Benedetto & Ballone, 2015). Additionally, we emphasize that this chapter is not meant as a comprehensive review of all enzyme/IL simulations to date. Rather, our aim is to provide a guide for researchers looking to start performing their own enzyme/IL simulations and thus we focus on best simulation practices, potential pitfalls and limitations, and common analysis techniques.

1.2 Experiments of Biocatalysis in ILs

Experiments of enzymes in ILs have been on the rise over the past decade or so, though, like its computational counterpart, the field is still very new compared to the study of enzymes in native or native-like environments. Difficulties arising from the use of nonaqueous solvents have prevented the use of many of the typical analysis techniques used to study enzymes. Many papers that have been published thus far that deal specifically with enzymes in ILs utilize techniques such as circular dichroism (CD) (Curto et al., 2014;

Ghaedizadeh et al., 2016), differential scanning calorimetry (Dabirmanesh et al., 2011), and fluorescence spectroscopy techniques (Ajloo, Sangian, Ghadamgahi, Evini, & Saboury, 2013; Dabirmanesh et al., 2011; Ghaedizadeh et al., 2016; Ghosh, Parui, Jana, & Bhattacharyya, 2015) to observe dynamics and conformational changes upon insertion in the IL. Dynamic light scattering has been used to observe enzyme aggregation or large-scale denaturation in ILs (Jaeger & Pfaendtner, 2013). Additionally, Kaar and coworkers have recently demonstrated the use of 2D NMR methods as a means to understand and improve enzyme stability in ILs (Nordwald, Armstrong, & Kaar, 2014). Common activity assays have also been used including calorimetry (Curto et al., 2014), activity and conformational stability assays (Nordwald & Kaar, 2013a, 2013b), and fluorescence quenching assays to find the extent of ion binding to the enzyme surface (Nordwald & Kaar, 2013a). Kinetic studies using ultraviolet–visible spectroscopy or other techniques have yielded Michaelis–Menten constants (Ajloo et al., 2013; Curto et al., 2014; Ghaedizadeh et al., 2016), and quantitative structure–activity relationship models have been developed to predict enzyme performance in ILs (Mai & Koo, 2014). To determine IL effects on enzyme activity, experiments by Kim and coworkers in 2014 investigated the lipase-catalyzed transesterification reaction of butyl alcohol with vinyl acetate in ILs (Kim, Ha, Sethaphong, Koo, & Yingling, 2014).

Directed evolution techniques have also recently been applied to enzyme/IL systems as a means of optimizing enzymes with respect to desired properties (Chen et al., 2013; Liu et al., 2013; Tee et al., 2008). For example, Carter and coworkers mutated the formate dehydrogenase from *Candida boidinii* with error-prone PCR to improve the enzyme's thermostability and tolerance to ILs (Carter, Bekhouche, Noiriel, Blum, & Doumèche, 2014). Wolski et al. evolved variants of cellulase enzyme Cel7A via biased clique shuffling, based on the standard DNA shuffling, to have increased stability in aqueous ILs to improve the biomass dissolution process for biofuels production (Wolski, Dana, Clark, & Blanch, 2016). Iterative saturation mutagenesis is another technique to direct the evolution of enzymes. Ultimately, the choice of technique depends on the desired enzyme property to be optimized (Carter et al., 2014).

1.3 Simulations of Biocatalysis in ILs

To our knowledge, the first paper of a molecular simulation study of an enzyme in an IL was published in 2008 (Micaêlo & Soares, 2008) followed by subsequent work in 2011 from other groups (Klahn, Lim, Seduraman, & Wu, 2011; Klahn, Lim, & Wu, 2011). Since that time, the number of classes of enzymes and types of ILs studied has grown significantly. For reference, Table 1 shows the broad range of combinations of enzymes and ILs, both aqueous and nonaqueous, studied up till now (n.b., this list may not be exhaustive but is meant to show range and depth both in the choice of enzyme and IL). The full names and chemical formulas of the ILs listed in Table 1 are given in Table 2. As Table 1 shows, although a wide variety of biomolecules and ILs has been studied, the large amount of prior experimental work has led to more common uses of lipases such as *Candida antarctica* lipase B (CALB) and *Candida rugosa* lipase (CRL), and imidazolium-based ILs (ie, [RMIM], where R is the alkyl side-chain defined in Table 2).

The majority of the studies published thus far has focused on characterizing biomolecular structure and dynamics in ILs with water as a reference to measure and learn how to control the stability and retained activity, respectively, of enzymes in nonaqueous media. Recently, examples of enzyme/IL simulations have begun to explicitly mimic advanced enzyme design procedures. Burney et al. (2015), following the experimental protocol developed by the Kaar group (Nordwald & Kaar, 2013a, 2013b), simulated the surface charge modification procedure and studied the effect of controlled surface charge modification on the long-range structure and properties of the IL near the enzyme surface. In many cases, as Table 1 shows, the role of water content has been explicitly examined by studying solute behavior in varying concentrations of aqueous IL solutions. Common practices of simulating enzymes in IL solutions, even in pure IL solvents, include retaining the crystallographic water molecules that surround and at times penetrate the structure of the enzyme, primarily because of their important role in upholding enzyme stability (Klahn, Lim, & Wu, 2011). A common method of analysis to gain mechanistic insight into enzyme stability in ILs is to track the diffusion of water from the enzyme surface over time in IL solutions compared to in pure water (Burney & Pfaendtner, 2013; Latif et al., 2014; Micaêlo & Soares, 2008). However, some studies have taken the approach of removing the surface crystallographic waters while retaining just the water molecules buried within the enzyme (Klahn, Lim, Seduraman, et al., 2011) or removing the structural waters altogether (Kim et al., 2014), citing the activation of the enzyme under anhydrous conditions.

Later in this chapter, we provide more detailed descriptions of common analysis tools and techniques to study enzymatic behavior in ILs with specific tips and suggestions to get started. However, to highlight the wide range of

Table 1 Overview of Enzyme/IL Systems Studied to Date with Molecular Simulations

Biomolecule	IL	% or Conc. in Water	References
Serine protease cutinase	[BMIM][NO$_3$], [BMIM][PF$_6$]	2.5, 5, 7.5, 10, 15, 25, 35, 50, 60, 75	Micaêlo and Soares (2008)
CALB	[BMIM][NO$_3$], [BMIM][BF$_4$], [BMIM][PF$_6$], [MOEMIM][BF$_4$], [BAGUA][BF$_4$], [BCGUA][BF$_4$], [MCGUA][NO$_3$], [DCGUA][NO$_3$]	100	Klahn, Lim, Seduraman, et al. (2011) and Klahn, Lim, and Wu (2011)
Xylanase	[EMIM][ACE], [EMIM][EtSO$_4$]	10, 20, 50	Jaeger and Pfaendtner (2013)
CRL	[BMIM][PF$_6$], [BMIM][NO$_3$]	100	Burney and Pfaendtner (2013)
Adenosine deaminase	[AMIM][Cl], [OMIM][Cl]	100	Ajloo et al. (2013)
CALB, CRL	[BMIM][PF$_6$], [BMIM][BF$_4$], [BMIM][Cl], [BMIM][TfO], [BMIM][Tf$_2$N]	5, 10, 15, 20, 50	Latif, Alif, Micaêlo, Rahman, and Basyaruddin (2014)
CALB	[BMIM][TfO], [BMIM][Cl]	100	Kim et al. (2014)
Cellulases[a]	[EMIM][ACE]	15, 50	Jaeger, Burney, and Pfaendtner (2015)
CRL, *Bos taurus* α-chymotrypsin	[BMIM][Cl], [EMIM][EtSO$_4$]	20	Burney, Nordwald, Hickman, Kaar, and Pfaendtner (2015)
Hen egg white lysozyme	[PMIM][Br]	1.5 M	Ghosh et al. (2015)
Renilla luciferase	[BMIM][PF$_6$], [BMIM][BF$_4$]	14 mM, 16 mM[b]	Ghaedizadeh et al. (2016)

[a]*Trichoderma viride, Thermotoga maritima, Pyrococcus horikoshii.*
[b]Pairwise for ILs.

Table 2 Abbreviations and Formulas for IL Cations and Anions Listed in Table 1

	Cation			Anion	
Abbrev.	Name	Formula	Abbrev.	Name	Formula
[EMIM]	1-Ethyl-3-methylimidazolium	$C_6H_{11}N_2^+$	[PF$_6$]	Hexafluorophosphate	PF_6^-
[BMIM]	1-Butyl-3-methylimidazolium	$C_8H_{15}N_2^+$	[NO$_3$]	Nitrate	NO_3^-
[OMIM]	1-Octyl-3-methylimidzolium	$C_{12}H_{23}N_2^+$	[BF$_4$]	Tetrafluoroborate	BF_4^-
[AMIM]	1-Allyl-3-methylimidzolium	$C_7H_{11}N_2^+$	[ACE]	Acetate	$CH_3CO_2^-$
[PMIM]	1-Methyl-3-pentylimidazolium	$C_9H_{17}N_2^+$	[EtSO$_4$]	Ethyl sulfate	$C_2H_6O_4S^-$
[MOEMIM]	1-Methoxyethyl-3-methylimidazolium	$C_7H_{13}N_2O^+$	[TfO]	Trifluoromethanesulfonate	$CF_3SO_3^-$
[BAGUA]	Acyclic butylpentamethylguanidinium	$C_{10}H_{24}N_3^+$	[Tf$_2$N]	Bis(trifluoromethanesulfonyl)imide	$C_2F_6NO_4S_2^-$
[MCGUA]	Cyclic tetramethylguanidinium	$C_6H_{13}N_4^+$	[Cl]	Chloride	Cl^-
[BCGUA]	Cyclic butyltrimethylguanidinium	$C_9H_{19}N_4^+$	[Br]	Bromide	Br^-
[DCGUA]	Cyclic decyltrimethylguanidinium	$C_{15}H_{31}N_4^+$			

examples and interesting analyses that atomistic molecular simulations uniquely enable, we describe a few unique examples in the remainder of this section. Some additional analysis methods to note include the use of principal component analysis to gain insights into the representative slow modes of enzymes in water and ILs (Jaeger & Pfaendtner, 2013), which suggested that the ILs can disrupt relative ordering and strength of various slow, correlated motions. Building on this was the observation that in glycoside hydrolases, transient behavior of hydrogen bonds and salt-bridges could be linked to enzyme stability (Jaeger et al., 2015).

Finally, there are growing examples of studying how ILs affect the structure and dynamics of substrate binding pockets in relation to enzyme activity. Kim and coworkers analyzed the conformational changes of the active site of CALB in various ILs and found a direct relationship with enzymatic activity (Kim et al., 2014). Depth profiles from the interior of the catalytic cavity to the enzyme surface changed in response to the IL type and differed notably from one solvent to the next. These findings were also verified with follow-up experiments. In another example of computational predictions that were validated for the case of biocatalysis in ILs, the prediction by Jaeger and Pfaendtner (2013) of the role of ILs as competitive inhibitors to enzyme substrates was later independently confirmed (Li et al., 2013). As the field of computational biocatalysis in ILs is still relatively new, many testing and validation studies are still needed to understand the accuracy of the force fields used. As others have noted, the tolerance of a given enzyme to any particular IL is generally unpredictable at this time (Benedetto & Ballone, 2015). Thus, it is an exciting step forward whenever computational predictions are later proved to be correct through experiments.

2. METHODS FOR SIMULATING BIOMOLECULES IN ILs

2.1 Choice of Force Field/Parameterization Process

Many different approaches exist for setting up simulations of enzymes in ILs. There are three main parts to the setup process—choice of force field to simulate the enzyme, choice of force field to simulate water, and parameterization of the force field for the IL. Whereas there are many choices for the former two, there are widely varying approaches to establishing the IL force field. Due to length considerations, we will not discuss the choice of biomolecular force field here beyond noting the general need for the class of the force field to have compatibility with both the water model and force field for the IL. Regarding the choice of water model, a wide range of force fields have been used to simulate both the water molecules present in aqueous IL solutions or the crystallographic water surrounding the enzyme in pure IL solvents. The most common choice of water model in studies published thus far are 3-site water models like SPC (Micaêlo & Soares, 2008), SPC/E (Ghosh et al., 2015), and TIP3P (Burney et al., 2015; Burney & Pfaendtner, 2013; Jaeger et al., 2015; Jaeger & Pfaendtner, 2013). We are aware of only a single study with a 4-site water model (Latif et al., 2014). While more computationally expensive than 3-site or 4-site water models, studies by Klahn et al. used the TIP5P water model (Klahn, Lim, Seduraman, et al., 2011; Klahn, Lim, & Wu, 2011), citing their previous experiments that found TIP3P overestimated the strength of water/IL interactions and led to incorrect mixing of [BMIM][PF$_6$], known to be a water-immiscible IL, with water; TIP5P correctly predicted no mixing of [BMIM][PF$_6$] with water in the simulations (Klähn, Stüber, Seduraman, & Wu, 2010).

The greatest variation in approach among the published studies on biocatalysis in ILs lies in the parameterization of the IL cations and anions. Generally, approaches fall into one of three categories: (1) use of or manual creation of a customized force field specifically refined for a particular IL, (2) use of a generic or universal force field combined with customized partial atomic charges obtained via quantum mechanics calculations, or (3) use of a self-contained computer program to generate force field parameters for the IL(s) of interest. Guiding the choice of how to proceed with IL parameterization are the major considerations of scope and scale of the researcher's study. If the goal is to test the stability of a particular enzyme in many ILs from different families (ie, no common cation or anion to all ILs) in a high-throughput manner, the parameterization method should be general and easily applied across the range of ILs tested, like in approach 2. Conversely, if the goal of the study is, for example, to reproduce experimental results of the behavior of an enzyme in a particular IL or family of ILs, an IL-specific (and thus more accurate) parameterization method would be appropriate, like in approach 1. A few studies thus far have utilized a united-atom model for the ILs: in the spirit of the first type of approach, Micaêlo and Soares developed a united-atom-based model for two imidazolium-based ILs in the framework of the GROMOS force field (Micaêlo & Soares, 2008), whereas Ghosh et al. and Ajloo et al. followed the third type of approach and generated united-atom IL force field parameters from software such as the Automated Force Field Builder and Repository (ATB and Repository) (Malde et al., 2011), also based on the GROMOS force field, and PRODRG2 (Schuttelkopf & van Aalten, 2004), respectively

(Ajloo et al., 2013; Ghosh et al., 2015). OPLS-based force fields have also been common: Latif et al. (2014) and Klahn, Lim, Seduraman, et al. (2011) and Klahn, Lim, and Wu (2011) used the IL family-transferable force fields developed by Canongia Lopes and Pádua (2004, 2006) and Canongia Lopes, Pádua, and Shimizu (2008) to model the ILs in their studies, and Kim et al. (2014) used the force field parameters developed for 68 different ILs by Sambasivarao and Acevedo (2009). Our own group, seeking to leverage the high-quality AMBER family force fields for proteins and carbohydrates, adopts the second approach and makes use of the fact that the general AMBER force field (GAFF) can reproduce many properties of neat ILs (Sprenger, Jaeger, & Pfaendtner, 2015). This has permitted seamless combining of GAFF-parameterized ILs with modern AMBER protein force fields like AMBER ff99SB (Hornak et al., 2006) and GLYCAM (Kirschner et al., 2008), the AMBER carbohydrate force field, to study many different biomolecule/IL systems (Burney et al., 2015; Burney & Pfaendtner, 2013; Jaeger et al., 2015; Jaeger & Pfaendtner, 2013; Jarin & Pfaendtner, 2014). For many ILs, GAFF already includes all the necessary bonded and nonbonded interactions with the exception of electrostatic point charges, which are obtained in a customized manner via a standard procedure. The developers of the CHARMM family of force fields recently published a generic force field, CGenFF (Vanommeslaeghe & MacKerell, 2012; Vanommeslaeghe, Raman, & MacKerell, 2012), which is an ideal candidate to further test the ability of a general force field to have widespread predictive power.

A common need, irrespective of the various parameterization methods, is accounting in some way for the slow dynamics of room temperature ILs, which if ignored may lead to poor configurational sampling and increase simulation convergence times (Latif et al., 2014; Micaêlo & Soares, 2008). The origin of this problem stems from the high viscosity of many ILs that can artificially dampen system dynamics, if not captured and corrected for in the IL force field, and can lead to wildly inaccurate predictions of IL transport properties. Micaêlo and Soares found it helpful to scale the masses of heavy atoms in their system in order to bring all of the characteristic vibrational frequencies more in line with each other (Micaêlo & Soares, 2008). They acknowledged that this had an effect on the overall dynamics of the system, but it left the equilibrium thermodynamic properties of the system unchanged. Another method to overcome the sluggish dynamics of ILs is to simply scale down the partial charge of each cation/anion. Studies have shown that classical force fields generally do a poor job of estimating the transport properties of ILs (eg, shear viscosity or self-diffusivity) when full charges are assigned to the ions (Zhang & Maginn, 2012). This is best explained by the fact that using full charges leads to a poor approximation of liquid-like behavior (Hurisso, Lovelock, & Licence, 2011; Men, Lovelock, & Licence, 2011; Zhang & Maginn, 2012), ie, charge screening by neighboring molecules and electron density effects by neighboring molecules of opposite charge are not well represented. It has been shown that scaled partial charges lead to significant improvements in the resulting transport properties of ILs (Zhang & Maginn, 2012). The use of GAFF combined with the Antechamber/RESP charge generation method and scaled IL partial charges by a factor of 0.8 has been used in some more recent biocatalysis/IL studies (Burney et al., 2015; Jaeger et al., 2015). The steps to carry out this particular parameterization method will be described in the next section, followed by a small case study to illustrate the effects of charge scaling in ILs on the conformational transitions of biomolecules. It should be noted that this method of performing quantum calculations on isolated ions or pairs of ions and then uniformly scaling the charges does sacrifice some accuracy; a much more accurate but computationally exhaustive approach is to use a QM/MM description of the liquid to derive the parameters for the Coulombic potential using the actual charge distributions in the liquid. This method has been employed by Klahn and coworkers to simulate CALB in a number of IL solvents (Klahn, Lim, Seduraman, et al., 2011; Klahn, Lim, & Wu, 2011).

2.2 System Setup Example Using GAFF/Antechamber Tools

There are many freely available programs to establish a workflow for creating the required force fields and starting structures for simulating enzymes in ILs. Fig. 1 shows a general flowchart for setting up an enzyme/IL system for use in a molecular dynamics (MD) simulation carried out with the GROMACS MD engine (Hess, Kutzner, van der Spoel, & Lindahl, 2008). We note that for any individual step, a number of different programs could be substituted, and interested readers are suggested to review the literature listed in Table 1 for alternative methods. Steps 1–5 are performed consecutively first for one ion, then for the other, but will be discussed here in tandem to avoid redundancy. The point charges for the cation/anion pair are created by drawing the chemical structures of the ions (step 1, completed in GaussView or a free program like Molden or Avogadro), and subsequent geometry optimization and calculation of the electrostatic potential around the molecule (in step 2, Gaussian (Frisch et al., 2009) conveniently works with the AmberTools

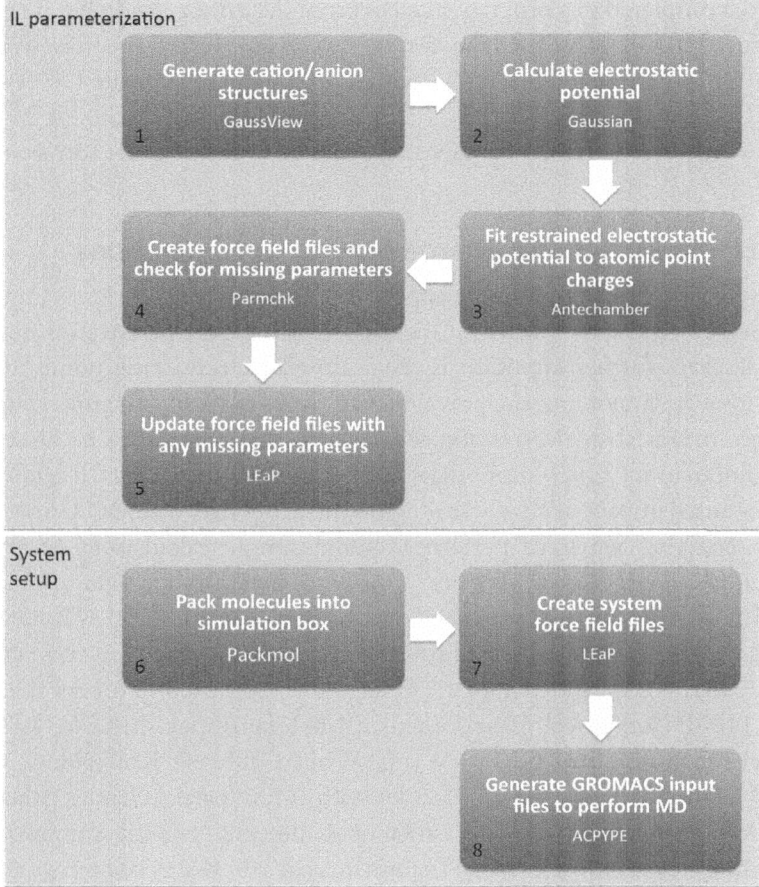

Fig. 1 Flow chart for developing MD simulation files for an enzyme/IL system to be simulated in GROMACS. This workflow is general and could be used with any MD engine compatible with modern force fields. (See the color plate.)

program in our workflow, but there are many other options for completing this step). Fitting the electrostatic potential leads to the determination of the atomic point charge on each atom (in step 3, Antechamber (Wang, Wang, Kollman, & Case, 2006) is a convenient program because it also assigns GAFF atom types). Combining the generated structural and electrostatic parameters with the equation of GAFF to describe the potential energies and their force derivatives creates force field files for the ions (a program like Parmchk can be used to identify missing parameters (step 4) that can be manually parameterized and incorporated in LEaP (Case et al., 2015) in step 5). After constructing the molecular structure file for the full system (step 6, completed in a program like Packmol (Martinez, Andrade, Birgin, & Martinez, 2009)), force field files for the system are created as before with the ions (step 7, completed in LEaP). Lastly, input files are generated to perform an MD simulation (in step 8, ACPYPE (Sousa da Silva & Vranken, 2012) is used to convert between the AMBER and GROMACS file formats).

2.3 Electrostatics and Charge Scaling Considerations

As noted earlier, many different approaches have been applied to set up simulations of biocatalysts in ILs. One aspect of the parameterization process that deserves further attention is the scaling of electrostatic point charges of IL ions. In IL systems, the prevalence of large numbers of significant point charges means that electrostatic interactions can contribute to molecular scale fluctuations much more than in purely aqueous systems. To illustrate the potential impact of charge scaling on the properties of an MD simulation of an enzyme, we have performed some simple calculations of alanine dipeptide in solutions of pure 1-butyl-3-methylimidazolium chloride ([BMIM][Cl]) with ion partial charges scaled to values of 1.0 (unscaled), 0.9, and 0.8. Additionally, a simulation is done in water for comparison. The AMBER03 force field (Case et al., 2005) was used to model the peptide, TIP3P (Jorgensen, Chandrasekhar, Madura, Impey, & Klein, 1983) to model the water, and GAFF (Wang, Wolf, Caldwell, Kollman, & Case, 2004) to model the organic molecules with RESP partial charges generated via Antechamber. Following a steepest-descent energy minimization of 10,000 steps, a short 1 ns NPT simulation at 300 K was run to equilibrate the system pressure, using a global stochastic thermostat (Bussi, Donadio, & Parrinello, 2007) and the Berendsen barostat (Berendsen, Postma, van Gunsteren, DiNola, & Haak, 1984). MD combined with the well-tempered version of metadynamics (MetaD) (Barducci, Bonomi, & Parrinello, 2011; Barducci, Bussi, & Parrinello, 2008; Laio & Parrinello, 2002) was used to construct free energy surfaces for each of the four simulations, shown in Fig. 2. The free energy has been projected onto the phi and psi dihedral collective variables (CVs), which were biased using PLUMED (Tribello, Bonomi, Branduardi, Camilloni, & Bussi, 2014) with Gaussian hill widths (σ) of 0.2 radians and hills added every 4 ps (τ) at an initial bias deposition rate of 0.075 kcal/mol/ps. A biasfactor (γ) of 9 was used in all simulations. Fig. 2 shows the free energy profile of alanine dipeptide in water (A) is very similar to that in [BMIM][Cl] with partial charges scaled by a factor of 0.8 (B). Scaling the charges by a factor of 0.9 (C) and by a factor of 1.0 (D) leads

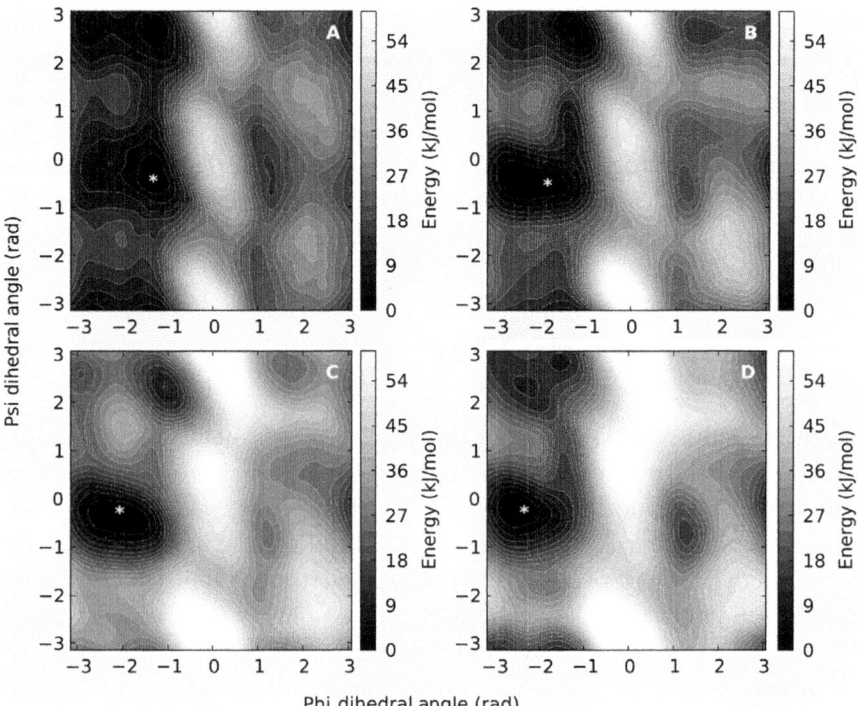

Fig. 2 Free energy surfaces for alanine dipeptide in (A) water, (B) [BMIM][Cl] with partial charges scaled by multiplying by 0.8, (C) [BMIM][Cl] with partial charges scaled by multiplying by 0.9, and (D) [BMIM][Cl] with partial charges scaled by multiplying by 1.0 (ie, not scaled). Asterisks indicate the α well in each of the two-state systems. (See the color plate.)

to increasingly higher barrier heights, hinting that longer transition times between states might exist for systems with less scaling of IL partial charges and that the main pathway(s) for changing states might differ as well. Despite these apparent differences in the free energy landscapes, all free energy surfaces show a two-state system with the same general locations of the low free energy wells, α (indicated with an asterisk) and β, across all simulations. It is interesting to note, however, that while the two-dimensional shape of the α well remains very similar across the four simulations, the shape of the β well changes considerably as the partial charges are less scaled, or essentially as the viscosity of the system increases. We also note that convergence was based on the free energy difference between the two wells over time, with the slow dynamics in the IL systems leading to longer convergence times.

To assess the rate of conformational transition in the different systems, we also initiated a series of simulations with the infrequent metadynamics approach (Tiwary & Parrinello, 2013) to estimate the rate of reaction of the $\alpha \to \beta$ transition of alanine dipeptide in the different solvents (n.b., the white asterisk in each plot in Fig. 2 represents the point in CV phase space at which state α is designated). These simulations employed well-tempered MetaD with hills now added every 80 ps at an initial bias deposition rate of 0.00375 kcal/mol/ps and postprocessed with the infrequent metadynamics algorithm. Calculation of an individual transition time over a barrier is calculated via Eq. (1),

$$t^{\text{eff}} = \sum_{i=0}^{N} \Delta t_i^{\text{MetaD}} e^{\beta V_{\text{bias}}(s,\, t)} \tag{1}$$

where Δt is the MetaD time step and V_{bias} is the instantaneous value of the MetaD bias potential at the ith point in the transition trajectory. The point in time at which alanine dipeptide leaves the α basin defines the interval of time in the so-called "transition trajectory." Details about the p value analysis can be found in the work of Salvalaglio, who suggested the approach for checking the validity of the calculations (Salvalaglio, Tiwary, & Parrinello, 2014). We also note our analysis and bootstrapping process followed our recent use of infrequent metadynamics to study chemical reaction rates (Fleming, Tiwary, & Pfaendtner, 2016).

Table 3 shows that all p values are above the recommended threshold of 0.05, thus we can conclude the transition times are uncorrelated and follow a Poisson distribution. The results show drastically different mean transition times, beyond an order of magnitude, are calculated depending on the degree of scaling of the ion partial charges. As predicted in the earlier discussion of the free energy profiles, the mean transition time generally increases as the degree of scaling is decreased from 0.8 to 1.0. In other words,

Table 3 Mean Escape Time Data for the Alanine Dipeptide $\alpha \to \beta$ Transition

Solvent	Charge Scaling	t^{eff} (ns)[a]	p Value[b]	Number of Events	τ (ps)[c]
Water	N/A	0.136 (0.016)	0.78	100	80
[BMIM][Cl]	0.8	9.00 (1.2)	0.99	100	80
[BMIM][Cl]	0.9	138 (17.9)	0.92	100	80
[BMIM][Cl]	1.0	149 (21.3)	0.08	100	80

[a]Mean transition times of 100 rare events, with standard deviations in parentheses.
[b]Kolmogorov–Smirnov significance (p) values.
[c]Time between successive Gaussian hills deposition in MetaD simulations.

the viscosity has a significant effect on the conformational transitions of biomolecules and should be treated with extreme care when constructing simulations of enzymes in IL solvents.

2.4 IL-Specific Aspects of Enzyme MD Simulations

Determining the force field, topology, and all input files for simulations of enzymes in ILs often proves to be the most time-consuming aspect (from the point of view of human effort) of the whole process. Once this is completed, the simulations can be performed with any parallel and efficient MD engine. Beyond the initial setup, there are a few potential considerations that merit further attention.

First, there is a common practice in the enzyme/protein simulation community to use a slow annealing from a low temperature near 10 K to the desired simulation temperature. Simulations with a barostat to mechanically equilibrate the system at a fixed pressure often follow this heating step. The IL systems are highly viscous and initial random configurations may lead to strong electrostatic interactions that could artificially pull the system far from equilibrium. To address this, we recommend a twofold remedy. First, simulations should be performed with multiple instances of generating the packed box of ILs (and water if desired). Second, in each system we suggest freezing all protein atoms and holding the system at a very high temperature (above 500 K) for ∼5 ns to equilibrate the solvent structure. The simulations can then be reliably quenched to a very low temperature and then slowly heated with an unfrozen protein according to the common community practice of annealing to room temperature for biomolecular studies.

Next, for simulations of enzyme/IL/water (ie, there are IL and water cosolvents together), the issue of water miscibility is a potential concern. There is a significant amount of experimental data to guide the selection of the system temperature and amount of water (in order to avoid phase separation). However, unless the force field has already been evaluated for its ability to predict experimental phase separation, it is important to remember the precise temperature and composition at which the phase split occurs. If concerns exist regarding miscibility, we suggest trial simulations of the water/IL mixture to evaluate this issue in detail. Finally, in light of the issue of slower timescales arising from high solvent viscosity, we also suggest extending MD simulations to longer timescales and/or using enhanced sampling approaches, which will be discussed further in the conclusions.

2.5 Typical Analysis Approaches

This section briefly highlights typical ways that other researchers have characterized MD simulations of enzyme/IL systems. We note that on a practical level, there is no reason any type of analysis based on studying average or time-dependent behavior of the enzyme, solvent, or both cannot be used in the context of MD simulations of nonaqueous biocatalysis. To date the most common way of characterizing enzyme/IL interactions is through the use of structural-based methods like those listed in Table 4, most of which are often calculated post priori vs on-the-fly to avoid computational overhead on MD simulations.

Some groups have made a point to focus on the thermodynamic aspects of enzyme/IL interactions as well. For instance, in 2011 Klahn and coworkers quantified the enthalpic changes that occurred upon solvation of an enzyme in IL compared to in water (Klahn, Lim, Seduraman, et al., 2011), as well as calculated the strengths of the Coulombic and Lennard–Jones interaction energies between the enzyme and IL ions (Klahn, Lim, Seduraman, et al., 2011; Klahn, Lim, & Wu, 2011). More recent papers by Burney et al. and Jaeger et al. directly quantified the entropic contributions to enzyme/IL interactions (Burney & Pfaendtner, 2013; Jaeger et al., 2015) by quantifying side-chain fluctuations. Jaeger and coworkers also quantified energetic contributions to enzyme/IL interactions by calculating

Table 4 Common Methods of Analysis for Enzyme/IL Systems

Method	What Is Calculated	What It Is Used for
Root mean square deviation (RMSD)	Mean deviation from a reference structure (usually the X-ray structure)	Assessing conformational changes and protein unfolding
Root mean square fluctuation (RMSF)	Residue-specific fluctuations averaged over portions of the simulation or the entire simulation	Assessing protein dynamics and backbone flexibility
Secondary structure	α-helix or β-sheet content	Assessing protein stability and loss or gain of structural motifs
Intra/intermolecular hydrogen bonds (H-bonds)	Number of hydrogen bonds between the enzyme and anions, enzyme and water in IL/water solutions, or between residues within the enzyme	Assessing protein stability and solvation in ILs and transient behavior of key nonbonded interactions

the internal energy of the protein, ie, summing up all of the molecular interactions within the protein (Jaeger et al., 2015).

A final type of analysis that is commonly performed on data from simulations of enzymes in ILs is the calculation of radial distribution functions (Ajloo et al., 2013; Burney et al., 2015; Ghosh et al., 2015) and spatial distribution functions (Latif et al., 2014; Micaêlo & Soares, 2008) of the IL ions and/or water molecules surrounding the enzyme surface. These types of analyses have provided insight into the behavior of ions and water, either solution or crystallographic, at the enzyme surface. In the case of Burney et al., this provided insightful information about the change in proportion of IL cations and anions around the enzyme surface with certain enzyme surface charge modifications (Burney et al., 2015). In the same paper, the authors also calculated charge distribution functions (ie, charge density of the IL ions as a function of the radial distance from the enzyme surface) to elucidate trends in solvent structure. Specifically, the goal of the analysis was to test the hypothesis that due to an anion-dominated surface layer created by enzyme charge modifications, the IL would organize in concentric spherical layers of alternating charge.

Since ILs have been shown to change the conformation of substrates, this is another potential area of investigation unique to enzyme/IL simulations, and so we briefly discuss here common methods of analysis applied to substrate/IL systems. Similar to the analysis methods for studying enzymes in ILs, many studies of substrates in ILs have focused on determining the solvent's structural and spatial distribution around the substrate to, for example, glean mechanistic aspects of cellulose solvation and dissolution in ILs (Cho, Gross, & Chu, 2011; Gross, Bell, & Chu, 2011; Gupta, Hu, & Jiang, 2011; Mostofian, Smith, & Cheng, 2011; Youngs, Hardacre, & Holbrey, 2007; Youngs et al., 2011). Thermodynamic analysis of the cellulose solvation process in ILs has revealed an entropic and energetic driving force for cellulose dissolution in ILs (Gross et al., 2011; Gross, Bell, & Chu, 2012), and more recently published papers have identified key structural and thermodynamic differences of bundles of cellulose vs individual chains in ILs (Li et al., 2015; Rabideau, Agarwal, & Ismail, 2013; Rabideau & Ismail, 2015; Zhao, Liu, Wang, & Zhang, 2013). New and interesting methods of analysis include analyzing the induced changes in the conformational free energy landscapes of sugar rings by ILs (Jarin & Pfaendtner, 2014), and performing dynamical calculations of mean lifetimes of different binding states of IL anions with hydroxyl groups on cellulose (Rabideau & Ismail, 2015).

3. PERSPECTIVE: CHALLENGES AND FUTURE DIRECTIONS

As we look to the future of molecular simulations of biocatalysts in ILs, a hurdle that will need to be overcome is the dearth of experimental data on enzyme/IL systems with which to validate computational models. Part of this problem may be that IL solvents often bring new challenges with standard experimental methods (eg, NMR, CD, crystallography). The importance of structural information is critical; experimental studies have played an invaluable role in continuing to refine modern force fields for protein simulations (eg, the identification in 2008 that many force fields incorrectly predicted alpha helicity; Best, Buchete, & Hummer, 2008). The absence of this data has been noted in the literature (Burney & Pfaendtner, 2013; Jaeger et al., 2015; Klahn, Lim, & Wu, 2011), and many groups have instead taken to comparing their simulation results to experimental data via indirect or qualitative measures. For example, Kim and coworkers found a direct linear correlation between experimentally observed reaction rates of a butyl acetate synthesis reaction and calculated (ie, simulated) nonbonded interaction energies between different IL solvents and the CALB enzyme that catalyzed the reaction (Kim et al., 2014). It has recently been suggested that the following experimental techniques could possibly be used to validate or invalidate computational results of enzymes in ILs: isothermal titration calorimetry to energetically describe substrate binding/unbinding to an enzyme in the presence of ILs, X-ray scattering or small-angle neutron scattering to observe large-scale enzyme structural changes, and Fourier transform infrared spectroscopy to monitor the evolution of enzyme secondary structure (Jaeger et al., 2015).

Another challenge of studying biocatalysis in ILs with computation is related to sampling. Due to strong interactions existing between the enzyme surface and IL ions (specifically anions), as well as the intrinsic properties of the solvent that lead to the previously discussed slow dynamics of many ILs, adequate sampling can sometimes be difficult to achieve (Burney & Pfaendtner, 2013). In turn, this can lead to complicated and potentially inaccurate calculations of equilibrium thermodynamic and/or dynamic system properties. The use of enhanced sampling techniques like parallel tempering, umbrella sampling, or metadynamics, has been proposed as a solution to this issue (Burney & Pfaendtner, 2013). Using special-purpose machines for MD simulations (eg, Anton (Shaw et al., 2008)) or hardware accelerators using platforms such as field-programmable gate arrays (FPGAs), graphics processing units (GPUs), the Cell Broadband Engine (CBE), and multicore processors (Sarkar, Majumder, Kalyanaraman, & Pande, 2010) is another potential solution.

The aforementioned areas for future focus and the proposed use of emerging hardware solutions to improve computational studies of biocatalysis in ILs highlight the potential growth of the field and the vast space for exploration and discovery. The field is growing in leaps and bounds every year, and we look forward to seeing the exciting new directions researchers take to inspire new experimental work and improve the predictive capabilities of enzyme/IL simulations.

REFERENCES

Ajloo, D., Sangian, M., Ghadamgahi, M., Evini, M., & Saboury, A. A. (2013). Effect of two imidazolium derivatives of ionic liquids on the structure and activity of adenosine deaminase. *International Journal of Biological Macromolecules, 55*, 47–61.

Barducci, A., Bonomi, M., & Parrinello, M. (2011). Metadynamics. *Wiley Interdisciplinary Reviews: Computational Molecular Science, 1*, 826–843.

Barducci, A., Bussi, G., & Parrinello, M. (2008). Well-tempered metadynamics: A smoothly converging and tunable free-energy method. *Physical Review Letters, 100*, 020603.

Benedetto, A., & Ballone, P. (2015). Room temperature ionic liquids meet biomolecules: A microscopic view of structure and dynamics. *ACS Sustainable Chemistry & Engineering, 4*, 392–412.

Berendsen, H. J. C., Postma, J. P. M., van Gunsteren, W. F., DiNola, A., & Haak, J. R. (1984). Molecular dynamics with coupling to an external bath. *The Journal of Chemical Physics, 81*, 3684–3690.

Best, R. B., Buchete, N.-V., & Hummer, G. (2008). Are current molecular dynamics force fields too helical? *Biophysical Journal, 95*, L07–L09.

Burney, P. R., Nordwald, E. M., Hickman, K., Kaar, J. L., & Pfaendtner, J. (2015). Molecular dynamics investigation of the ionic liquid/enzyme interface: Application to engineering enzyme surface charge. *Proteins: Structure, Function, and Bioinformatics, 83*, 670–680.

Burney, P. R., & Pfaendtner, J. (2013). Structural and dynamic features of *Candida rugosa* lipase 1 in water, octane, toluene, and ionic liquids BMIM-PF6 and BMIM-NO3. *The Journal of Physical Chemistry. B, 117*, 2662–2670.

Bussi, G., Donadio, D., & Parrinello, M. (2007). Canonical sampling through velocity rescaling. *The Journal of Chemical Physics, 126*, 014101.

Canongia Lopes, J. N., & Pádua, A. A. H. (2004). Molecular force field for ionic liquids composed of triflate or bistriflylimide anions. *The Journal of Physical Chemistry. B, 108*, 16893–16898.

Canongia Lopes, J. N., & Pádua, A. A. H. (2006). Molecular force field for ionic liquids III: Imidazolium, pyridinium, and phosphonium cations; chloride, bromide, and dicyanamide anions. *The Journal of Physical Chemistry. B, 110*, 19586–19592.

Canongia Lopes, J. N., Pádua, A. A. H., & Shimizu, K. (2008). Molecular force field for ionic liquids IV: Trialkylimidazolium and alkoxycarbonyl-imidazolium cations; alkylsulfonate and alkylsulfate anions. *The Journal of Physical Chemistry. B, 112*, 5039–5046.

Carter, J. L. L., Bekhouche, M., Noiriel, A., Blum, L. J., & Doumèche, B. (2014). Directed evolution of a formate dehydrogenase for increased tolerance to ionic liquids reveals a new site for increasing the stability. *ChemBioChem, 15*, 2710–2718.

Case, D. A., Berryman, J. T., Betz, R. M., Cerutti, D. S., Cheatham, T. E., 3rd., Darden, T. A., ... Kollman, P. A. (2015). *Amber 2015*. San Francisco, CA: University of California.

Case, D. A., Cheatham, T. E., 3rd., Darden, T., Gohlke, H., Luo, R., Merz, K. M., Jr., ... Woods, R. J. (2005). The Amber biomolecular simulation programs. *Journal of Computational Chemistry, 26*, 1668–1688.

Chen, Z., Pereira, J. H., Liu, H., Tran, H. M., Hsu, N. S. Y., Dibble, D., ... Sale, K. L. (2013). Improved activity of a thermophilic cellulase, Cel5A, from *Thermotoga maritima* on ionic liquid pretreated switchgrass. *PLoS One, 8*, e79725.

Cho, H. M., Gross, A. S., & Chu, J.-W. (2011). Dissecting force interactions in cellulose deconstruction reveals the required solvent versatility for overcoming biomass recalcitrance. *Journal of the American Chemical Society, 133*, 14033–14041.

Curto, V. F., Scheuermann, S., Owens, R. M., Ranganathan, V., MacFarlane, D. R., Benito-Lopez, F., & Diamond, D. (2014). Probing the specific ion effects of biocompatible hydrated choline ionic liquids on lactate oxidase biofunctionality in sensor applications. *Physical Chemistry Chemical Physics, 16*, 1841–1849.

Dabirmanesh, B., Daneshjou, S., Sepahi, A. A., Ranjbar, B., Khavari-Nejad, R. A., Gill, P., ... Khajeh, K. (2011). Effect of ionic liquids on the structure, stability and activity of two related alpha-amylases. *International Journal of Biological Macromolecules, 48*, 93–97.

Fleming, K. L., Tiwary, P., & Pfaendtner, J. (2016). New approach for investigating reaction dynamics and rates with ab initio calculations. *The Journal of Physical Chemistry A, 120*, 299–305.

Fort, D. A., Remsing, R. C., Swatloski, R. P., Moyna, P., Moyna, G., & Rogers, R. D. (2007). Can ionic liquids dissolve wood? Processing and analysis of lignocellulosic materials with 1-n-butyl-3-methylimidazolium chloride. *Green Chemistry, 9*, 63–69.

Frisch, M. J., Trucks, G. W., Schlegel, H. B., Scuseria, G. E., Robb, M. A., Cheeseman, J. R., ... Fox, D. J. (2009). *Gaussian 09*. Wallingford, CT: Gaussian, Inc.

Ghaedizadeh, S., Emamzadeh, R., Nazari, M., Rasa, S. M. M., Zarkesh-Esfahani, S. H., & Yousefi, M. (2016). Understanding the molecular behaviour of *Renilla* luciferase in imidazolium-based ionic liquids, a new model for the α/β fold collapse. *Biochemical Engineering Journal, 105*(Pt. B), 505–513.

Ghosh, S., Parui, S., Jana, B., & Bhattacharyya, K. (2015). Ionic liquid induced dehydration and domain closure in lysozyme: FCS and MD simulation. *The Journal of Chemical Physics, 143*, 125103.

Gross, A. S., Bell, A. T., & Chu, J.-W. (2011). Thermodynamics of cellulose solvation in water and the ionic liquid 1-butyl-3-methylimidazolim chloride. *The Journal of Physical Chemistry. B, 115*, 13433–13440.

Gross, A. S., Bell, A. T., & Chu, J.-W. (2012). Entropy of cellulose dissolution in water and in the ionic liquid 1-butyl-3-methylimidazolim chloride. *Physical Chemistry Chemical Physics, 14*, 8425–8430.

Guo, Z., & Xu, X. (2006). Lipase-catalyzed glycerolysis of fats and oils in ionic liquids: A further study on the reaction system. *Green Chemistry, 8*, 54–62.

Gupta, K. M., Hu, Z., & Jiang, J. (2011). Mechanistic understanding of interactions between cellulose and ionic liquids: A molecular simulation study. *Polymer, 52*, 5904–5911.

Ha, S. H., Lan, M. N., Lee, S. H., Hwang, S. M., & Koo, Y.-M. (2007). Lipase-catalyzed biodiesel production from soybean oil in ionic liquids. *Enzyme and Microbial Technology, 41*, 480–483.

Hess, B., Kutzner, C., van der Spoel, D., & Lindahl, E. (2008). GROMACS 4: Algorithms for highly efficient, load-balanced, and scalable molecular simulation. *Journal of Chemical Theory and Computation, 4*, 435–447.

Hornak, V., Abel, R., Okur, A., Strockbine, B., Roitberg, A., & Simmerling, C. (2006). Comparison of multiple Amber force fields and development of improved protein backbone parameters. *Proteins*, *65*, 712–725.

Hurisso, B. B., Lovelock, K. R. J., & Licence, P. (2011). Amino acid-based ionic liquids: Using XPS to probe the electronic environment via binding energies. *Physical Chemistry Chemical Physics*, *13*, 17737–17748.

Jaeger, V., Burney, P., & Pfaendtner, J. (2015). Comparison of three ionic liquid-tolerant cellulases by molecular dynamics. *Biophysical Journal*, *108*, 880–892.

Jaeger, V. W., & Pfaendtner, J. (2013). Structure, dynamics, and activity of xylanase solvated in binary mixtures of ionic liquid and water. *ACS Chemical Biology*, *8*, 1179–1186.

Jarin, Z., & Pfaendtner, J. (2014). Ionic liquids can selectively change the conformational free-energy landscape of sugar rings. *Journal of Chemical Theory and Computation*, *10*, 507–510.

Jorgensen, W. L., Chandrasekhar, J., Madura, J. D., Impey, R. W., & Klein, M. L. (1983). Comparison of simple potential functions for simulating liquid water. *The Journal of Chemical Physics*, *79*, 926–935.

Kim, H. S., Ha, S. H., Sethaphong, L., Koo, Y. M., & Yingling, Y. G. (2014). The relationship between enhanced enzyme activity and structural dynamics in ionic liquids: A combined computational and experimental study. *Physical Chemistry Chemical Physics*, *16*, 2944–2953.

Kirschner, K. N., Yongye, A. B., Tschampel, S. M., Gonzalez-Outeirino, J., Daniels, C. R., Foley, B. L., & Woods, R. J. (2008). GLYCAM06: A generalizable biomolecular force field. Carbohydrates. *Journal of Computational Chemistry*, *29*, 622–655.

Klahn, M., Lim, G. S., Seduraman, A., & Wu, P. (2011a). On the different roles of anions and cations in the solvation of enzymes in ionic liquids. *Physical Chemistry Chemical Physics*, *13*, 1649–1662.

Klahn, M., Lim, G. S., & Wu, P. (2011b). How ion properties determine the stability of a lipase enzyme in ionic liquids: A molecular dynamics study. *Physical Chemistry Chemical Physics*, *13*, 18647–18660.

Klähn, M., Stüber, C., Seduraman, A., & Wu, P. (2010). What determines the miscibility of ionic liquids with water? Identification of the underlying factors to enable a straightforward prediction. *The Journal of Physical Chemistry. B*, *114*, 2856–2868.

Laio, A., & Parrinello, M. (2002). Escaping free-energy minima. *Proceedings of the National Academy of Sciences of the United States of America*, *99*, 12562–12566.

Latif, M., Alif, M., Micaêlo, N. M., Rahman, A., & Basyaruddin, M. (2014). Influence of anion-water interactions on the behaviour of lipases in room temperature ionic liquids. *RSC Advances*, *4*, 48202–48211.

Li, H., Kankaanpää, A., Xiong, H., Hummel, M., Sixta, H., Ojamo, H., & Turunen, O. (2013). Thermostabilization of extremophilic dictyoglomus thermophilum GH11 xylanase by an N-terminal disulfide bridge and the effect of ionic liquid [emim]OAc on the enzymatic performance. *Enzyme and Microbial Technology*, *53*, 414–419.

Li, Y., Liu, X., Zhang, S., Yao, Y., Yao, X., Xu, J., & Lu, X. (2015). Dissolving process of a cellulose bunch in ionic liquids: A molecular dynamics study. *Physical Chemistry Chemical Physics*, *17*, 17894–17905.

Liu, H., Zhu, L., Bocola, M., Chen, N., Spiess, A. C., & Schwaneberg, U. (2013). Directed laccase evolution for improved ionic liquid resistance. *Green Chemistry*, *15*, 1348–1355.

Mai, N. L., & Koo, Y.-M. (2014). Quantitative prediction of lipase reaction in ionic liquids by QSAR using COSMO-RS molecular descriptors. *Biochemical Engineering Journal*, *87*, 33–40.

Malde, A. K., Zuo, L., Breeze, M., Stroet, M., Poger, D., Nair, P. C., … Mark, A. E. (2011). An automated force field topology builder (ATB) and repository: Version 1.0. *Journal of Chemical Theory and Computation*, *7*, 4026–4037.

Martinez, L., Andrade, R., Birgin, E. G., & Martinez, J. M. (2009). PACKMOL: A package for building initial configurations for molecular dynamics simulations. *Journal of Computational Chemistry*, *30*, 2157–2164.

Men, S., Lovelock, K. R. J., & Licence, P. (2011). X-ray photoelectron spectroscopy of pyrrolidinium-based ionic liquids: Cation-anion interactions and a comparison to imidazolium-based analogues. *Physical Chemistry Chemical Physics*, *13*, 15244–15255.

Micaêlo, N. M., & Soares, C. M. (2008). Protein structure and dynamics in ionic liquids. Insights from molecular dynamics simulation studies. *The Journal of Physical Chemistry. B*, *112*, 2566–2572.

Moniruzzaman, M., Nakashima, K., Kamiya, N., & Goto, M. (2010). Recent advances of enzymatic reactions in ionic liquids. *Biochemical Engineering Journal*, *48*, 295–314.

Mostofian, B., Smith, J., & Cheng, X. (2011). The solvation structures of cellulose microfibrils in ionic liquids. *Interdisciplinary Sciences, Computational Life Sciences*, *3*, 308–320.

Nordwald, E. M., Armstrong, G. S., & Kaar, J. L. (2014). NMR-guided rational engineering of an ionic-liquid-tolerant lipase. *ACS Catalysis*, *4*, 4057–4064.

Nordwald, E. M., & Kaar, J. L. (2013a). Mediating electrostatic binding of 1-butyl-3-methylimidazolium chloride to enzyme surfaces improves conformational stability. *The Journal of Physical Chemistry. B*, *117*, 8977–8986.

Nordwald, E. M., & Kaar, J. L. (2013b). Stabilization of enzymes in ionic liquids via modification of enzyme charge. *Biotechnology and Bioengineering*, *110*, 2352–2360.

Rabideau, B. D., Agarwal, A., & Ismail, A. E. (2013). Observed mechanism for the breakup of small bundles of cellulose Iα and Iβ in ionic liquids from molecular dynamics simulations. *The Journal of Physical Chemistry. B*, *117*, 3469–3479.

Rabideau, B. D., & Ismail, A. E. (2015). Mechanisms of hydrogen bond formation between ionic liquids and cellulose and the influence of water content. *Physical Chemistry Chemical Physics*, *17*, 5767–5775.

Salvalaglio, M., Tiwary, P., & Parrinello, M. (2014). Assessing the reliability of the dynamics reconstructed from metadynamics. *Journal of Chemical Theory and Computation*, *10*, 1420–1425.

Sambasivarao, S. V., & Acevedo, O. (2009). Development of OPLS-AA force field parameters for 68 unique ionic liquids. *Journal of Chemical Theory and Computation*, *5*, 1038–1050.

Sarkar, S., Majumder, T., Kalyanaraman, A., & Pande, P. P. (2010). Hardware accelerators for biocomputing: A survey. In *Proceedings of 2010 IEEE international symposium on circuits and systems*, (pp. 3789–3792), Paris: IEEE. http://dx.doi.org/10.1109/ISCAS.2010.5537736.

Schuttelkopf, A. W., & van Aalten, D. M. F. (2004). PRODRG: A tool for high-throughput crystallography of protein-ligand complexes. *Acta Crystallographica Section D*, *60*, 1355–1363.

Shaw, D. E., Deneroff, M. M., Dror, R. O., Kuskin, J. S., Larson, R. H., Salmon, J. K., ... Wang, S. C. (2008). Anton, a special-purpose machine for molecular dynamics simulation. *Communications of the ACM*, *51*, 91–97.

Sousa da Silva, A. W., & Vranken, W. F. (2012). ACPYPE—AnteChamber PYthon Parser interfacE. *BMC Research Notes*, *5*, 1–8.

Sprenger, K. G., Jaeger, V. W., & Pfaendtner, J. (2015). The general amber force field (GAFF) can accurately predict thermodynamic and transport properties of many ionic liquids. *The Journal of Physical Chemistry. B*, *119*, 5882–5895.

Tee, K. L., Roccatano, D., Stolte, S., Arning, J., Jastorff, B., & Schwaneberg, U. (2008). Ionic liquid effects on the activity of monooxygenase P450 BM-3. *Green Chemistry*, *10*, 117–123.

Tiwary, P., & Parrinello, M. (2013). From metadynamics to dynamics. *Physical Review Letters*, *111*, 230602.

Tribello, G. A., Bonomi, M., Branduardi, D., Camilloni, C., & Bussi, G. (2014). PLUMED 2: New feathers for an old bird. *Computer Physics Communications*, *185*, 604–613.

Vanommeslaeghe, K., & MacKerell, A. D. (2012). Automation of the CHARMM general force field (CGenFF) I: Bond perception and atom typing. *Journal of Chemical Information and Modeling*, *52*, 3144–3154.

Vanommeslaeghe, K., Raman, E. P., & MacKerell, A. D. (2012). Automation of the CHARMM general force field (CGenFF) II: Assignment of bonded parameters and partial atomic charges. *Journal of Chemical Information and Modeling*, *52*, 3155–3168.

Wang, J., Wang, W., Kollman, P. A., & Case, D. A. (2006). Automatic atom type and bond type perception in molecular mechanical calculations. *Journal of Molecular Graphics and Modelling*, *25*, 247–260.

Wang, J., Wolf, R. M., Caldwell, J. W., Kollman, P. A., & Case, D. A. (2004). Development and testing of a general amber force field. *Journal of Computational Chemistry*, *25*, 1157–1174.

Wolski, P. W., Dana, C. M., Clark, D. S., & Blanch, H. W. (2016). Engineering ionic liquid-tolerant cellulases for biofuels production. *Protein Engineering, Design & Selection*, *29*, 117–122.

Youngs, T. G. A., Hardacre, C., & Holbrey, J. D. (2007). Glucose solvation by the ionic liquid 1,3-dimethylimidazolium chloride: A simulation study. *The Journal of Physical Chemistry. B*, *111*, 13765–13774.

Youngs, T. G. A., Holbrey, J. D., Mullan, C. L., Norman, S. E., Lagunas, M. C., D'Agostino, C., … Hardacre, C. (2011). Neutron diffraction, NMR and molecular dynamics study of glucose dissolved in the ionic liquid 1-ethyl-3-methylimidazolium acetate. *Chemical Science*, *2*, 1594–1605.

Zhang, Y., & Maginn, E. J. (2012). A simple AIMD approach to derive atomic charges for condensed phase simulation of ionic liquids. *The Journal of Physical Chemistry. B*, *116*, 10036–10048.

Zhao, Y., Liu, X., Wang, J., & Zhang, S. (2013). Effects of anionic structure on the dissolution of cellulose in ionic liquids revealed by molecular simulation. *Carbohydrate Polymers*, *94*, 723–730.

CHAPTER SEVENTEEN

The MOD-QM/MM Method: Applications to Studies of Photosystem II and DNA G-Quadruplexes

M. Askerka*, J. Ho*,[1], E.R. Batista[†], J.A. Gascón[‡], V.S. Batista*,[2]

*Yale University, New Haven, CT, United States
[†]Los Alamos National Laboratory, Los Alamos, NM, United States
[‡]University of Connecticut, Storrs, CT, United States
[2]Corresponding author: e-mail address: victor.batista@yale.edu

Contents

1. Introduction	444
2. Methods	449
2.1 QM/MM Methodology	449
2.2 MOD-QM/MM Method	450
3. EXAFS Simulations	461
3.1 Post-QM/MM Refinement Model of the OEC of PSII	462
4. MOD-QM/MM Models of DNA Quadruplexes	464
5. Conclusions	468
Acknowledgments	469
References	469

Abstract

Quantum mechanics/molecular mechanics (QM/MM) hybrid methods are currently the most powerful computational tools for studies of structure/function relations and catalytic sites embedded in macrobiomolecules (eg, proteins and nucleic acids). QM/MM methodologies are highly efficient since they implement quantum chemistry methods for modeling only the portion of the system involving bond-breaking/forming processes (QM layer), as influenced by the surrounding molecular environment described in terms of molecular mechanics force fields (MM layer). Some of the limitations of QM/MM methods when polarization effects are not explicitly considered include the approximate treatment of electrostatic interactions between QM and MM layers. Here, we review recent advances in the development of computational protocols that allow for rigorous modeling of electrostatic interactions in biomacromolecules and structural

---

[1] Institute of High Performance Computing, 1 Fusionopolis Way, #16-16 Connexis North, Singapore 138632.

refinement, beyond the common limitations of QM/MM hybrid methods. We focus on photosystem II (PSII) with emphasis on the description of the oxygen-evolving complex (OEC) and its high-resolution extended X-ray absorption fine structure spectra (EXAFS) in conjunction with Monte Carlo structural refinement. Furthermore, we review QM/MM structural refinement studies of DNA G4 quadruplexes with embedded monovalent cations and direct comparisons to NMR data.

1. INTRODUCTION

The development of computer architectures and efficient algorithms has allowed computational chemistry to establish itself as a powerful and valuable tool for general purpose applications, including a wide range of structural and mechanistic studies of large biomolecular systems (eg, proteins and nucleic acids). In particular, hybrid quantum mechanics/molecular mechanics (QM/MM) methods have become available in popular software packages and are routinely applied in studies of catalytic complexes embedded in enzymatic macromolecular structures (Bakowies & Thiel, 1996; Dapprich, Komaromi, Byun, Morokuma, & Frisch, 1999; Field, Bash, & Karplus, 1990; Humbel, Sieber, & Morokuma, 1996; Luber et al., 2011; Maseras & Morokuma, 1995; Murphy, Philipp, & Friesner, 2000b; Pal et al., 2013; Philipp & Friesner, 1999; Shoji, Isobe, & Yamaguchi, 2015a; Shoji et al., 2015b; Singh & Kollman, 1986; Svensson et al., 1996; Vreven & Morokuma, 2000a, 2000b; Warshel & Levitt, 1976). QM/MM methods significantly expand the scope of quantum mechanical calculations to much larger systems by limiting the quantum mechanical description (typically based on density functional theory (DFT)) to a subsystem where chemical reactivity is localized (QM layer), while the "spectator" region is usually treated with an inexpensive model chemistry such as classical molecular mechanics force fields, or a lower level of electronic structure theory (eg, semiempirical methods). This partitioning scheme ensures highly efficient, yet accurate, modeling of electrostatic interactions as exploited by the Moving-Domain QM/MM (MOD-QM/MM) methodology (Gascon, Leung, Batista, & Batista, 2006; Ho et al., 2014; Menikarachchi & Gascon, 2008; Sandberg, Rudnitskaya, & Gascón, 2012) and the studies of the oxygen-evolving complex (OEC) of photosystem II (PSII) reviewed in this chapter.

Technical differences between the various QM/MM methods concern whether the energy is evaluated according to additive or subtractive schemes, and whether the boundaries between QM and MM layers are treated with link atoms, pseudopotentials, or localized bond orbitals (Senn & Thiel, 2009; van der Kamp & Mulholland, 2013). Other differences include the level of quantum chemistry and MM force field and different approaches for the description of electrostatic interactions between the QM and MM layers (eg, mechanical embedding, electronic embedding, and polarized embedding schemes). In particular, the mechanical embedding approach couples in vacuo QM electronic densities with point charges in the MM layer, as described by classical electrostatic interactions. In the electronic embedding approach, the atom-centered point charges in the MM layer can polarize the QM electronic density. Finally, higher-level corrections to include the mutual polarization between the MM and QM layers can be included in the polarized embedding scheme (Devries et al., 1995; Eichler, Kolmel, & Sauer, 1997; Gao & Xia, 1992; Thompson, 1996; Vreven, Mennucci, da Silva, Morokuma, & Tomasi, 2001), and in the MOD-QM/MM scheme (Gascon, Leung, et al., 2006; Ho et al., 2014; Menikarachchi & Gascon, 2008).

The MOD-QM/MM method partitions the system into molecular domains and obtains the geometry and electrostatic properties of the whole system from an iterative self-consistent treatment of the constituent molecular domains (Gascon, Leung, et al., 2006; Ho et al., 2014; Menikarachchi & Gascon, 2008; Sandberg et al., 2012). Self-consistently optimized structures and electrostatic potential (ESP) atomic charges are obtained for each domain modeled as a QM layer polarized by the distribution of atomic charges in the surrounding fragments. The resulting computational task scales linearly with the size of the system, bypassing the exponential demand of memory and computational resources typically required by "brute-force" full QM calculations. Computations for several benchmark systems that allow for full QM calculations have shown that the MOD-QM/MM method produces ab initio quality structures and ESPs in quantitative agreement with full QM results (Gascon, Leung, et al., 2006; Ho et al., 2014; Menikarachchi & Gascon, 2008; Sandberg et al., 2012). The work reviewed in this chapter illustrates the MOD-QM/MM method as applied to the description of macromolecules of biological interest, including PSII and DNA quadruplexes, in conjunction with structural refinement based on simulations of X-ray absorption spectra and direct comparisons with experimental data.

The key aspect of MOD-QM/MM is the description of polarization effects in the MM layer of QM/MM calculations. In that respect, a number of methods were previously proposed (Mayhall & Raghavachari, 2012;

Saha & Raghavachari, 2015). One of them includes a standard atomic dipole description where dipole moments are induced in each atom of the MM layer, defined as the product of the atomic polarizability and the electric field at the position of the atom (Houjou, Inoue, & Sakurai, 2001; Illingworth et al., 2006; Thompson, 1996; Thompson & Schenter, 1995; Warshel, 1976). Atomic polarizabilities are therefore required as input parameters. The approach has been implemented at the "ab initio" level in quantum chemistry packages such as GAMESS (Schmidt et al., 1993) and Q-CHEM (Shao et al., 2015) where it is called an "effective fragment potential" method (Chen & Gordon, 1996; Day et al., 1996; Gordon et al., 2001). The polarization of the fragments is described by standard distributed multipoles, using atomic polarizabilities obtained in preliminary ab initio calculations as input parameters. Another related approach is the explicit polarization (X-Pol) method, which is an approximate fragment-based molecular orbital method. This approach incorporates many-body polarization at a cost that scales linearly with the number of fragments but must be augmented with empirical Lennard-Jones potentials (Gao, 1998; Xie, Orozco, Truhlar, & Gao, 2009).

Polarization effects can be accounted for at the MM level by using polarizable force fields (Banks et al., 1999; Bernardo, Ding, Kroghjespersen, & Levy, 1994; Burnham & Xantheas, 2002; Corongiu, 1992; Halgren & Damm, 2001; Huang, Lopes, Roux, & MacKerell, 2014; Maple et al., 2005; Palmo, Mannfors, Mirkin, & Krimm, 2003; Ren & Ponder, 2003; Schworer et al., 2013; Stern et al., 1999; Stern, Rittner, Berne, & Friesner, 2001; Vanommeslaeghe & MacKerell, 2015; Yu, Hansson, & van Gunsteren, 2003), although such methods are computationally expensive. One of the early classical methods to describe polarization effects has been the "fluctuating charge model," studied by Rappe and Goddard (Rappe & Goddard, 1991), Berne (Bader & Berne, 1996; New & Berne, 1995; Rick & Berne, 1996; Rick, Stuart, Bader, & Berne, 1995; Rick, Stuart, & Berne, 1994; Stuart & Berne, 1996), Parr and Yang (Parr & Yang, 1989), Friesner and Berne (Stern et al., 2001), and Field (Field, 1997). In this model, the system is broken down into fragments; whereas the total charge of the fragment is conserved, the charges on the atoms are calculated self-consistently through minimizing the total energy of the system with respect to the atomic charges. This self-consistent procedure leads to the charges, for which the partial derivatives of the total energy E with respect to the atomic charges—electronegativities—are equal. Similar approaches include the chemical potential equalization proposed by York and Yang (York, 1995; York & Yang, 1996) and the Drude oscillator models implemented by Voth (Lobaugh & Voth, 1994), Chandler (Thompson, Schweizer, & Chandler, 1982), Cao and Berne (Cao & Berne, 1992), and Sprik and Klein (Sprik & Klein, 1988).

Semiempirical models, based on the equalization of atomic electronegativities (Borgis & Staib, 1995; Field, 1997; Gao, 1997), obtain the self-consistent density matrices of the constituent fragments by neglecting the off-diagonal elements associated with atoms of different fragments. The interaction between the fragments can be added back by including Lennard-Jones potential term as well as electrostatic interactions based on Mulliken charges (Field, 1997; Gao, 1997). The Mulliken charges can be then improved by rescaling, using empirical parameters obtained from comparisons of the computed and experimental properties of the system (Gao, 1997). The charges can be also calculated using a semiempirical charge model (CM), developed by Cramer and Truhlar (Storer, Giesen, Cramer, & Truhlar, 1995). However, the charges cannot be improved to reproduce ab initio quality ESPs, as in the MOD-QM/MM method since they would be too small and would require empirical scaling (Field, 1997; Gao, 1997).

Semiempirical fluctuating CMs also differ from MOD-QM/MM in the treatment of the boundaries between QM and MM layers. MOD-QM/MM was specifically designed to model ESPs of extended systems (eg, proteins, RNA, DNA) where it is necessary to cut covalent bonds to define the boundaries between QM and MM layers. When combined with accurate calculations of ESP atomic charges, the MOD-QM/MM method allows one to obtain ab initio quality ESPs and refined structures of extended systems.

The MOD-QM/MM approach usually converges within a few iterations of self-consistent calculations of polarized charges and optimized geometries of the constituent domains, and therefore has convergence properties similar to those of "linear scaling" techniques where the whole system is treated at the QM level (Babu & Gadre, 2003; Baldwin, Kampf, & Pecoraro, 1993; Burant, Scuseria, & Frisch, 1996; Daniels & Scuseria, 1999; Greengard & Rokhlin, 1987; Li, Nunes, & Vanderbilt, 1993; Schwegler & Challacombe, 1996; Strain, Scuseria, & Frisch, 1996; Stratmann, Burant, Scuseria, & Frisch, 1997; Stratmann, Scuseria, & Frisch, 1996; Strout & Scuseria, 1995; White & Headgordon, 1994; White, Johnson, Gill, & HeadGordon, 1996). In such linear scaling methods, the high computational cost is avoided by neglecting the two electron integrals between orbitals that are situated at long distances from each other, leading to sparse Fock matrices. Full diagonalization is circumvented and the correlation energy is computed in a less expensive way (Ko et al., 2003; Saebo & Pulay, 1993), or by using localized orbitals (Subotnik & Head-Gordon, 2005). Another linear scaling method is the "divide-and-conquer" technique, developed by Yang and coworkers (Lee, York, & Yang, 1996; Yang, 1991; Yang & Lee, 1995). Such a method divides the density matrix into fragments that are treated by conventional quantum chemistry techniques to ultimately obtain the properties of the whole system. Later, Merz and coworkers developed a semiempirical version of divide-and-conquer (Dixon & Merz, 1996, 1997; Ermolaeva, van der Vaart, & Merz, 1999), Exner and Mezey proposed the adjustable density-matrix assembler approach (Exner & Mezey, 2002, 2003, 2004), Zhang and coworkers developed the molecular fractionation with conjugated caps approach (Gao, Zhang, Zhang, & Zhang, 2004; Kurisu, Zhang, Smith, & Cramer, 2003; Mei, Zhang, & Zhang, 2005; Zhang, Xiang, & Zhang, 2003; Zhang & Zhang, 2003), Gadre and coworkers developed the molecular tailoring approach (Babu & Gadre, 2003; Babu, Ganesh, Gadre, & Ghermani, 2004; Gadre, Shirsat, & Limaye, 1994), Kitaura and coworkers proposed the fragment molecular-orbital FMO method (Fedorov & Kitaura, 2004a, 2004b; Kitaura, Ikeo, Asada, Nakano, & Uebayasi, 1999; Kitaura, Sugiki, Nakano, Komeiji, & Uebayasi, 2001), Li and coworkers proposed the localized molecular-orbital assembler approach (Li & Li, 2005; Li, Li, & Fang, 2005), Mayhall and Raghavachari proposed the multilayer molecules-in-molecules method that is similar in spirit to the ONIOM approach (Mayhall & Raghavachari, 2011) (vide infra), while Collins and Bettens introduced the combined and systematic fragmentation approach (Collins, Cvitkovic, & Bettens, 2014).

The MOD-QM/MM approach does not have any kind of limitation with regard to the level of quantum chemistry applied to describe the QM layer. Typically, it has been implemented by using DFT, which allows for an efficient description of electron correlation at the computational cost of single-determinant calculations. MOD-QM/MM with DFT has been successfully applied to PSII and other biological systems (Gascon, Leung, Batista, & Batista, 2006; Gascon, Sproviero, & Batista, 2006; Gascon, Sproviero, McEvoy, Brudvig, & Batista, 2008; Menikarachchi & Gascon, 2008; Sandberg et al., 2012; Sproviero, Gascon, McEvoy, Brudvig, & Batista, 2006a, 2006b, 2007, 2008a, 2008b, 2008c, 2008d). This includes modeling charge redistribution, polarization effects and changes in the ligation scheme of metal centers induced by oxidation state transitions, and bond breaking and forming processes associated with chemical reactivity and proton transfer. The resulting MOD-QM/MM model systems have allowed for direct comparison with high-resolution spectroscopic data, including FTIR, NMR, extended X-ray absorption fine structure spectra (EXAFS), and EPR spectroscopy (Sproviero et al., 2007; Sproviero, Gascon, et al., 2008a, 2008b, 2008c, 2008d, 2008b).

This chapter reviews the MOD-QM/MM as a rigorous yet practical methodology suited to study a wide range of enzymes and biological systems, including DNA, RNA, and proteins with catalytic sites beyond the capabilities of molecular mechanics force fields. The presentation is organized as follows. First, we review the QM/MM methodology, including a detailed discussion of its implementation to metalloenzymes and DNA. Then, we review the post-QM/MM refinement scheme based on simulations of X-ray absorption spectra and direct comparisons with experimental measurements. The applications review benchmark studies, illustrating the capabilities of the MOD-QM/MM methodology and the post-QM/MM analysis of model structures through direct comparisons with NMR and EXAFS spectroscopic data.

2. METHODS

2.1 QM/MM Methodology

The MOD-QM/MM method is a general approach that can be applied in conjunction with any electronic embedding QM/MM methods (Chung, Hirao, Li, & Morokuma, 2012; Gascon, Leung, et al., 2006; Menikarachchi & Gascon, 2008; Sproviero, Gascon, et al., 2008b; Vreven et al., 2006). It has been implemented in the ONIOM (Chung et al., 2012; Vreven et al., 2006) electronic embedding method in the Gaussian program (Frisch et al., 2004). The ONIOM method is a subtractive QM/MM approach. When implemented in term of two layers, the whole system is partitioned into a "high-level" layer generally treated with QM methods and a "low-level" layer, described by MM force fields (Fig. 1). The total energy is given by the following extrapolation,

$$E = E^{\text{MM,high+low}} + E^{\text{QM,high}} - E^{\text{MM,high}} \qquad (1)$$

where $E^{\text{MM,high+low}}$ is the energy of the full system computed at the molecular mechanics level of theory, while $E^{\text{QM,high}}$ and $E^{\text{MM,high}}$ are the energies of the high-level layer calculated by using QM and MM levels of theory, respectively. In this extrapolation scheme, electrostatic interactions

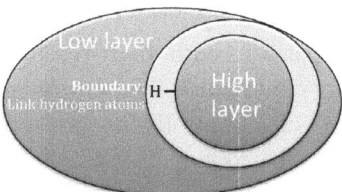

Fig. 1 QM/MM fragmentation of the system into a reduced QM (high) layer and the surrounding molecular environment described as an MM (low) layer.

computed at the MM level in $E^{MM,high}$ and $E^{MM,high+low}$ cancel, therefore, giving a quantum mechanical description of polarization of the high layer due to the electrostatic influence of the surrounding environment. However, when implemented with standard nonpolarizable force fields based on distributions of single point atomic charges (eg, Amber (Cornell et al., 1995), CHARMM (Vanommeslaeghe & MacKerell, 2015)), the method does not include the self-consistent polarization of the MM layer due to the distribution of charge in the QM layer. Such a polarization effect often corresponds to small corrections of single point atomic charges of about 10–15%. However, adding all of the corrections over the whole interface between the QM and MM layers often amounts to significant energy corrections of 10–15 kcal mol^{-1}. Neglecting such corrections during energy minimization, thus, can lead to structural rearrangement artifacts in both of the QM and MM layers. The MOD-QM/MM approach is a simple methodology for addressing these important corrections.

2.2 MOD-QM/MM Method

In the MOD-QM/MM method, the system is partitioned into n molecular domains using a simple space-domain decomposition scheme (Gascon, Leung, et al., 2006; Ho et al., 2014; Menikarachchi & Gascon, 2008). A two-layer QM/MM scheme is used to optimize the geometry of each molecular domain and to compute the electrostatic potential (RESP) atomic charges of each fragment as influenced by the surrounding molecular environment. Mutual polarization interactions between the constituent molecular domains are thus included and the whole cycle of QM/MM calculations is iterated several times until a self-consistent geometry and point-charge model of the ESP are obtained.

Fig. 2 illustrates the application of the MOD-QM/MM method on a guanine G-DNA quadruplex, where each domain corresponds to one of the guanine nucleotides. The self-consistent solution is usually obtained within a few (four to five) iterations, which makes the computational cost

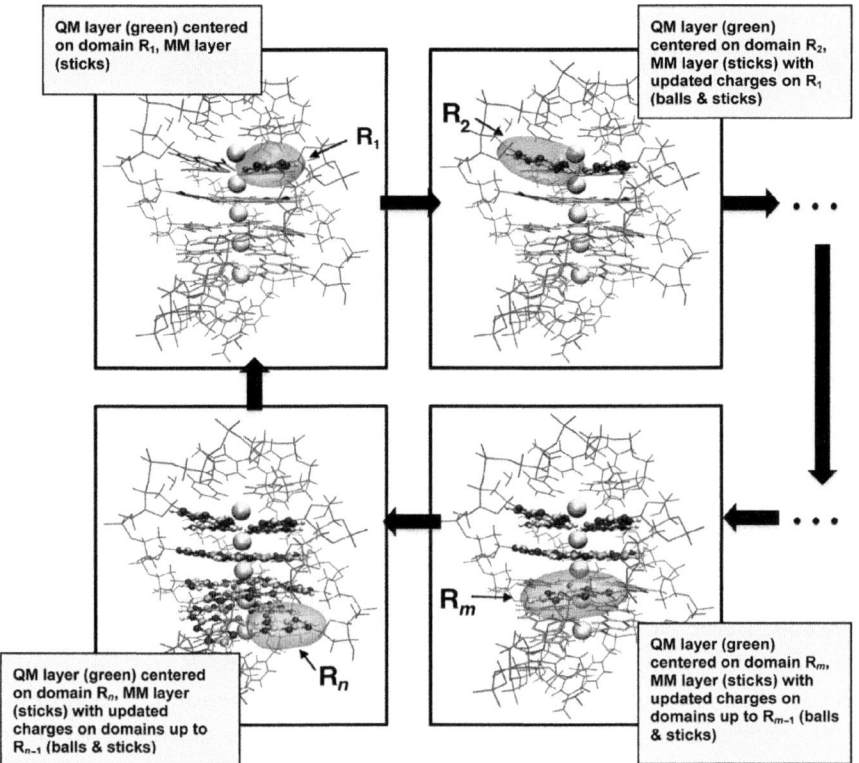

Fig. 2 MOD-QM/MM iterative cycle of QM/MM calculations with QM layers (in *green* (*gray* in the print version)) defined on individual molecular domains along the different steps of the cycle (Gascon, Leung, et al., 2006). *Balls* and *sticks* represent molecular domains with ESP atomic charges updated in previous steps of the cycle. *Ho, J., Newcomer, M.B., Ragain, C.M., Gascon, J.A., Batista, E.R., Loria, J.P., & Batista, V.S. (2014). The MoD-QM/MM structural refinement method: Characterization of hydrogen bonding in the Oxytricha nova G-quadruplex.* Journal of Chemical Theory and Computation, 10, 5125–5135. Copyright American Chemical Society 2014.

scale linearly with the system size. This makes MOD-QM/MM much more applicable when compared to the quantum-chemical treatment of the complete system.

The atomic charges obtained by the MOD-QM/MM iterative procedure typically correct the charges of popular molecular mechanics force fields (Cornell et al., 1995; Dykstra, 1993) by only 10–20%. As mentioned before, however, these small corrections across multiple domains can amount to a significant change in energy (10–15 kcal/mol). Those corrections are comparable to the energetics of changes in protonation states, hydrogen-bonding pattern, or oxidation state transitions of redox-active cofactors. Therefore, it is important to properly account for electrostatic interactions that could otherwise lead to artifact structural rearrangements across multiple domains over the entire QM/MM interface.

Ab initio quality molecular electrostatic potentials are obtained from the updated distribution of atomic charges. However, the charges are generally not transferable to very different configurations of the system since polarization effects are configuration dependent. Exceptions are cases where the conformation remains close to the optimized structure, such as in studies of rigid DNA quadruplexes and intermediate structures of the OEC of PSII shown to be very similar at very low temperatures as evidenced by X-ray absorption data (Haumann et al., 2005), the studies of enzyme catalysis with electron transfer and proton transport (Ilan, Tajkhorshid, Schulten, & Voth, 2004; Lobaugh & Voth, 1996; Okamura & Feher, 1992; Popovic & Stuchebrukhov, 2004; Sham, Muegge, & Warshel, 1998, 1999; Warshel, 1981; Warshel & Aqvist, 1991), ion channels (Aqvist & Luzhkov, 2000; Bliznyuk, Rendell, Allen, & Chung, 2001; Ilan et al., 2004; Luzhkov & Aqvist, 2000), docking and ligand binding (Cho, Guallar, Berne, & Friesner, 2005; Grater, Schwarzl, Dejaegere, Fischer, & Smith, 2005; Lappi et al., 2004; Norel, Sheinerman, Petrey, & Honig, 2001; Sheinerman & Honig, 2002; Simonson, Archontis, & Karplus, 1997, 1999; Vasilyev & Bliznyuk, 2004; Wang, Donini, Reyes, & Kollman, 2001), macromolecular assembly (Norel et al., 2001; Sheinerman, Norel, & Honig, 2000; Wang et al., 2001), and signal transduction (Gentilucci, Squassabia, & Artali, 2007; Green, Dennis, Fam, Tidor, & Jasanoff, 2006; Hurley, 2006; Mulgrew-Nesbitt et al., 2006). Therefore, the range of potential applications of MOD-QM/MM is wide and addresses the need for descriptions based on ab initio quality ESPs in biomolecular systems (Exner & Mezey, 2002, 2003, 2004, 2005; Gao et al., 2004; Kitaura et al., 2001; Li et al., 2005; Nakano et al., 2000, 2002; Szekeres, Exner & Mezey, 2005).

2.2.1 MOD-QM/MM Dangling Bonds

The interface between high and low layers often has to be drawn across covalent bonds. To address the electron density of "dangling" bonds at the QM–MM interface, a number of methods were developed (Lin & Truhlar, 2005). One of the earliest approaches is the link H-atom scheme that completes the covalency of the high layer with capping hydrogen atoms (Eurenius, Chatfield, Brooks, & Hodoscek, 1996; Field et al., 1990; Singh & Kollman, 1986). When the QM–MM interface is defined by breaking C–C bond, the hydrogen atom link is most suitable due to the similarity in electronegativities between hydrogen and carbon atoms. However, other atoms or functional groups can be used as well (Antes & Thiel, 1999; Chen, Zhang, & Zhang, 2004; Gao et al., 2004; Zhang, Chen, & Zhang, 2003). For instance, when it is necessary to simulate electron donation or withdrawal, using halogen-like atoms—pseudohalogens—have proved useful (Zhang, Lee, & Yang, 1999). Other methods to treat the QM/MM boundary use frozen orbitals (Gao, Amara, Alhambra, & Field, 1998; Murphy, Philipp, & Friesner, 2000a; Murphy et al., 2000b; Philipp & Friesner, 1999), pseudopotentials (Bergner, Dolg, Kuchle, Stoll, & Preuss, 1993; Estrin, Tsoo, & Singer, 1991; Zhang et al., 1999), or electron density-derived schemes (Ferre & Angyan, 2002), which does not require insertion of extra atoms. In particular, the delocalized Gaussian MM method (Das et al., 2002) is implemented in the CHARMM/GAMESS-UK and CHARMM/GAMESS-US software programs (Brooks et al., 1983). A comparative study was carried out by Karplus and coworkers, who showed that link atoms and frozen orbitals method give comparable results and neither is systematically better than the other (Reuter, Dejaegere, Maigret, & Karplus, 2000). In practice, the Gaussian ONIOM implementations of MOD-QM/MM use hydrogen link atoms (Dapprich et al., 1999; Humbel et al., 1996; Maseras & Morokuma, 1995; Svensson et al., 1996; Vreven et al., 2001; Vreven & Morokuma, 2000a, 2003), while calculations based on the Qsite package (Murphy et al., 2000a, 2000b; Philipp & Friesner, 1999) use frozen orbitals. However, MOD-QM/MM in general is not limited to any particular treatment of dangling bonds. While neither Gaussian nor Qsite software packages have not yet implemented the MOD-QM/MM method as a built-in algorithm that could be invoked with a keyword, the software can be run as a free standalone software package in conjunction with Gaussian or Qsite, as described in the user guide posted at http://gascon.chem.uconn.edu/software.

2.2.2 MOD-QM/MM ESP Charges

The ESP atomic charges of individual molecular fragments obtained by the MOD-QM/MM method (Gascon, Leung, et al., 2006) are computed according to a least-squares fit procedure similar to the Besler–Merz–Kollman method, implemented in Gaussian (Besler, Merz, & Kollman, 1990). In the least squares fit procedure, the following function is minimized:

$$\gamma(q_1, q_2, \ldots, q_N) = \sum_{j=1}^{m} \left(V_j - E_j\right)^2 + \sum_{k} \lambda_k g_k(q_1, q_2, \ldots, q_N), \quad (2)$$

where V_j is the ab initio ESP at a space point j and $E_j = \sum_{k=1}^{N} q_k/r_{jk}$ and λ_k represents the Lagrange multiplier imposing the constraint. In the Besler–Merz–Kollman method, a single constraint is used, which ensures that the sum of the atomic charges equals the total charge of the fragment:

$$g_1(q_1, q_2, \ldots, q_N) = \sum_{K=1}^{N} q_k - q_{tot} = 0,$$ where N is the number of atoms in the QM layer. In MOD-QM/MM approach, such a scheme would lead to charge leakage to the capping fragments and, therefore, would change the total charge of the system along subsequent iterations. Therefore, the MOD-QM/MM method introduces another constraint to ensure that the sum of the charges of link atoms is equal to zero: $g_1(q_1, q_2, \ldots, q_N) = \sum_{K=1}^{N-M} q_k = 0$, where $N - M$ is the number of link atoms (Fig. 3).

In the MOD-QM/MM method, the ESP charges are obtained by considering a QM layer with N atoms, including atoms whose charges would be fitted (region 1), atoms whose charges are not fitted (region 2), and $N - M$ link H-atoms. Often, nearest neighbor residues are included as capping

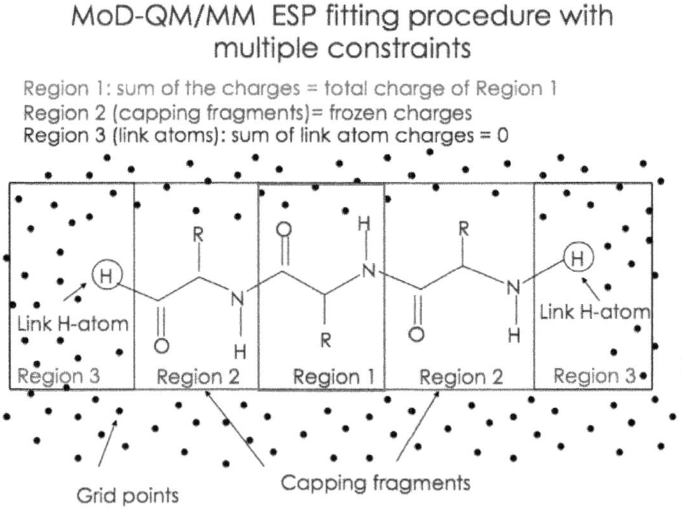

MoD-QM/MM ESP fitting procedure with multiple constraints

Region 1: sum of the charges = total charge of Region 1
Region 2 (capping fragments) = frozen charges
Region 3 (link atoms): sum of link atom charges = 0

Fig. 3 Charge constraints for fitting the electrostatic potential of a domain (region 1) according to the MOD-QM/MM method (Gascon, Leung, et al., 2006), when capped by nearest neighbor residues (region 2) and link hydrogen atoms.

fragments (region 2) to separate the domain whose charges would be fitted (region 1) from the QM/MM interface (Fig. 3) and avoid overpolarization effects. The total charge of the QM layer is $Q = Q_1 + Q_2$, where $Q_1 = \sum_{i=1}^{M-1} q_i + q_M$ is the net charge of the fragment and $Q_2 = \sum_{i=M+1}^{N-1} q_i + q_N$ is set equal to zero. The ESP at position r_j, due to all point charges in the QM region, is $u_j = \sum_{i=1}^{N} q_i/r_{ji}$ where $r_{ji} = |r_j - r_i|$.

Considering the conditions imposed for Q_1 and Q_2 the ESP becomes,

$$u_j = \sum_{i=1}^{M-1} \left[\frac{q_i}{r_{ji}} - \frac{q_i}{r_{jM}}\right] + \frac{Q_1}{r_{jM}} + \sum_{i=M+1}^{N-1} \left[\frac{q_i}{r_{ji}} - \frac{q_i}{r_{jN}}\right] + \frac{Q_2}{r_{jN}}. \tag{3}$$

Substituting $F_{jik} = (1/r_{ji} - 1/r_{jk})$ and $K_{jk} = 1/r_{jk}$, Eq. (3) can be rewritten as follows:

$$u_j = \sum_{i=1}^{M-1} q_i F_{jiM} + \sum_{i=M+1}^{N-1} q_i F_{jiN} + Q_1 K_{jM} + Q_2 K_{jN}. \tag{4}$$

To compute the ESP atomic charges, a least-square minimization procedure of the error function $\chi^2 = \sum_{j=1}^{N_g} (u_j - U_j)^2$ is performed, where U_j is the QM/MM ESP at grid point j and u_j is the corresponding ESP as defined by the distribution of point charges. The sum is typically carried over the set of N_g grid points, associated with four layers at 1.4, 1.6, 1.8, and 2.0 times the van der Waals radii around the QM region span the N_g grid points (Fig. 3). The minimum of χ^2 is obtained by the following gradient:

$$\frac{\partial \chi^2}{\partial q_k} = -\sum_{j=1}^{N_g} 2(u_j - U_j) \frac{\partial u_j}{\partial q_k} = 0, \tag{5}$$

for all q_k in the set $(q_1, q_2, \ldots, q_{M-1}, q_{M+1}, \ldots, q_{N-1})$. Since $\frac{\partial u_j}{\partial q_k} = F_{jks}$, where s corresponds to M or N depending on whether $k < M$ or $M < k < N$, respectively, we can rewrite Eq. (5):

$$\sum_{j=1}^{N_g}\left[U_j-\left(Q_1K_{jM}+Q_2K_{jN}\right)\right]F_{jks} = \sum_{i=1}^{M-1}q_i\sum_{j=1}^{N_g}F_{jiM}F_{jks} \\ -\sum_{i=M+1}^{N-1}q_i\sum_{j=1}^{N_g}F_{jiN}F_{jks}.$$
(6)

Or, in matrix notation:

$$c = q\begin{bmatrix} F_1 & 0 \\ 0 & F_2 \end{bmatrix},$$
(7)

where

$$c^k = \sum_{j=1}^{N_g}\left[U_j-\left(Q_1K_{jM}+Q_2K_{jN}\right)\right]F_{jks},$$
$$q = (q_1, q_2, \ldots, q_{M-1}, q_{M+1}, \ldots, q_{N-1}),$$

$F_1^{lk} = \sum_{j=1}^{N_g} F_{jlM}F_{jks}$, and $F_2^{lk} = \sum_{j=1}^{N_g} F_{jlN}F_{jks}$. Note that vector c and matrix F are only functions of distances between atomic positions. The grid points r_{jk}, the partial charges Q_1 and Q_2, and the ESP are evaluated at grid points $j = 1, \ldots, N_g$.

2.2.3 MOD-QM/MM Optimization

In MOD-QM/MM, the total system is divided into molecular domains and an iterative cycle of QM/MM calculations is performed. In each step, a different molecular domain is considered as a QM layer, for which a geometry relaxation is performed and the ESP atomic charges are updated. The rest of the system is treated at the MM level with atomic charges defined as updated along the MOD-QM/MM iterations. The iterative scheme was initially developed for ab initio computations of protein ESPs (Gascon, Leung, et al., 2006) and was recently generalized for self-consistent structural refinement of extended systems (Ho et al., 2014).

Geometry optimization is typically carried out using the Newton–Raphson method (Fan & Ziegler, 1991; Versluis & Ziegler, 1988), and the Hessian matrix is updated with the Broyden–Fletcher–Goldfarb–Shanno strategy (Schlegel, 1987). Initially, the geometry of the first domain (R_1) is optimized taking into account the surrounding environment. In its minimum energy configuration, the ESP charges of the QM layer are computed and used to update the charges of that domain when treated at the MM level in subsequent steps of the iterative cycle. The procedure is applied analogously to all domains, and the cycle is iterated multiple times until the geometry and distribution of atomic charges converge self-consistently as influenced by mutual polarization effects. Recently, the approach was applied to characterize the hydrogen-bonded geometry of the DNA *Oxytricha nova* guanine quadruplex (G4), showing results comparable to the full QM calculations (Ho et al., 2014).

ONIOM-EE Optimization: Interactions between QM and MM layers in MOD-QM/MM are often dominated by the electrostatic terms. In the ONIOM electronic embedding (EE) scheme (Vreven et al., 2006; Vreven & Morokuma, 2000b; Vreven, Morokuma, Farkas, Schlegel, & Frisch, 2003), the electrostatic interactions are included as one-electron operators of the QM Hamiltonian. The atom-centered partial point charges in the QM Hamiltonian give rise to the ESP due to the MM layer. Other types of interactions between QM and MM layers, such as bonding and non-bonding Van der Waals terms, are also retained at the MM level.

Optimization according to the ONIOM implementation of MOD-QM/MM is performed by following the total energy gradient:

$$\frac{\partial E^{ONIOM}}{\partial q} = \frac{\partial E_X^{QM}}{\partial q} \cdot \mathbf{J} + \frac{\partial E_{X+Y}^{MM}}{\partial q} - \frac{\partial E_X^{MM}}{\partial q} \cdot \mathbf{J} \qquad (8)$$

where $\mathbf{J}$ is the Jacobian necessary to transform the coordinates of the reduced system (X) into the coordinates of the complete system $(X + Y)$. Initially, the coordinates of the reduced system are frozen, leaving only the second term of the expansion. This leads to optimization of the full system at MM level through microiterations. Then, a QM energy minimization step is performed to update the coordinates of the reduced system, based on the forces calculated according with Eq. (8). Microiterations are repeated until the total energy converges to a minimum with respect to all nuclear coordinates.

In the ONIOM electronic embedding (EE) method, $\partial E_X^{QM}/\partial \mathbf{q}^Y \neq 0$ and $\partial E_X^{MM}/\partial \mathbf{q}^Y \neq 0$, that is why the coordinates of the system cannot be rigorously separated into $\mathbf{q}^X$ and $\mathbf{q}^Y$, making the abovementioned microiteration scheme not applicable. In the ONIOM implementation, this issue is addressed by approximating the electrostatic interactions between QM and MM layers during the microiterations, as described by the distribution of ESP charges computed for the initial configuration. After the microiterations are completed, with essentially a frozen distribution of atomic charges, the wave function of the QM layer is reevaluated giving rise to new ESP charges for subsequent microiterations. The QM wave function is recomputed and the MM geometry is reoptimized several times until a self-consistent solution is obtained. Only after that, the QM energy minimization step is performed, and the coordinates of the QM layer are updated. This double self-consistency process is repeated until convergence is reached.

2.2.4 MOD-QM/M Model of the OEC of Photosystem II

MOD-QM/MM models of the OEC of PSII have been originally based on the 1S5L X-ray crystal structure of cyanobacterium PSII (Ferreira, Iverson, Maghlaoui, Barber, & Iwata, 2004) and more recently refined in light of recent breakthroughs in conventional and femtosecond X-ray crystallography (Askerka, Vinyard, Wang, Brudvig, & Batista, 2015; Askerka, Wang, Brudvig, & Batista, 2014; Li, Siegbahn, & Ryde, 2015; Luber et al., 2011; Pal et al., 2013). The models included the Mn_3CaO_4Mn OEC cluster and all amino acid residues with α-carbons within 15 Å from any atom in the OEC, resulting in a total of 1987 atoms. Water ligands were added to complete the coordination spheres of the Mn and Ca atoms, while keeping the displacement of the amino acid residues at minimum. This led to the characteristic coordination numbers of 5 or 6, 6 for Mn^{III}, Mn^{IV} and 8 for Ca, respectively. The presence of at least one water molecule bound to Ca^{2+} is suggested by ^{18}O isotope exchange measurements (Hillier & Wydrzynski, 2004, 2008). The resulting structure of the OEC hydration shell is consistent with pulsed EPR experiments, revealing the presence of several exchangeable deuterons near the Mn cluster in the S_0, S_1, and S_2 states (Britt et al., 2004).

MOD-QM/MM computations of the OEC of PSII required high-quality initial guesses of the QM layer electronic structure, including a proper description of spin states in the OEC cluster as directly compared to spectroscopic measurements. These were originally obtained by using the ligand field theory as implemented in Jaguar 5.5 (Schrodinger, 2003, Jaguar, Portland, OR) and most recently according to the spin fragment approach as implemented in Gaussian 09. Once the initial state was properly defined, it was used within the ONIOM approach as the QM layer of MOD-QM/MM calculations (region X), including the Mn_3CaO_4Mn OEC complex, water molecules, hydroxide and chloride ions in the OEC vicinity, as well as the directly ligated carboxylate groups of D1-D189, CP43-E354, D1-A344, D1-E333, D1-D170, D1-D342, and the imidazole ring of D1-H332 amino acid residues (Fig. 4). The QM/MM interface was defined by breaking C–C bonds of side chains in the abovementioned amino acid residues. Capping of the QM layer was based on the link hydrogen atom scheme. The molecular structure beyond the

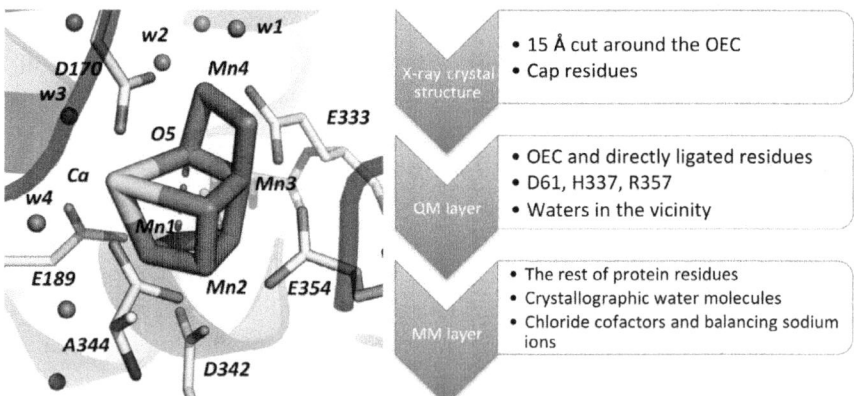

Fig. 4 *Left*: Structural model of the OEC of PSII (QM layer shown explicitly, MM layer shown as cartoon). *Right*: Scheme of QM/MM model preparation. (See the color plate.)

QM layer was treated using Amber classical force field (Cornell et al., 1995). A protein buffer shell with amino acid residues with α-carbons within 15–20 Å from any atom in the OEC ion cluster was also added with harmonic constraints to preserve the natural shape of the system even in the absence of the rest of the protein complex and membrane.

An important aspect of the OEC cluster is the antiferromagnetic coupling between manganese centers. These were simulated using broken-symmetry states (Noodleman, 1981; Noodleman & Case, 1992; Noodleman & Davidson, 1986; Noodleman, Peng, Case, & Mouesca, 1995) in conjunction with DFT. The QM layer was prepared in various spin states and the spin coupling patterns were stabilized by ligand arrangement. Upon optimization, the resulting structures are analyzed in terms of the total energy as well as by comparison of structural and electronic features to experimental data from high-resolution spectroscopy (EPR).

2.2.5 Post-QM/MM Structural Refinement Based on Spectroscopic Data

Typically, potential energy surfaces (PESs) of biological systems are very rugged, with many local minima of comparable in energy. It is, therefore, often difficult to find a subset of configurations with maximum statistical weight by energy minimization. However, it is possible to implement a post-QM/MM structural refinement scheme that addresses the likelihood of energetically similar structures in terms of their consistency with the high-resolution spectroscopy (for instance, EXAFS) (Pal et al., 2013; Sproviero, Gascon, et al., 2008c).

The refinement algorithm starts with a reference configuration of the system (eg, the QM/MM minimum energy structure) and iteratively adjusts the nuclear coordinates to minimize the mean square deviation between the simulated and experimental spectra by simulated annealing Monte Carlo (MC). Each MC step consists of an attempt to move all atoms of choice according to the Metropolis algorithm. The probability of accepting the new position is given by: $P=\min[1, e^{-\Delta F/kT}]$, where k is the Boltzmann constant, T is the temperature (which in this case is varied exponentially from an initial value to 0 K), and ΔF denotes the difference in the cost function, defined as the sum of squared differences between experimental and calculated spectra, when comparing the displaced and undisplaced configurations (Fig. 5). In previous applications, the cost function has been calculated by taking the sum of the square deviations between the experimental and calculated EXAFS spectra. Normally, the simulation is performed by harmonically constraining all the surrounding ligands and relaxing the metal oxide core of the OEC, restricting the displacements to those of the atoms in the OEC and surrounding ligands known to be responsible for affecting the

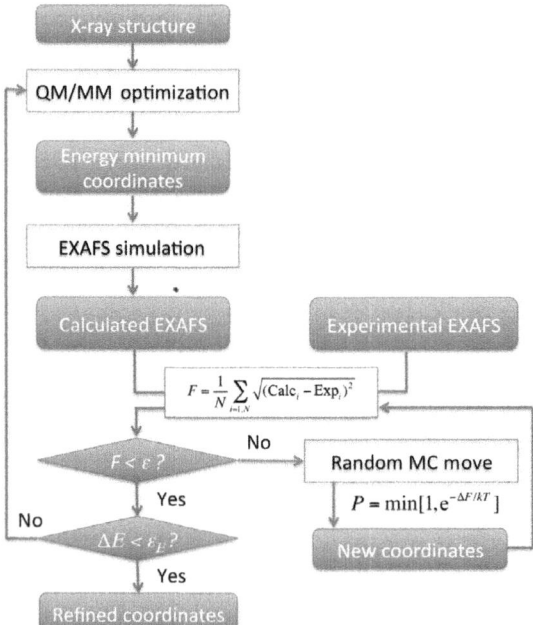

Fig. 5 Post-QM/MM structural refinement scheme based on EXAFS calculations and simulated annealing MC (Pal et al., 2013).

experimental EXAFS of S-state intermediates. To harmonically constrain the atoms, an additional penalty of $k^*||r-r_{int}||^2$ is added to the cost function, where $r-r_{int}$ is the displacement vector with respect to the initial position r_{int}. The EXAFS spectra for a given configuration of the system are computed by calls to the external program FEFF8 (version 8.2) (Ankudinov, Bouldin, Rehr, Sims, & Hung, 2002a; Ankudinov, Ravel, Rehr, & Conradson, 1998; Bouldin, Sims, Hung, Rehr, & Ankudinov, 2001).

It is important to note that even though the post-QM/MM structural refinement algorithm has been implemented for EXAFS spectra simulations, the ΔF minimization procedure could also be based on other spectroscopic data whenever experimental data are available and the calculated spectra can be reliably and efficiently computed. Examples include calculations of EPR data or NMR chemical shifts.

3. EXAFS SIMULATIONS

Ab initio simulations of EXAFS spectra involve solving the multi-scattering problem associated with the photodetached electrons emitted upon X-ray absorption by the Mn centers. The quantum mechanical interference of outgoing photoelectrons with the backscattering waves from atoms surrounding the Mn ions gives rise to oscillations of X-ray absorption intensities as a function of the kinetic energy (or momentum) of the photodetached electron. The Fourier transform of these oscillations provides information on the Mn–Mn distances, the coordination of Mn ions, and changes in the Mn coordination sphere as determined by changes in the oxidation state of the metal centers. Calculations are typically carried out according to the real-space Green function approach, as implemented in the program FEFF8 (version 8.2) (Ankudinov et al., 2002a, 1998; Bouldin et al., 2001), using the theory of the oscillatory structure due to multiple scattering originally proposed by Kronig (Kronig, 1931; Kronig & Penney, 1931) and worked out in detail by Sayers (Sayers, Stern, & Lytle, 1971), Stern (Stern, 1974), Lee and Pendry (Lee & Pendry, 1975), and by Ashley and Doniach (Ashley & Doniach, 1975). In FEFF, the EXAFS spectrum is calculated according to the following equation:

$$\chi(k) = \sum_j \frac{N_j f_j(k) e^{-2k^2\sigma_j^2}}{kR_j^2} \sin\left[2kR_j + \delta_j(k)\right], \tag{9}$$

where $f(k)$ and $\delta(k)$ are scattering properties (amplitudes and phase shifts) of the atoms neighboring the excited atom, N is the number of neighboring atoms, R is the distance to the neighboring atom, and σ^2 is a measure of the disorder in the neighbor distance.

EXAFS using Mn atoms as absorption centers has been particularly important in structural characterization of the OEC in PSII (Dau, Grundmeier, Loja, & Haumann, 2008; Dau & Haumann, 2008; Haumann et al., 2005; Yano et al., 2006). Unlike X-ray diffraction experiments, the X-ray intensities used for EXAFS do not induce radiation damage (Davis & Pushkar, in press; Galstyan, Robertazzi, & Knapp, 2012; Luber et al., 2011; Pal et al., 2013; Pantazis, Ames, Cox, Lubitz, & Neese, 2012) to highly charged (III–IV) Mn (Yano et al., 2005). The onset of radiation damage can be precisely determined and controlled by monitoring the absorber metal K-edge position, allowing us to use as a reference an (almost) intact metal cluster within the protein. In addition, EXAFS provides metal-to-metal and metal-to-ligand distances with high accuracy ($\sim$0.02 Å) and a resolution of $\sim$0.1 Å (Askerka et al., 2015) and can be performed in solution allowing for capturing the intermediate (S_n states) in the PSII catalytic cycle (Haumann et al., 2005). The latter analysis in combination with highly accurate computational tools to predict the EXAFS spectrum (Ankudinov, Bouldin, Rehr, Sims, & Hung, 2002b; Newville, 2001) has been extensively exploited by the computational groups as it allows for direct comparisons to DFT and hybrid QM/MM models of the PSII intermediates with the experimental spectroscopic data (Askerka et al., 2015, 2014; Li et al., 2015; Luber et al., 2011; Pal et al., 2013).

3.1 Post-QM/MM Refinement Model of the OEC of PSII

Structural refinement proved to be particularly useful when dealing with subtle changes in structure that may affect the EXAFS spectrum, such as those produced by changes in protonation states of the μ-oxo bridges in the PSII OEC. For example, in Pal et al. (2013), it was important to determine the protonation states of the O4 and O5 μ-oxo bridges in the S_0 state of PSII. It is known that S_0 to S_1 transition in the catalytic cycle includes release of the proton; however, its position in S_0 state remained known. To understand the nature of deprotonation proton, a preliminary MC screening was conducted using simulated annealing to fit the S_0 EXAFS data, as based on the geometry of the previously reported S_1 model. The model was composed of the metal oxide cluster, the functional groups of the residues which are directly coordinated, and the hydrogen-bonded water network around

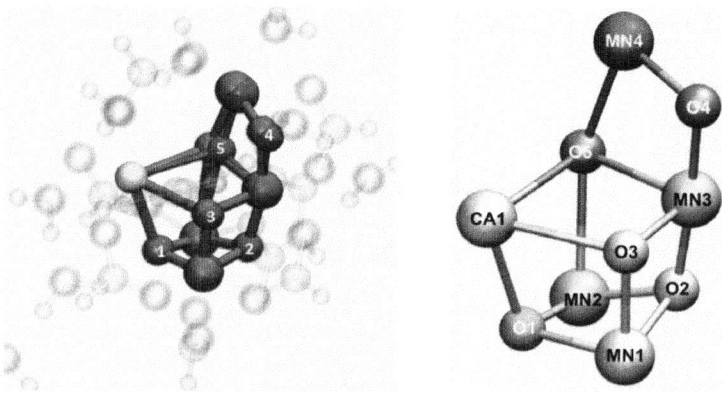

Fig. 6 Monte Carlo (MC)-refined OEC cluster of the S_0 state (in *red* (*gray* in the print version)) and S_1 state (in *purple* (*dark gray* in the print version)) structures (*left*) and the crystal structure B-factors for the atoms in the OEC with most certain positions colored *green* (*light gray* in the print version) and least certain (Mn4 and corresponding bridges) colored *red* (*gray* in the print version) (*right*) (Pal et al., 2013). *Pal, R., Negre, C.F.A., Vogt, L., Pokhrel, R., Ertem, M.Z., Brudvig, G.W., & Batista, V.S. (2013). S_0-state model of the oxygen-evolving complex of photosystem II, Biochemistry, 52(44), 7703–7706. http://dx.doi.org/10.1021/bi401214v. Copyright American Chemical Society 2013.*

the OEC. An overlay of the OEC geometry for the MC-prescreened S_0 state and the S_1 state is shown in Fig. 6.

MC refinement showed that the main difference in the S_0 EXAFS when compared to S_1 is in positions of Mn3, Mn4, O4, and O5 atoms. These results suggested that the proton may originate from the O5 or O4 μ-oxo bridge in the S_0 state. Since the S_0 to S_1 transition also involves an electron release, the movement of Mn3 and Mn4 in the MC refinement indicated that one of those Mn centers may be responsible for oxidation. Interestingly, this is in agreement with the reported B-factors of the atoms in X-ray crystal structure at 1.9 Å resolution (PDB: 3ARC) (Umena, Kawakami, Shen, & Kamiya, 2011). The O4 and O5 atoms exhibit the highest displacement among all the bridging μ-oxos (see Fig. 6, left), which could originate from radiation damage and partial reduction of the resting S_1 state to S_0. Therefore, the next step was to explore the protonation of O4 and O5 sites using the QM/MM model of the S_0 state:

Fig. 7 shows that the 8–11.6 $Å^{-1}$ region of the experimental EXAFS is more consistent with O5 protonated compared to O4 protonated. While both of the refined QM/MM models for both O5-H and O4-H fit well, RMSD of the $CaMn_4O_5$ cluster for O5-H is much lower (0.05 Å) than the RMSD for O4-H (0.12 Å), indicating that smaller refinements are needed

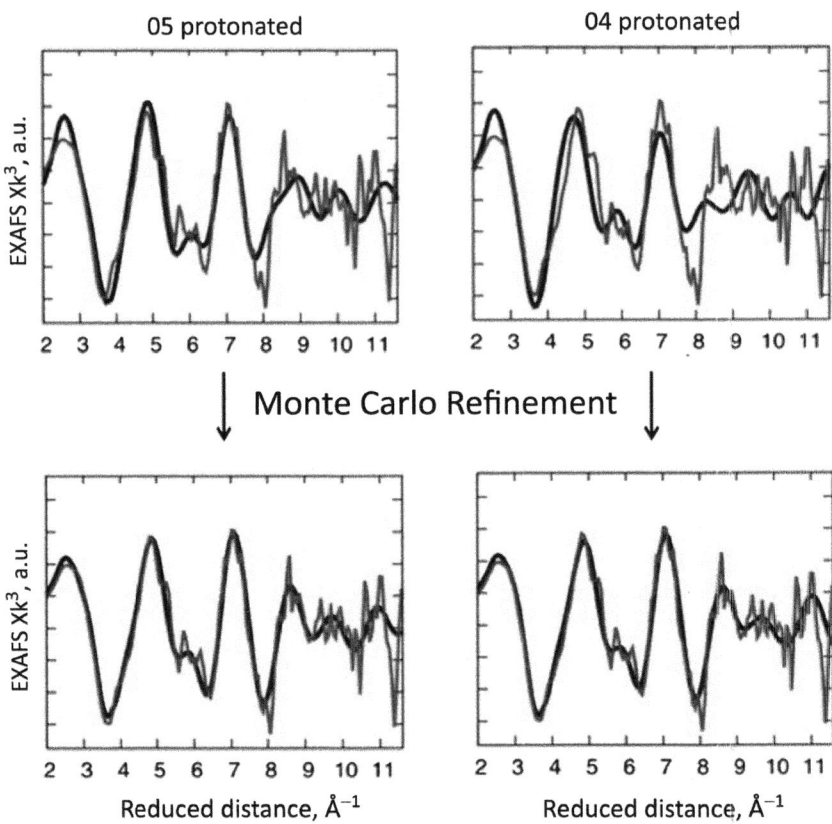

Fig. 7 EXAFS spectra of the S_0 state of the OEC with O5 (*left*) and O4 (*right*) protonated as well as corresponding MC refined spectra (*bottom*) (Pal et al., 2013). Pal, R., Negre, C.F.A., Vogt, L., Pokhrel, R., Ertem, M.Z., Brudvig, G.W., & Batista, V.S. (2013). S_0-state model of the oxygen-evolving complex of photosystem II, Biochemistry, 52(44), 7703–7706. http://dx.doi.org/10.1021/bi401214v.

to get a good match to experimental EXAFS. Upon inspection, we find that the O5-H structure shows minimal movements, whereas O4-H undergoes more significant structural changes during the MC refinement. This serves as another proof that QM/MM O5-H structure is most likely representative of the actual S_0 state observed by EXAFS measurements.

4. MOD-QM/MM MODELS OF DNA QUADRUPLEXES

MOD-QM/MM applications have produced various structural models that could be partially validated by direct comparisons with experimental data, including NMR, EXAFS, X-ray diffraction, and calorimetry measurements (Gascon, Leung, et al., 2006; Gascon et al., 2008; Menikarachchi & Gascon, 2008; Sproviero et al., 2006a; Sproviero, Gascon, McEvoy, Brudvig, & Batista, 2006b; Sproviero et al., 2007; Sproviero, Gascon, et al., 2008a, 2008b, 2008c, 2008d; Sproviero, Gascon, et al., 2008b). Recently, a benchmark MOD-QM/MM study has focused on the DNA *Oxytricha nova* guanine quadruplex (Ho et al., 2014), including direct comparisons to high-resolution X-ray diffraction (Haider, Parkinson, & Neidle, 2002, 2003) and NMR data (Gill, Strobel, & Loria, 2006). The DNA quadruplex has a central channel containing five potassium ions (K^+ shown in yellow in Fig. 8) in between layers of guanine quartets (shown in orange in Fig. 8) that are essential for stabilizing the quadruplex structure. The presence of these ions introduces significant polarization effects that are not adequately described by static molecular mechanics force fields. This is manifested by a very loosely bound potassium ion in the thymine loops (shown in gray in Fig. 8) of the quadruplex, as well as convergence to the incorrect bifurcated hydrogen bonding geometry of the guanine quartets (Gkionis et al., 2014; Song, Ji, & Zhang, 2013).

For comparison, Fig. 9 shows a rigid scan of the PES for a two-layer model where a potassium ion is displaced stepwise along the z-direction. The comparison of PES profiles shows significant deviations between results

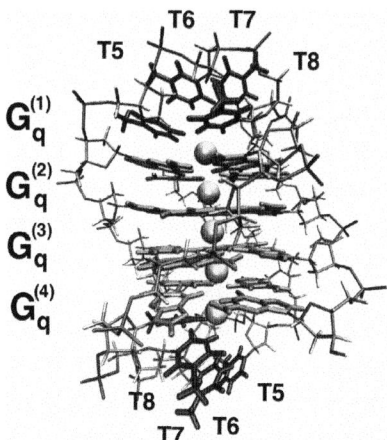

Fig. 8 1JPQ crystal structure of the DNA *Oxytricha nova* guanine quadruplex. *Ho, J., Newcomer, M.B., Ragain, C.M., Gascon, J.A., Batista, E.R., Loria, J.P., & Batista, V.S. (2014). MoD-QM/MM structural refinement method: Characterization of hydrogen bonding in the Oxytricha nova G-Quadruplex,* Journal of Chemical Theory and Computation, *10(11), 5125–5135. http://pubs.acs.org/doi/full/10.1021/ct500571k. Copyright American Chemical Society 2014.*

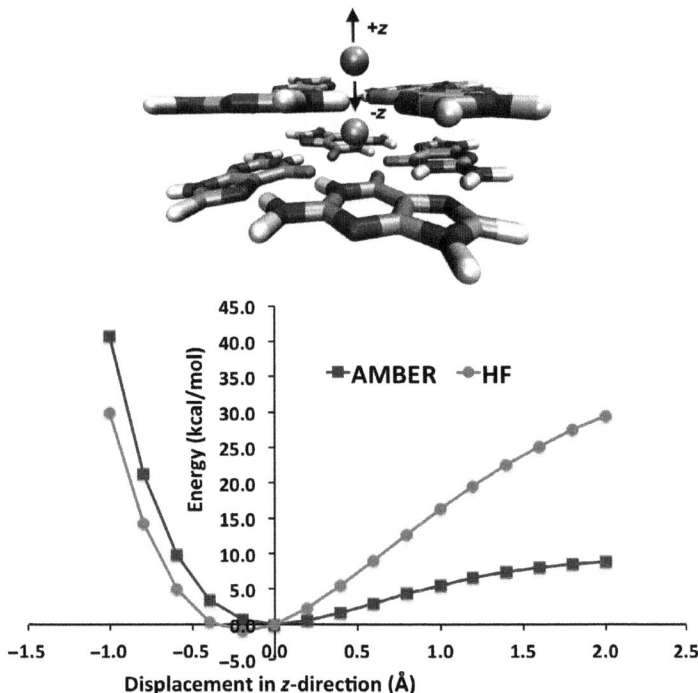

Fig. 9 Benchmark calculations of electrostatic interactions for a potassium ion dragged through the internal channel of a two-layer model of the DNA guanine quadruplex, as described by HF and Amber, revealing significant polarization effects missed by non-polarizable molecular mechanics force fields. (See the color plate.)

obtained with HF and the Amber molecular mechanics force field. In particular, the force field substantially underestimates the binding energy of the potassium ion because it does not account explicitly for mutual polarization effects. Therefore, electrostatic stabilization of the monovalent cations by the guanine quartets is underestimated.

The main goal of the MOD-QM/MM study has been to explore the refinement of the G-quadruplex geometry that is sensitive to electrostatic interactions between the embedded K^+ ions and the polarized guanine moieties, and to validate the resulting MOD-QM/MM models through direct comparison of ^{1}H NMR chemical shifts (Ho et al., 2014) which are very sensitive to the detailed configuration of hydrogen bonding in the folded quadruplex configuration. Therefore, it was important to describe electrostatic interactions and polarization effects accurately since K^+ ions polarize the guanine moieties and affect the hydrogen bonds and stacking interactions that regulate the NMR chemical shifts. As shown in Fig. 10, the

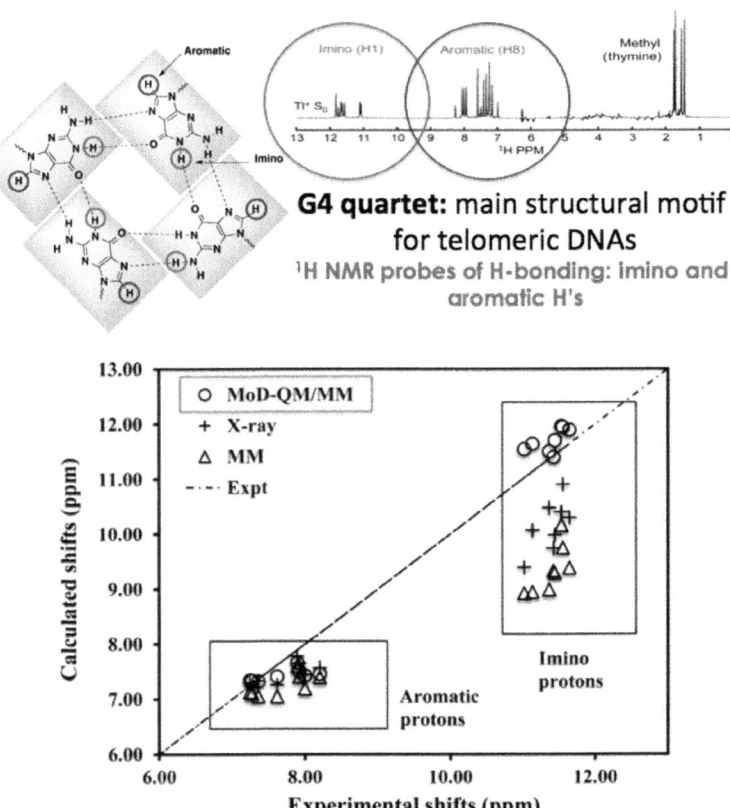

Fig. 10 Comparison of experimental and calculated ^{1}H NMR chemical shifts, obtained at the ONIOM(ωB97XD/6-31G(d,p):AMBER)-EE level of theory for MOD-QM/MM structural models (*circles*), and structures predicted by a classical molecular mechanics (AMBER) force field (*triangles*) and the X-ray structure (*crosses*). Copyright American Chemical Society 2014.

MOD-QM/MM refined structure gives a significant improvement in the description of the chemical shifts when compared to those predicted by popular molecular mechanics force fields (Ho et al., 2014).

The MOD-QM/MM model of the DNA quadruplex (Haider et al., 2002) was prepared from the 1JPQ X-ray crystal model by using the standard protocol of adding hydrogen atoms and solvating the resulting structure in a periodic box of water molecules. Classical molecular dynamics simulations equilibrated the solvated system while constraining the quadruplex structure. Production run calculations were based on an ensemble of solvated configurations, composed of the quadruplex and water molecules within 15 Å from any atom in the G4 quadruplex. For each representative configuration of the quadruplex, molecular domains were defined as individual guanine nucleotides within quartets. Accordingly, the hydrogen bonding interactions between the nucleotides were described at the MM level, albeit using MOD-QM/MM-restrained ESP charges. Then, the MOD-QM/MM iterative algorithm was implemented where each of the 16 nucleotides is treated as the QM layer in an ONIOM (ωB97XD/6-31G(d):Amber) calculation while keeping fixed all other domains. The geometry of the QM layer was relaxed in the presence of the surrounding charges, and the restrained ESP charges were calculated on the converged geometry. The cycle then proceeded to the next domain, using the atomic charges and geometry updated from the previous QM/MM calculation, and the cycle was complete when the algorithm iterated over all 16 domains and was iterated until the charges and geometry of each domain were converged below a desired threshold tolerance value.

Proton NMR chemical shifts were then calculated using the gauge-independent atomic orbital (GIAO) method at the DFT QM/MM (ωB97XD/6-31G(d):Amber) level for each of the solvated configurations after structural refinement based on the MOD-QM/MM method. To account for the shielding effects between the guanine quartets, the QM layer included all 16 nucleotides. The NMR chemical shifts were then obtained as an average over solvated configurations, sampled from classical MD simulations. The overall average yielded proton NMR chemical shifts for the imino- and aromatic protons that are within 0.5 ppm from experimental data.

5. CONCLUSIONS

Establishing a simple and reliable approach to describe electrostatic interactions in complex molecular systems, including the influence of polarization effects on structural models and mechanisms, has been a long-standing challenge in computational enzymology. In this chapter, we have shown that the recently developed MOD-QM/MM hybrid method provides a rigorous yet practical approach to obtain ab initio quality models based on a simple space domain decomposition scheme and the iterative self-consistent treatment of the constituent molecular fragments, according to state-of-the-art QM/MM methods.

The MOD-QM/MM method has been successfully applied to complex biological systems, including important applications to PSII, human aldose reductase enzyme, and guanine DNA quadruplexes, demonstrating the capabilities of the method as applied to systems where electrostatic interactions are critical. The post-QM/MM refinement methodology allows to assess structures of similar energy and to explore subtle structural changes, including those induced by monovalent cations or variable protonation states, as directly compared to high-resolution spectroscopic data. The method has been developed and implemented using EXAFS computations and simulated annealing MC minimization, where the nuclear coordinates of the model are refined for best agreement between calculated and experimental spectra. While implemented in conjunction with calculations of EXAFS spectra, the methodology could be implemented more generally with other types of spectroscopic techniques (eg, FTIR, NMR) whenever experimental data are available and the calculated spectra can be reliably and efficiently computed.

We have shown that the MOD-QM/MM approach can be applied to address fundamental questions on the role of electrostatic interactions on structural refinement of PSII and DNA quadruplexes that for many years remained largely elusive to rigorous first-principles examinations. We found that the resulting QM/MM methodology, where polarization effects are explicitly considered, is a powerful tool when applied in conjunction with high-resolution spectroscopy. Therefore, it is natural to anticipate that the MPD-QM/MM approach will continue to make important contributions on enzymology-based structural and mechanistic characterization.

ACKNOWLEDGMENTS

V.S.B. acknowledges an allocation of DOE high-performance computing time from NERSC and the support from the NSF CHE-1465108 grant. Development of the MOD-QM/MM approach was partially funded by the NIH grant 1R01GM10621-01A1. We also thank Dr. Christian Negre for contributions to EXAFS-refinement methodology and Dr. Leslie Vogt for contribution to the PSII model preparation.

REFERENCES

Ankudinov, A. L., Bouldin, C. E., Rehr, J. J., Sims, J., & Hung, H. (2002a). Parallel calculation of electron multiple scattering using Lanczos algorithms. *Physical Review B: Condensed Matter*, *65*(10), 104107.

Ankudinov, A. L., Bouldin, C. E., Rehr, J. J., Sims, J., & Hung, H. (2002b). Parallel calculation of electron multiple scattering using Lanczos algorithms. *Physical Review B*, *65*(10), 104107. http://dx.doi.org/10.1103/Physrevb.65.104107.

Ankudinov, A. L., Ravel, B., Rehr, J. J., & Conradson, S. D. (1998). Real-space multiple-scattering calculation and interpretation of x-ray-absorption near-edge structure. *Physical Review B: Condensed Matter*, *58*(12), 7565–7576.

Antes, I., & Thiel, W. (1999). Adjusted connection atoms for combined quantum mechanical and molecular mechanical methods. *Journal of Physical Chemistry A*, *103*(46), 9290–9295.

Aqvist, J., & Luzhkov, V. (2000). Ion permeation mechanism of the potassium channel. *Nature (London), 404*(6780), 881–884.

Ashley, C. A., & Doniach, S. (1975). Theory of extended X-ray absorption-edge fine-structure (EXAFS) in crystalline solids. *Physical Review B: Condensed Matter, 11*(4), 1279–1288.

Askerka, M., Vinyard, D. J., Wang, J. M., Brudvig, G. W., & Batista, V. S. (2015). Analysis of the radiation-damage-free X-ray structure of photosystem II in light of EXAFS and QM/MM data. *Biochemistry, 54*(9), 1713–1716. http://dx.doi.org/10.1021/acs.biochem.5b00089.

Askerka, M., Wang, J., Brudvig, G. W., & Batista, V. S. (2014). Structural changes in the oxygen-evolving complex of photosystem II induced by the S1 to S2 transition: A combined XRD and QM/MM study. *Biochemistry, 53*(44), 6860–6862. http://dx.doi.org/10.1021/bi5011915.

Babu, K., & Gadre, S. R. (2003). Ab initio quality one-electron properties of large molecules: Development and testing of molecular tailoring approach. *Journal of Computational Chemistry, 24*(4), 484–495.

Babu, K., Ganesh, V., Gadre, S. R., & Ghermani, N. E. (2004). Tailoring approach for exploring electron densities and electrostatic potentials of molecular crystals. *Theoretical Chemistry Accounts, 111*(2–6), 255–263.

Bader, J. S., & Berne, B. J. (1996). Solvation energies and electronic spectra in polar, polarizable media: Simulation tests of dielectric continuum theory. *Journal of Chemical Physics, 104*(4), 1293–1308.

Bakowies, D., & Thiel, W. (1996). Hybrid models for combined quantum mechanical and molecular mechanical approaches. *Journal of Physical Chemistry, 100*(25), 10580–10594.

Baldwin, M. J., Kampf, J. W., & Pecoraro, V. L. (1993). The linear MnII complex $Mn_3(5-NO_2-Salimh)_2(Oac)_4$ provides an alternative structure type for the carboxylate shift in proteins. *Journal of the Chemical Society, Chemical Communications*, (23), 1741–1743.

Banks, J. L., Kaminski, G. A., Zhou, R. H., Mainz, D. T., Berne, B. J., & Friesner, R. A. (1999). Parametrizing a polarizable force field from ab initio data. I. The fluctuating point charge model. *Journal of Chemical Physics, 110*(2), 741–754.

Bergner, A., Dolg, M., Kuchle, W., Stoll, H., & Preuss, H. (1993). Ab-Initio energy-adjusted pseudopotentials for elements of groups 13-17. *Molecular Physics, 80*(6), 1431–1441.

Bernardo, D. N., Ding, Y. B., Kroghjespersen, K., & Levy, R. M. (1994). An anisotropic polarizable water model—Incorporation of all-atom polarizabilities into molecular mechanics force-fields. *Journal of Physical Chemistry, 98*(15), 4180–4187.

Besler, B. H., Merz, K. M., & Kollman, P. A. (1990). Atomic charges derived from semi-empirical methods. *Journal of Computational Chemistry, 11*(4), 431–439.

Bliznyuk, A. A., Rendell, A. P., Allen, T. W., & Chung, S. H. (2001). The potassium ion channel: Comparison of linear scaling semiempirical and molecular mechanics representations of the electrostatic potential. *Journal of Physical Chemistry B, 105*(50), 12674–12679.

Borgis, D., & Staib, A. (1995). A semiempirical quantum polarization model for water. *Chemical Physics Letters, 238*(1–3), 187–192.

Bouldin, C., Sims, J., Hung, H., Rehr, J. J., & Ankudinov, A. L. (2001). Rapid calculation of X-ray absorption near edge structure using parallel computation. *X-Ray Spectrometry, 30*(6), 431–434.

Britt, R. D., Campbell, K. A., Peloquin, J. M., Gilchrist, M. L., Aznar, C. P., Dicus, M. M., ... Messinger, J. (2004). Recent pulsed EPR studies of the photosystem II oxygen-evolving complex: Implications as to water oxidation mechanisms. *Biochimica et Biophysica Acta, 1655*, 158–171.

Brooks, B. R., Bruccoleri, R. E., Olafson, B. D., States, D. J., Swaminathan, S., & Karplus, M. (1983). Charmm—A program for macromolecular energy, minimization, and dynamics calculations. *Journal of Computational Chemistry, 4*(2), 187–217.

Burant, J. C., Scuseria, G. E., & Frisch, M. J. (1996). A linear scaling method for Hartree-Fock exchange calculations of large molecules. *Journal of Chemical Physics*, *105*(19), 8969–8972.

Burnham, C. J., & Xantheas, S. S. (2002). Development of transferable interaction models for water. III. Reparametrization of an all-atom polarizable rigid model (TTM2-R) from first principles. *Journal of Chemical Physics*, *116*(4), 1500–1510.

Cao, J., & Berne, B. J. (1992). Many-body dispersion forces of polarizable clusters and liquids. *Journal of Chemical Physics*, *97*(11), 8628–8636.

Chen, W., & Gordon, M. S. (1996). Energy decomposition analyses for many-body interaction and applications to water complexes. *Journal of Physical Chemistry*, *100*(34), 14316–14328.

Chen, X. H., Zhang, D. W., & Zhang, J. Z. H. (2004). Fractionation of peptide with disulfide bond for quantum mechanical calculation of interaction energy with molecules. *Journal of Chemical Physics*, *120*(2), 839–844.

Cho, A. E., Guallar, V., Berne, B. J., & Friesner, R. (2005). Importance of accurate charges in molecular docking: Quantum mechanical/molecular mechanical (QM/MM) approach. *Journal of Computational Chemistry*, *26*(9), 915–931.

Chung, L. W., Hirao, H., Li, X., & Morokuma, K. (2012). The ONIOM method: Its foundation and applications to metalloenzymes and photobiology. *Wiley Interdisciplinary Reviews: Computational Molecular Science*, *2*(2), 327–350. http://dx.doi.org/10.1002/wcms.85.

Collins, M. A., Cvitkovic, M. W., & Bettens, R. P. A. (2014). The combined fragmentation and systematic molecular fragmentation methods. *Accounts of Chemical Research*, *47*(9), 2776–2785.

Cornell, W. D., Cieplak, P., Bayly, C. I., Gould, I. R., Merz, K. M., Ferguson, D. M., ... Kollman, P. A. (1995). A 2nd generation force-field for the simulation of proteins, nucleic-acids, and organic-molecules. *Journal of the American Chemical Society*, *117*(19), 5179–5197.

Corongiu, G. (1992). Molecular-dynamics simulation for liquid water using a polarizable and flexible potential. *International Journal of Quantum Chemistry*, *42*(5), 1209–1235.

Daniels, A. D., & Scuseria, G. E. (1999). What is the best alternative to diagonalization of the Hamiltonian in large scale semiempirical calculations? *Journal of Chemical Physics*, *110*(3), 1321–1328.

Dapprich, S., Komaromi, I., Byun, K. S., Morokuma, K., & Frisch, M. J. (1999). A new ONIOM implementation in Gaussian 98. Part I. The calculation of energies, gradients, vibrational frequencies and electric field derivatives. *Journal of Molecular Structure*, *461*, 1–21.

Das, D., Eurenius, K. P., Billings, E. M., Sherwood, P., Chatfield, D. C., Hodoscek, M., & Brooks, B. R. (2002). Optimization of quantum mechanical molecular mechanical partitioning schemes: Gaussian delocalization of molecular mechanical charges and the double link atom method. *Journal of Chemical Physics*, *117*(23), 10534–10547.

Dau, H., Grundmeier, A., Loja, P., & Haumann, M. (2008). On the structure of the manganese complex of photosystem II: Extended-range EXAFS data and specific atomic-resolution models for four S-states. *Philosophical Transactions of the Royal Society, B: Biological Sciences*, *363*(1494), 1237–1243. http://dx.doi.org/10.1098/Rstb.2007.2220.

Dau, H., & Haumann, M. (2008). The manganese complex of photosystem II in its reaction cycle—Basic framework and possible realization at the atomic level. *Coordination Chemistry Reviews*, *252*(3–4), 273–295. http://dx.doi.org/10.1016/J.Ccr.2007.09.001.

Davis, K. M., & Pushkar, Y. (2015). Structure of the oxygen evolving complex of Photosystem II at room temperature. *The Journal of Physical Chemistry. B*, *119*(8), 3492–3498.

Day, P. N., Jensen, J. H., Gordon, M. S., Webb, S. P., Stevens, W. J., Krauss, M., ... Cohen, D. (1996). An effective fragment method for modeling solvent effects in quantum mechanical calculations. *Journal of Chemical Physics*, *105*(5), 1968–1986.

Devries, A. H., Vanduijnen, P. T., Juffer, A. H., Rullmann, J. A. C., Dijkman, J. P., Merenga, H., & Thole, B. T. (1995). Implementation of reaction field methods in quantum-chemistry computer codes (Vol 16, Pg 37, 1995). *Journal of Computational Chemistry, 16*(11), 1445–1446.

Dixon, S. L., & Merz, K. M. (1996). Semiempirical molecular orbital calculations with linear system size scaling. *Journal of Chemical Physics, 104*(17), 6643–6649.

Dixon, S. L., & Merz, K. M. (1997). Fast, accurate semiempirical molecular orbital calculations for macromolecules. *Journal of Chemical Physics, 107*(3), 879–893.

Dykstra, C. E. (1993). Electrostatic interaction potentials in molecular-force fields. *Chemical Reviews, 93*(7), 2339–2353.

Eichler, U., Kolmel, C. M., & Sauer, J. (1997). Combining ab initio techniques with analytical potential functions for structure predictions of large systems: Method and application to crystalline silica polymorphs. *Journal of Computational Chemistry, 18*(4), 463–477.

Ermolaeva, M. D., van der Vaart, A., & Merz, K. M. (1999). Implementation and testing of a frozen density matrix—Divide and conquer algorithm. *Journal of Physical Chemistry A, 103*(12), 1868–1875.

Estrin, D. A., Tsoo, C., & Singer, S. J. (1991). Accurate nonlocal electron argon pseudopotential for condensed phase simulation. *Chemical Physics Letters, 184*(5–6), 571–578.

Eurenius, K. P., Chatfield, D. C., Brooks, B. R., & Hodoscek, M. (1996). Enzyme mechanisms with hybrid quantum and molecular mechanical potentials. 1. Theoretical considerations. *International Journal of Quantum Chemistry, 60*(6), 1189–1200.

Exner, T. E., & Mezey, P. G. (2002). *Ab initio*-quality electrostatic potentials for proteins: An application of the ADMA approach. *Journal of Physical Chemistry A, 106*(48), 11791–11800.

Exner, T. E., & Mezey, P. G. (2003). *Ab initio* quality properties for macromolecules using the ADMA approach. *Journal of Computational Chemistry, 24*(16), 1980–1986.

Exner, T. E., & Mezey, P. G. (2004). The field-adapted ADMA approach: Introducing point charges. *Journal of Physical Chemistry A, 108*(19), 4301–4309.

Exner, T. E., & Mezey, P. G. (2005). Evaluation of the field-adapted ADMA approach: Absolute and relative energies of crambin and derivatives. *Physical Chemistry Chemical Physics, 7*(24), 4061–4069.

Fan, L. Y., & Ziegler, T. (1991). The influence of self-consistency on nonlocal density functional calculations. *The Journal of Chemical Physics, 94*, 6057–6063.

Fedorov, D. G., & Kitaura, K. (2004a). The importance of three-body terms in the fragment molecular orbital method. *Journal of Chemical Physics, 120*(15), 6832–6840.

Fedorov, D. G., & Kitaura, K. (2004b). On the accuracy of the 3-body fragment molecular orbital method (FMO) applied to density functional theory. *Chemical Physics Letters, 389*(1–3), 129–134.

Ferre, N., & Angyan, J. G. (2002). Approximate electrostatic interaction operator for QM/MM calculations. *Chemical Physics Letters, 356*(3–4), 331–339.

Ferreira, K. N., Iverson, T. M., Maghlaoui, K., Barber, J., & Iwata, S. (2004). Architecture of the photosynthetic oxygen-evolving center. *Science, 303*(5665), 1831–1838.

Field, M. J. (1997). Hybrid quantum mechanical molecular mechanical fluctuating charge models for condensed phase simulations. *Molecular Physics, 91*(5), 835–845.

Field, M. J., Bash, P. A., & Karplus, M. (1990). A combined quantum-mechanical and molecular mechanical potential for molecular-dynamics simulations. *Journal of Computational Chemistry, 11*(6), 700–733.

Frisch, M. J., Trucks, G. W., Schlegel, H. B., Scuseria, G. E., Robb, M. A., Cheeseman, J. R., ... Pople, J. A. (2004). *Gaussian 03, revision B.04 (version 03, revision B.04)*. Wallingford, CT: Gaussian, Inc.

Gadre, S. R., Shirsat, R. N., & Limaye, A. C. (1994). Molecular tailoring approach for simulation of electrostatic properties. *Journal of Physical Chemistry*, *98*(37), 9165–9169.

Galstyan, A., Robertazzi, A., & Knapp, E. W. (2012). Oxygen-evolving Mn cluster in photosystem II: The protonation pattern and oxidation state in the high-resolution crystal structure. *Journal of the American Chemical Society*, *134*(17), 7442–7449. http://dx.doi.org/10.1021/ja300254n.

Gao, J. L. (1997). Toward a molecular orbital derived empirical potential for liquid simulations. *Journal of Physical Chemistry B*, *101*(4), 657–663.

Gao, J. L. (1998). A molecular-orbital derived polarization potential for liquid water. *Journal of Chemical Physics*, *109*(6), 2346–2354. http://dx.doi.org/10.1063/1.476802.

Gao, J. L., Amara, P., Alhambra, C., & Field, M. J. (1998). A generalized hybrid orbital (GHO) method for the treatment of boundary atoms in combined QM/MM calculations. *Journal of Physical Chemistry A*, *102*(24), 4714–4721.

Gao, J. L., & Xia, X. F. (1992). A priori evaluation of aqueous polarization effects through Monte-Carlo QM-MM simulations. *Science*, *258*(5082), 631–635.

Gao, A. M., Zhang, D. W., Zhang, J. Z. H., & Zhang, Y. K. (2004). An efficient linear scaling method for ab initio calculation of electron density of proteins. *Chemical Physics Letters*, *394*(4–6), 293–297.

Gascon, J. A., Leung, S. S. F., Batista, E. R., & Batista, V. S. (2006). A self-consistent space-domain decomposition method for QM/MM computations of protein electrostatic potentials. *Journal of Chemical Theory and Computation*, *2*(1), 175–186.

Gascon, J. A., Sproviero, E. M., & Batista, V. S. (2006). Computational studies of the primary phototransduction event in visual rhodopsin. *Accounts of Chemical Research*, *39*(3), 184–193.

Gascon, J. A., Sproviero, E. M., McEvoy, J. P., Brudvig, G. W., & Batista, V. S. (2008). Ligation of the C-terminus of the D1-polypeptide of photosystem II to the oxygen evolving complex of photosystem II. In J. F. Allen, E. Gautt, J. H. Golbeck, & B. Osmond (Eds.), *Photosynthesis: Energy from the sun* (pp. 363–368). New York: Springer.

Gentilucci, L., Squassabia, F., & Artali, R. (2007). Re-discussion of the importance of ionic interactions in stabilizing ligand-opioid receptor complex and in activating signal transduction. *Current Drug Targets*, *8*(1), 185–196.

Gill, M. L., Strobel, S. A., & Loria, J. P. (2006). Crystallization and characterization of the thallium form of the Oxytricha nova G-quadruplex. *Nucleic Acids Research*, *34*(16), 4506–4514.

Gkionis, K., Kruse, H., Platts, J. A., Mladek, A., Koca, J., & Sponer, J. (2014). Ion Binding to quadruplex DNA stems. Comparison of MM and QM descriptions reveals sizable polarization effects not included in contemporary simulations. *Journal of Chemical Theory and Computation*, *10*(3), 1326–1340.

Gordon, M. S., Freitag, M. A., Bandyopadhyay, P., Jensen, J. H., Kairys, V., & Stevens, W. J. (2001). The effective fragment potential method: A QM-based MM approach to modeling environmental effects in chemistry. *Journal of Physical Chemistry A*, *105*(2), 293–307.

Grater, F., Schwarzl, S. M., Dejaegere, A., Fischer, S., & Smith, J. C. (2005). Protein/ligand binding free energies calculated with quantum mechanics/molecular mechanics. *Journal of Physical Chemistry B*, *109*(20), 10474–10483.

Green, D. F., Dennis, A. T., Fam, P. S., Tidor, B., & Jasanoff, A. (2006). Rational design of new binding specificity by simultaneous mutagenesis of calmodulin and a target peptide. *Biochemistry*, *45*(41), 12547–12559.

Greengard, L., & Rokhlin, V. (1987). A fast algorithm for particle simulations. *Journal of Computational Physics*, *73*(2), 325–348.

Haider, S., Parkinson, G. N., & Neidle, S. (2002). Crystal structure of the potassium form of an Oxytricha nova G-quadruplex. *Journal of Molecular Biology*, *320*(2), 189–200.

Haider, S. M., Parkinson, G. N., & Neidle, S. (2003). Structure of a G-quadruplex-ligand complex. *Journal of Molecular Biology, 326*(1), 117–125.

Halgren, T. A., & Damm, W. (2001). Polarizable force fields. *Current Opinion in Structural Biology, 11*(2), 236–242.

Haumann, M., Muller, C., Liebisch, P., Iuzzolino, L., Dittmer, J., Grabolle, M., ... Dau, H. (2005a). Structural and oxidation state changes of the photosystem II manganese complex in four transitions of the water oxidation cycle (S0 - > S1, S1 - > S2, S2 - > S3, and S3,S4 - > S0) characterized by X-ray absorption spectroscopy at 20 K and room temperature. *Biochemistry, 44*(6), 1894–1908.

Hillier, W., & Wydrzynski, T. (2004). Substrate water interactions within the photosystem II oxygen evolving complex. *Physical Chemistry Chemical Physics, 6*(20), 4882–4889.

Hillier, W., & Wydrzynski, T. (2008). O-18-water exchange in photosystem II: Substrate binding and intermediates of the water splitting cycle. *Coordination Chemistry Reviews, 252*(3–4), 306–317. http://dx.doi.org/10.1016/J.Ccr.2007.09.004.

Ho, J., Newcomer, M. B., Ragain, C. M., Gascon, J. A., Batista, E. R., Loria, J. P., & Batista, V. S. (2014). MoD-QM/MM structural refinement method: Characterization of hydrogen bonding in the Oxytricha nova G-Quadruplex. *Journal of Chemical Theory and Computation, 10*(11), 5125–5135.

Houjou, H., Inoue, Y., & Sakurai, M. (2001). Study of the opsin shift of bacteriorhodopsin: Insight from QM/MM calculations with electronic polarization effects of the protein environment. *Journal of Physical Chemistry B, 105*(4), 867–879.

Huang, J., Lopes, P. E. M., Roux, B., & MacKerell, A. D. (2014). Recent advances in polarizable force fields for macromolecules: Microsecond simulations of proteins using the classical drude oscillator model. *Journal of Physical Chemistry Letters, 5*(18), 3144–3150. http://dx.doi.org/10.1021/jz501315h.

Humbel, S., Sieber, S., & Morokuma, K. (1996). The IMOMO method: Integration of different levels of molecular orbital approximations for geometry optimization of large systems: Test for n-butane conformation and SN2 reaction: RCl+Cl-. *The Journal of Chemical Physics, 105*(5), 1959–1967.

Hurley, J. H. (2006). Membrane binding domains. *Biochimica et Biophysica Acta, 1761*(8), 805–811.

Ilan, B., Tajkhorshid, E., Schulten, K., & Voth, G. A. (2004). The mechanism of proton exclusion in aquaporin channels. *Proteins: Structure, Function, and Bioinformatics, 55*(2), 223–228.

Illingworth, C. J. R., Gooding, S. R., Winn, P. J., Jones, G. A., Ferenczy, G. G., & Reynolds, C. A. (2006). Classical polarization in hybrid QM/MM methods. *Journal of Physical Chemistry A, 110*(20), 6487–6497.

Kitaura, K., Ikeo, E., Asada, T., Nakano, T., & Uebayasi, M. (1999). Fragment molecular orbital method: An approximate computational method for large molecules. *Chemical Physics Letters, 313*(3–4), 701–706.

Kitaura, K., Sugiki, S. I., Nakano, T., Komeiji, Y., & Uebayasi, M. (2001). Fragment molecular orbital method: Analytical energy gradients. *Chemical Physics Letters, 336*(1–2), 163–170.

Ko, C., Levine, B., Toniolo, A., Manohar, L., Olsen, S., Werner, H. J., & Martinez, T. J. (2003). Ab initio excited-state dynamics of the photoactive yellow protein chromophore. *Journal of the American Chemical Society, 125*(42), 12710–12711.

Kronig, R. D. (1931). The quantum theory of dispersion in metallic conductors—II. *Proceedings of the Royal Society of London. Series A, Containing Papers of a Mathematical and Physical Character, 133*(821), 255–265.

Kronig, R. D., & Penney, W. G. (1931). Quantum mechanics of electrons on crystal lattices. *Proceedings of the Royal Society of London. Series A, Containing Papers of a Mathematical and Physical Character, 130*(814), 499–513.

Kurisu, G., Zhang, H. M., Smith, J. L., & Cramer, W. A. (2003). Structure of the cytochrome b(6)f complex of oxygenic photosynthesis: Tuning the cavity. *Science*, *302*(5647), 1009–1014.

Lappi, A. K., Lensink, M. F., Alanen, H. I., Salo, K. E. H., Lobell, M., Juffer, A. H., & Ruddock, L. W. (2004). A conserved arginine plays a role in the catalytic cycle of the protein disulphide isomerases. *Journal of Molecular Biology*, *335*(1), 283–295.

Lee, P. A., & Pendry, J. B. (1975). Theory of the extended X-ray absorption fine structure. *Physical Review B: Condensed Matter*, *11*(8), 2795–2811.

Lee, T. S., York, D. M., & Yang, W. T. (1996). Linear-scaling semiempirical quantum calculations for macromolecules. *Journal of Chemical Physics*, *105*(7), 2744–2750.

Li, W., & Li, S. H. (2005). A localized molecular-orbital assembler approach for Hartree-Fock calculations of large molecules. *Journal of Chemical Physics*, *122*(19), 194109.

Li, S. H., Li, W., & Fang, T. (2005). An efficient fragment-based approach for predicting the ground-state energies and structures of large molecules. *Journal of the American Chemical Society*, *127*(19), 7215–7226.

Li, X. P., Nunes, R. W., & Vanderbilt, D. (1993). Density-matrix electronic-structure method with linear system-size scaling. *Physical Review B*, *47*(16), 10891–10894.

Li, X. C., Siegbahn, P. E. M., & Ryde, U. (2015). Simulation of the isotropic EXAFS spectra for the S-2 and S-3 structures of the oxygen evolving complex in photosystem II. *Proceedings of the National Academy of Sciences of the United States of America*, *112*(13), 3979–3984. http://dx.doi.org/10.1073/pnas.1422058112.

Lin, H., & Truhlar, D. G. (2005). Redistributed charge and dipole schemes for combined quantum mechanical and molecular mechanical calculations. *Journal of Physical Chemistry A*, *109*(17), 3991–4004.

Lobaugh, J., & Voth, G. A. (1994). A path-integral study of electronic polarization and nonlinear coupling effects in condensed-phase proton-transfer reactions. *Journal of Chemical Physics*, *100*(4), 3039–3047.

Lobaugh, J., & Voth, G. A. (1996). The quantum dynamics of an excess proton in water. *Journal of Chemical Physics*, *104*(5), 2056–2069.

Luber, S., Rivalta, I., Umena, Y., Kawakami, K., Shen, J.-R., Kamiya, N., … Batista, V. S. (2011a). S_1-state model of the O_2-evolving complex of photosystem II. *Biochemistry*, *50*(29), 6308–6311. http://dx.doi.org/10.1021/bi200681q.

Luzhkov, V. B., & Aqvist, J. (2000). A computational study of ion binding and protonation states in the KcsA potassium channel. *Biochimica et Biophysica Acta, Protein Structure and Molecular Enzymology*, *1481*(2), 360–370.

Maple, J. R., Cao, Y. X., Damm, W. G., Halgren, T. A., Kaminski, G. A., Zhang, L. Y., & Friesner, R. A. (2005). A polarizable force field and continuum solvation methodology for modeling of protein-ligand interactions. *Journal of Chemical Theory and Computation*, *1*(4), 694–715.

Maseras, F., & Morokuma, K. (1995). Imomm—A new integrated *ab-initio* plus molecular mechanics geometry optimization scheme of equilibrium structures and transition-states. *Journal of Computational Chemistry*, *16*(9), 1170–1179.

Mayhall, N. J., & Raghavachari, K. (2011). Molecules-in-molecules: An extrapolated fragment-based approach for accurate calculations on large molecules and materials. *Journal of Chemical Theory and Computation*, *7*(5), 1336–1343. http://dx.doi.org/10.1021/ct200033b.

Mayhall, N. J., & Raghavachari, K. (2012). Many-overlapping-body (MOB) expansion: A generalized many body expansion for nondisjoint monomers in molecular fragmentation calculations of covalent molecules. *Journal of Chemical Theory and Computation*, *8*(8), 2669–2675. http://dx.doi.org/10.1021/ct300366e.

Mei, Y., Zhang, D. W., & Zhang, J. Z. H. (2005). New method for direct linear-scaling calculation of electron density of proteins. *Journal of Physical Chemistry A*, *109*(1), 2–5.

Menikarachchi, L. C., & Gascon, J. A. (2008). Optimization of cutting schemes for the evaluation of molecular electrostatic potentials in proteins via moving-domain QM/MM. *Journal of Molecular Modeling, 14*(6), 479–487.

Mulgrew-Nesbitt, A., Diraviyam, K., Wang, J. Y., Singh, S., Murray, P., Li, Z. H., ... Murray, D. (2006). The role of electrostatics in protein-membrane interactions. *Biochimica et Biophysica Acta, Molecular and Cell Biology of Lipids, 1761*(8), 812–826.

Murphy, R. B., Philipp, D. M., & Friesner, R. A. (2000a). Frozen orbital QM/MM methods for density functional theory. *Chemical Physics Letters, 321*(1–2), 113–120.

Murphy, R. B., Philipp, D. M., & Friesner, R. A. (2000b). A mixed quantum mechanics/molecular mechanics (QM/MM) method for large-scale modeling of chemistry in protein environments. *Journal of Computational Chemistry, 21*(16), 1442–1457.

Nakano, T., Kaminuma, T., Sato, T., Akiyama, Y., Uebayasi, M., & Kitaura, K. (2000). Fragment molecular orbital method: Application to polypeptides. *Chemical Physics Letters, 318*(6), 614–618.

Nakano, T., Kaminuma, T., Sato, T., Fukuzawa, K., Akiyama, Y., Uebayasi, M., & Kitaura, K. (2002). Fragment molecular orbital method: Use of approximate electrostatic potential. *Chemical Physics Letters, 351*(5–6), 475–480.

New, M. H., & Berne, B. J. (1995). Molecular-dynamics calculation of the effect of solvent polarizability on the hydrophobic interaction. *Journal of the American Chemical Society, 117*(27), 7172–7179.

Newville, M. (2001). IFEFFIT: Interactive XAFS analysis and FEFF fitting. *Journal of Synchrotron Radiation, 8*, 322–324. http://dx.doi.org/10.1107/S0909049500016964.

Noodleman, L. (1981). Valence bond description of anti-ferromagnetic coupling in transition-metal dimers. *The Journal of Chemical Physics, 74*, 5737–5743.

Noodleman, L., & Case, D. A. (1992). Density functional theory of spin polarization and spin coupling in iron-sulfur clusters. *Advances in Inorganic Chemistry, 38*, 423–470.

Noodleman, L., & Davidson, E. R. (1986). Ligand spin polarization and antiferromagnetic coupling in transition-metal dimers. *Chemical Physics, 109*, 131–143.

Noodleman, L., Peng, C. Y., Case, D. A., & Mouesca, J. M. (1995). Orbital interactions, electron delocalization and spin coupling in iron-sulfur clusters. *Coordination Chemistry Reviews, 144*, 199–244.

Norel, R., Sheinerman, F., Petrey, D., & Honig, B. (2001). Electrostatic contributions to protein-protein interactions: Fast energetic filters for docking and their physical basis. *Protein Science, 10*(11), 2147–2161.

Okamura, M. Y., & Feher, G. (1992). Proton-transfer in reaction centers from photosynthetic bacteria. *Annual Review of Biochemistry, 61*, 861–896.

Pal, R., Negre, C. F. A., Vogt, L., Pokhrel, R., Ertem, M. Z., Brudvig, G. W., & Batista, V. S. (2013). S_0-state model of the oxygen-evolving complex of photosystem II. *Biochemistry, 52*(44), 7703–7706. http://dx.doi.org/10.1021/bi401214v.

Palmo, K., Mannfors, B., Mirkin, N. G., & Krimm, S. (2003). Potential energy functions: From consistent force fields to spectroscopically determined polarizable force fields. *Biopolymers, 68*(3), 383–394.

Pantazis, D. A., Ames, W., Cox, N., Lubitz, W., & Neese, F. (2012). Two interconvertible structures that explain the spectroscopic properties of the oxygen-evolving complex of photosystem II in the S_2 state. *Angewandte Chemie, International Edition, 51*(39), 9935–9940. http://dx.doi.org/10.1002/anie.201204705.

Parr, R. G., & Yang, W. (1989). *In density-functional theory of atoms and molecules*. Oxford: Oxford University Press.

Philipp, D. M., & Friesner, R. A. (1999). Mixed ab initio QM/MM modeling using frozen orbitals and tests with alanine dipeptide and tetrapeptide. *Journal of Computational Chemistry, 20*(14), 1468–1494.

Popovic, D. M., & Stuchebrukhov, A. A. (2004). Electrostatic study of the proton pumping mechanism in bovine heart cytochrome c oxidase. *Journal of the American Chemical Society, 126*(6), 1858–1871.

Rappe, A. K., & Goddard, W. A. (1991). Charge equilibration for molecular-dynamics simulations. *Journal of Physical Chemistry, 95*(8), 3358–3363.

Ren, P. Y., & Ponder, J. W. (2003). Polarizable atomic multipole water model for molecular mechanics simulation. *Journal of Physical Chemistry B, 107*(24), 5933–5947.

Reuter, N., Dejaegere, A., Maigret, B., & Karplus, M. (2000). Frontier bonds in QM/MM methods: A comparison of different approaches. *Journal of Physical Chemistry A, 104*(8), 1720–1735.

Rick, S. W., & Berne, B. J. (1996). Dynamical fluctuating charge force fields: The aqueous solvation of amides. *Journal of the American Chemical Society, 118*(3), 672–679.

Rick, S. W., Stuart, S. J., Bader, J. S., & Berne, B. J. (1995). Fluctuating charge force-fields for aqueous-solutions. *Journal of Molecular Liquids, 65–6*, 31–40.

Rick, S. W., Stuart, S. J., & Berne, B. J. (1994). Dynamical fluctuating charge force-fields—Application to liquid water. *Journal of Chemical Physics, 101*(7), 6141–6156.

Saebo, S., & Pulay, P. (1993). Local treatment of electron correlation. *Annual Review of Physical Chemistry, 44*, 213–236.

Saha, A., & Raghavachari, K. (2015). Analysis of different fragmentation strategies on a variety of large peptides: Implementation of a low level of theory in fragment-based methods can be a crucial factor. *Journal of Chemical Theory and Computation, 11*(5), 2012–2023. http://dx.doi.org/10.1021/ct501045s.

Sandberg, D. J., Rudnitskaya, A. N., & Gascón, J. A. (2012). QM/MM prediction of the stark shift in the active site of a protein. *Journal of Chemical Theory and Computation, 8*(8), 2817–2823.

Sayers, D. E., Stern, E. A., & Lytle, F. W. (1971). New technique for investigating non-crystalline structures—Fourier analysis of extended X-ray—Absorption fine structure. *Physical Review Letters, 27*(18), 1204–1207.

Schlegel, H. B. (1987a). Optimization of equilibrium geometries and transition structures. *Advances in Chemical Physics, 67*, 249–285.

Schmidt, M. W., Baldridge, K. K., Boatz, J. A., Elbert, S. T., Gordon, M. S., Jensen, J. H., ... Montgomery, J. A. (1993). General atomic and molecular electronic-structure system. *Journal of Computational Chemistry, 14*(11), 1347–1363.

Schwegler, E., & Challacombe, M. (1996). Linear scaling computation of the Hartree-Fock exchange matrix. *Journal of Chemical Physics, 105*(7), 2726–2734.

Schworer, M., Breitenfeld, B., Troster, P., Bauer, S., Lorenzen, K., Tavan, P., & Mathias, G. (2013). Coupling density functional theory to polarizable force fields for efficient and accurate Hamiltonian molecular dynamics simulations. *Journal of Chemical Physics. 138*(24). http://dx.doi.org/10.1063/1.4811292. Artn 244103.

Senn, H. M., & Thiel, W. (2009). QM/MM methods for biomolecular systems. *Angewandte Chemie, International Edition, 48*(7), 1198–1229.

Sham, Y. Y., Muegge, I., & Warshel, A. (1998). Effect of protein relaxation on charge-charge interaction and dielectric constants in protein. *Biophysical Journal, 74*(4), 1744–1753.

Sham, Y. Y., Muegge, I., & Warshel, A. (1999). Simulating proton translocations in proteins: Probing proton transfer pathways in the Rhodobacter sphaeroides reaction center. *Proteins: Structure, Function, and Genetics, 36*(4), 484–500.

Shao, Y. H., Gan, Z. T., Epifanovsky, E., Gilbert, A. T. B., Wormit, M., Kussmann, J., ... Head-Gordon, M. (2015). Advances in molecular quantum chemistry contained in the Q-Chem 4 program package. *Molecular Physics, 113*(2), 184–215. http://dx.doi.org/10.1080/00268976.2014.952696.

Sheinerman, F. B., & Honig, B. (2002). On the role of electrostatic interactions in the design of protein-protein interfaces. *Journal of Molecular Biology, 318*(1), 161–177.

Sheinerman, F. B., Norel, R., & Honig, B. (2000). Electrostatic aspects of protein-protein interactions. *Current Opinion in Structural Biology, 10*(2), 153–159.

Shoji, M., Isobe, H., & Yamaguchi, K. (2015a). QM/MM study of the S-2 to S-3 transition reaction in the oxygen-evolving complex of photosystem II. *Chemical Physics Letters, 636*, 172–179. http://dx.doi.org/10.1016/j.cplett.2015.07.039.

Shoji, M., Isobe, H., Yamanaka, S., Umena, Y., Kawakami, K., Kamiya, N., … Yamaguchi, K. (2015b). Theoretical modelling of biomolecular systems I. Large-scale QM/MM calculations of hydrogen-bonding networks of the oxygen evolving complex of photosystem II. *Molecular Physics, 113*(3–4), 359–384. http://dx.doi.org/10.1080/00268976.2014.960021.

Simonson, T., Archontis, G., & Karplus, M. (1997). Continuum treatment of long-range interactions in free energy calculations. Application to protein-ligand binding. *Journal of Physical Chemistry B, 101*(41), 8349–8362.

Simonson, T., Archontis, G., & Karplus, M. (1999). A Poisson-Boltzmann study of charge insertion in an enzyme active site: The effect of dielectric relaxation. *Journal of Physical Chemistry B, 103*(29), 6142–6156.

Singh, U. C., & Kollman, P. A. (1986). A combined abinitio quantum-mechanical and molecular mechanical method for carrying out simulations on complex molecular systems applications to the $Ch_3Cl + Cl^-$ exchange reaction and gas phase protonation of polyethers. *Journal of Computational Chemistry, 7*(6), 718–730.

Song, J., Ji, C., & Zhang, J. Z. H. (2013). The critical effect of polarization on the dynamical structure of guanine quadruplex DNA. *Physical Chemistry Chemical Physics, 15*(11), 3846–3854.

Sprik, M., & Klein, M. L. (1988). A polarizable model for water using distributed charge sites. *Journal of Chemical Physics, 89*(12), 7556–7560.

Sproviero, E. M., Gascon, J. A., McEvoy, J. P., Brudvig, G. W., & Batista, V. S. (2006a). Characterization of synthetic oxomanganese complexes and the inorganic core of the O_2-evolving complex in photosystem—II: Evaluation of the DFT/B3LYP level of theory. *Journal of Inorganic Biochemistry, 100*(4), 786–800.

Sproviero, E. M., Gascon, J. A., McEvoy, J. P., Brudvig, G. W., & Batista, V. S. (2006b). QM/MM models of the O_2-evolving complex of photosystem II. *Journal of Chemical Theory and Computation, 2*(4), 1119–1134.

Sproviero, E. M., Gascon, J. A., McEvoy, J. P., Brudvig, G. W., & Batista, V. S. (2007). Structural models of the oxygen-evolving complex of photosystem II. *Current Opinion in Structural Biology, 17*, 173–180.

Sproviero, E. M., Gascon, J. A., McEvoy, J. P., Brudvig, G. W., & Batista, V. S. (2008a). Computational insights into the O_2-evolving complex of photosystem II. *Photosynthesis Research, 97*, 91–114.

Sproviero, E. M., Gascon, J. A., McEvoy, J. P., Brudvig, G. W., & Batista, V. S. (2008b). Computational studies of the O2-evolving complex of photosystem II and biomimetic oxomanganese complexes. *Coordination Chemistry Reviews, 252*, 395–415.

Sproviero, E. M., Gascon, J. A., McEvoy, J. P., Brudvig, G. W., & Batista, V. S. (2008c). A model of the oxygen evolving center of photosystem II predicted by structural refinement based on EXAFS simulations. *Journal of the American Chemical Society, 130*, 6728–6730.

Sproviero, E. M., Gascon, J. A., McEvoy, J. P., Brudvig, G. W., & Batista, V. S. (2008d). QM/MM study of the catalytic cycle of water splitting in photosystem II. *Journal of the American Chemical Society, 130*, 3428–3442.

Stern, E. A. (1974). Theory of extended X-ray-absorption fine-structure. *Physical Review B: Condensed Matter, 10*(8), 3027–3037.

Stern, H. A., Kaminski, G. A., Banks, J. L., Zhou, R. H., Berne, B. J., & Friesner, R. A. (1999). Fluctuating charge, polarizable dipole, and combined models: Parameterization from ab initio quantum chemistry. *Journal of Physical Chemistry B, 103*(22), 4730–4737.

Stern, H. A., Rittner, F., Berne, B. J., & Friesner, R. A. (2001). Combined fluctuating charge and polarizable dipole models: Application to a five-site water potential function. *Journal of Chemical Physics, 115*(5), 2237–2251.

Storer, J. W., Giesen, D. J., Cramer, C. J., & Truhlar, D. G. (1995). Class IV charge models—A new semiempirical approach in quantum-chemistry. *Journal of Computer-Aided Molecular Design, 9*(1), 87–110.

Strain, M. C., Scuseria, G. E., & Frisch, M. J. (1996). Achieving linear scaling for the electronic quantum coulomb problem. *Science, 271*(5245), 51–53.

Stratmann, R. E., Burant, J. C., Scuseria, G. E., & Frisch, M. J. (1997). Improving harmonic vibrational frequencies calculations in density functional theory. *Journal of Chemical Physics, 106*(24), 10175–10183.

Stratmann, R. E., Scuseria, G. E., & Frisch, M. J. (1996). Achieving linear scaling in exchange-correlation density functional quadratures. *Chemical Physics Letters, 257*(3–4), 213–223.

Strout, D. L., & Scuseria, G. E. (1995). A quantitative study of the scaling properties of the Hartree-Fock method. *Journal of Chemical Physics, 102*(21), 8448–8452.

Stuart, S. J., & Berne, B. J. (1996). Effects of polarizability on the hydration of the chloride ion. *Journal of Physical Chemistry, 100*(29), 11934–11943.

Subotnik, J. E., & Head-Gordon, M. (2005). A localized basis that allows fast and accurate second-order Moller-Plesset calculations. *The Journal of Chemical Physics, 122*(3), 034109.

Svensson, M., Humbel, S., Froese, R. D. J., Matsubara, T., Sieber, S., & Morokuma, K. (1996). ONIOM: A multilayered integrated MO + MM method for geometry optimizations and single point energy predictions. A test for Diels-Alder reactions and $Pt(P(t-Bu)_3)_2 + H_2$ oxidative addition. *Journal of Physical Chemistry, 100*(50), 19357–19363.

Szekeres, Z., Exner, T., & Mezey, P. G. (2005). Fuzzy fragment selection strategies, basis set dependence and HF-DFT comparisons in the applications of the ADMA method of macromolecular quantum chemistry. *International Journal of Quantum Chemistry, 104*(6), 847–860.

Thompson, M. A. (1996). QM/MMpol: A consistent model for solute/solvent polarization. Application to the aqueous solvation and spectroscopy of formaldehyde, acetaldehyde, and acetone. *Journal of Physical Chemistry, 100*(34), 14492–14507.

Thompson, M. A., & Schenter, G. K. (1995). Excited-states of the bacteriochlorophyll-B dimer of Rhodopseudomonas-viridis—A QM/MM study of the photosynthetic reaction-center that includes MM polarization. *Journal of Physical Chemistry, 99*(17), 6374–6386.

Thompson, M. J., Schweizer, K. S., & Chandler, D. (1982). Quantum-theory of polarization in liquids—Exact solution of the mean spherical and related approximations. *Journal of Chemical Physics, 76*(2), 1128–1135.

Umena, Y., Kawakami, K., Shen, J.-R., & Kamiya, N. (2011). Crystal structure of oxygen-evolving photosystem II at a resolution of 1.9 angstrom. *Nature, 473*(7345), 55–60.

van der Kamp, M. W., & Mulholland, A. J. (2013). Combined quantum mechanics/molecular mechanics (QM/MM) methods in computational enzymology. *Biochemistry, 52*(16), 2708–2728.

Vanommeslaeghe, K., & MacKerell, A. D. (2015). CHARMM additive and polarizable force fields for biophysics and computer-aided drug design. *Biochimica et Biophysica Acta-General Subjects, 1850*(5), 861–871. http://dx.doi.org/10.1016/j.bbagen.2014.08.004.

Vasilyev, V., & Bliznyuk, A. (2004). Application of semiempirical quantum chemical methods as a scoring function in docking. *Theoretical Chemistry Accounts, 112*(4), 313–317.

Versluis, L., & Ziegler, T. (1988). The determination of molecular-structures by density functional theory—The evaluation of analytical energy gradients by numerical-integration. *The Journal of Chemical Physics, 88*, 322–328.

Vreven, T., Byun, K. S., Komaromi, I., Dapprich, S., Montgomery, J. A., Morokuma, K., & Frisch, M. J. (2006). Combining quantum mechanics methods with molecular mechanics methods in ONIOM. *Journal of Chemical Theory and Computation, 2*(3), 815–826.

Vreven, T., Mennucci, B., da Silva, C. O., Morokuma, K., & Tomasi, J. (2001). The ONIOM-PCM method: Combining the hybrid molecular orbital method and the polarizable continuum model for solvation. Application to the geometry and properties of a merocyanine in solution. *The Journal of Chemical Physics, 115*(1), 62–72.

Vreven, T., & Morokuma, K. (2000a). On the application of the IMOMO (integrated molecular orbital plus molecular orbital) method. *Journal of Computational Chemistry, 21*(16), 1419–1432.

Vreven, T., & Morokuma, K. (2000b). The ONIOM (our own N-layered integrated molecular orbital plus molecular mechanics) method for the first singlet excited (S_1) state photoisomerization path of a retinal protonated Schiff base. *The Journal of Chemical Physics, 113*(8), 2969–2975.

Vreven, T., & Morokuma, K. (2003). Investigation of the S_0 -> S_1 excitation in bacteriorhodopsin with the ONIOM(MO: MM) hybrid method. *Theoretical Chemistry Accounts, 109*(3), 125–132.

Vreven, T., Morokuma, K., Farkas, O., Schlegel, H. B., & Frisch, M. J. (2003). Geometry optimization with QM/MM, ONIOM, and other combined methods. I. Microiterations and constraints. *Journal of Computational Chemistry, 24*(6), 760–769.

Wang, W., Donini, O., Reyes, C. M., & Kollman, P. A. (2001). Biomolecular simulations: Recent developments in force fields, simulations of enzyme catalysis, protein-ligand, protein-protein, and protein-nucleic acid noncovalent interactions. *Annual Review of Biophysics and Biomolecular Structure, 30*, 211–243.

Warshel, A. (1976). Bicycle-pedal model for the first step in the vision process. *Nature (London), 260*, 679–683.

Warshel, A. (1981). Electrostatic basis of structure-function correlation in proteins. *Accounts of Chemical Research, 14*(9), 284–290.

Warshel, A., & Aqvist, J. (1991). Electrostatic energy and macromolecular function. *Annual Review of Biophysics and Biophysical Chemistry, 20*, 267–298.

Warshel, A., & Levitt, M. (1976). Theoretical studies of enzymic reactions—Dielectric, electrostatic and steric stabilization of carbonium-ion in reaction of lysozyme. *Journal of Molecular Biology, 103*(2), 227–249.

White, C. A., & Headgordon, M. (1994). Derivation and efficient implementation of the fast multipole method. *Journal of Chemical Physics, 101*(8), 6593–6605.

White, C. A., Johnson, B. G., Gill, P. M. W., & HeadGordon, M. (1996). Linear scaling density functional calculations via the continuous fast multipole method. *Chemical Physics Letters, 253*(3–4), 268–278.

Xie, W., Orozco, M., Truhlar, D. G., & Gao, J. (2009). X-Pol potential: An electronic structure-based force field for molecular dynamics simulation of a solvated protein in water. *Journal of Chemical Theory and Computation, 5*(3), 459–467. http://dx.doi.org/10.1021/ct800239q.

Yang, W. T. (1991). Direct calculation of electron-density in density-functional theory. *Physical Review Letters, 66*(11), 1438–1441.

Yang, W. T., & Lee, T. S. (1995). A density-matrix divide-and-conquer approach for electronic-structure calculations of large molecules. *Journal of Chemical Physics, 103*(13), 5674–5678.

Yano, J., Kern, J., Sauer, K., Latimer, M. J., Pushkar, Y., Biesiadka, J., ... Yachandra, V. K. (2006). Where water is oxidized to dioxygen: Structure of the photosynthetic Mn_4Ca cluster. *Science, 314*(5800), 821–825. http://dx.doi.org/10.1126/science.1128186.

Yano, J., Pushkar, Y., Glatzel, P., Lewis, A., Sauer, K., Messinger, J., ... Yachandra, V. (2005). High-resolution Mn EXAFS of the oxygen-evolving complex in photosystem II: Structural implications for the Mn_4Ca cluster. *Journal of the American Chemical Society, 127*(43), 14974–14975.

York, D. M. (1995). A generalized formulation of electronegativity equalization from density functional theory. *International Journal of Quantum Chemistry: Quantum Chemistry Symposium, 29*, 385–394.

York, D. M., & Yang, W. T. (1996). A chemical potential equalization method for molecular simulations. *Journal of Chemical Physics, 104*(1), 159–172.

Yu, H. B., Hansson, T., & van Gunsteren, W. F. (2003). Development of a simple, self-consistent polarizable model for liquid water. *Journal of Chemical Physics, 118*(1), 221–234.

Zhang, D. W., Chen, X. H., & Zhang, J. Z. H. (2003). Molecular caps for full quantum mechanical computation of peptide-water interaction energy. *Journal of Computational Chemistry, 24*(15), 1846–1852.

Zhang, Y. K., Lee, T. S., & Yang, W. T. (1999). A pseudobond approach to combining quantum mechanical and molecular mechanical methods. *Journal of Chemical Physics, 110*(1), 46–54.

Zhang, D. W., Xiang, Y., & Zhang, J. Z. H. (2003). New advance in computational chemistry: Full quantum mechanical ab initio computation of streptavidin-biotin interaction energy. *Journal of Physical Chemistry B, 107*(44), 12039–12041.

Zhang, D. W., & Zhang, J. Z. H. (2003). Molecular fractionation with conjugate caps for full quantum mechanical calculation of protein-molecule interaction energy. *Journal of Chemical Physics, 119*(7), 3599–3605.

AUTHOR INDEX

Note: Page numbers followed by "*f*" indicate figures, "*t*" indicate tables, and "*s*" indicate schemes.

A

Abascal, J.L.F., 265
Abel, R., 426–427
Abeysinghe, T., 300–308
Acevedo, O., 92–93, 426–427
Acharya, C., 89–90
Adams, J.B., 201
Adams, P.D., 253–254
Adamson, R.D., 396–397
Agafonov, R.V., 214
Agarwal, A., 435
Agarwal, P.K., 271–273, 307–308, 379
Agirre, J., 167–169
Agmon, N., 344
Agrawal, N., 296, 298, 300–302, 311
Ahlrichs, R., 124
Ahn, N.G., 379
Aitchison, E.W., 71
Ajloo, D., 420–421, 423*t*, 426–427, 435
Åkerfeldt, K.S., 129
Akiyama, Y., 452
Alagona, G., 33–34
Alanen, H.I., 452
Albaret, T., 201, 205–207
Alber, T., 306–307
Alberts, I.L., 238–239
Alberty, R.A., 59–60
Alder, B.J., 271
Alexandrova, A.N., 320–338
Alexov, E., 256
Alhambra, C., 126, 193–194, 260–261, 266, 273–274, 348–349, 361, 363, 378–381, 452–453
Ali, M.E., 43–45
Alif, M., 422, 423*t*, 425–428, 435
Allemann, R.K., 272–273, 305–308, 379–381
Allen, J.F., 448–449, 464–465
Allen, M.P., 258–259
Allen, T.W., 452
Al-Shawi, M.K., 190

Altenberg, G.A., 189–190
Althoff, E.A., 254
Altun, A., 398–399
Alvarez, S., 228, 230, 233–234
Amara, P., 126, 193–194, 260–261, 266, 348–349, 452–453
Amaya, M.F., 162–163
Ames, G.F.-L., 189–190
Ames, W., 462
Amrein, B.A., 342
Andersson, K.K., 134
Andrade, R., 428–430
André, I., 129
Andrejic, M., 140–141, 142*f*
Andrews, L.D., 223–225, 233–234, 236–239, 241–242
Angyan, J.G., 452–453
Ankudinov, A.L., 460–462
Antes, I., 260–261, 266, 348–349, 452–453
Antonini, E., 45
Antoniou, D., 288, 295–296, 305–306, 310–311, 361
Antony, J., 124, 267, 394*f*, 407–408
Appleman, J.R., 296
Aqvist, J., 220, 254, 272–273, 452
Aragones, J.L., 265
Aranda, J., 67
Arantes, G.M., 199–200
Archontis, G., 32, 220–221, 390, 452
Ardèvol, A., 160–180
Arkin-Ojo, O., 201, 205–207
Armstrong, G.S., 420–421
Arning, J., 421
Arora, K., 305–306
Artacho, E., 391–392
Artali, R., 452
Asada, T., 109, 447–448
Ascenzi, P., 42, 45
Ashley, C.A., 461–462
Ashworth, J., 254
Askerka, M., 444–469

483

Assfeld, X., 126, 260–261, 266, 342–345, 348–349
Athokpam, B., 401
Atwell, S., 189–190
Austin, R.H., 41–42
Autschbach, J., 396–398
Ayers, P.W., 92–93
Aznar, C.P., 458
Azuri, A., 272–273

B

Babu, K., 447–448
Bacic, Z., 392
Bacskay, G.B., 393
Bader, J.S., 446–447
Bae, S.-H., 307–308
Baer, R., 396–398
Baichwal, V., 189–190
Baiya, S., 175
Baker, D., 214, 223–225
Baker, N., 256
Bakowies, D., 444
Bala, P., 34
Baldridge, K.K., 445–446
Balduz, J.L., 396
Baldwin, M.J., 447–448
Ball, S.E., 20–21
Ballone, P., 43–45, 420, 425
Bandaria, J.N., 223–225
Bandyopadhyay, P., 445–446
Banerjee, R., 135, 140
Banks, J.L., 446–447
Baptista, A.M., 256
Barber, J., 458
Barducci, A., 63–64, 160, 220–221, 430–431
Barriocanal, J.A., 131
Bartels, C., 63, 269–270
Bartlett, P.A., 223–225
Bash, P.A., 34, 58–59, 107–108, 218–219, 260–261, 265–266, 342, 363, 444, 452–453
Bashford, D., 45, 124, 193–194, 216–217, 229
Basu, P., 140
Basyaruddin, M., 422, 423t, 425–428, 435
Bathelt, C.M., 20–21
Batista, E.R., 444–469, 451f, 454f

Batista, V.S., 134–135, 141, 444–469, 451f, 454f, 460f, 463–464f
Bauer, S., 446–447
Bauman, N.P., 140
Bayly, C.I., 264–265, 449–452, 458–459
Beall, E., 189–190
Beauchamp, K.A., 221, 264–265
Becke, A.D., 123, 217–218, 226, 394f, 395–397, 407–408
Becker, O.M., 34, 45
Beer, M., 128–129, 400
Beeson, K.W., 41–42
Begley, T.P., 296
Beglov, D., 258–259
Behiry, E.M., 305–306, 379–381
Bekhouche, M., 421
Bekker, C., 401
Bell, A.T., 92–93, 435
Bellott, M., 45, 193–194, 216–217, 229
Benabdelhak, H., 190–191
Benedetti, P., 296
Benedetto, A., 420, 425
Benighaus, T., 218–219, 222, 230, 258–259
Benito-Lopez, F., 420–421
Benkovic, S.J., 214, 221, 223–225, 288, 292, 305–308, 310, 379–381, 390
Bennett, C.H., 76–78, 86, 130–131
Benoit, M., 369, 392
Benschoten, A.H.V., 253–254
Benson, R.W., 23
Bentzien, J., 81, 84–85
Berendsen, H.J.C., 33–34, 143, 265, 430–431
Berg, B.A., 58–59, 63
Bergner, A., 452–453
Berlow, R.B., 306–307
Berman, H.M., 322–323
Bernard, S., 201, 207
Bernard, T., 162–163
Bernardo, D.N., 446–447
Bernasconi, C.F., 375
Berne, B.J., 273, 369–370, 390–392, 401–404, 446–447, 452
Berneche, S., 218–219, 229
Bernstein, N., 343–344
Berryman, J.T., 428–430
Bertran, J., 342
Bertrn, J., 400

Besler, B.H., 453–454
Best, R.B., 264–265, 436
Betker, J., 214
Bettens, R.P.A., 447–448
Betz, R.M., 428–430
Bevilacqua, P.C., 67
Bhabha, G., 379
Bhat, T.N., 322–323
Bhattacharyya, K., 420–421, 423t, 425–427, 435
Biarnés, X., 160–161, 163, 169–172, 176–178
Biesiadka, J., 462
Billeter, S.R., 271–273
Billings, E.M., 265, 452–453
Bilton, P., 186–187
Bingaman, J.L., 67
Birgin, E.G., 428–430
Bjelic, S., 254
Blaha-Nelson, D., 342
Blanch, H.W., 421
Blanchet, C., 190–191
Blank, I.D., 266–267
Blight, M.A., 191
Bliznyuk, A.A., 363, 452
Blomberg, L.M., 43
Blomberg, M.R.A., 43–45, 120, 140, 217–218, 220
Blomberg, R., 214
Blum, L.J., 421
Blumberger, J., 145–146
Boatz, J.A., 445–446
Bobyr, E., 232–233
Böckmann, M., 343–344
Bocola, M., 421
Bodnarchuk, M.S., 254
Boehr, D.D., 221, 223–225, 253–254, 307–308, 379
Bolhuis, P.G., 166–167, 221, 270–271
Bolognesi, M., 42, 45
Bonomi, M., 160, 220–221, 430–431
Boone, A., 218–219, 259
Borden, W.T., 295
Bordwell, F.G., 375
Boresch, S., 76–98, 131–132
Borgis, D., 447
Borodkin, V., 169, 172, 176
Borowski, T., 120, 217–218, 220, 390

Borst, P., 186–187
Boulanger, E., 222, 258–259, 264–265
Bouldin, C.E., 460–462
Bourassa, J.L., 43
Bouzida, D., 2–3, 116, 205, 269–270
Bowers, K.E., 4–11
Bowler, D.R., 92–93, 391–392
Bowman, A.L., 76–77
Bowman, G.R., 221
Boxer, G., 221
Boxer, S.G., 360, 393, 408–409
Boyd, D.B., 218–219, 260, 360–361, 363
Boyle, W.J., 375
Brändas, E., 342
Branduardi, D., 430–431
Brauman, J., 45–47
Braun, R., 349–350
Braun-Sand, S., 196–197
Breeze, M., 426–427
Breitenfeld, B., 446–447
Britt, R.D., 458
Brodie, B.B., 20–21
Bromberg, A., 252–253, 260
Brooks, B.R., 33–34, 38–39, 45, 49–50, 66, 84–87, 89–90, 92–93, 98, 193–194, 203, 229–230, 232–233, 342, 452–453
Brooks, C.L., 38–39, 49–50, 66, 89–90, 98, 193–194, 203, 229, 258–259, 264–265, 288, 305–308
Brooks, P.R., 31–32
Brouillette, C.G., 189–190
Brown, S.T., 89–90, 98, 106–109, 203
Broyde, S., 106–107
Bruccoleri, R.E., 33–34, 45, 203, 230, 452–453
Bruckner, S., 77–78, 86
Brudvig, G.W., 444, 448–450, 458–459, 460f, 462–465, 463–464f
Brunger, A.T., 225, 238f, 240, 254, 258–259
Brunk, E., 215, 218
Brunori, M., 45
Brüschweiler, R., 58–59, 63
Bruskin, E.J., 403–404
Buchete, N.-V., 436
Buettner, A., 254–256
Buldyrev, S.V., 321
Bulo, R.E., 343–344
Burant, J.C., 447–448

Burger, S.K., 80–81, 92–93, 232–233
Burke, K., 123, 395–396
Burney, P.R., 422–428, 423t, 434–437
Burnham, C.J., 201, 446–447
Burton, N.A., 126–127
Buschiazzo, A., 162–163
Bussi, G., 63–64, 403, 430–431
Bylaska, E.J., 80–81
Bystroff, C., 305–306
Byun, K.S., 126–127, 364–365, 444, 449–450, 452–453, 457

C

Caaveiro, J.M.M., 407
Cabedo Martinez, A.I., 84–85
Caines, M.E., 163–165
Caldararu, O., 141, 142f
Caldwell, J.W., 430–431
Camilloni, C., 430–431
Cammi, R., 124, 226–228
Campbell, K.A., 458
Campomanes, P., 201
Campopiano, D.J., 186–187
Cannon, B.R., 256
Canongia Lopes, J.N., 426–427
Cantarel, B.L., 162–163
Cao, J., 361–362, 366–367, 390–391, 403–404, 409, 446–447
Cao, L., 58–71
Cao, Y.X., 446–447
Cao, Z., 106–107, 342
Car, R., 197–198, 363, 390–391
Caratzoulas, S., 310–311
Carbone, I., 63–64
Carey, F.A., 23–24
Carfi, A., 325
Carosati, E., 296
Carpenter, B.K., 18–20
Carpenter, E.P., 186–187
Carreras, C.W., 296
Carroll, D.L., 203
Carroll, K.S., 214
Carroll, M.J., 379
Carter, E.A., 63
Carter, J.L.L., 421
Case, D.A., 33–34, 90, 106–109, 111–112, 124, 216–217, 264–265, 390, 399, 428–431, 459

Casey, P.J., 4–11
Castaneda, C.A., 256
Castro-Lopez, J., 169, 172, 176
Catlow, C.R.A., 128
Cavanagh, J., 134, 253–254, 256–257
Cave-Ayland, C., 84–85
Cembran, A., 289, 305–308, 370, 379, 381–382
Ceperley, D.M., 370
Ceriotti, M., 369, 392, 401, 403–404, 406, 409
Cerutti, D.S., 264–265, 428–430
Cha, Y., 375
Chai, J.-D., 396–397
Chakraborty, A., 92–93
Chakraborty, S., 143
Chakravorty, D.K., 7
Challacombe, M., 111, 447–448
Chan, G.K.-L., 217–218, 266
Chandler, D., 58, 221, 270–271, 362, 366–368, 446–447
Chandrasekhar, J., 45, 76–77, 80–81, 265, 408, 430–431
Chang, C.J., 409–410
Chang, M.C.Y., 409–410
Chatfield, D.C., 265, 452–453
Chaudret, R.F., 263
Cheatham, T.E.I., 90, 106–107, 111–112, 264–265, 399, 428–431
Cheatum, C.M., 223–225, 379–380
Cheeseman, J.R., 226, 349–350, 428–430, 449–450
Chen, B., 390–391
Chen, G., 134–135, 141
Chen, H.N., 20–21, 58–59
Chen, J., 143–145, 186–189, 223–225
Chen, M., 80–81, 86–87, 190
Chen, M.E., 58–61, 67–69, 71
Chen, N., 106–107, 421
Chen, O., 45–47
Chen, P.-Y., 369, 392
Chen, S.-L., 43–45, 140
Chen, W., 445–446
Chen, W.Q., 23–24
Chen, X.H., 452–453
Chen, Y.K., 360–361, 363
Chen, Z., 421
Cheng, C.-L., 34–35

Cheng, X.L., 58–71, 435
Chiarotti, G.L., 201, 207
Chipot, C., 77–79, 85–86
Cho, A.E., 452
Cho, H.M., 435
Cho, Y.J., 214
Chodera, J.D., 77, 216–217
Choi, K.Y., 407
Chong, L., 132–133
Choutko, A., 342
Chowdhury, S., 408
Christov, C., 342
Chu, J.-W., 435
Chu, Z.T., 140, 264–265, 271–273
Chuang, Y.-Y., 273–274
Chung, L.W., 320, 342, 449–450
Chung, S.H., 452
Ciccotti, G., 58–59, 62–63, 92–93, 195–197, 201
Cieplak, P., 264–265, 449–452, 458–459
Cioloboc, D., 141, 142*f*
Cisneros, G.A., 80–81
Claeyssens, F., 34, 76–77, 217–218, 222
Clair, J.L.S., 214
Clark, D.S., 421
Clark, M.A., 128, 399–400, 408
Clarkson, M.W., 306–307
Cleland, W.W., 223, 290–292, 301, 371–373, 390–391
Clemente, F.R., 254
Clementi, C., 268
Clore, G.M., 254
Cobb, N., 140
Cohen, A.J., 217–218, 396
Cohen, D., 445–446
Cohen, S., 20–21
Cole, D.J., 262–263
Collins, E.J., 379
Collins, M.A., 447–448
Collins, S.J., 126–127
Collman, J., 45–47
Colombi Ciacchi, L., 343–344
Colombo, M.C., 201, 205–206
Colonna, F., 396–397
Conde, M.M., 265
Conners, K., 189–190

Conrads, T., 140
Conradson, S.D., 460–462
Constantin, L.A., 395
Contreras-Garcia, J., 396
Cook, P.F., 290–292, 301
Cooper, A.M., 130
Cooper, J., 253, 256–257
Corchado, J.C., 34, 273–274, 361, 363, 378–381
Corcoran, G.B., 20–21
Corminboeuf, C., 395–396
Cornell, W.D., 264–265, 449–452, 458–459
Corongiu, G., 446–447
Cossi, M., 124
Coulson, C.A., 41
Coutinho, P.M., 162–163
Cowtan, K.D., 167–169
Cox, N., 462
Cox, S.R., 109–110
Craig, I.R., 390–391, 409
Cramer, C.J., 122–123, 447
Cramer, W.A., 447–448
Crawford, T.D., 90
Crehuet, R., 67
Cremer, P.S., 396–397
Criddle, M.P., 84–85
Crooks, G.E., 84–85
Crowley, M.F., 108–109, 166–167
Cruywagen, J.J., 225, 233–234, 236–238
Csajka, F.S., 270–271
Csányi, G., 201, 205–207, 343–344
Csonka, G.I., 395
Cui, G.L., 5–11, 14
Cui, Q., 32, 34, 76–77, 86–87, 91–92, 127, 199–200, 214–242, 258–259, 263, 364–365, 391–392, 398–399
Cui, Q.A., 364–365
Cui, Q.J., 126–127
Cukier, R.I., 409–410
Cummins, P.L., 363
Cundari, T.R., 273–274
Cuny, J., 401, 403
Curchod, B.F.E., 396–398
Curto, V.F., 420–421
Cvitkovic, M.W., 447–448
Czekster, C.M., 305–306
Czerminski, R., 92–93

D

da Silva, C.O., 444–445, 452–453
Dabin, J., 165, 169–171
Dabirmanesh, B., 420–421
D'Agostino, C., 435
Dama, J.F., 63–64, 66–67, 174
Damager, I., 162–163
Dametto, M., 288, 305–306, 310–311
Damm, W.G., 258–259, 264–265, 446–447
Damrauer, N.H., 409–410
Dana, C.M., 421
Danenberg, P.V., 296
Daneshjou, S., 420–421
Daniels, A.D., 447–448
Daniels, C.R., 169, 426–427
Danielsson, J., 35–36, 38, 42
Danielsson, U., 41
Dapprich, S., 126–127, 444, 449–450, 452–453, 457
Darden, T.A., 90, 106–107, 109–112, 124–125, 216–219, 258–259, 264–265, 399, 428–431
Dargan, P.I., 20–21
Darve, E., 63
Das, A.K., 31–51
Das, D., 265, 452–453
Das, R., 264–265
Das, S., 252–276
Dasgupta, S., 32–33
Dassa, E., 186–189
Dau, H., 452, 462
Davidson, A.L., 186–190, 223–225
Davidson, D.G., 20–21
Davidson, E.R., 459
Davies, G.J., 160–165, 167–169
Davis, A.M., 256–257
Davis, C., 343–345, 348, 352–353, 353f
Davis, D.C., 20–21
Davis, K.M., 462
Dawson, R.J.P., 186–187
Dawson, W.M., 263, 288, 307–308
Day, P.N., 445–446
Dayal, P., 34–35
de Groot, B.L., 259
de Groot, M.J., 20–21
de Ruiter, A., 76–77
De Vita, A., 201, 205–207, 343–344
De Vivo, M., 160, 166–167
de Vries, A.H., 126–128
DeChancie, J., 214
Degnan, S.C., 306–307
Dejaegere, A., 132–133, 266, 452–453
Delcey, M.G., 120–122, 147–149
Deleage, G., 190–191
Dellago, C., 84, 166–167, 221, 270–271
Demma, M., 10–11
Deneroff, M.M., 436–437
Deng, H., 233–234
Deng, Y.Q., 220–221
Dennis, A.T., 452
Deumens, E., 109
Devries, A.H., 444–445
Dewar, M.J.S., 124, 193–194, 217–218, 364–365
Dewilde, S., 42
Dey, B.K., 92–93
Di Pietro, A., 190–191
Diamond, D., 420–421
Dibble, D., 421
Dickson, B.M., 58–59
Dicus, M.M., 458
Dijkman, J.P., 444–445
DiLabio, G.A., 395–396
Dinev, Z., 165, 169–171
Ding, F., 321, 337
Ding, Y.B., 446–447
DiNola, A., 430–431
Dinur, U., 33–34
Diraviyam, K., 452
Dittmer, J., 452, 462
Dixit, M., 252–276
Dixon, S.L., 447–448
Doemer, M., 201
Dokholyan, N.V., 321, 337
Dolenc, J., 342
Dolence, J.M., 4–5
Dolg, M., 452–453
Dölker, N., 135–136
Doltsinis, N.L., 343–344
Donadio, D., 430–431
Dong, G., 120–122, 121–122f, 140, 147–149
Doniach, S., 461–462
Donini, O., 452
Doren, D.J., 131
Doron, D., 217–218, 221, 257, 263–264, 268, 270–273

Doumèche, B., 421
Doxsee, K., 45–47
Doyle, L., 254
Drakenberg, T., 134
Dror, R.O., 2–3, 220–221, 264–265, 436–437
Drula, E., 162–163
Duan, Y., 408
Duarte, C.M., 254
Duarte, F., 342
Ducros, V.M., 161–162
Duggan, B.M., 307–308
Duke, R.E., 106–107, 111–112, 264–265, 399
Dunbrack, R.L., 45, 193–194, 216–217, 229
Dupuis, M., 80–81
Dupuis, R., 369, 392
Duster, A., 342–353
Dutta, S., 223–225
Dykstra, C.E., 451–452
Dyson, H.J., 221, 223–225, 253–254, 305–306, 379
Dzierlenga, M.W., 361

E

Eastham, W.N., 20–21
Eastwood, M.P., 220–221, 264–265
Effio, A., 112
Ehrlich, S., 124, 147, 394f, 407–408
Eichenberger, A.P., 342
Eichler, U., 444–445
Eisenberg, H., 396–397
Eisenstein, L., 41–42
Eitan, R., 252–276
Ekiert, D.C., 379
Elber, R., 58–59, 92–93
Elbert, S.T., 445–446
Eliasson, J., 128–129
Ellingson, B.A., 273–274
Ellison, F.O., 32–33
Elsner, J., 76–77, 217–218
Elstner, M.E., 32, 34, 76–77, 217–219, 221–222, 229, 343–344, 364–365, 391–392, 395–396
Elstner, T., 34
Emamzadeh, R., 420–421, 423t
Embry, A.C., 6–11

Emtage, S., 189–190
Engel, H., 263–264, 272–273
Ensing, B., 65, 67, 160, 166–167, 176, 343–344
Epifanovsky, E., 89–90, 98, 445–446
Ercolessi, F., 201
Eriksson, L.A., 128–129, 143
Erion, M.D., 86–87, 131
Erion, R., 306–307
Ermolaeva, M.D., 447–448
Ernzerhof, M., 123
Ertem, M.Z., 444, 458–459, 460f, 462–463, 463–464f
Essex, J.W., 84–85, 254
Estrin, D.A., 452–453
Eubanks, L.M., 4–5
Eurenius, K.P., 265, 452–453
Evangelista, F.A., 90
Evanseck, J.D., 45, 193–194
Evenseck, J.D., 216–217, 229
Evini, M., 420–421, 423t, 426–427, 435
Ewig, C.S., 33–34
Exner, M., 43
Exner, T.E., 256, 447–448, 452
Eyring, H., 32–33, 267–268

F

Fafarman, A.T., 407
Fainberg, A.H., 112
Fairbrother, W.J., 134, 253–254, 256–257
Fajer, M.I., 58–71
Fam, P.S., 452
Fan, K.-N., 193
Fan, L.Y., 456–457
Fan, Y., 289, 305–308, 370, 379, 381–382
Fang, D., 215, 220–222
Fang, T., 447–448, 452
Fang, W., 169, 172, 176, 401
Fanourgakis, G.S., 409
Farkas, O., 457
Farrokhnia, M., 120–122, 145–149
Fast, P.L., 273–274
Fedorov, D.G., 447–448
Feher, G., 452
Feierberg, I., 272–273
Feig, M., 193–194, 229, 264–265
Feng, Z., 322–323

Fenn, T.D., 225, 232–233, 238–241, 238f, 253–254, 407
Ferenczy, G.G., 445–446
Ferenczy, G.J., 126
Ferguson, D.M., 264–265, 449–452, 458–459
Fermann, J.T., 90
Fernandes, P.A., 120, 122–123
Fernandez-Ramos, A., 273–274, 361, 366
Ferrari, S., 296
Ferré, N., 126, 452–453
Ferreira, K.N., 458
Ferrenberg, A.M., 116
Ferrer, S., 288, 296–298, 300, 304–305, 342
Ferrin-O'Connell, I., 189–190
Fersht, A., 112–113, 220–221
Feynman, R.P., 271–273, 361–362, 366–368, 401
Field, M
Field, M.J., 34, 45, 58–59, 67, 107–108, 126, 193–194, 216–219, 229, 260–261, 265–266, 342, 348–349, 360–361, 363, 444, 446–447, 452–453
Fierke, C.A., 4–11, 292, 305–306
Finer-Moore, J.S., 296, 300–302, 304–305
Finke, R.G.J., 135
Fischer, A., 58–59, 92–93, 195–197, 201
Fischer, M., 253–254
Fischer, S., 132–133, 452
Fitzpatrick, P.F., 254–256, 263–264, 272–273, 360, 370, 375–378
Flaig, D., 128–129, 266–267, 400
Fleming, K.L., 431–432
Fleurat-Lessard, P., 343–344
Florián, J., 81, 84–85, 265
Fock, V.A., 123
Foley, B.L., 169, 426–427
Foloppe, N., 20–21
Fort, D.A., 420
Foster, T.J., 169
Fowler, P.W., 186–187
Fox, D.J., 428–430
Fox, S.J., 400
Fox, T., 84–85, 400
Francis, K., 256, 305–312, 306s, 310f
Franzen, S., 43–45
Fraser, J.S., 253–256, 306–307
Frauenfelder, H., 41–42

Frauenheim, T., 34, 76–77, 217–218, 222, 364–365, 395–396
Freddolino, P.L., 264–265
Freier, E., 221–222
Freindorf, M., 80–81, 127, 218–219, 364–365
Freitag, M.A., 445–446
French, S.A., 128
Frey, P.A., 252, 390–391
Fried, S.D., 360, 393, 407–409
Friedlander, A.M., 67
Friedman, J., 42
Friesner, R.A., 126–128, 218, 260, 398–399, 444, 446–447, 452–453
Frisch, M.J., 226, 349–350, 428–430, 444, 447–450, 452–453, 457
Froese, R.D.J., 125, 128, 444, 452–453
Frushicheva, M.P., 81
Fu, H.W., 6–11
Fukui, K., 178–179, 201
Fukuzawa, K., 452
Furche, F., 124
Furlani, T.R., 127

G

Gadre, S.R., 447–448
Gaikwad, S.M., 176–178
Gakhar, L., 257
Gallinari, P., 325
Gallup, N.M., 320–338
Galstyan, A., 462
Gam, J., 379
Gan, Z.T., 89–90, 98, 445–446
Gandour, R.D., 360
Ganesh, V., 447–448
Ganguly, A., 67
Gao, A.M., 447–448, 452–453
Gao, J., 32, 80–81, 126, 160, 193–194, 196–197, 199–200, 203, 214, 218–219, 254–261, 263–266, 268–269, 271–274, 289, 295, 305–308, 342, 348–349, 360–382, 390, 445–446
Gao, J.L., 58–59, 106, 108–110, 214, 218–219, 221–222, 256, 259, 263–264, 271–273, 360–361, 363, 444–447, 452–453
Gao, M., 2–4
Garcia-Moreno, B., 256

Garcia-Viloca, M., 67, 160, 197, 214, 256–257, 263, 271–274, 295–298, 360–361, 363, 365, 367–368, 370, 378–381
Garret, B.C., 267, 273–274
Garrett, B.C., 80–81, 366–367
Garza, C., 342–353
Gascón, J.A., 342, 444–469, 451*f*, 454*f*
Gaudet, R., 189–190
Gaus, M., 32, 215, 229, 364–365, 391–392
Gautt, E., 448–449, 464–465
Geacintov, N.E., 106–107
Gedeck, P., 38–39
Geissler, P.L., 221, 270–271
Gekko, K., 307–308
Gelin, B.R., 2–3
Genheden, S., 84–85, 130–133
Gentilucci, L., 452
George, A.M., 193
Geourjon, C., 190–191
Gerlt, J.A., 390–391
Germann, M., 35
Gervasio, F.L., 166–167
Gerwert, K., 200–201, 221–222
Ghadamgahi, M., 420–421, 423*t*, 426–427, 435
Ghaedizadeh, S., 420–421, 423*t*
Ghanem, M., 306–307
Ghasemi, S.A., 92–93
Ghermani, N.E., 447–448
Gheyi, T., 189–190
Ghio, C., 33–34
Ghosh, A.K., 298, 303–305, 305*f*
Ghosh, N., 34, 86–87, 218–222, 342, 391–392, 398–399
Ghosh, S., 420–421, 423*t*, 425–427, 435
Ghysels, A., 92–93
Gibbs, R.A., 4–5
Giese, T.J., 218–219
Giesen, D.J., 447
Gilbert, A.T.B., 89–90, 92, 98, 445–446
Gilbert, H.J., 165
Gilchrist, M.L., 458
Gill, M.L., 464–465
Gill, P.M.W., 89–90, 92, 111–112, 396–397, 420–421, 447–448
Gillan, M.J., 366–367, 391–392
Gillette, J.R., 20–21

Gilliland, G., 322–323
Gkionis, K., 464–465
Glatzel, P., 462
Glowacki, D.R., 263, 288, 307–308
Goddard, W.A., 32–33, 446–447
Goedecker, S., 92–93, 391–392
Goerigk, L., 147
Goerner, C., 254–256
Goetz, A.W., 393, 399–400, 408
Goez, A., 229
Gohlke, H., 90, 264–265, 430–431
Golaconda Ramulu, H., 162–163
Golbeck, J.H., 448–449, 464–465
Goldstein, H., 222–223
Golubkov, P.A., 216–217
Gompf, S., 190
Gonzalez-Lafont, A., 67, 198–199
Gonzalez-Outeirino, J., 169, 426–427
Goodey, N.M., 308, 310, 379–381
Gooding, S.R., 445–446
Goodrow, A., 92–93
Goosen, T.C., 20–21
Görbitz, C.H., 134
Gordon, M.S., 445–446
Görling, A., 396–397
Goto, M., 420
Götz, A.W., 128, 166–167, 343–344
Gould, I.R., 264–265, 449–452, 458–459
Govender, K., 264
Govind, N., 396–398
Goyal, P., 67, 215, 218–219, 229
Grabolle, M., 452, 462
Grant, I.M., 76–77
Gräter, F., 132–133, 452
Gready, J.E., 363
Greatbanks, S.P., 126–127
Green, D.F., 452
Green, M.T., 20–21
Greengard, L., 447–448
Grey, D.T., 76–77
Griffith, W.P., 225, 236–238
Grimme, S., 124, 140, 147, 267, 394*f*, 395–396, 407–408
Grob, S., 45–47
Grochowski, P., 34
Groenhof, G., 259
Gromova, A.V., 379
Gronenborn, A.M., 254

Gross, A.S., 435
Grotendorst, J., 342, 350–351
Groves, J.T., 43, 396–397
Grubmüller, H., 2–3, 259
Grundmeier, A., 462
Grunwald, E., 112
Grutter, M.G., 214
Guaitoli, G., 296
Guallar, V., 218, 260, 342, 398–399, 452
Guertin, M., 42, 45
Guest, M.F., 128
Guggino, W.B., 189–190
Guidoni, L., 363
Guillot, B., 265
Gumbart, J., 349–350
Gunner, M.R., 218–219
Gunsalus, I.C., 41–42
Gunsteren, W.F.V., 264–265
Guo, H., 106–107, 263
Guo, J.N., 305–306, 379–381
Guo, Z., 420
Gupta, K.M., 435
Gurevic, I., 298, 300
Guvench, O., 169

H

Ha, N.-C., 407
Ha, S.H., 420–422, 423t, 425–427, 436
Haak, J.R., 430–431
Habershon, S., 366–367, 390–391, 401, 409
Habibollahzadeh, D., 364–365
Haddad Momeni, M., 166–167
Hadt, R.G., 143
Hagler, A.T., 33–34
Haider, S.M., 464–465, 467–468
Hakki, Z., 175
Halbert, T., 45–47
Halgren, T.A., 258–259, 264–265, 446–447
Hall, J., 140
Halperin, I., 214
Hammes, G.G., 59–60, 252, 268
Hammes-Schiffer, S., 67, 160, 214, 221, 256, 271–273, 288–289, 305–308, 306s, 311–312, 342, 360–361, 379, 390, 409–410
Han, G.W., 223–225
Hancock, S.M., 163–165
Handy, N.C., 396–397

Hansen, J.A., 140
Hansen, N., 130–131
Hansson, T., 446–447
Haranczyk, M., 218
Hardacre, C., 435
Harris, T.K., 390–391
Harris, W., 86–87
Harrison, D.H.T., 390–391
Hartman, H.L., 6–11
Hartree, D.R., 123
Hartsough, D.S., 112
Harvey, J.N., 20–21, 34, 76–77, 130, 140, 217–218, 222, 267, 342
Hatcher, E., 89–90, 169
Haugk, M., 76–77, 217–218
Haumann, M., 452, 462
Havranek, J.J., 254
Haworth, N.L., 393
Hay, B.P., 135
Hay, P.J., 226
Hay, S., 221, 289
Hayatsu, H., 303
Haynes, P.D., 391–392
Head-Gordon, M., 89–90, 92–93, 98, 123, 396–397, 445–448
Head-Gordon, T., 216–217
Healy, E.F., 124, 193–194, 217–218, 364–365
Heaton-Burgess, T., 396
Hedegård, E.D., 120–122, 147–149
Hedman, B., 143
Hegeman, A.D., 252
Hehre, W.J., 32, 197
Heide, L., 11
Heimdal, J., 80–81, 84–85, 120–122, 128–129, 131–134, 143–149
Heine, T., 230
Hengge, A.C., 223
Henkelman, G., 58–59, 92–93
Henrissat, B., 162–163
Henry, D.J., 393
Henry, F., 89–90, 92
Henzler-Wildman, K.A., 221, 253–254, 306–307
Herbert, J.M., 92
Herman, M.F., 403–404
Hermann, J.C., 92–93
Hermans, J., 33–34, 265

Hernandez, R., 361–362
Herold, S., 43
Heroux, A., 254–256, 263–264, 272–273, 360, 370, 377–378
Herschlag, D., 214, 223–225, 228, 232–234, 236–242, 238f, 407
Hersleth, H.-P., 134
Hertel, M., 254–256
Hess, B., 428–430
Heyden, A., 92–93, 343–348, 352–353, 353f
Heymann, B., 2–3
Hibbs, A.R., 271–273, 361–362, 366–368, 401
Hickman, K., 422, 423t, 425–428, 435
Hightower, K.E., 4–6, 10–11
Hille, R., 140
Hillier, I.H., 126
Hillier, W., 458
Hilvert, D., 214
Himo, F.J., 120, 128, 217–218, 220
Hindle, S.A., 126
Hinsen, K., 267
Hirao, H., 342, 449–450
Hirao, K., 396–397
Hirschfelder, J., 45
Hirschi, J.S., 222–223
Hitzenberger, M., 343–344
Ho, J., 444–469
Ho, M.-H., 4–5
Hobza, P., 395–396
Hocky, G.M., 63–64, 66–67
Hodgson, K.O., 143
Hodoscek, M., 81, 83–85, 89–90, 92–93, 95–97, 232–233, 265, 342, 452–453
Hofacker, M., 190
Hofer, T.S., 343–344
Hoffman, M., 126–127, 222
Hoffmann, M., 34
Hohenberg, P., 123, 395
Hohenstein, E.G., 90
Holbrey, J.D., 435
Holden, Z.C., 92
Holland, I.B., 186–187, 189–193
Hollenberg, P.F., 20–21
Hollenstein, K., 186–187
Hollfelder, F., 223
Holm, C., 166–167

Holtz, K.M., 223–229, 228f, 232–234, 235f, 238–239
Hong, B., 292, 296, 298, 300–305
Hong, G., 40–41, 143
Honig, B., 124, 252–253, 452
Hooft, R.W., 269–270
Hopmann, K.H., 128
Horenstein, N.A., 162–163, 166–167
Hornak, V., 426–427
Horner, J.H., 20–21
Horning, E.C., 20–21
Hosseinzadeh, P., 143
Hou, G.H., 215, 218–219, 229, 233–234
Houghton, K.T., 20–21
Houjou, H., 445–446
Houk, K.N., 66–67, 360–361, 363
Hoyle, S., 76–77
Hrovat, D.A., 295
Hsiao, Y.-W., 134–135
Hsu, N.S.Y., 421
Hu, H., 58–59, 80–81, 86–87, 92–93, 106, 132, 218–219, 222, 232–233, 259–260, 342
Hu, L., 120–122, 126–130, 145–149, 400
Hu, L.H., 216–217
Hu, P., 106–107
Hu, W.-P., 273–274
Hu, X., 92–93
Hu, Z., 435
Huang, C.C., 4–5, 10–11
Huang, H., 58–59
Huang, J., 446–447
Huang, L., 216–217
Huang, X., 256, 306s, 311–312
Huang, Y., 256
Hub, J.S., 259
Huber, T., 63
Hudson, P.S., 76–98
Hugosson, H.W., 201, 205–206
Humbel, S., 125, 128, 444, 452–453
Hummel, M., 425
Hummer, G., 67, 84, 92–93, 197, 436
Hünenberger, P.H., 169, 259
Hung, H., 460–462
Hunke, S., 186–187
Huo, S., 132–133
Hurisso, B.B., 427–428
Hurley, J.H., 452

Hurst, S., 20–21
Hurtado-Guerrero, R., 169, 172, 176
Husberg, C., 126
Huskey, W.P., 374–375
Hutter, J., 43–45
Hwang, J.-K., 271–273, 362, 366–368
Hwang, M.J., 33–34
Hwang, S.M., 420
Hyland, R., 20–21
Hynes, J.T., 63, 374–375

I

Iannuzzi, M., 174–175
Ichihara, S., 307–308
Iftimie, R., 131
Iglesias-Fernández, J., 165–171, 175–178
Iikura, H., 396–397
Ikeo, E., 447–448
Ilan, B., 452
Illingworth, C.J.R., 445–446
Im, S.C., 20–21
Im, W., 218–219, 229
Impey, R.W., 45, 265, 408, 430–431
Inoue, Y., 445–446
Iordanov, T., 271–273
Iriyama, K., 307–308
Irle, S., 229
Isborn, C.M., 390–410
Ishida, T., 257
Islam, Z., 288–313, 305f
Ismail, A.E., 435
Isobe, H., 444
Isom, D.G., 256
Isralewitz, B., 2–4
Ito, S., 215, 220–222
IUPAC, 161
Iuzzolino, L., 452, 462
Ivanov, I., 390–391
Iverson, T.M., 458
Ives, E.L., 43
Iwata, S., 458
Iyengar, S.S., 197–198
Izvekov, S., 201, 205–206

J

Jaeger, V.W., 420–428, 423t, 434–436
Jamet, H., 132–133
Jana, B., 420–421, 423t, 425–427, 435

Jana, D.F., 395–396
Janas, E., 190
Jang, S., 366–367, 390–391, 409
Jarin, Z., 426–427, 435
Jarzynski, C., 4, 77–79, 84–86
Jasanoff, A., 452
Jastorff, B., 421
Javier Ruiz-Pernía, J., 288, 307–308
Jencks, W.P., 223
Jenewein, S., 189–193
Jensen, F., 122–123
Jensen, J.H., 221–222, 445–446
Jensen, K.P., 43–45, 135, 136–139f, 139–140
Ji, C., 464–465
Jiang, J., 435
Jiang, L., 214, 254
Jiao, D., 216–217
Jin, G., 41–42
Johns, K., 165
Johnson, B.G., 447–448
Johnson, E.R., 186–187, 395–396
Johnson, K.A., 292, 305–306
Johnson, M.A., 396–397
Johnston, H.S., 32–33
Johnston, P.G., 296
Jollow, D.J., 20–21
Jones, A., 343–344
Jones, B.C., 20–21
Jones, G.A., 445–446
Jones, P.M., 193, 325
Jones, T.A., 133
Jönsson, B., 129
Jonsson, H., 58–59, 366–367
Jorgensen, W.L., 33–34, 45, 58–59, 76–77, 80–81, 92–93, 264–265, 408, 430–431
Josephine, H.R., 60–61, 65
Jost, M., 16–20
Juffer, A.H., 444–445, 452
Jumpertz, T., 189–193
Jung, Y., 89–90, 98, 106–109, 203
Jungnickel, G., 76–77, 217–218

K

Kaar, J.L., 420–422, 423t, 425–428, 435
Kabsch, W., 333–334
Kaduk, B., 34–35
Kairys, V., 445–446

Kalsi, S.S., 20–21
Kalyanaraman, A., 436–437
Kamerlin, S.C.L., 81, 84–85, 132, 196, 207, 218, 221–222, 268–269, 342
Kaminski, G.A., 446–447
Kaminuma, T., 452
Kamiya, N., 420, 444, 458, 462–463
Kampf, J.W., 447–448
Kanaan, N., 296–298
Kandt, C., 193
Kankaanpää, A., 425
Kantrowitz, E.R., 223–229, 228f, 232–234, 235f, 238–239
Kapil, V., 409
Kapral, R., 63
Karplus, M., 2–3, 32–34, 38–39, 42, 45, 49–50, 58–59, 63, 66, 76–77, 89–90, 92–93, 98, 107–108, 160, 203, 214, 216–221, 229–230, 252–254, 258–261, 263, 265–266, 269–270, 273–274, 342, 360, 363–365, 379, 390, 444, 452–453
Kastenholz, M.A., 259
Kästner, J., 3, 130
Kathmann, S.M., 80–81
Kato, M., 58–59, 214, 264–265
Kaukonen, M., 120–122, 132–133, 143–145, 147–149
Kawakami, K., 444, 458, 462–463
Kaxiras, E., 76–77, 364–365, 395–396
Kazemi, M., 220
Ke, Z., 106–107
Kearns, F.L., 76–98
Keedy, D.A., 253–254
Kepp, K.P., 43–45
Kerdcharoen, T., 343–344
Kermode, J.R., 343–344
Kern, D., 214, 221, 306–307
Kern, J., 462
Kerns, S.J., 214, 221
Kesvatera, T., 129
Khajeh, K., 420–421
Khan, M.M.T., 190
Khare, D., 188
Khavari-Nejad, R.A., 420–421
Khavrutskii, I.V., 67, 140
Khersonsky, O., 214
Kim, H.S., 420–422, 423t, 425–427, 436
Kim, M.-C., 396

Kim, M.-S., 407
Kim, S., 41–42
Kim, Y., 34
Kipp, D.R., 222–223
Kirk, M., 140
Kirkwood, J.G., 2–3, 63, 76, 86, 130–131
Kirschner, K.N., 169, 426–427
Kiss, G., 214
Kitamura, Y., 343–344
Kitaura, K., 447–448, 452
Klähn, M., 196–197, 421–422, 423t, 425–428, 434–436
Klamt, A., 124, 323–324
Klein, M.L., 4–5, 45, 65, 67, 160, 166–167, 176, 265, 362, 367–368, 390–391, 408, 430–431, 446–447
Klepeis, J.L., 264–265
Kleywegt, G.J., 133, 256–257
Klimes, J., 92–93
Klinman, J.P., 221, 266–267, 288–289, 305–307, 360, 374–375, 379–380, 390–393
Klippenstein, S.J., 361, 366
Knapp, E.W., 462
Knight, C., 201, 205–206
Knott, B.C., 166–167
Ko, C., 447–448
Koca, J., 67, 464–465
Koenigs, L.L., 23–24
Kofke, D.A., 77–78
Koga, N., 109
Kohen, A., 217–218, 221–225, 256–257, 263–264, 268, 270–273, 288–313, 305f, 306s, 310f, 360, 379–381, 390–391, 403
Kohn, W., 123, 395
Kollman, P.A., 2–3, 33–34, 80–81, 116, 128, 132–133, 205, 269–270, 363, 399, 408, 428–431, 444, 449–454, 458–459
Kollmann, P.A., 342
Kolmel, C.M., 444–445
Komáromi, I., 126–127, 444, 449–450, 452–453, 457
Komeiji, Y., 447–448, 452
Kong, J., 92–93, 127
Kongsted, J., 120–122, 126, 147–149
König, G., 76–78, 81–85, 88–90, 92, 131–132
König, P.H., 34, 126–127, 222

Konowalow, D., 45
Koo, Y.-M., 420–422, 423t, 425–427, 436
Koslover, E.F., 92–93
Koslowski, A., 222
Kotting, C., 200–201
Kovach, I.M., 390–391
Kovrigin, E.L., 268
Kowalczyk, T., 34–35
Kowalski, K., 80–81
Koyano, Y., 109, 343–344
Kozlowski, P.M., 140
Kozmon, S., 67
Kraisler, E., 396
Kralova, B., 163
Kramer, C., 38–39
Krauss, M., 445–446
Kraut, D.A., 214
Kraut, J., 305–306
Kremer, K., 166–167
Kresge, A., 375
Krieg, H., 124, 394f, 407–408
Kries, H., 214
Krimm, S., 446–447
Kroghjespersen, K., 446–447
Kronig, R.D., 461–462
Kronik, L., 396–398
Kroon, J., 269–270
Krueger, J., 303–304
Kruse, H., 464–465
Krylov, A.I., 111–112
Kubař, T., 343–344
Kuchle, W., 452–453
Kuczera, K., 32, 42
Kuemmel, S., 396
Kuhanek, P., 67
Kuhn, B., 132–133
Kulik, H.J., 217–218, 263, 266–267, 391–398
Kumano, T., 11
Kumar, D., 20–21
Kumar, M., 140
Kumar, S., 2–3, 116, 205, 269–270
Kumari, M., 67
Kumbhar, S., 343–344
Kunc, K., 43–45
Kundu, S., 89–90
Kuntz, D.A., 179
Kuo, I.F.W., 66–67

Kurashige, Y., 217–218
Kurisu, G., 447–448
Kuritz, N., 397–398
Kusalik, P.G., 401
Kuskin, J.S., 436–437
Kussmann, J., 89–90, 98, 106–109, 203, 445–446
Kutzner, C., 428–430
Kuwajima, S., 258–259
Kuzuyama, T., 11, 14
Kwiecien, R.A., 140

L

Labarre, M., 42
Labow, B.I., 223
Ladner, R.D., 296
Lagunas, M.C., 435
Laino, T., 66–67, 343–344
Laio, A., 58–59, 63, 67, 163, 166–167, 172, 174–176, 201, 205–207, 430–431
Lakner, C., 60–61, 65
Lambert, A.R., 214
Lambry, J.C., 42
Lan, M.N., 420
Lans, I., 67
Lansing, J.C., 305–306
Lappi, A.K., 452
Larkin, J.D., 89–90, 232–233, 342
Larson, R.H., 436–437
Lassila, J.K., 214, 225, 228, 232–233
Latif, M., 422, 423t, 425–428, 435
Latimer, M.J., 462
Layfield, J.P., 256, 271, 289, 306s, 311–312, 360–361
Le, H.V., 10–11
Lee Woodcock, H., 81, 83–85, 90–91, 95–97
Lee, A.L., 301–302, 304–305, 307–308, 310–311, 379
Lee, C., 226
Lee, C.T., 123
Lee, C.Y., 269, 273–274
Lee, F.S., 264–265
Lee, J.J., 232–233, 379
Lee, M.C., 408
Lee, P.A., 461–462
Lee, S.H., 420

Lee, T.-S., 107–108, 260–261, 266, 348–349, 379, 447–448, 452–453
Lee, W., 407
Legler, P.M., 67
Lei, J., 106–107
Lei, M., 221, 253–254
Leimkuhler, B., 343–344
Leitgeb, M., 76–77
Lennartz, C., 398–399
Lensink, M.F., 452
Lenz, H.-J., 296
Leone, S.R., 396–397
Lesniak, P.J.B., 225, 236–238
Lesyng, B., 34
Leuhr, N., 217–218
Leung, S.S.F., 444–445, 448–450, 451*f*, 453–454, 454*f*, 456, 464–465
Lev, B., 343
Lever, G., 262–263
Levine, B., 447–448
Levitt, M., 33–34, 58–59, 76–77, 106, 126, 128, 193, 196, 218, 252–253, 260–261, 264–265, 320, 342, 363, 398–399, 444
Levy, C., 289
Levy, M., 396–397
Levy, R.M., 446–447
Lewinson, O., 186–187, 189–190
Lewis, A., 462
Leys, D., 289
Li, G., 91–92, 127, 220–221
Li, H.Z., 60–61, 65, 77, 80–81, 86–87, 425
Li, J.-L., 120–122, 140–141, 143–149, 193
Li, L., 306–307
Li, M., 106–107
Li, S.H., 447–448, 452
Li, S.M., 16–20
Li, W., 140, 447–448, 452
Li, X., 11, 134–135, 141, 197–198, 320, 342, 401, 449–450
Li, X.C., 458, 462
Li, X.P., 447–448
Li, Y., 106–116, 435
Li, Z.H., 452
Liakos, D.G., 267
Liao, R.-Z., 120, 128–129, 137–138, 217–218, 220, 266–267
Licence, P., 427–428
Liebisch, P., 452, 462

Liechty, A., 45–47
Liedl, K.R., 343–344
Lifson, S., 33–34
Lightstone, F.C., 2–25
Lim, G.S., 421–422, 423*t*, 425–428, 434–436
Lim, M., 41–42
Limaye, A.C., 447–448
Limbach, H.H., 360, 374–375, 403
Lim-Wilby, M., 254–256
Lin, H., 120, 122–123, 126–127, 342–353, 353*f*, 398–399, 452–453
Lin, J., 188
Lin, Y.-l., 263–264, 272–273
Lin, Y.-L., 364–365, 370
Lin, Y.S., 264–265
Lindahl, E., 428–430
Lindh, R., 120–122, 147–149
Lindorff-Larsen, K., 2–3, 220–221, 264–265
Lindqvist, Y., 225
Linse, S., 129
Lior-Hoffmann, L., 106–107
Lipkowitz, K.B., 218–219, 260, 273–274, 360–361, 363
Liu, C.T., 256, 306*s*, 311–312
Liu, F., 392–393
Liu, H., 80–81, 132, 272–273, 288–289, 305–308, 310, 421
Liu, H.B., 58–59, 214
Liu, H.Y., 110–111
Liu, J., 143, 167
Liu, M.Y., 360–361, 363, 396
Liu, X., 143–145, 435
Liu, Y.-P., 63–64, 273–274
Liu, Z.W., 160, 166–167
Livshits, E., 396–397
Lluch, J.M., 67
Lobaugh, J., 446–447, 452
Lobell, M., 452
Locher, K.P., 186–187
Loffler, P., 360–361
Logan, D., 134
Loja, P., 462
Lombard, V., 162–163
Lonardi, A., 169
London, F., 32–33
Lonsdale, R., 20–21, 76–77, 130, 262–263, 267, 342

Loos, M., 199–200
Lopes, P.E.M., 264–265, 446–447
Lopez, X., 199–200
Lorant, F., 32–33
Lorenzen, K., 446–447
Loria, J.P., 268, 306–307, 444–445, 450, 456–457, 464–467
Lotrich, V., 395–396
Lovell, S.C., 256–257
Lovelock, K.R.J., 427–428
Loveridge, E.J., 263, 288, 305–308, 379–381
Lovick, H.M., 214
Lpez-Canut, V., 400
Lu, G., 188
Lu, J., 218–219
Lu, M., 106–107
Lu, N., 77–78
Lu, X., 215, 218–222, 229, 435
Lu, Y., 143
Lu, Z.Y., 80–81, 92–93, 232–233
Luber, S., 444, 458, 462
Lubitz, W., 462
Luehr, N., 263, 266–267, 391–398
Luk, L.Y.P., 263, 288, 307–308
Luo, H.-B., 106–107
Luo, R., 90, 264–265, 430–431
Luque, F.J., 80–81, 364–365
Lutz, S., 379
Luzhkov, V., 80–81, 131, 131f, 272–273, 452
Luzum, C., 301–302, 304–305
Lv, C., 58–59, 67–71, 86–87
Lynch, G.C., 273–274
Lytle, F.W., 461–462

M

Ma, B.Y., 214
Ma, S., 58–59, 214, 218, 268, 289, 295, 305–308, 360–361, 365–366, 370, 379–382
MacFarlane, D.R., 420–421
MacKenzie, G., 126
Mackenzie, L.F., 166–167
MacKerell, A.D., 38–39, 45, 49–50, 66, 86–87, 89–90, 98, 124–125, 193–194, 203, 216–217, 229, 264–265, 426–427, 446–447, 449–450

Mackie, I.D., 395–396
MacPherson, I.S., 60–61, 65
Mader, M.M., 223–225
Madura, J.D., 45, 265, 408, 430–431
Maeda, S., 92–93
Maghlaoui, K., 458
Maginn, E.J., 427–428
Mahoney, M.W., 265
Mai, N.L., 420–421
Maigret, B., 126, 266, 452–453
Mainz, D.T., 446–447
Majewski, J.A., 396–397
Major, D.T., 58–59, 214, 217–218, 221, 252–276, 360–362, 364–372, 375–378
Majumder, T., 436–437
Makarov, D.E., 366–367
Makmal, A., 396
Malcolm, N., 400
Malde, A.K., 426–427
Maley, F., 292, 303–305
Mallajosyula, S.S., 169
Manabe, S., 161–162
Manby, F.R., 34, 76–77, 80–81, 217–218, 222
Mannfors, B., 446–447
Manohar, L., 447–448
Manolopoulos, D.E., 366–367, 390–392, 401–403, 406, 409
Manzoni, F., 134
Maple, J.R., 33–34, 446–447
Mar, B.D., 396–398
Maragakis, P., 220–221, 264–265
Maragliano, L., 58–59, 92–93, 195–197, 201
Marcus, R.A., 289
Mark, A.E., 143, 264–265, 426–427
Markham, K.A., 308, 311, 379–381
Markland, T.E., 360, 366–367, 369, 390–410
Marks, G.T., 390–391
Marks, P.A., 324–325
Marqueling, A.L., 43
Marsalek, O., 369, 392, 409
Marsh, E.N.G., 135, 140
Marshall, N.M., 143
Mart, S., 400
Marti, S., 222, 296–298, 342
Martin, B.D., 135
Martin, J.L., 42

Martinez, J.M., 428–430
Martinez, L., 428–430
Martínez, T.J., 217–218, 263, 266–267, 391–400, 408, 447–448
Marverti, G., 296
Marx, D., 343–344, 390–391
Maseras, F., 135–136, 365, 444, 452–453
Masia, M., 343
Massiah, M.A., 390–391
Massova, I., 132–133
Mata, R.A., 76–77, 140–141, 142f, 143–145, 217–218, 222
Mathias, G., 446–447
Matsubara, T., 125, 128, 444, 452–453
Mattu, M., 325
Matuschek, M., 16–20
Mauldin, R.V., 307–308, 379
Maupin, C.M., 58–59, 201, 205–206
Maurer, P., 201, 205–206
Mavri, J., 272–273
Maxwell, D.S., 264–265
Mayhall, N.J., 445–448
Mazack, M.J.M., 360–361
McCammon, J.A., 2–3, 34, 220–221, 269, 273–274, 390
McCracken, J.A., 311
McCullagh, M., 66–67
McElheny, D., 253–254, 305–308, 379
McEvoy, J.P., 448–450, 459, 464–465
McGibbon, R., 392–393
McKenzie, R.H., 401
McMurry, J.E., 296
McNeish, J., 34–35
McQuarrie, D., 32
Meadows, C.W., 306–307
Medina, M., 67
Meheut, M., 369, 392
Mehler, E.L., 256
Mei, Y., 447–448
Meier, K., 342
Men, S., 427–428
Meng, Y.L., 214
Menikarachchi, L.C., 342, 444–445, 448–450, 464–465
Mennucci, B., 124, 226–228, 342, 444–445, 452–453
Merenga, H., 444–445

Merz, K.M., 2–25, 80–81, 90, 112, 218, 264–265, 342, 360–361, 430–431, 447–454, 458–459
Messina, M., 366–367
Messinger, J., 458, 462
Metz, S., 130
Meuwly, M., 31–51, 216–217
Mezei, M., 269–270
Mezey, P.G., 447–448, 452
Mhashal, A.R., 252–276
Miao, Y.P., 14
Micaêlo, N.M., 421–422, 423t, 425–428, 435
Michaelides, A., 92–93, 401
Michel, C., 343–344
Mie, G., 38–39
Miertus, S., 200–201
Mihai, C., 296, 298, 300–302
Mikkelsen, K.V., 126
Mikulskis, P., 130–131
Milani, M., 42, 45
Mildvan, A.S., 390–391
Milet, A., 132–133
Millam, J.M., 197–198
Miller, B.G., 58–59
Miller, B.T., 89–90, 342
Miller, J.A., 361, 366
Miller, T.F., 366–367, 390–391, 409
Miller, W.H., 366–367, 392
Milletti, F., 256
Mills, G., 366–367
Mills, J.H., 223–225
Min, D., 60–61, 63–65, 77, 86–87
Mincer, J.S., 310–311
Mirkin, N.G., 446–447
Mishra, S., 36, 43
Mitchell, J.R., 20–21
Mittal, J., 264–265
Mittl, P.R.E., 214
Miyazaki, T., 391–392
Mladek, A., 464–465
Mo, Y., 203, 365
Mobley, D.L., 77
Mohan, M.S., 190
Molina, P.A., 221–222
Moliner, V., 160, 222, 288, 296–298, 300, 304–305, 342, 400
Møller, C., 123

Molnar, L.F., 89–90, 98, 106–109, 203
Momany, F.A., 109–110
Monard, G., 218, 342
Monari, A., 260–261, 266, 342
Mones, L., 343–344
Moniruzzaman, M., 420
Monnat, R.J., 254
Montgomery, J.A., 126–127, 393, 445–446, 449–450, 457
Moore, P.B., 160, 166–167, 343–344
Moradi, M., 193
Morales, M.A., 401
Moran, D., 86–87
Moras, G., 343–344
Mori, T., 229
Mori-Sánchez, P., 217–218, 396
Morokuma, K., 92–93, 125–128, 140, 320, 342–344, 365, 444–445, 449–450, 452–453, 457
Morris, G.M., 254–256
Morris, H., 305–306
Morrone, J.A., 390–391
Mosey, N.J., 34–35
Mostofi, A.A., 391–392
Mostofian, B., 435
Mouesca, J.M., 459
Moyna, G., 420
Moyna, P., 420
Mu, Y.Q., 4–5
Muegge, I., 452
Mulgrew-Nesbitt, A., 452
Mulholland, A.J., 20–21, 34, 76–77, 80–81, 92–93, 130, 132, 140, 217–218, 222, 260, 262–263, 267, 342, 398–399, 444–445
Mullan, C.L., 435
Muller, C., 452, 462
Muller, R.P., 81, 84–85
Mundy, C.J., 66–67
Murdzek, J., 296
Murphy, R.B., 127–128, 444, 452–453
Murray, C.J., 375
Murray, D., 452
Murray, P., 452
Musaev, D.G., 140

N

Nadassy, K., 238–239
Nagaoka, M., 109, 343–344
Nagaraju, M., 186–207
Nagel, Z.D., 221, 289, 379–380
Nagy, T., 35–36
Naidoo, K.J., 264
Nair, P.C., 426–427
Nakano, T., 447–448, 452
Nakashima, K., 420
Nam, K., 58–59, 108–110, 199–200, 214, 218, 221, 258–259, 268, 270–273, 360–361, 365–366
Nasertorabi, F., 223–225
Nauser, T., 43
Nayak, S., 395–396
Naylor, G.J., 60–61, 65
Nazari, M., 420–421, 423t
Neaton, J.B., 397–398
Nedd, S., 337
Neese, F., 222, 267, 349–350, 462
Negre, C.F.A., 444, 458–459, 460f, 462–463, 463–464f
Neidle, S., 464–465, 467–468
Nelson, S.D., 23–24
Neuhauser, D., 396–397
Neumark, D.M., 31–32
New, M.H., 446–447
Newcomb, M., 20–21
Newcomer, M.B., 444–445, 450, 456–457, 464–467
Newstead, S., 186–187
Newville, M., 462
Ng, S.-L., 189–190
Nguyen, T., 162–163
Nicoll, R.M., 126
Nielsen, J.E., 390
Nielsen, S.O., 343–344
Nielson, J.B., 401
Nieto, J., 160–161, 163, 169–172
Nikaido, K., 189–190
Niklasson, A.M.N., 111
Nikolic-Hughes, I., 232–233
Nilsson, K., 120–122, 133–134
Nilsson, L., 38–39, 49–50, 66, 89–90, 98, 193–194, 203, 229, 254
Nin-Hill, A., 160–180
Nishiyama, M., 11
Nitoker, N., 271–273
Nocera, D.G., 409–410
Noel, J.P., 11, 14

Noiriel, A., 421
Noodleman, L., 459
Nordlander, E., 141
Nordwald, E.M., 420–422, 423t, 425–428, 435
Norel, R., 452
Norman, S.E., 435
Northrop, D.B., 292–293, 311–312
Northrup, S.H., 269, 273–274
Noskov, S.Y., 343
Nowotny, M., 197
Nowotyny, M., 67
Numao, S., 179
Nunes, R.W., 447–448
Nussinov, R., 214
Nutt, D.R., 34–36
Nydegger, M.W., 223–225
Nymeryer, H., 58–59, 63

O

O'Brien, P.J., 223
Ochsenfeld, C., 89–90, 98, 106–109, 128–129, 203, 400
Ochterski, J.W., 393
Offen, W.A., 165
Ogilby, P.R., 126
Oh, B.-H., 407
Ohmae, E., 307–308
Ohno, K., 92–93
Ojamo, H., 425
Ojeda-May, P., 186–207
Okamoto, T., 109
Okamoto, Y., 58–59, 63, 66–67, 215, 220–222
Okamura, M.Y., 452
Oksanen, E., 134
Okur, A., 89–90, 342, 426–427
Olafson, B.D., 33–34, 45, 203, 230, 452–453
Olah, J., 20–21
Oldham, M.L., 188
Oliveira, A.F., 230
Olivucci, M., 126
Oloo, E.O., 193
Olsen, L., 120–122, 133–134
Olsen, S., 447–448
Olsson, M.A., 131–132
Olsson, M.H.M., 58–59, 120–122, 128, 140, 143, 160, 214, 272–273
Olssson, M.H.A., 143
O'Mara, M.L., 193
Oostenbrink, C., 76–77, 143, 264–265
Oppeneer, P.M., 43–45
Ordejón, P., 391–392
Orelle, C., 186–191
Orlando, A., 58–59
Orozco, M., 80–81, 264–265, 360–361, 364–365, 445–446
Orville, A.M., 254–256, 263–264, 272–273, 360, 370, 377–378
Osmond, B., 448–449, 464–465
Osted, A., 126
Oswald, C., 186–187, 189–191
Ott, M., 253–254
Otten, R., 214
Oude Elferink, R., 186–187
Ouellet, H., 42
Ouellet, Y., 42, 45
Ovcharenko, E., 189–190
Ovchinnikov, V., 215, 220–222
Owens, R.M., 420–421
Oyeyemi, O.A., 379

P

Pachov, D.V., 214
Padmakumar, R., 135
Pádua, A.A.H., 426–427
Pais, J.E., 4–7, 10–11
Pal, R., 444, 458–459, 460f, 462–463, 463–464f
Palmer, A.G., 134, 253–254, 256–257
Palmo, K., 446–447
Pan, A.C., 58–59
Pan, L.L., 2–25
Pande, P.P., 436–437
Pande, V.S., 77–78, 86, 214, 216–217, 221, 223–225, 236–239, 241–242, 264–265, 392–393
Paneth, P., 140
Pang, J., 289
Pang, X., 215
Pantazis, D.A., 462
Parkinson, C.J., 393
Parkinson, G.N., 464–465, 467–468
Parks, J.M., 80–81, 92–93, 232–233

Parr, C., 32–33
Parr, R.G., 123, 197, 217–218, 226, 396, 446–447
Parrinello, M., 43–45, 58–59, 63–64, 66–67, 160, 163, 166–167, 174–176, 197–198, 201, 220–221, 363, 390–392, 401, 403, 430–432
Parrish, R.M., 90
Parson, W.W., 160
Parui, S., 420–421, 423t, 425–427, 435
Pastor, R.W., 86–87
Patey, G.N., 62–63, 111
Pauling, L., 32–33, 305–306
Payne, M.C., 201, 205–207, 343–344, 391–392
Pear, M.R., 269, 273–274
Pearson, A.D., 223–225
Peck, A., 223–225, 236–239, 241–242
Pecoraro, V.L., 447–448
Pedersen, L., 109–110, 124–125, 218–219, 258–259
Peguiron, A., 343–344
Peloquin, J.M., 458
Pendry, J.B., 461–462
Peng, C.Y., 459
Pengthaisong, S., 175
Penney, W.G., 461–462
Peraro, M.D., 4–5
Perdew, J.P., 123, 395–396
Pereira, J.H., 421
Perissinotti, L.L., 379–380
Perrin, C.L., 401
Pesce, A., 42, 45
Peter, R.M., 23–24
Peters, B., 92–93
Peters, K.S., 112
Petersen, L., 171
Petersen, M.K., 58–59
Petersson, G.A., 393
Petit, C.M., 307–308, 379
Petrella, R.J., 38–39, 49–50, 66, 89–90, 98, 193–194, 203, 229
Petrey, D., 452
Petrich, J.W., 42
Petrik, I., 143
Petsko, G.A., 253–254
Pettie, A.E., 23–24
Pezeshki, S., 342–350, 352–353, 353f

Pfaendtner, J., 420–437, 423t
Phatak, P., 221–222
Philipp, D.M., 126–128, 444, 452–453
Philipsen, P., 230
Phillips, J.C., 349–350
Phung, Q.M., 120–122, 147–149
Piana, S., 2–3, 220–221, 264–265
Pickett, J.S., 4–11
Piecuch, P., 140
Pierdominici-Sottile, G., 162–163, 166–167
Pierloot, K., 120–122, 143, 147–149
Pinkas, D.M., 214
Pisliakov, A.V., 264–265
Pittock, C., 400
Planas, A., 160–165, 169–172, 176–178
Plapp, B.V., 310–311
Platts, J.A., 464–465
Plazinski, W., 169
Plesset, M.S., 123
Plimpton, S., 408
Plotnikov, N.V., 81, 84–85, 132, 207, 222
Poger, D., 426–427
Pohorille, A., 63, 77–79, 85, 401–402
Pokhrel, R., 444, 458–459, 460f, 462–463, 463–464f
Polanyi, M., 32–33
Pollock, E.L., 370
Polyak, I., 222
Ponder, J.W., 89–90, 106–109, 216–217, 265, 342, 349–350, 390, 446–447
Pontiggia, F., 214
Pople, J.A., 123, 197, 226, 396–397, 449–450
Popovic, D.M., 452
Porezag, D., 76–77, 217–218
Post, C.B., 58–59
Postma, J.P.M., 33–34, 265, 430–431
Potter, W.Z., 20–21
Poulsen, T.D., 126, 197, 256–257
Poulter, C.D., 4–5, 10–11
Poyart, C., 42
Prasad, R.B., 196
Prat-Resina, X., 83, 218–220, 342, 391–392, 402
Prescott, L.F., 20–21
Preuss, H., 452–453
Privett, H.K., 214
Procko, E., 189–190

Proctor, E.A., 337
Pu, J., 58–59, 126, 186–207, 214, 218, 266, 268–269, 271–274, 289, 295, 360–361, 365–366, 378–382
Pudney, C.R., 289
Pulay, P., 447–448
Pullman, B., 265
Pushkar, Y., 462

Q

Qian, H.J., 229
Quapp, W., 92–93
Quiocho, F.A., 188, 223–225

R

Rabideau, B.D., 435
Radisky, E.S., 10–11
Radkiewicz, J.L., 288, 306–308
Radom, L., 197, 393
Ragain, C.M., 444–445, 450, 456–457, 464–467
Raghavachari, K., 123, 445–448
Rahman, A., 265, 422, 423t, 425–428, 435
Raich, L., 160–180
Rajagopalan, P.T.R., 307–308, 379–381
Raman, E.P., 169, 216–217, 426–427
Ramesh, S.G., 401
Ramos, M.J., 120, 122–123
Ranaghan, K.E., 34, 76–77, 92–93, 217–218, 222, 262–263, 398–399
Rance, M., 253–254, 256–257
Rancurel, C., 162–163
Randolf, B.R., 343–344
Ranganathan, V., 420–421
Ranjbar, B., 420–421
Rappe, A.K., 446–447
Rappoport, D., 124
Rasa, S.M.M., 420–421, 423t
Ravel, B., 460–462
Reddy, M.R., 86–87, 131
Redler, R.L., 337
Rees, D.C., 186–187
Refaely-Abramson, S., 396–398
Rehr, J.J., 460–462
Reilly, P.J., 171
Remsing, R.C., 420
Ren, P.Y., 216–217, 265, 446–447
Ren, W., 58–59, 65–66, 92–93, 196–197, 201
Rendell, A.P., 363, 452
Rendic, S., 22–24, 25t
Repasky, M.P., 76–77
Resing, K.A., 379
Ressl, S., 241
Retegan, M., 132–133
Reuter, N., 266, 452–453
Reuwer, J.F., 294–295, 374–375
Reyes, C.M., 132–133, 452
Reyes, J.Y., 35
Reynolds, C.A., 445–446
Riahi, S., 128
Riccardi, D., 86–87, 127, 218–222, 229, 342, 391–392, 398–399
Richard, C., 42
Richard, R.M., 92
Richard, S.B., 11, 14
Richardson, D.C., 256–257
Richardson, J.S., 256–257
Richmond, G., 396–397
Rick, S.W., 446–447
Ridder, L., 76–77, 92–93
Ridge, D.P., 45–47
Rinaldi, D., 126
Ringe, D., 253–254
Riniker, S., 342
Rishavy, M.A., 371–373
Rittle, J., 20–21
Rittner, F., 446–447
Rivail, J.-L., 126, 260–261, 266, 342–345, 348–349
Rivalta, I., 444, 458, 462
Robb, M.A., 226, 349–350, 428–430, 449–450
Robertazzi, A., 462
Roberts, D.W., 23
Roca, M., 67, 263, 288, 296–298, 307–308
Roccatano, D., 421
Rock, W., 223–225
Rod, T.H., 80–81, 120–122, 126–127, 131–132, 131f, 147–149, 222, 288, 306–308
Rode, B.M., 343–344
Roe, A.L., 23
Rogers, R.D., 420
Rohrdanz, M.A., 268

Rohrig, U.F., 363
Roitberg, A.E., 109, 162–163, 166–167, 426–427
Rokhlin, V., 447–448
Román-Meléndez, G.D., 140
Rooklin, D.W., 106–107
Roothaan, C.C.J., 203
Rose, D.R., 162–163, 165, 179
Rosenberg, J.M., 2–3, 116, 205, 269–270
Ross, G.A., 254
Rossi, I., 217–218, 263
Rosta, E., 40–41, 67, 92–93, 196–197
Roston, D., 214–242, 289, 379–380
Röthlisberger, D., 214, 254
Rothlisberger, U., 126, 131, 172, 201, 205–206, 215, 218, 363
Rotskoff, J., 63–64, 66–67
Roux, B., 38–39, 49–50, 58–59, 66, 89–90, 98, 116, 193–194, 203, 214, 216–221, 229, 258–259, 267, 343, 446–447
Rovira, C., 43–45, 160–180
Rowley, C.N., 128, 343
Rubach, J.K., 310–311
Rubinstein, A., 263, 272–273, 370
Rudack, T., 200–201
Ruddock, L.W., 452
Rudnitskaya, A.N., 444–445, 448–449
Rudolph, J., 92–93
Ruiz-Lopez, M., 343–345
Ruiz-Pernía, J.J., 67, 263, 342
Rulíšek, L., 120–122, 134–135, 143–149
Rullmann, J.A.C., 444–445
Ruzsinszky, A., 395
Rydberg, P., 134
Ryde, U., 43–45, 80–81, 84–85, 120–149, 121–122f, 131f, 136–139f, 142f, 216–217, 222, 400, 458, 462
Rye, C.S., 163

S

Sabin, J.R., 342
Saboury, A.A., 420–421, 423t, 426–427, 435
Sadeghian, K., 266–267
Saderholm, M.J., 5–6
Sadler, P.J., 186–187
Saebo, S., 447–448
Saha, A., 445–446
Sakane, S., 131
Sakurai, M., 445–446
Salahub, D., 131
Sale, K.L., 421
Sali, A., 254
Salmon, J.K., 436–437
Salo, K.E.H., 452
Salomon-Ferrer, R., 216–217
Salvalaglio, M., 431–432
Salzman, J., 241
Salzner, U., 396–397
Sambasivarao, S.V., 426–427
Sampson, C., 84–85
Sanchez, M.L., 273–274, 361, 378, 380–381
Sanchez-Martinez, M., 67
Sandberg, D.J., 444–445, 448–449
Sandgren, M., 166–167
Sangian, M., 420–421, 423t, 426–427, 435
Sankaran, B., 189–190
Sansom, M.S.P., 342
Santi, D.V., 296
Santry, D.P., 197
Sanyal, B., 43–45
Sapienza, P.J., 301–302, 304–305, 307–308, 310–311, 379
Sarkar, S., 436–437
Sato, S., 32–33
Sato, T., 452
Satoh, H., 161–162
Sauer, J., 444–445
Sauer, K., 462
Saunders, M.G., 66–67
Saunders, W.H., 374
Sauter, N.K., 253–254
Sautet, P., 343–344
Savin, A., 396–397
Sayers, D.E., 461–462
Scandolo, S., 201, 207
Schaad, L.J., 294–295, 374–375
Schaefer, I., 89–90, 92
Schaefer, M., 258–259
Schaefer, P., 86–87, 218–220, 229, 342, 391–392, 398–399
Schenter, G.K., 80–81, 267, 366–367, 445–446
Scheuermann, S., 420–421
Schfer, A., 398–399

Schlegel, H.B., 197–198, 226, 349–350, 428–430, 449–450, 456–457
Schleyer, P.v.R., 197
Schlitter, J., 200–201
Schmidt, M.W., 445–446
Schmidt, T., 396
Schmitt, L., 186–187, 190–193
Schmitt, U.W., 58–59
Schneider, E., 186–187
Schneider, G., 225
Schneider, S., 266–267
Schnell, J.R., 305–308
Schnieders, M.J., 216–217
Schofield, J., 131
Schowen, R.L., 360, 374–375
Schramm, V.L., 178–179, 214–216, 221–225, 254–256, 270–271, 295–296, 305–307, 360
Schreckenbach, G., 128
Schrepfer, P., 254–256
Schröder, C., 76–77
Schulten, K., 2–4, 222, 452
Schultz, P.G., 223–225
Schulzke, C., 140
Schuttelkopf, A.W., 426–427
Schütz, M., 123
Schüürmann, G., 124, 323–324
Schwaneberg, U., 421
Schwans, J.P., 407
Schwartz, S.D., 214, 221, 270–271, 288, 295–296, 305–306, 310–311, 361
Schwarzl, S.M., 132–133, 452
Schwegler, E., 447–448
Schweizer, K.S., 446–447
Schworer, M., 446–447
Scoles, G., 395–396
Scott, R.A., 134–135
Scrocco, E., 200–201
Scrutton, N.S., 221, 272–273, 390–391
Scuseria, G.E., 123, 197–198, 226, 349–350, 428–430, 447–450
Sderhjelm, P., 400
Seabra, G.D.M., 109
Seduraman, A., 421–422, 423t, 425–428, 434–435
Segal, G.A., 197
Seidl, A., 396–397
Seifert, G., 76–77, 217–218

Selzer, T., 288, 306–308, 379
Sen, A., 292–293, 307–308, 310f, 311
Senior, A.E., 189–190
Senn, H.M., 58–59, 106, 120, 122–123, 125–127, 130–131, 215, 218, 260, 262, 320, 342, 360–361, 398–399, 444–445
Sepahi, A.A., 420–421
Sethaphong, L., 420–422, 423t, 425–427, 436
Sevastik, R., 128
Sezer, D., 58–59
Shaik, S., 20–21, 398–399
Shakhnovich, E.I., 321
Sham, L.J., 395
Sham, Y.Y., 452
Shao, Y., 89–90, 92–93, 98, 106–109, 127, 203
Shao, Y.H., 232–233, 445–446
Sharifzadeh, S., 397–398
Sharma, P.K., 58–59, 140, 196, 214
Sharma, S., 189–190, 217–218, 223–225
Sharp, K., 124
Shashidhara, K.S., 176–178
Shaw, D.E., 2–3, 220–221, 264–265, 436–437
Shaw, K.E., 132
Sheinerman, F.B., 452
Shelton, G.R., 295
Shen, J.-R., 444, 458, 462–463
Sheng, X., 20–21
Sheppard, D., 92–93
Sherwood, D., 253, 256–257
Sherwood, P., 126–128, 265, 342, 350–351, 452–453
Shi, W., 178–179
Shi, Y., 106–107
Shiga, M., 343
Shim, J., 89–90, 264–265
Shimada, H., 268
Shimanovich, R., 43
Shimizu, K., 426–427
Shirsat, R.N., 447–448
Shirts, M.R., 77–78, 86
Shirvanyants, D., 321, 337
Shleev, S., 145–149
Shoichet, B.K., 253–254
Shoji, M., 444
Shokes, J.E., 134–135

Shrimpton, P.J., 307–308, 379
Shukla, D., 214
Shyamala, V., 189–190
Sieber, S., 125, 128, 444, 452–453
Siebrand, W., 295
Siegbahn, P.E.M., 43–45, 120, 134–136, 140, 217–218, 220, 272–273, 390, 458, 462
Siegel, J.B., 214
Sigala, P.A., 407
Sikkema, J., 343–344
Sikorski, R.S., 308, 311, 379–381
Sim, E., 396
Simmerling, C.L., 106–107, 111–112, 399, 426–427
Simmonett, A.C., 90
Simon, I., 343–344
Simonson, T., 32, 220–221, 390, 452
Sims, J., 460–462
Singer, S.J., 452–453
Singh, P., 288–313, 310*f*
Singh, S., 452
Singh, U.C., 33–34, 86–87, 128, 342, 363, 444, 452–453
Singleton, S.F., 379
Sisto, A., 393
Sivak, D.A., 253–254
Sixta, H., 425
Skelton, N.J., 134, 253–254, 256–257
Skylaris, C.-K., 84–85, 391–392, 400
Smedarchina, Z., 295
Smith, D.A., 20–21, 265
Smith, G.K., 106–107
Smith, J., 435
Smith, J.C., 132–133, 452
Smith, J.L., 447–448
Smith, S.F., 80–81
Snawder, J.E., 23
Snider, M.J., 160, 252
Snijders, J.G., 143
Soares, C.M., 421–422, 423*t*, 425–428, 435
Söderhjelm, P., 84–85, 120–122, 126–134, 147–149
Sokalski, W.A., 92–93
Soler, J.M., 391–392
Solomon, E.I., 134–135
Soloviov, M., 43–45, 50
Solt, I., 343–344

Song, J., 464–465
Song, Y., 201, 205–207, 223–225
Sosa Vazquez, X.A., 396–398
Souaille, M., 116
Sours, K.M., 379
Sousa da Silva, A.W., 428–430
Sousa, S.F., 120, 122–123
Sparta, M., 321, 337–338
Speciale, G., 161–162
Spencer, H.T., 296
Spiess, A.C., 421
Spiwok, V., 163
Sponer, J., 464–465
Sprenger, K.G., 420–437
Sprik, M., 62–63, 143–145, 362, 367–368, 446–447
Sproviero, E.M., 134–135, 141, 448–450, 459, 464–465
Squassabia, F., 452
Srebro, M., 396–398
Srnec, M., 120–122, 143–145, 147–149
Staib, A., 447
Stanley, H.E., 321
Stanton, C.L., 66–67
Starke, K., 141
Staroverov, V.N., 123
States, D.J., 33–34, 45, 203, 218–219, 230, 452–453
Stathis, D., 266–267
Stauffacher, C.V., 225
Stec, B., 223–229, 228*f*, 232–234, 235*f*, 238–239
Stefan, L., 307–308
Stehle, T., 16–20
Stein, T., 396–398
Steinfels, E., 190–191
Steitz, T.A., 191
Stepan, J., 67
Stern, E.A., 461–462
Stern, H.A., 446–447
Stevens, W.J., 445–446
Stewart, J.J.P., 124, 193–194, 203, 217–218, 363–365
Stick, R.V., 163–165
Stillinger, F.H., 265
Stivers, E.C., 294–295, 374–375
Stockfisch, T.P., 33–34
Stoddard, B.L., 254

Stoddart, J.F., 161–162
Stojkovic, V., 257, 379–380
Stoll, H., 452–453
Stolte, S., 421
Storer, J.W., 447
Story, R.M., 191
Stote, R., 34, 45, 218–219
Straatsma, T.P., 220–221
Strain, M.C., 447–448
Strambi, A., 126
Strasser, R., 266–267
Stratmann, R.E., 447–448
Strickland, C.L., 10–11
Strid, Å., 128–129
Strobel, S.A., 464–465
Strockbine, B., 426–427
Strodel, P., 222
Stroet, M., 426–427
Stroud, R.M., 296, 300
Strout, D.L., 447–448
Strutzenberg, T.S., 298, 300, 304–305, 305f
Stuart, S.J., 446–447
Stubbs, M.T., 191
Stüber, C., 425–426
Stuchebrukhov, A.A., 409–410, 452
Stutz, A.E., 163–165
Subotnik, J.E., 447–448
Sugiki, S.I., 447–448, 452
Suhai, S., 34, 395–396
Sulpizi, M., 143–145
Sumner, S., 120–122, 128–130, 147–149
Sumowski, C.V., 128–129, 400
Sun, J., 395
Sun, N., 143
Sun, Q., 266
Sun, R., 63–64, 66–67
Sundberg, R.J., 23–24
Sunden, F., 223–225, 236–239, 241–242
Sure, R., 147
Suslick, K., 45–47
Sussman, D., 254
Sutcliffe, M.J., 289, 390–391
Svensson, M., 125, 128, 444, 452–453
Swain, C.G., 294–295, 374–375
Swaminathan, S., 33–34, 45, 203, 230, 452–453
Swanson, J.M.J., 58–59
Swanwick, R.S., 307–308, 379

Swart, M., 143
Swatloski, R.P., 420
Swendsen, R.H., 2–3, 116, 205, 269–270
Szefczyk, B., 92–93
Szekeres, Z., 452

T

Tajkhorshid, E., 34, 189–190, 193, 349–350, 452
Takenaka, N., 343–344
Tal, N., 189–190
Tampe, R., 190
Tankrathok, A., 175
Tao, J., 123
Tao, P., 232–233
Tao, Y., 134–135
Tarcz, S., 16–20
Tarling, C.A., 163–165
Tautermann, C.S., 84–85, 400
Tavan, P., 2–3, 446–447
Tavernelli, I., 126, 201, 396–398
Tawada, Y., 396–397
Teague, S.J., 256–257
Tee, K.L., 421
Tello, M., 11
ten Brink, T., 256
Terrell, R., 92–93
Terstegen, F., 398–399
Terwilliger, T.C., 253–254
Tew, D.P., 396–397
Thai, V., 221, 253–254
Tharp, S., 307–308
Théry, V., 126
Thiel, S., 131
Thiel, W., 20–21, 76–77, 106, 120, 122–127, 129–131, 137–138, 197, 203, 215, 217–219, 222, 230, 258–267, 272–273, 320, 342, 348–350, 360–361, 398–399, 444–445, 452–453
Thirumalai, D., 273, 390–392, 401–402
Thole, B.T., 444–445
Thompson, A.J., 161–162, 165, 169–171
Thompson, M.A., 193, 196, 444–446
Thompson, M.J., 446–447
Thompson, S.J., 23–24
Thomson, J.S., 20–21
Thorpe, I.F., 288, 307–308
Tian, B., 128–129

Tian, S., 143
Tichy, S.E., 377
Tidor, B., 32, 452
Tieleman, D.P., 193
Tildesley, D.J., 258–259
Tirado-Rives, J., 33–34, 264–265
Tironi, I.G., 2–3
Titmuss, S.J., 363
Titov, A., 392–393
Tivanski, A.V., 305–306
Tiwary, P., 431–432
Tochowicz, A., 296
Tomasi, J., 124, 200–201, 226–228, 444–445, 452–453
Tong, X., 35
Toniolo, A., 447–448
Topaler, M., 366–367
Torda, A.E., 63
Torras, J., 109
Torrie, G.M., 2–3, 82, 166–167, 220–221, 269, 369–370
Torrier, G.M., 196–197, 205
Tosatti, E., 201, 207
Toulouse, J., 396–397
Tozer, D.J., 396
Trager, W.F., 23–24
Tran, H.M., 421
Tratnyek, P.G., 80–81
Trauenheim, T., 126–127
Tribello, G.A., 430–431
Trickey, S.B., 109
Troster, P., 446–447
Trucks, G.W., 226, 349–350, 428–430, 449–450
Truhlar, D., 58–59
Truhlar, D.G., 34, 58–59, 106, 120, 122–123, 126, 160, 193–194, 197–199, 214, 217–218, 221–222, 256–257, 262–269, 271–274, 289, 295, 342–350, 360–361, 363, 365–366, 378–382, 390, 398–399, 445–447, 452–453
Truhlar, D.G.J., 126–127
Truhlar, D.J., 295
Truong, T.N., 198–199
Trygubenko, S.A., 92–93
Tsai, M.-K., 80–81
Tsang, J.E., 306–307
Tschampel, S.M., 169, 426–427
Tsoo, C., 452–453
Tsuneda, T., 396–397
Tuckerman, M.E., 392
Tucks, G.W., 123
Tull, D., 165
Tun, I., 400
Tuñón, I., 67, 222, 296–298, 342
Turney, J.M., 90
Turunen, O., 425
Tuttle, T., 76–77
Tvaroska, I., 67, 163
Tymczak, C.J., 111

U

Uberuaga, B.P., 58–59
Ucisik, M.N., 9–11
Uebayasi, M., 447–448, 452
Ufimtsev, I.S., 217–218, 263, 266–267, 391–398, 408
Umena, Y., 444, 458, 462–463
Urbatsch, I.L., 189–190

V

Vaishnav, Y.N., 20–21
Valdez, C.E., 337
Valiev, M., 80–81
Valleau, J.P., 2–3, 62–63, 82, 111, 166–167, 196–197, 205, 220–221, 269, 369–370
Valley, M.P., 254–256, 263–264, 272–273, 360, 370, 375–378
van Aalten, D.M., 169, 172, 176
van Aalten, D.M.F., 426–427
van den Bedem, H., 253–256
Van den Bosch, M., 143
van der Does, C., 190
van der Kamp, M.W., 218, 260, 263, 444–445
van der Spoel, D., 428–430
van der Vaart, A., 447–448
van Duin, A.C.T., 32–33
van Eijck, B.P., 269–270
van Gunsteren, W.F., 33–34, 63, 130–131, 265, 342, 430–431, 446–447
van Lenthe, E., 230
van Rijn, J., 254–256
Van Severen, M.-C., 141
Van Voorhis, T., 34–35
Vancoillie, S., 120–122, 147–149

Vanden-Eijnden, E., 58–59, 65–66, 92–93, 195–197, 201, 221
Vanderbilt, D., 447–448
VandeVondele, J., 131, 143–145, 172, 409
Vanduijnen, P.T., 444–445
VanEtten, R.L., 225
Vangunsteren, W.F., 2–3
Vanicek, J., 392
Vannini, A., 325
Vanommeslaeghe, K., 89–90, 169, 216–217, 426–427, 446–447, 449–450
Vardi-Kilshtain, A., 263–264, 271–273
Vargas, M.C., 395–396
Várnai, C., 343–344
Varrot, A., 161–162
Vasilyev, V., 452
Vega, C., 265
Versluis, L., 456–457
Vihko, P., 225
Villa, A., 264–265
Villa, E., 349–350
Villa, J., 34, 366–367
Villafranca, J.E., 296
Vincent, M.A., 126–127
Vinyard, D.J., 458, 462
Viragh, C., 390–391
Visscher, L., 343–344
Vivo, M.D., 4–5
Vocadlo, D.J., 161–165
Vogl, P., 396–397
Vogt, L., 444, 458–459, 460f, 462–463, 463–464f
Voityuk, A.A., 76–77
Volpari, C., 325
von Glehn, P., 140
Von Lilienfeld, O.A., 126
von Ragu Schleyer, P., 395–396
Voth, G.A., 58–59, 63–64, 66–67, 174, 197–198, 201, 205–206, 361–362, 366–367, 390–391, 409, 446–447, 452
Vranken, W.F., 428–430
Vreven, T., 126–127, 444–445, 449–450, 452–453, 457

W

Wadt, W.R., 226
Wahiduzzaman, M., 230
Wainwright, T.E., 271
Wakata, A., 222–223
Waldman, M., 33–34
Wales, D.J., 92–93, 220, 262–263
Walker, B., 401
Walker, R.C., 108–109, 128, 216–217, 343–344, 393, 399–400, 408
Wall, M.E., 253–254
Waller, M.P., 343–344
Wallqvist, A., 67, 392, 402–403
Wallrapp, F., 254–256, 342
Walter, T., 58–59
Wang, B., 7, 9–11, 14
Wang, F., 201, 205–207
Wang, J., 106–107, 111–112, 399, 428–431, 435, 458, 462
Wang, J.M., 458, 462
Wang, J.Y., 452
Wang, L., 106–107, 307–308, 310–311, 360, 379–381, 390–410
Wang, L.-P., 34–35, 392–393
Wang, N.X., 393
Wang, Q., 272–273
Wang, S., 106–116, 436–437
Wang, W., 349–350, 428–430, 452
Wang, W.-N., 193
Wang, X., 167
Wang, Y., 20–21, 266, 360–361, 363
Wang, Z., 288, 295–298, 300–306, 305f
Wanko, M., 222
Ward, A.B., 254
Waring, W.S., 20–21
Warshel, A., 33–34, 40–41, 58–59, 76–77, 80–81, 84–85, 106, 126, 128, 131–132, 131f, 140, 143, 160, 193, 196–197, 207, 214, 218, 220–222, 252–253, 258–261, 263–265, 268–269, 271–273, 288–289, 305–308, 310, 320, 342, 362–363, 366–368, 398–399, 444–446, 452
Waskell, L., 20–21
Watanabe, H.C., 343–344
Wataya, Y., 303
Watney, J.B., 305–308, 379
Watt, E.D., 268, 306–307
Watts, A.G., 162–163
Webb, S.P., 271–273, 445–446
Weber, J., 189–190
Wehenkel, A., 162–163
Wei, D., 131

Weigend, F., 124
Weinan, E., 92–93, 221
Weiner, P., 33–34
Weiner, S.J., 33–34
Weiss, R.M., 34, 40–41, 263
Weissig, H., 322–323
Weitman, M., 262–264, 272–273
Wen, P.-C., 189–190, 193
Weng, J.-W., 193
Werner, H.-J., 76–77, 123, 217–218, 222, 447–448
Wesolowski, T., 80–81
Westbrook, J., 322–323
Weyand, S.A., 34–35
Wheeler, R.A., 77, 86–87
White, A., 162–163, 165
White, C.A., 447–448
White, J.K., 81, 83–85, 95–97
Whittleton, S.R., 396
Wiedenmann, A., 189–191
Wiersma-Koch, H.I., 232–233
Wigner, E., 58
Williams, D.E., 109–110
Williams, J.A., 20–21
Williams, R.J., 175
Williams, S.J., 161–162, 165, 169–171
Willitsch, S., 35
Willmer, D., 379–380
Wilson, A., 393
Wilson, I.A., 254, 379
Wilson, K.S., 167–169
Wilson, M.A., 253–254
Wilson, P.M., 296
Windsor, W.T., 10–11
Winfield, S.A., 343–344
Wing, C., 306–307
Winn, P.J., 445–446
Winstein, S., 112
Witek, H.A., 230
Withers, S.G., 163, 165, 179
Wittenberg, B., 42
Wittenberg, J., 42
Wodak, S.J., 238–239
Wodrich, M.D., 395–396
Wolf, R.M., 430–431
Wolf, S., 221–222
Wolfenden, R., 58–59, 160, 252

Wolfson, H., 214
Wolf-Watz, M., 253–254
Wollacott, A.M., 214
Wolski, P.W., 421
Wong, K.F., 288, 305–308, 379
Wong, K.-Y., 360, 370, 375–376
Wood, D.M., 20–21
Wood, R.H., 131
Woodcock, H.L., 76–98, 342
Woods, C.J., 80–81, 84–85, 132
Woods, R.J., 90, 426–427, 430–431
Woolf, T.B., 77–78
Word, J.M., 256–257
Wormit, M., 89–90, 98, 445–446
Wriggers, W., 220–221
Wright, P.E., 221, 223–225, 305–306, 379
Wrodnigg, T.M., 163–165
Wu, C., 216–217, 408
Wu, D., 58–71
Wu, P., 421–422, 423t, 425–428, 434–436
Wu, Q., 34–35
Wu, R., 106–107, 342
Wu, X., 217–218, 263–264, 272–273, 395–396
Wu, Y.J., 58–59
Wu, Z., 10–11
Wydrzynski, T., 458

X

Xantheas, S.S., 446–447
Xia, F., 200–201
Xia, X., 80–81, 265, 363–366, 377–378, 444–445
Xiang, Y., 58–59, 214, 447–448
Xie, D., 106–107
Xie, L., 392–393
Xie, W., 264–265, 360–361, 445–446
Xie, X.L., 16–20
Xing, J., 34
Xiong, G., 408
Xiong, H., 425
Xu, D., 167
Xu, J., 435
Xu, J.C., 58–59
Xu, L., 360
Xu, X., 420
Xu, Y., 360

Y

Yachandra, V.K., 462
Yadav, A., 271–273
Yahashiri, A., 292–293, 310–311
Yamada, K., 109
Yamaguchi, K., 444
Yamanaka, S., 444
Yanagisawa, S., 396–397
Yanai, T., 217–218, 396–397
Yang, A.-S., 124
Yang, J., 343–344
Yang, S., 218–219, 221–222
Yang, W., 58–71, 77, 80–81, 86–87, 92–93, 132, 197, 226, 259–261, 266, 342, 348–349, 396, 446–447
Yang, W.J., 132
Yang, W.T., 58–59, 92–93, 106–108, 110–111, 123, 217–219, 222, 226, 232–233, 446–448, 452–453
Yang, Y., 2–25, 76–77, 86–87, 218, 220, 342, 391–392, 398–399
Yano, J., 462
Yao, X., 435
Yao, Y., 435
Yezdimer, E.M., 131
Yingling, Y.G., 420–422, 423t, 425–427, 436
Yongye, A.B., 169, 426–427
York, D.M., 76–77, 108–110, 124–125, 199–200, 218–219, 258–259, 263, 364–365, 370, 375–378, 446–448
Yosa Reyes, J., 35–36
Yosa, J., 36
Young, J., 190
Youngs, T.G.A., 435
Yousefi, M., 420–421, 423t
Yu, G., 128
Yu, H., 76–77, 86–87, 218, 220–222, 342, 391–392, 398–399
Yu, H.B., 446–447
Yu, Y., 143
Yue, W., 395

Z

Zaitseva, J., 189–193
Zalatan, J.G., 223, 225, 228, 238f, 240–241
Zanghellini, A., 214, 254
Zarkesh-Esfahani, S.H., 420–421, 423t
Zechel, D.L., 161–163, 165
Zeng, X., 92–93
Zhang, D.W., 447–448, 452–453
Zhang, H.M., 20–21, 447–448
Zhang, J., 266–267, 392–393
Zhang, J.Z.H., 360–361, 363, 447–448, 452–453, 464–465
Zhang, L.Y., 446–447
Zhang, M., 225
Zhang, S., 67, 435
Zhang, W., 408
Zhang, Y., 106–116, 126, 132, 260–261, 266, 342, 348–350, 396, 427–428, 447–448, 452–453
Zhao, Q., 392–393
Zhao, Y., 262–264, 435
Zhechkov, L., 230
Zheng, L., 71, 80–81, 86–87
Zheng, W., 92–93, 268
Zheng, X., 396
Zheng, Y.J., 80–81
Zhong, S., 89–90
Zhou, H.X., 58–59, 63
Zhou, H.Y., 34
Zhou, J., 106–107
Zhou, M., 225
Zhou, N., 106–107
Zhou, R.H., 215, 446–447
Zhou, Y., 106–116, 186–207
Zhu, L., 421
Zhu, X., 218–219, 264–265
Ziegler, T., 456–457
Zienau, J., 215, 218–219, 230, 258–259
Zinovijev, K., 67
Zinovjev, K., 67
Zocher, G., 16–20
Zoebisch, E.G., 124, 193–194, 217–218, 364–365
Zoghbi, M.E., 189–190
Zuckerman, D.M., 58–59, 63
Zuo, L., 426–427
Żurek, J., 20–21
Zwanzig, R.W., 2–3, 76–78, 84, 86, 130–131, 270, 361–362

SUBJECT INDEX

Note: Page numbers followed by "*f*" indicate figures, "*t*" indicate tables, and "*s*" indicate schemes.

A

Ab initio molecular dynamics (AIMD) simulations, 390–393
   nuclear quantum effects (NQEs) in, 401–409
Ab initio path integral molecular dynamics (AI-PIMD) simulations, 390–392
Ab initio QM/MM molecular dynamics (aiQM/MM-MD)
   PME potential with periodic boundary condition, 108–110
   simulation, 106–107
Active-site model, of alkaline phosphatase, 227–228f, 231f
Active-site optimization, transition state of alkaline phosphatase, 226–228
Acylation reaction, for serine protease, 112–113, 114f
Adaptive biasing force (ABF) method, 63
Adaptive partitioning (AP) scheme, 344
   applications
      parallel scalability, 352
      parameters for AP calculations, 351
      solvent exchange of protein active site, 352–353
   energy-based, 346–348
   force-based, 348
Adaptive solvation scheme, 343–344
Adenosine triphosphate (ATP)-binding cassette (ABC) transporters, 186–187
   ATP hydrolysis, 189–191
   chemomechanical coupling cycles in, 187–189
   H-Loop His in catalysis, 191–193
Adenosine triphosphate (ATP) hydrolysis, 189–190, 196
   existing SE(-SRP) methods, 199–200
   in HlyB, 193–194, 194f
   potential energy surfaces (PESs) of, 194–195, 195f

Adiabatic reactive molecular dynamics (ARMD), 35–38, 37f
   multisurface, 38–40, 39f
AdoCbl coenzyme, 138f
AIMD simulations. *See* Ab initio molecular dynamics (AIMD) simulations
AI-PIMD simulations. *See* Ab initio path integral molecular dynamics (AI-PIMD) simulations
aiQM/MM-Ewald method, 109
aiQM/MM-MD. *See* ab initio QM/MM molecular dynamics (aiQM/MM-MD)
Alchemical mutations, 221
Alkaline phosphatase (AP)
   active-site model of, 227–228f, 231f
   hydrolysis of phosphate esters in, 224f
   transition state of, 223–241
α-isomers, 161f
α-mannosidase
   active site of, 161f, 168f
   glycosidic bond cleavage in, 177f
AP. *See* Alkaline phosphatase (AP)
AP scheme. *See* Adaptive partitioning (AP) scheme
ARMD. *See* Adiabatic reactive molecular dynamics (ARMD)
Aromatic prenyltransferase NphB, 11–16, 12f
Aspergillus fumigatus prenyltransferase (FtmPT1), 16–20

B

BAR. *See* Bennett's acceptance ratio (BAR)
BDE. *See* Bond dissociation energies (BDE)
Bennett's acceptance ratio (BAR), 77–78
Besler–Merz–Kollman method, 453–454
BEST. *See* Boundary-based-on-exchange symmetry-theory (BEST)
β-isomers, 161f
Big-QM approach, 128–130
Biocatalysis, in ionic liquids (ILs), 420–425

513

Biocatalysts accelerate reaction rates, 252
Biological hydrogen bond network, electronic and nuclear quantum fluctuations in, 406–409
Biological system, electronic quantum effect in, 392–400
Biomolecule, in ionic liquids, 420
Bisection sampling technique, 370
Blue copper proteins, 143–147
BO-MD. *See* Born–Oppenheimer molecular dynamics (BO-MD)
Bond dissociation energies (BDE), 135–136
Born–Oppenheimer ab initio QM/MM molecular dynamics (aiQM/MM-MD), 106–107
Born–Oppenheimer aiQM/MM-MD simulation, 111
Born–Oppenheimer approximation, 362–363
Born–Oppenheimer molecular dynamics (BO-MD), 321
  simulation, 111–112
Boundary-based-on-exchange symmetry-theory (BEST), 343
BQCP, 370
Broyden–Fletcher–Goldfarb–Shanno strategy, 456–457
Buffered-force (BF) scheme, 343–344
Butane torsional rotation
  QM-NBB, 95
  QM-NEW, 96

C

Carbohydrate structure, 160–162
Catalytic effect, 139f
Catalytic mechanism
  glycoside hydrolases, 162–163
  MD simulation of, 167–179
Centroid molecular dynamics (CMD), 390–391
Chain-of-states (COS) path optimization, 67
CHARMM program, 66
Chemical linchpin, 191–193
Chemical reaction, 31–32, 34, 40–41
Circular dichroism (CD), 420–421
Classical free energy simulations, 269–271
Classical MD equilibration, 169–171

CMD. *See* Centroid molecular dynamics (CMD)
Collective variables (CVs), 60–63, 62f
  metadynamics simulation, 174–176
Computational methods, 215–223
  calculation type, 220–222
  levels of theory, 216–219
Condensed-phase system, 364–365
Conductor-like solvent model (COSMO), 124
Conformational catalytic itinerary, 163–167
Covalent connection, QM and MM region, 266
CPMD, 172
Crystal, optimized geometries of small-molecule, 227t
Cytochrome P450s (CYP), 20–24, 22f
  acetaminophen metabolism, 24f
  APAP-CYPs regioselectivity determination, 25t

D

DAD. *See* H-donor and H-acceptor distance (DAD)
Denitrification, 43–45
Density functional theory (DFT), 123, 321, 391–392
  error in size-dependent ionization potential, 398f
Density functional tight binding (DFTB), 32
Density function theory (DFT), 197
DFTB. *See* Density functional tight binding (DFTB)
D-glucose, 161f
Dihydrofolate reductase (DHFR), 288, 379
  case study, 305–312
  catalyze reaction, 306s, 310f
  competitive KIE measurement, 311–312
  network of coupled motions in, 306–311
Dimensionality limit, 64–65
Discrete molecular dynamics (DMD)
  region, setting, 326–330, 326–327f
Discrete path-integral method, 367
Divide-and-conquer technique, 447–448
DNA quadruplexes, MOD-QM/MM models of, 464–468, 465f
Dual-topology molecule, 89
Dynamic light scattering, 420–421

Subject Index

E

EE. See Electrostatic embedding (EE)
Effective fragment potential method, 445–446
Electronic fluctuation, in biological hydrogen bond networks, 406–409
Electronic quantum effect, in biological system, 392–400
 basis sets, 393–395
 density functionals and role of exact exchange, 395–398
 QM regions in QM/MM calculations, 398–400
Electrostatic effect, 137–138
Electrostatic embedding (EE), 126
Electrostatic energy components, 138f
Electrostatic potential fitting (ESP) charges, 109–110
Empirical force fields, 32–34, 39–41
Empirical valence bond (EVB)
 approach, 263
 method, 40–41
 model, 76–77
Energy-based AP schemes, 346–348
Energy component, in glutamate mutase, 135–140, 136f
Energy gap reaction, 268–269
Enzymatic catalytic mechanism, 160
Enzymatic glycoside hydrolysis, 164f
Enzymatic kinetic isotope effect, protein dynamics on, 378–382
Enzyme
 active sites, 92
 inverting, 163
 mo oxo-transfer, 140–143
Enzyme catalysis, 291–292, 296
 dominant factor in, 360
Enzyme reaction, 106–108, 112–113, 116
 metadynamics-based, 65–67
 rates, 59–60, 267–269
Enzyme simulation protocol, 114–116, 115f
Enzyme system modeling, 253–254, 268
 boundary conditions, 257–259
 mutant forms, 256–257
 protein structure, 256–257
 protonation state and tautomers, 256
 silico enzyme-modeling project, 253–259

Escherichia coli CLC (EcCLC), 352–353
Escherichia coli (ecDHFR), 305–307
 network of coupled promoting motions in, 307f
EVB. See Empirical valence bond (EVB)
EXAFS simulations, 461–464
 post-QM/MM refinement model, 462–464
Exchange-hole dipole model (XDM), 395–396
Explicit polarization (X-Pol) method, 445–446
Extended Lagrangian (EL) technique, 197

F

Farnesylation reaction, simulation, 9–11
Farnesyl-pyrophosphate (FPP), conformational activation of, 5–7
Farnesyltransferase (FTase)
 Mg^{2+} binding site in, 7–9, 8f
 protein, 4–11
FEP. See Free energy perturbation (FEP)
FES. See Free energy simulation (FES); Free energy surface (FES)
Feynman path-integral method, 271–272
Feynman path-integral simulation, 366–367
Force-based AP scheme, 348
Force fields (FFs), 34–35, 39–40, 50–51
 empirical, 32–34, 39–41
 parameterization, 425–428
 parameters, 38, 43, 45, 50
Force-matched multistate empirical valence bond (FM-MS-EVB) method, 205–206
Free energy, 106–107, 112–113, 113–114f
Free energy perturbation (FEP), 77–78, 79f, 81, 270
 method, 404
 theory, 361–362, 369
Free energy simulation (FES), 76–79, 130–131
 alchemical, 86–92, 87f
 classical, 269–271
 direct vs. indirect, 79–81
 quantum, 271–274
Free energy surface (FES), 176–178

G

GAC mechanism, 194–195
Gas-phase reactions, 268
Gauge independent atomic orbital (GIAO) method, 468
Gaussian and polynomial functions (GAPOs), 39–40
Gaussian function, 63–64
Gaussian-shaped functions, 174
General amber force field (GAFF)/ antechamber tools, 428–430
Generalized ensemble (GE) sampling, 58–59
  string optimization, OTPRW method, 67–69
  vs. traditional importance sampling, 62–64
Generalized gradient approximation (GGA), 395
Generalized hybrid orbital (GHO), 266
Generalized Langevin equation (GLE), 402–403
Generalized solvent boundary potential (GSBP), 229
GHO. See Generalized hybrid orbital (GHO)
GLE. See Generalized Langevin equation (GLE)
Global hybrid technique, 396–397
Glutamate mutase, energy components in, 135–140, 136f
Glycosidase. See Glycoside hydrolases (GHs)
Glycoside hydrolases (GHs), 160
  catalytic mechanism, 162–163
  electrostatic region, 173f
  reaction mechanism, 162–163
Glycoside hydrolysis, 164f
Glycosidic bond cleavage, 177f
Glycosylation reaction
  in golgi α-mannosidase, 178–179
  substrate-assisted, 69–70, 69f
Golgi α-mannosidase
  glycosylation reaction in, 178–179
  reaction pathway of, 178t
Graphical processing units (GPUs), 391–392
Grid-based hybrid approach, 361
GROMACS MD engine, 428–430, 429f
GSBP. See Generalized solvent boundary potential (GSBP)
G115T mutant, regioselectivity of, 18–20

H

Haemolysin A (HlyA), 190
Haemolysin B (HlyB), 190
  ATP hydrolysis in, 193–194, 194f
  HlyB-NBD, active site, 192f
  reaction, QM/MM minimum free energy paths (MEFPs) study, 195–196
Hartree–Fock (HF) method, 123, 322
H-donor and H-acceptor distance (DAD)
  sampling, 289, 299f
  frequency, 300
High-performance computing (HPC) system, 2–3
His–His dyads, 329–330
H-link-atom approach, 126–127
H-Loop His, in catalysis, 191–193
Human DHFR (hsDHFR), 310–311
Hybrid DFT method, 123
Hybrid QM and discrete molecular dynamics (QM/DMD) engine, 323
Hybrid quantum mechanical/molecular mechanical (QM/MM) method, 76–77, 260–261, 260f, 320
Hydride transfer, 299f
  measurements of KIEs on, 301–303, 301s, 303f
  PESs for, 299f
Hydrogen isotope, Swain–Schaad relationship, 294
Hydrolysis
  ATP, 189–191
  enzymatic glycoside, mechanisms for, 164f
  of phosphate ester in alkaline phosphatase, 224f
  Wang–Landau metadynamics simulation, 61f

I

ILs. See Ionic liquids (ILs)
Inhibitor, TS analogue (TSA), 223
Inosine monophosphate dehydrogenase (IMPDH), 60–61
Ionic liquids (ILs), 420
  biocatalysis in, 420–425
  cations and anions, 424t
  molecular simulations, 423t

motivation for studying biomolecules in, 420
simulating biomolecules in, methods for
　electrostatics and charge scaling considerations, 430–433
　force field/parameterization process, 425–428
　GAFF/antechamber tools, 428–430
　IL-specific aspects of enzyme MD simulations, 433
　typical analysis approaches, 434–435, 434t

K

Kabsch RMSD, 333–334, 334f
KIEs. See Kinetic isotope effects (KIEs)
Kinetic complexity, 292
Kinetic isotope effects (KIEs), 288–290, 360, 368–371
　double-mutant, 308–309
　enzymatic, 378–382
　KIE_{obs} vs. KIE_{int}, 290–292, 291f
　measurement
　　on hydride transfer, 301–303, 301s, 303f
　　on proton abstraction, 303–305, 305f
　for non-hydrogen atoms, 371–373
KSI^{D40N} enzyme mutant, 406–407, 407–408f, 409

L

Langevin equations, 402–403
Learn-on-the-fly (LOTF) approach, 206–207
Liquid scintillation counting (LSC) analysis, 301–302
Long-range corrected (LC) hybrid, 396–397

M

Maltose transporter, 188, 188f
MbNO, rebinding dynamics in, 41–42
MD. See Molecular dynamics (MD)
Mechanical embedding (ME), 126
Mechanical linchpin, 192f
MEPs. See Minimum energy pathways (MEPs)
Message passing interface (MPI), 399
Metadynamics-based enzyme reaction study, 65–67

Metadynamics simulation, 174–176
Metalloenzyme, 214–215, 217–218, 241–242, 320–321, 324–325
Metropolis–Hastings Monte Carlo approach, 132
MFEP. See Minimum free energy path (MFEP)
Microiterative optimization method, 110–111
Minimum energy path (MEP), 70, 220
Minimum free energy path (MFEP), 65–70
　HlyB reaction, 195–196
Modified PAP (mPAP)
　scheme, 352–353
　simulation, 353f
MOD-QM/MM method. See Moving-Domain QM/MM (MOD-QM/MM) methodology
Molecular dynamics (MD), 2–3, 124–125, 166–167. See also Adiabatic reactive molecular dynamics (ARMD)
　algorithm, 2–3
　equilibration, classical, 169–171
　simulation, 2–3, 113–114f, 115, 428–430, 429f
　of catalytic mechanism, 167–179
　method, 58–59
　steered, 4
Molecular ESPs, 452
Molecular mechanics (MM), 106, 120, 216–217
　approach, 128
　boundary across covalent bonds, 107–108, 107f, 109–110f
　energy components, 137f
　free-energy calculations, 130–133
　investigations, 147–149
　methods, 124–125
　region, selection method, 264–265
Molecular mechanics combined with Poisson–Boltzmann and surface area solvation (MM/PBSA), 132–133
Møller–Plesset second-order perturbation theory, 123
Monte Carlo method, 124–125
Monte Carlo simulation, 361–362
Mo oxo-transfer enzymes, 140–143

Moving-Domain QM/MM (MOD-QM/MM) methodology, 444–461, 451*f*
  dangling bonds, 452–453
  ESP charges, 453–456, 454*f*
  model
    of DNA quadruplexes, 464–468
    of OEC of photosystem II, 458–459, 459*f*
  optimization, 456–458
  post-QM/MM structural refinement, 459–461, 460*f*
MPI. *See* Message passing interface (MPI)
Mulliken charges, 447
Multicopper oxidases, 143–147
Multiscale modeling, 253
Multiscale QM/MM method, RP–FM, 200–207
  generic procedure, 201–203, 202*f*
  proton transfer, 203–205
  strategy, 200–201
Multisurface ARMD, 38–40, 39*f*
Mutant forms, 256–257

N

NBD. *See* Nucleotide-binding domain (NBD)
Nitroalkane oxidase (NAO), 377–378
NO detoxification reaction, in trHbN, 42–43
Non-Boltzmann Bennett acceptance ratio (NBB) approach, 131–132
Nonbonded interaction, QM and MM region, 265
Non-hydrogen atoms, kinetic isotope effects (KIEs) for, 371–373
Northrop method, 292–296
NphB
  aromatic prenyltransferase, 11–16, 12*f*
  catalysis, 12–13, 14–15*f*
NQE. *See* Nuclear quantum effects (NQE)
Nuclear quantum effect (NQE), 253, 271–274, 390–391
  in AIMD simulations, incorporating, 401–409
  enzyme enhancing, 377–378
Nuclear quantum fluctuation, in biological hydrogen bond networks, 406–409

Nucleotide-binding domain (NBD), 186–187, 189–190
Number-adaptive scheme, 343–344

O

ONIOM
  electronic embedding (EE) optimization, 457
  method, 449–450
  model, 365
Optimal potential (OP) method, 207
Optimization
  MOD-QM/MM, 456–458
  ONIOM electronic embedding (EE), 457
Order parameter, 60–61
OTPRW
  method, 67–69
  substrate-assisted glycosylation reaction, 69–70, 69*f*
  switching mechanism in, 66*f*

P

PAP scheme. *See* Permuted adaptive partitioning (PAP) scheme
Particle exchange, 344–345
Particle mesh Ewald (PME) method, 109–110
Path-integral free energy perturbation (PI-FEP) method, 368–371
Pathintegral (PI) method, 271–273, 272*f*
Path integral molecular dynamics (PIMD) simulations, 390–391, 401–402
  convergence using generalized Langevin equations, 402–403
  extracting thermodynamic isotope effects from, 403–406
Path-integral simulation, 361–362, 366–367, 369–370
PBC. *See* Periodic boundary conditions (PBC)
PDB. *See* Protein Data Bank (PDB)
Periodic boundary conditions (PBC), 258–259, 258*f*
Permuted adaptive partitioning (PAP) scheme, 343–344, 346, 348
  implementation in QMMM, 349–351, 350*f*
PES. *See* Potential energy surface (PES)

Phosphate ester hydrolysis reaction, transition state, 238f
Phosphoryl transfer, 223, 224f, 234–235, 240–241
Photoactive yellow protein (PYP), 399–400, 400f
Photosystem II, MOD-QM/M model of OEC of, 458–459, 459f
PI-FEP method. *See* Path-integral free energy perturbation (PI-FEP) method
PIMD simulation. *See* Path integral molecular dynamics (PIMD) simulation
Plastocyanin
  QM systems of, 144–145f
  redox potential, 146t
PMF. *See* Potential of mean force (PMF)
Polarizable continuum model (PCM), 124
Polarization effect, 446–447
Potential energy function, 363
Potential energy of the QM/MM (PBC) system, 108–109
Potential energy surface (PES), 32, 220, 360–361, 363–366, 459
  of ATP hydrolysis, 194–195, 195f
  covalent connections between QM and MM region, 266
  DAD sampling frequency, 300
  for H-transfer, 299f
  hybrid quantum mechanical–molecular mechanical approach, 260–261
  MM region selection method, 264–265
  nonbonded interactions between QM and MM region, 265
  QM region
    selection method, 262–264
    size, 266–267
Potential of mean force (PMF), 366–368
  construction, 116
  of proton transfer, 205, 206f
  simulation, 2–3
PPVs. *See* Protein promoting vibrations (PPVs)
Product state (PS), 254–256
Protein Data Bank (PDB), 322–323, 325, 328f

Protein dynamics, on enzymatic KIEs, 378–382
Protein farnesyltransferase, 4–11
Protein promoting vibrations (PPVs), 310–311
Proton abstraction, 299f
  KIEs measurement on, 303–305, 305f
Proton transfer
  PMFs of, 205, 206f
  QM/MM setup and computer programs, 203
  RP–FM
    base on gas-phase reaction path, 203–205, 204f
    for solution-phase, 205
PS. *See* Product state (PS)
Pseudobond approach, QM/MM boundary problem, 107–108, 107f
Puckering coordinates, 179
Pyranose, 162f

Q

Q-chem–amber interface, 111–113
Q-chem–tinker interface, 112–113
QM/MM. *See* Quantum mechanics/molecular mechanics (QM/MM)
QM-NBB. *See* Quantum mechanics-non-Boltzmann BAR method (QM-NBB)
QM-NEW method. *See* Quantum mechanics-nonequilibrium work (QM-NEW) method
QTCP, 131f, 132–133, 143–147, 146t
QTST. *See* Quantum transition state theory (QTST)
Quantized classical path (QCP), 272–273
Quantum-classical boundary effect, 260–261
Quantum effect, enzyme enhancing nuclear, 377–378
Quantum free energy simulation
  PI methods, 272–273, 272f
  quantum simulation methods, 271–272
  semiclassical methods, 273–274
  wavefunction (WF) methods, 273
Quantum mechanics (QM), 217–218, 360–361
  approach, 120, 128

Quantum mechanics (QM) (*Continued*)
  calculations, 120, 125–128
  free-energy calculations, 130–133
  investigations, 147–149
  methods, 106, 123–124
    boundary across covalent bonds, 107–108, 107f, 109–110f
    on-the-fly reclassification, 344
    region
      selection method, 262–264
      size, 266–267
Quantum mechanics and discrete molecular dynamics (QM/DMD)
  blueprint of algorithm, 331–333, 332f
  convergence analysis, 335–337
  energy, 336f
  engine, 321
    advantages, 321–323
    full iteration, 333–335
    initial iteration, 333
    method, 323–324, 324f
    permissions, copyrights, and utilities, 337–338
    protein crystal structure, 325–326, 326–327f
    setting, 324–331
    task, 337
Quantum mechanics/molecular mechanics (QM/MM)
  algorithm, 343–344
  buffer zone, 344–345
  calculations, 218, 398–400
  covalent connections between, 266
  docking of NO_2-ethane, 255f
  hybrid, 76–77, 260–261, 260f
  interactions, 218–219, 219f
  interpolating energy and gradients, 345–348
  many-body expansion, 346, 350–351
  MD setup and equilibration, 171–174
  MD simulation of catalytic mechanism, 167–179
  metadynamics, 160
    simulation of GHs, 170f
  method, 449–450, 450f
  minimum free energy paths (MEFPs) study of HlyB reaction, 195–196
  nonbonded interactions between, 265

optimization, transition state of AP, 229–230
PAP and SAP implementation in, 349–351, 350f
PES scan, 230–233, 232f
PMF simulation, 14–16
potential energy study, 193–194
primary subsystem (PS), 342
schemes, 321–323
secondary subsystem (SS), 342
setting zeros of energy, 348–349, 349f
simulations, 160, 218–219, 309
software packages with multiscale capabilities, 275–276t
Quantum mechanics-non-Boltzmann BAR method (QM-NBB), 82–84, 83f, 88–89, 94f
  butane torsional rotation, 95
Quantum mechanics-nonequilibrium work (QM-NEW) method, 84–85, 86f, 90
  butane torsional rotation, 96
Quantum-refinement approach, 133–134
Quantum transition state theory (QTST), 366, 368

R

Range-separated hybrid (RSH) technique, 396–397
RBOPs. *See* Reaction bond order parameters (RBOPs)
RCD method. *See* Reaction coordinate driving (RCD) method
Reactant state (RS), 254–256
Reaction bond order parameters (RBOPs), 60–61
Reaction coordinate (RC), 2–3
Reaction coordinate driving (RCD) method, 110–111
Reaction mechanism, and conformational itinerary, 176–179
Reaction path–force matching (RP–FM), multiscale QM/MM method, 200–207
  generic procedure, 201–203, 202f
  proton transfer, 203–205
  strategy, 200–201
Reaction profile, 92–97

Subject Index

Reactive molecular dynamics (RMD), 34.
 See also Adiabatic reactive molecular dynamics (ARMD)
Receptor tyrosine kinase (RTK) signal transduction pathway, 4–5
Reference-potential method, 131, 131f
Ring-polymer molecular dynamics (RPMD), 390–391
RMD. See Reactive molecular dynamics (RMD)
Room temperature ionic liquids (RTILs), 420
RPMD. See Ring-polymer molecular dynamics (RPMD)
RS. See Reactant state (RS)
RTILs. See Room temperature ionic liquids (RTILs)
Rule of geometric mean (RGM), 374–376

S

SAC. See Substrate-assisted catalysis (SAC)
SAP scheme. See Sorted adaptive-partitioning (SAP) scheme
SBC. See Stochastic boundary conditions (SBC)
Schrödinger equation, 32–33, 123
Semiclassical method, 273–274
Semiempirical model, 447
Semiempirical QM/MM (SQM/MM), 76–77
Serine protease, acylation reaction for, 112–113, 114f
Silico enzyme-modeling project, 253–259
Simulation
 Born–Oppenheimer aiQM/MM-MD, 111
 Born–Oppenheimer molecular dynamics, 111–112
 classical free energy, 269–271
 molecular dynamics, 113–114f, 115
 protocol, 114–116, 115f
 QM/MM, 309
Sizeconsistent multipartitioning scheme, 343–344
Sorted adaptive-partitioning (SAP) scheme, 343–344, 347–350
 implementation in QMMM, 349–351, 350f

Specific reaction parameters (SRPs), 198–199, 202f, 204f
Split valence basis set (SVP), 124
SRPs. See Specific reaction parameters (SRPs)
SSE. See Swain–Schaad exponent (SSE)
Standard crystallographic refinement, 133
Steered molecular dynamics (MD) method, 4
Stochastic boundary conditions (SBC), 258–259, 258f
Stoddart's diagram, 166f
Substrate-assisted catalysis (SAC), 191–193
Substrate-assisted glycosylation (SAG) reaction, OTPRW, 69–70, 69f
Swain–Schaad exponent (SSE), 295, 374–376

T

Tautomers, 256
TeraChem electronic structure program, 391–395, 399, 406
Tert-butyl chloride, dissociation of, 112, 113f
Thermal reservoir, 58
Thermodynamic cycle, 131f
Thermodynamic integration (TI), 270
Thermodynamic isotope effect, 403–406
Thymidylate synthase (TSase), 288
 case study, 296–305
 chemical mechanism, 297s
 KIEs measurement
  on hydride transfer, 301–303, 301s, 303f
  on proton abstraction, 303–305, 305f
TI. See Thermodynamic integration (TI)
TMD. See Transmembrane domain (TMD)
Transition path sampling (TPS), 221, 270–271
Transition state (TS), 58, 215–216, 254–256
 of AP, 223–241
  active-site optimizations, 226–228
  comparison with experiment, 236–240
  computational benchmarks, 226
  QM/MM optimization, 229–230
  QM/MM PES scan, 230–233, 232f
  tests of protonation states, 233–235
  TSA binding, 223–226
 phosphate ester hydrolysis reaction, 238f
 vs. TS analogue (TSA), 240–241

Transition state analogue (TSA)
   binding, 223–226
   geometriesa of calculated and experimental, 237t
   inhibitors, 223
   structures, 236f
   vs. TSs, 240–241
Transition state theory (TST), 267, 366
Transmembrane domain (TMD), 186–188
trHbN
   competitive ligand binding in, 43–49
   NO detoxification reaction in, 42–43
TS. See Transition state (TS)
TSA. See Transition state analogue (TSA)
TSase. See Thymidylate synthase (TSase)
TST. See Transition state theory (TST)

U
Umbrella sampling method, 3

V
Variational transition state theory (VTST) method, 361
Variational transition state theory with multidimensional tunneling (VTST/MT), 273–274

W
Wang–Landau flat-histogram procedure, 63–64
Wang–Landau metadynamics, 61f, 65–66
Water, tert-butyl chloride in, 112, 113f
Wavefunction (WF) methods, 273
WT FtmPT1, regioselectivity of, 18

Z
Zero point energy (ZPE), 390–391

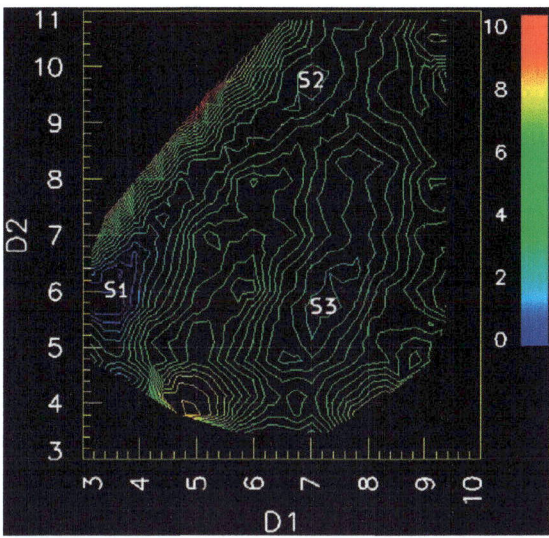

Y. Yang et al., **Fig. 6** 2D Free energy profile (kcal/mol) of 1,6-DHN binding in NphB as a function of D_1 and D_2 (Å). *Reprint with permission from Cui, G., Li, X., & Merz, K. M., Jr. (2007). Understanding the substrate selectivity and the product regioselectivity of Orf2-catalyzed aromatic prenylations. Biochemistry, 46(5), 1303–1311. doi:10.1021/bi062076z. Copyright (2007) American Chemical Society.*

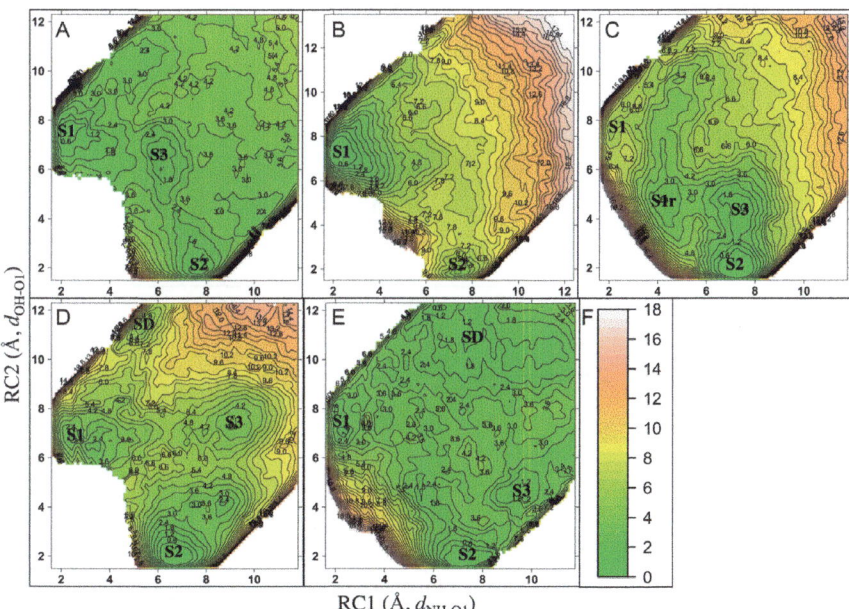

Y. Yang et al., **Fig. 12** Free energy profiles for CYP–acetaminophen binding. (A) CYP3A4, (B) CYP2E1, (C) CYP2A6, (D) CYP1A2, (E) CYP2C9. Topography energy legend with energy values (kcal/mol) represented by defined color is given in (F). Binding states, S1, S2, S3, S1r, and SD (not discussed in this chapter) are labeled. *Reprinted from Yang, Y., Wong, S. E., & Lightstone, F. C. (2014). Understanding a substrate's product regioselectivity in a family of enzymes: A case study of acetaminophen binding in cytochrome P450s. PloS One, 9(2), e87058. doi:10.1371/journal.pone.0087058.*

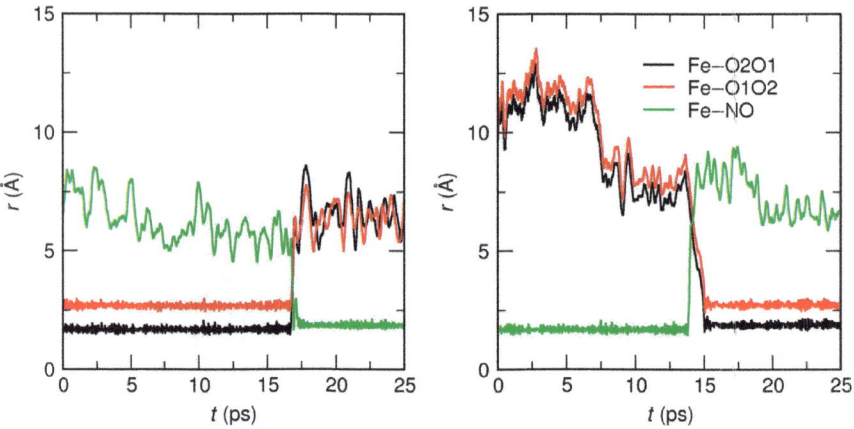

A.K. Das and M. Meuwly, Fig. 4 Variation of distance between ligands and the metal during the ligand exchange reaction.

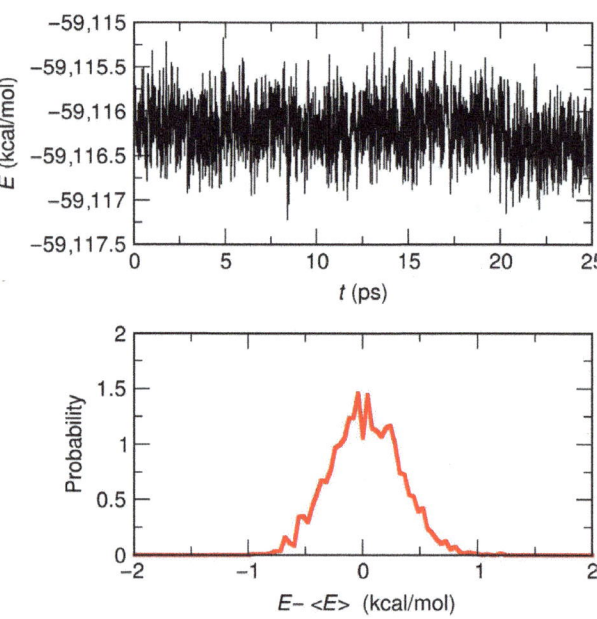

A.K. Das and M. Meuwly, Fig. 5 Variation of total energy along the trajectory (*upper panel*) together with the distribution around the mean (*lower panel*).

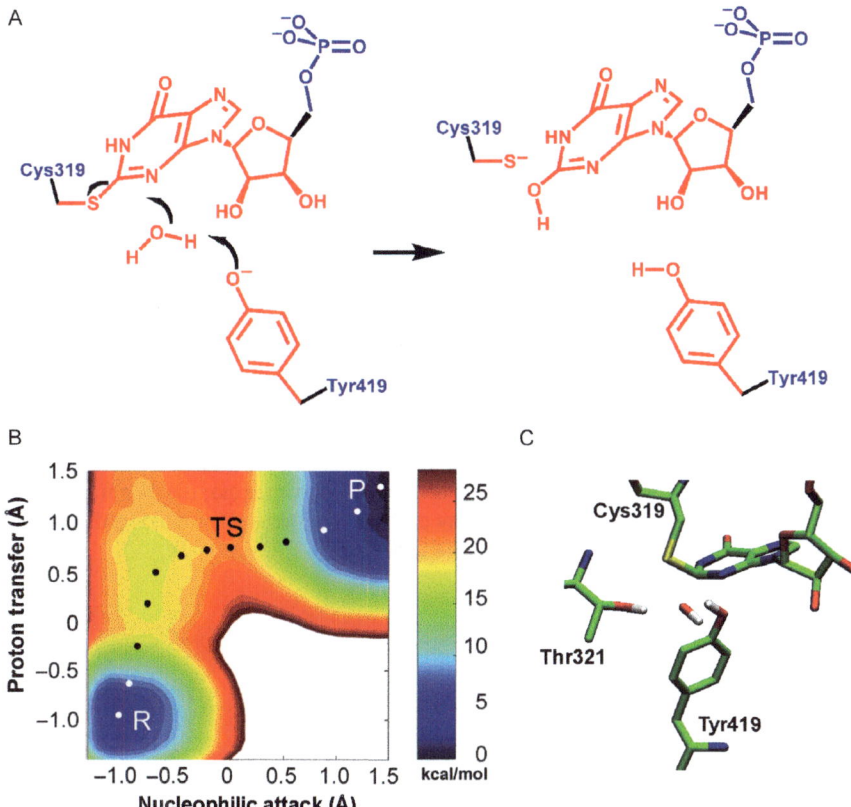

D. Wu et al., Fig. 1 Wang–Landau (flat-histogram) metadynamics simulation of the hydrolysis step in the IMPDH Arg418Gln variant. Figure was originally published in PLoS Biology (open-access, doi:10.1371/journal.pbio.0060206.g003). (A) The proposed mechanism on the hydrolysis of E-XMP* with Tyr419 acting as the general base. (B) The free energy landscape of the Tyr419 pathway in the Ar418Gln variant. P, product; R, reactant; and TS, transition state. (C) The corresponding transition state structure.

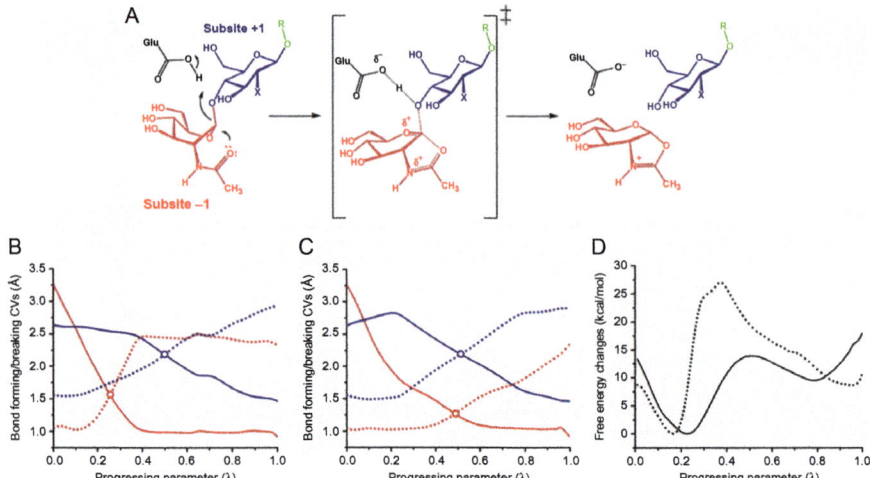

D. Wu et al., Fig. 4 The OTPRW simulation of the substrate-assisted glycosylation reaction in OfHex1. (A) The proposed mechanism on the substrate-assisted glycosylation reaction in OfHex1. (B) The chemical order parameter changes along the initial minimum energy path. *Red*: The CVs for the proton transfer process. *Blue*: The CVs for the nucleophilic substitution process. (C) The chemical order parameter changes along the OTPRW optimized minimum free energy path. *Red*: The CVs for the proton transfer process. *Blue*: The CVs for the nucleophilic substitution process. (D) The free energy changes along the initial minimum energy path (the *dotted line*) and along the OTPRW optimized minimum free energy path (the *solid line*).

Y. Zhou et al., Fig. 1 Illustration of the pseudobond approach in the treatment of the QM/MM boundary problem.

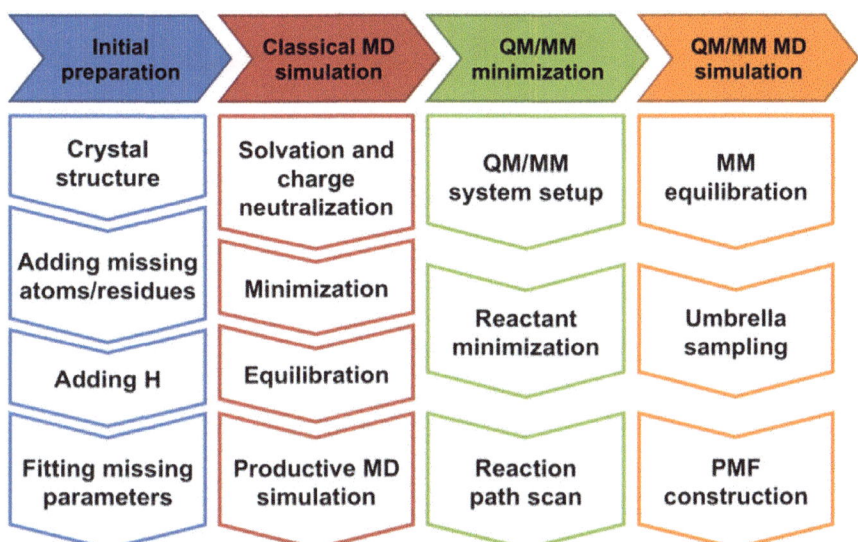

Y. Zhou et al., Fig. 6 Illustration of our enzyme simulation protocol.

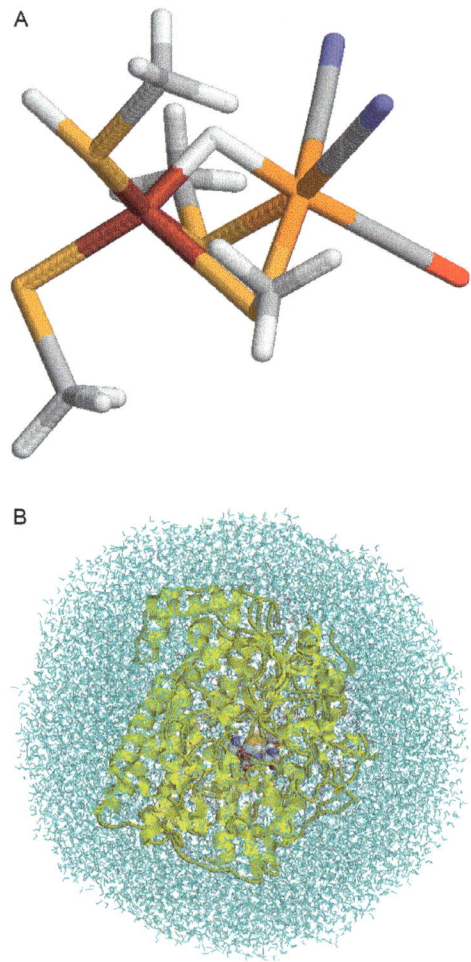

U. Ryde, Fig. 1 Examples of (A) QM-cluster, (B) QM/MM (the QM system is shown as *balls*), and

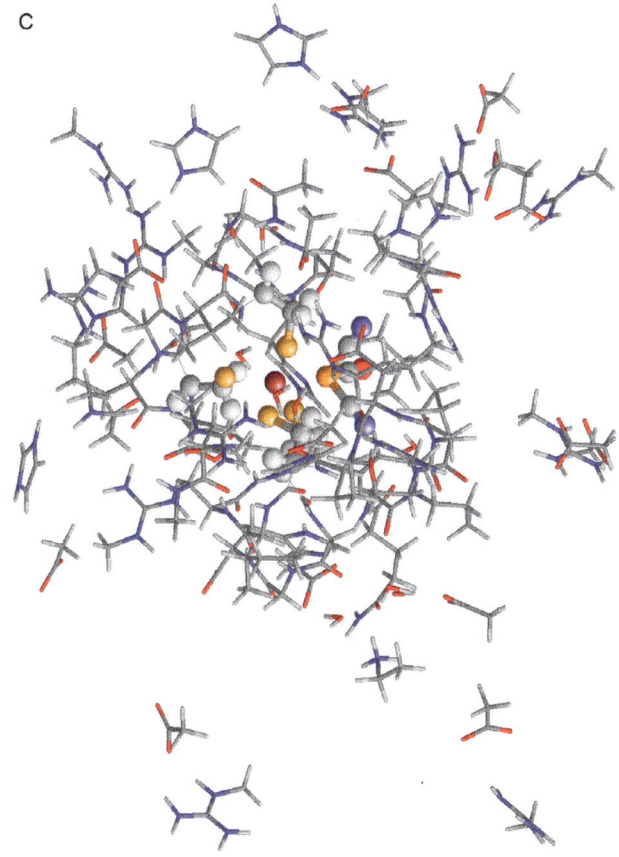

U. Ryde, Fig. 1—Cont'd (C) big-QM models for [NiFe] hydrogenase (Dong & Ryde, 2016).

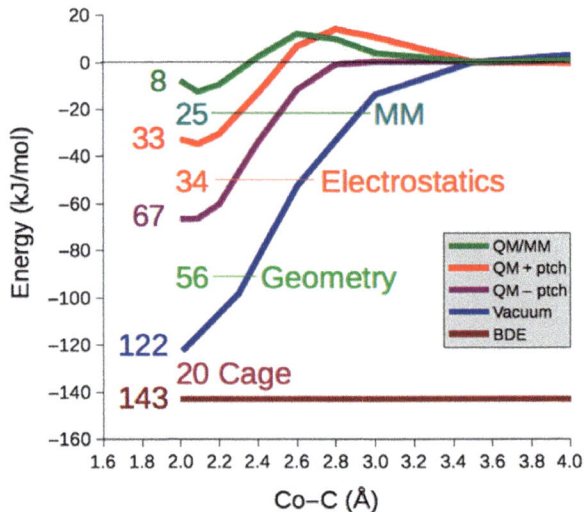

U. Ryde, Fig. 3 Energy components for glutamate mutase (Jensen & Ryde, 2005). The *colored numbers* to the *left* are the energies for the five curves at Co–C=2.0 Å (using the energy at Co–C=3.5 Å as the reference). The *numbers* to the *right* are the pairwise differences between these numbers. All energies are in kJ/mol. The components are further explained in the main text.

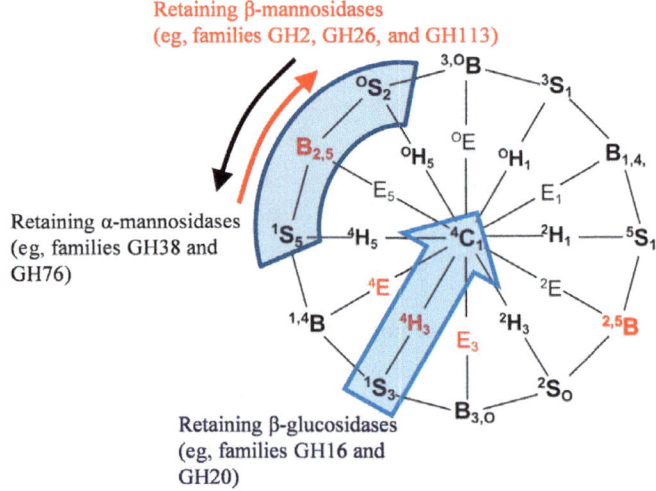

L. Raich et al., Fig. 4 Stoddart's diagram (centered at $^4C_1$) illustrating the proposed itineraries followed by selected GH families.

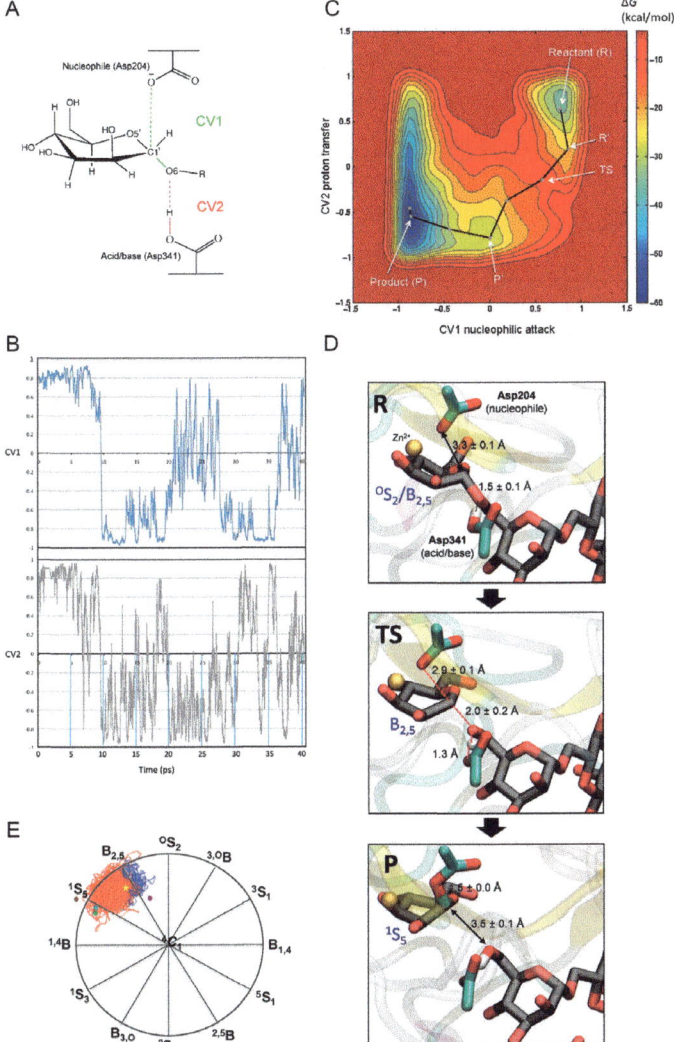

L. Raich et al., Fig. 9 Results obtained for the QM/MM metadynamics simulation of glycosidic bond cleavage in α-mannosidase II. (A) Collective variables definition. (B) Evolution of the CVs during the metadynamics simulation. (C) Reconstructed free energy landscape. The minimal free energy pathway is shown by *gray dots* connected with *black lines*. Contour lines are separated by 4 kcal/mol intervals. (D) Structures of the reactants (R), transition state (TS), and glycosyl-enzyme intermediate (P). The Zn ligands have been omitted for clarity. (E) Metadynamics simulation, mapped onto a Cremer–Pople sphere. The conformations visited before the TS are shown in *blue*, while those visited after the TS are shown in *red*. The average TS conformation is shown with a *yellow star*. The experimentally observed conformation for the Michaelis complex in a mutated (D341N) enzyme is shown with a *purple dot* (PDB 3BUP). Three conformations corresponding to glycosyl-enzyme intermediate structures are shown as *brown* (PDB 1QX1), *green* (PDB 1QWU), and *light blue* (PDB 1QWN) *dots*. Adapted with permission from Petersen, L., Ardèvol, A., Rovira, C., & Reilly, P. J. (2010). Molecular mechanism of the glycosylation step catalyzed by Golgi alpha-mannosidase II: A QM/MM metadynamics investigation. *Journal of the American Chemical Society, 132*, 8291–8300. Copyright 2010 American Chemical Society.

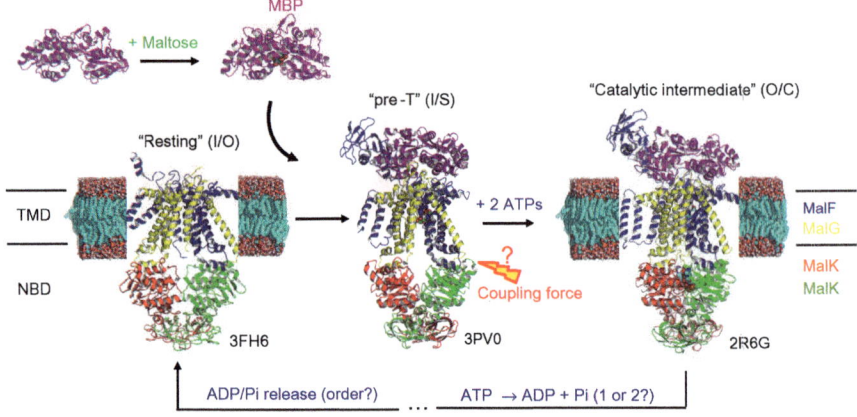

Y. Zhou et al., Fig. 1 Crystal structures captured for key intermediate states in the working cycle of maltose transporter. The TMD/NBD conformations are given in parentheses (TMD: I or O, for inward- or outward-facing conformations; NBD: O, S, or C, for open, semiopen, or closed conformations).

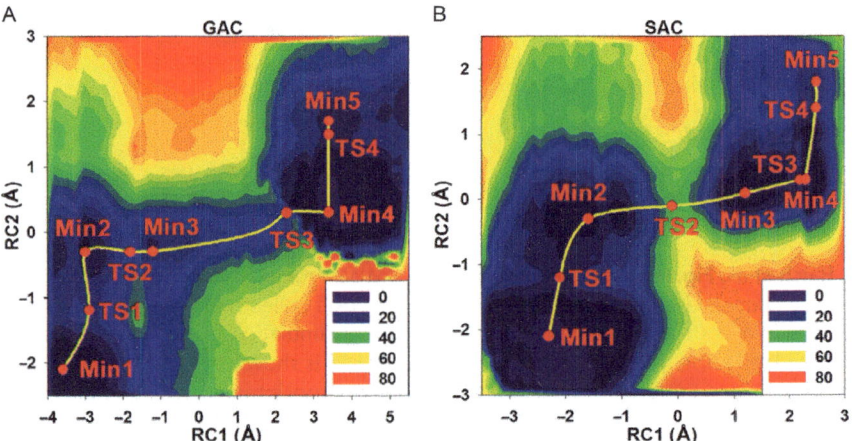

Y. Zhou et al., Fig. 4 2D potential energy surfaces (AM1/CHARMM) of ATP hydrolysis catalyzed by HlyB-NBD: (A) the GAC and (B) the SAC mechanisms. The highest barrier found in the GAC mechanism (22.1 kcal/mol; GAC:TS3) is substantially lower than that in the SAC mechanism (32.1 kcal/mol; SAC:TS2). *Reproduced from Zhou, Y., Ojeda-May, P., & Pu, J. (2013). H-loop histidine catalyzes ATP hydrolysis in the E. coli ABC-transporter HlyB. Physical Chemistry Chemical Physics, 15, 15811, doi: 10.1039/C3CP50965F with permission from the PCCP Owner Societies.*

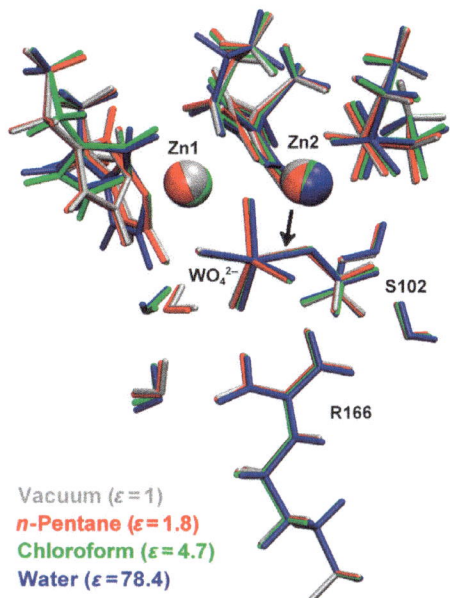

D. Roston and Q. Cui, Fig. 4 Structures of the active-site model of AP with tungstate using different solvent models (indicated by *color*). Structures were optimized at the B3LYP/6-31+G* level, using an effective core potential for the metal atoms. The bond from the W to the nucleophilic serine alkoxide (O_{nuc}) is indicated by the *arrow*. Alpha carbons were frozen to their crystal structure (Stec et al., 2000) positions during the optimizations.

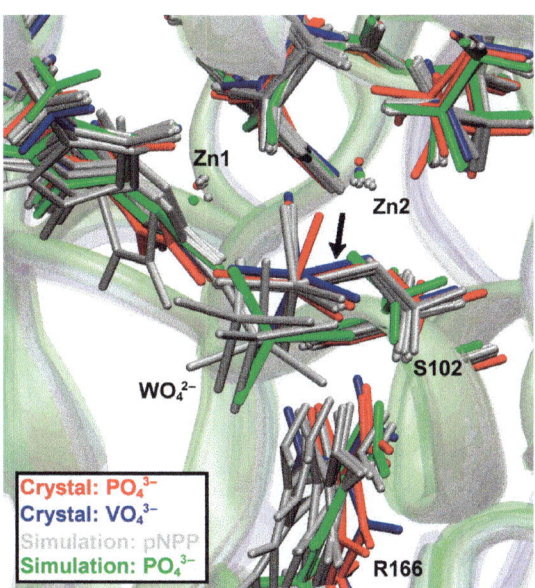

D. Roston and Q. Cui, Fig. 5 Optimized structures from B3LYP/MM calculations of tungstate in the active site of AP. Starting structures were obtained from crystal structures and simulations with other ligands as indicated by *color*. In cases where the W–O_{nuc} distance is less than 2.5 Å, a bond is drawn between the two atoms, indicated by the *arrow*.

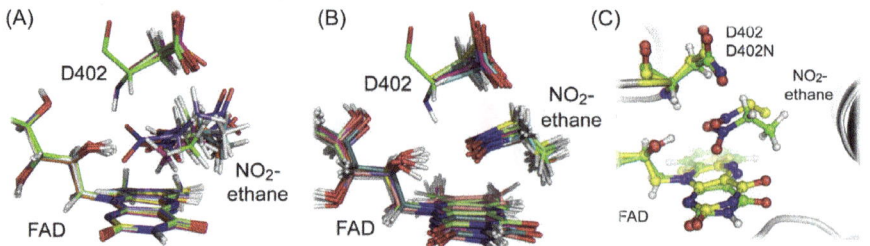

M. Dixit et al., Fig. 1 QM/MM docking of NO$_2$-ethane in the active site of nitroalkane oxidase. (A) Randomly generated configurations of NO$_2$-ethane. (B) Final docked poses of NO$_2$-ethane. (C) Comparison of a docked pose (*green* carbons) and the crystal structure of the D402N mutant enzyme (*yellow* carbons).

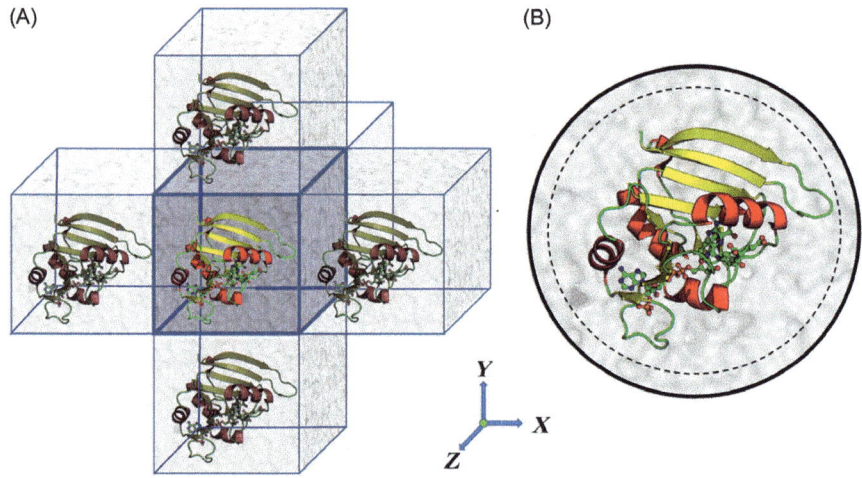

M. Dixit et al., Fig. 2 Possible setups for enzyme systems (ie, dihydrofolate reductase): (A) Periodic boundary conditions where the enzyme is immersed in an orthorhombic cell, which is surrounded by an infinite number of replicas of the unit cell. (B) Stochastic boundary conditions where the enzyme is immersed in a spherical water droplet. Within the *dotted circle*, the atoms are propagated by unrestrained MD, whereas beyond it the system is restrained and propagated with Langevin dynamics.

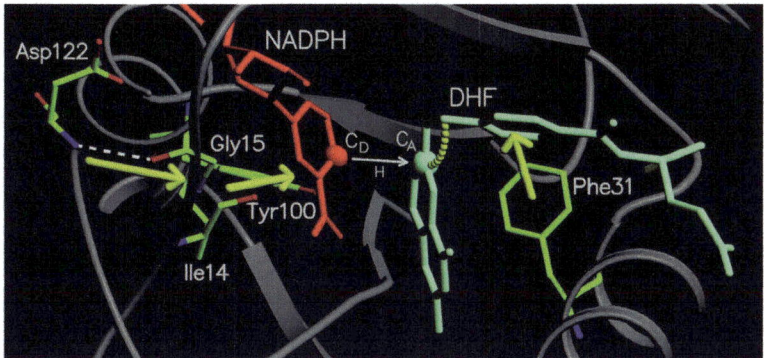

P. Singh et al., Fig. 6 Diagram of a portion of the network of coupled promoting motions in ecDHFR. The *yellow arrows* and *arc* indicate the coupled promoting motions. From Agarwal, P. K., Billeter, S. R., Rajagopalan, P. T., Benkovic, S. J., & Hammes-Schiffer, S. (2002). Network of coupled promoting motions in enzyme catalysis. Proceedings of the National Academy of Sciences of the United States of America, 99, 2794–2799, with permission from Proceedings of the National Academy of Sciences of the United States of America.

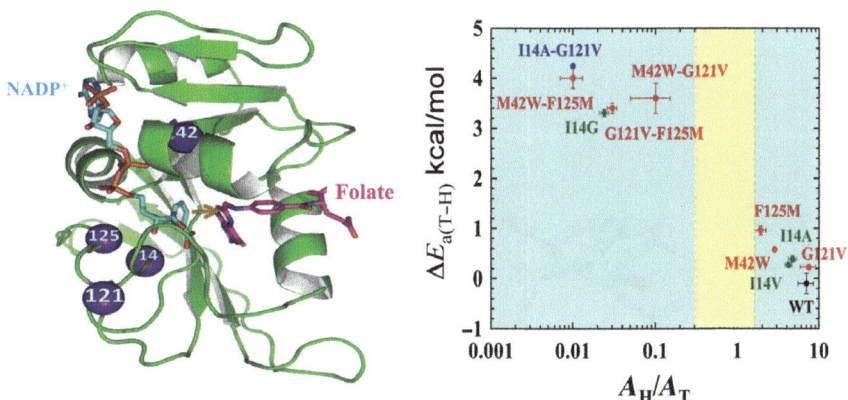

P. Singh et al., Fig. 8 Roles of active site and distal residues on the DHFR-catalyzed reaction. *Left panel*: Structure of WT-ecDHFR (PDB Code 1RX2), with folate in *magenta* and NADP$^+$ in *light blue*. A *yellow arrow* marks the hydride's path from C4 of the nicotinamide to C6 of the folate, and the residues studied in Singh et al. (2014) are marked as *purple spheres*. *Right panel*: Presentation of the temperature-dependence parameters of KIE$_{int}$ for WT (*black*), distal (*red*), and active site I14 (*green*) mutants of DHFR, where error bars represent standard deviation. For each point, the ordinate is the isotope effect on activation energy (ΔE_a), and the abscissa is the isotope effect on the Arrhenius preexponential factor (A_H/A_T), both of which were determined from a nonlinear regression of KIE$_{int}$ to the Arrhenius equation. The *yellow block* represents the semiclassical range of the Arrhenius preexponential factor (0.3–1.7) (Kohen, 2006). Reprinted from Singh, P., Francis, K., & Kohen, A. (2015). Network of remote and local protein dynamics in dihydrofolate reductase catalysis. ACS Catalysis, 5, 3067–3073, with permission from the American Chemical Society.

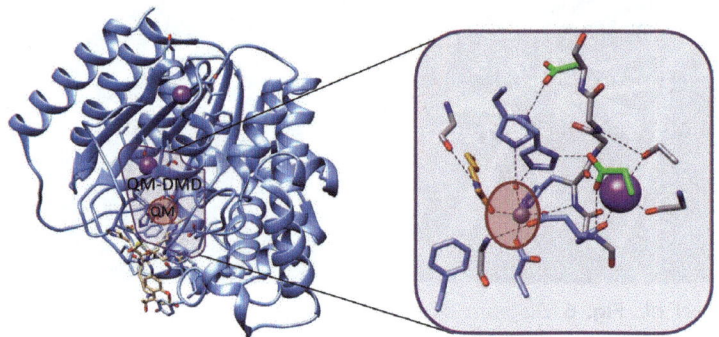

N.M. Gallup and A.N. Alexandrova, Fig. 1 The *purple* and *red* regions represent the QM–DMD boundary, QM–DMD region, and QM region. The rest of the protein constitutes the DMD-only region. The *inset* on the *right* is the cutout of the QM–DMD region.

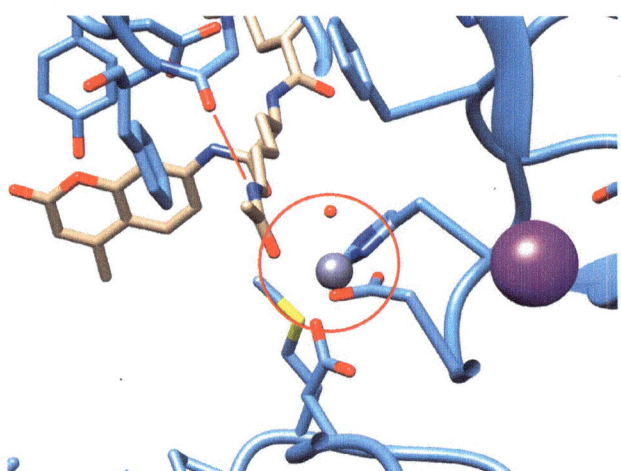

N.M. Gallup and A.N. Alexandrova, Fig. 5 *Red marks* indicate potential targets for DMD constraints. The *red line* connects a hydrogen bond between the substrate and the backbone of the protein, while the *red circle* highlights the coordinating ligands around the active site metal.

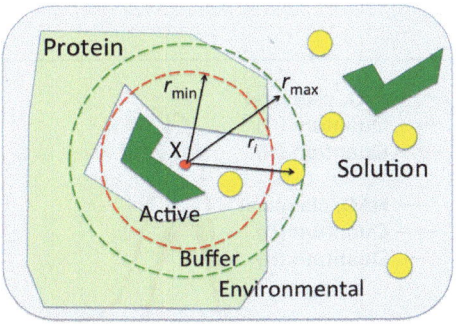

A. Duster et al., Fig. 1 Illustration of adaptive-partitioning setups. The protein is displayed as a *light-green polygon*, substrate molecules as *dark-green polygons*, and solvent molecules as *yellow spheres*. The active zone is defined as a sphere of radius r_{min} (*dashed red circle*) centered at the active-site center X (*small red circle*). The environmental zone is outside the sphere of radius r_{max} (*dashed green circle*). The thin shell between r_{min} and r_{max} represents the buffer zone. The distance r_i is measured between the active-site center and the ith group, which may be a solvent molecule, a substrate molecule, a cofactor molecule, or a residue side-chain of the protein.

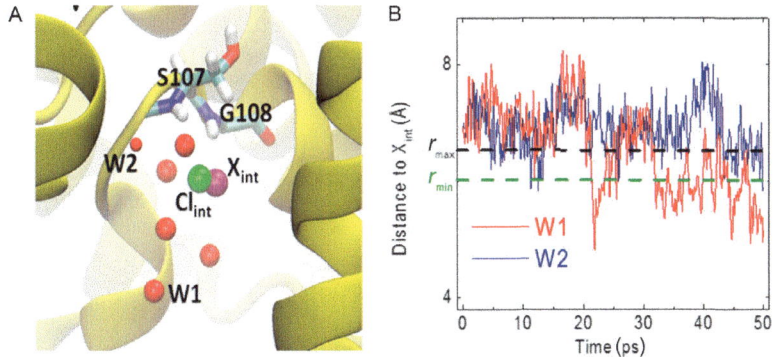

A. Duster et al., Fig. 4 (A) Snapshot from the mPAP simulation. The protein is shown as cartoon in *yellow*, with two key residues S107 and G108 in *sticks* (N: *blue*, C: *cyan*, O: *red*, and H: *white*). The bound Cl⁻ (Cl$_{int}$) and pseudoatom (X$_{int}$) are indicated as *green* and *purple spheres*, respectively, and water in the active and buffer zones as *large and small red spheres*, respectively. W1 and W2 are two water molecules, whose distances to X$_{int}$ are plotted in (B) over simulation time, where the *green* and *black dashed lines* indicate the active-buffer and buffer-environmental boundaries, respectively. *Reprinted with permission from Pezeshki, S., Davis, C., Heyden, A., & Lin, H. (2014). Adaptive-partitioning QM/MM dynamics simulations: 3. Solvent molecules entering and leaving protein binding sites.* Journal of Chemical Theory and Computation 10, *4765–4776. Copyright 2014 American Chemical Society.*

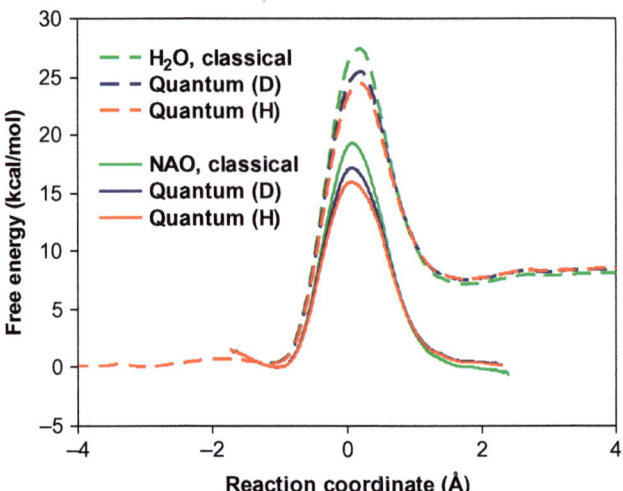

J. Gao, Fig. 3 Classical (*blue*) and quantum mechanical potential of mean force for the proton (*red*) and deuteron (*green*) transfer from nitroethane to acetate ion in water (*dashed curves*), and to Asp402 (*solid curves*) in the active site of nitroalkane oxidase.

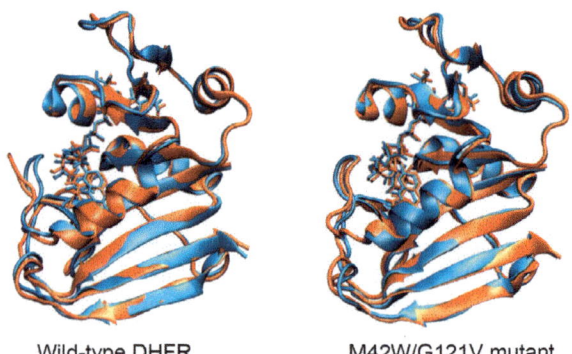

J. Gao, Fig. 4 Superposition comparison of snapshots of the Michaelis complex (*blue*) and transition state (*brown*) structures for the wild-type enzyme dihydrofolate reductase (*left*) and the M42W/G121V double mutant (*right*).

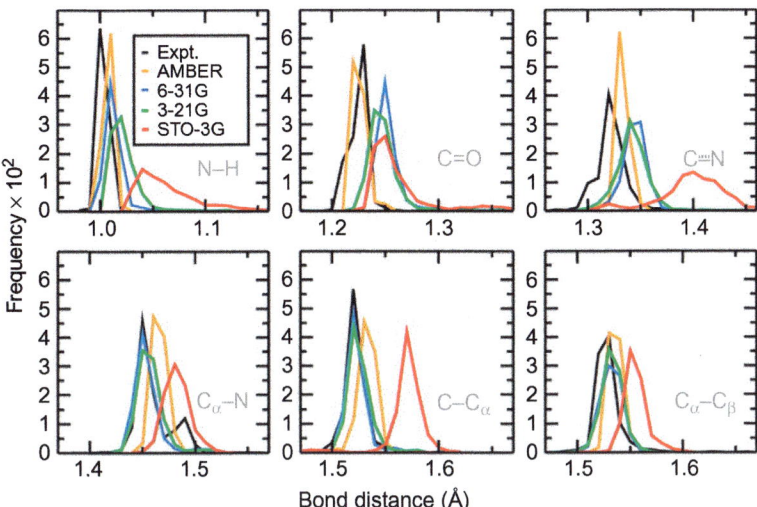

L. Wang et al., Fig. 2 Distributions of bond length for experimental structures (*black*) compared to those obtained from structural optimization with the classical AMBER force field (*orange*), ωPBEh/STO-3G (*red*), ωPBEh/3-21G (*green*), and ωPBEh/6-31G (*blue*) for a collection of 58 proteins. *Reproduced from Kulik, H. J., Luehr, N., Ufimtsev, I. S., & Martinez, T. J. (2012). Ab initio quantum chemistry for protein structures.* The Journal of Physical Chemistry B, 116 *(41), 12501–12509. doi: 10.1021/jp307741u with permission from American Chemical Society.*

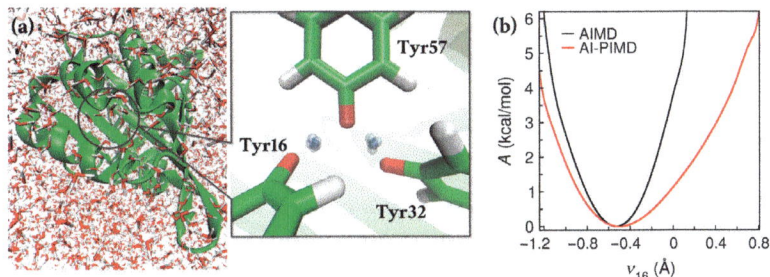

L. Wang et al., Fig. 5 (A) A snapshot of KSI^{D40N} from the AI-PIMD simulation. Its active-site hydrogen bond network is enlarged, in which *green*, *red*, and *white* represent carbon, oxygen, and hydrogen atoms, respectively. The *blue spheres* are the full ring-polymer representation of the protons. For clarity, all the other atoms are shown as their centroids. (B) The free energy surface of the proton movement along ν_{16}, as calculated from AIMD and AI-PIMD simulations.

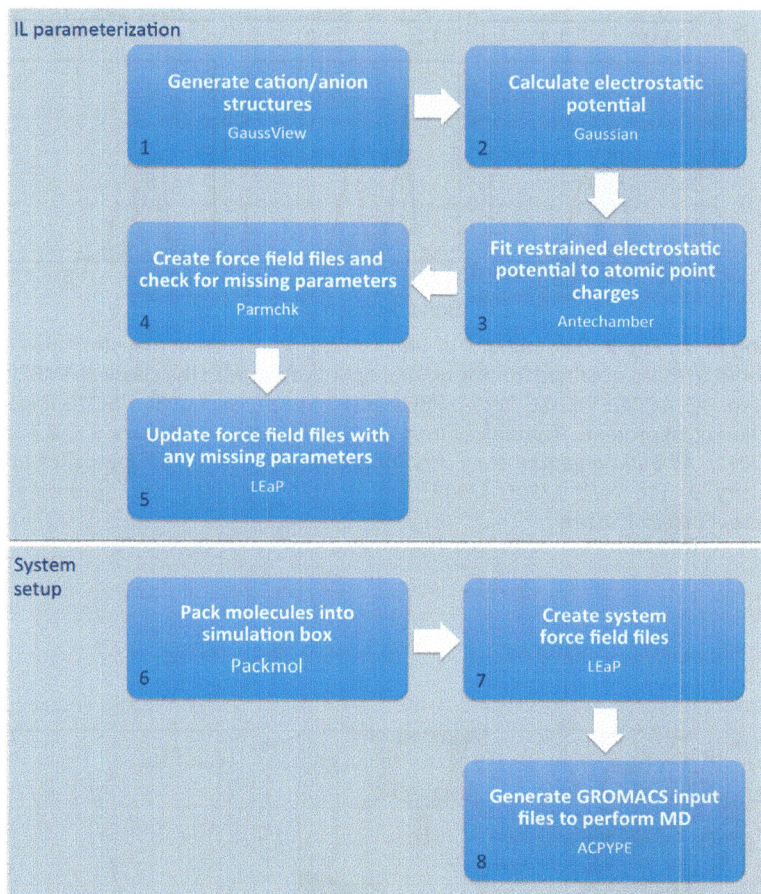

K.G. Sprenger and J. Pfaendtner, Fig. 1 Flow chart for developing MD simulation files for an enzyme/IL system to be simulated in GROMACS. This workflow is general and could be used with any MD engine compatible with modern force fields.

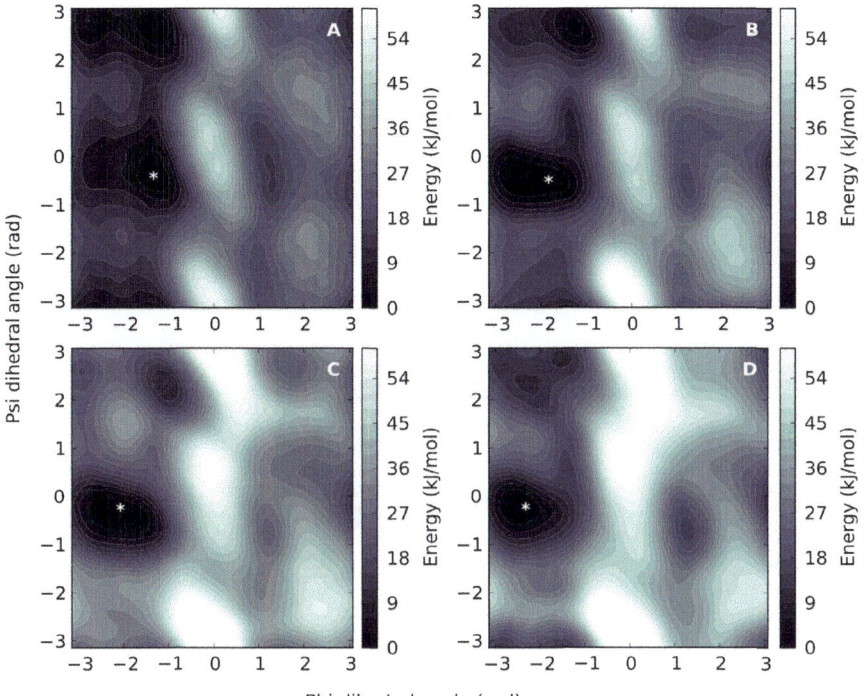

K.G. Sprenger and J. Pfaendtner, Fig. 2 Free energy surfaces for alanine dipeptide in (A) water, (B) [BMIM][Cl] with partial charges scaled by multiplying by 0.8, (C) [BMIM][Cl] with partial charges scaled by multiplying by 0.9, and (D) [BMIM][Cl] with partial charges scaled by multiplying by 1.0 (ie, not scaled). Asterisks indicate the α well in each of the two-state systems.

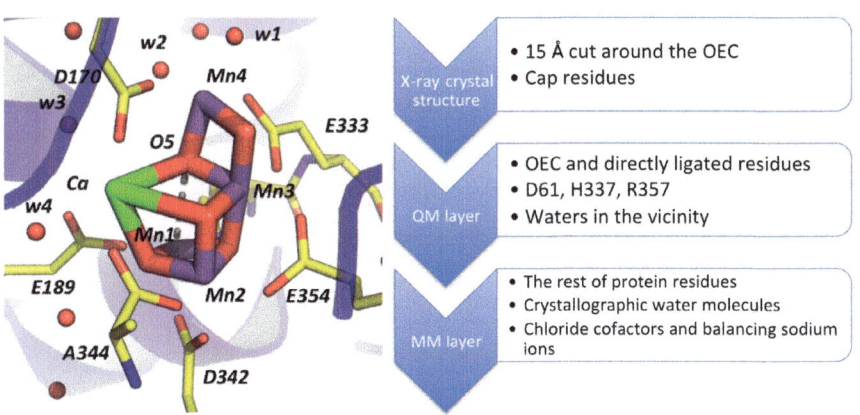

M. Askerka et al., Fig. 4 *Left*: Structural model of the OEC of PSII (QM layer shown explicitly, MM layer shown as cartoon). *Right*: Scheme of QM/MM model preparation.

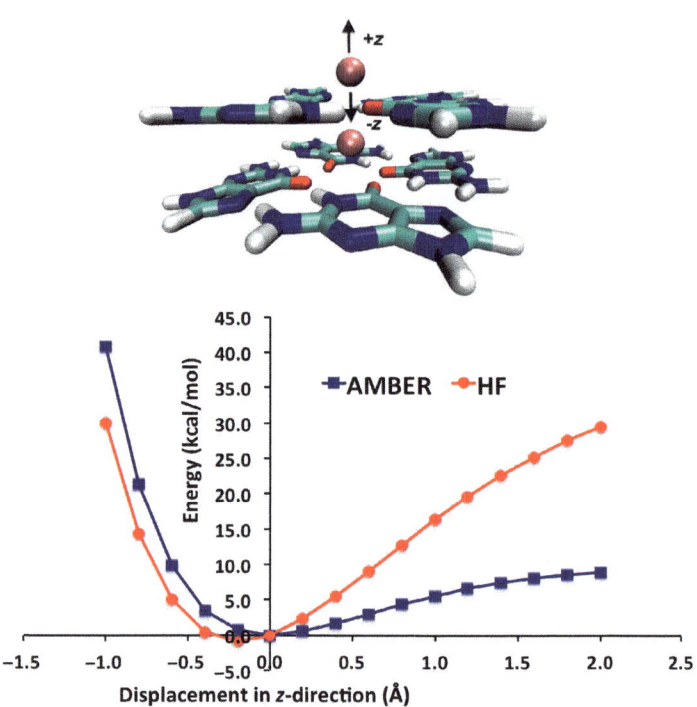

M. Askerka et al., Fig. 9 Benchmark calculations of electrostatic interactions for a potassium ion dragged through the internal channel of a two-layer model of the DNA guanine quadruplex, as described by HF and Amber, revealing significant polarization effects missed by nonpolarizable molecular mechanics force fields.

Edwards Brothers Malloy
Ann Arbor MI. USA
August 15, 2016